Inorganic Rings and Cages

To my Parents and my Sister

Inorganic Rings and Cages

Fred Armitage

Lecturer in Inorganic Chemistry,
Queen Elizabeth College, Kensington

Edward Arnold London
Crane, Russak New York

First published 1972 by
Edward Arnold (Publishers) Limited, London

Distributed in the United States of America by
Crane, Russak & Company Inc.,
52 Vanderbilt Avenue,
New York, New York 10017

Library of Congress Catalog Card Number 72-76951
ISBN: 0 8448 0004 X

Printed in Great Britain

Preface

The purpose of writing this book is to bring together under a general heading, many of the interesting and important facets of Main Group Inorganic Chemistry not comprehensively covered in general textbooks. These topics are often historically surveyed, but are up to date, and modern theories are used to interpret the structural data and chemical properties. Many of the topics provide the platform for final year lecture courses, while more general material gleaned from different Chapters could be expanded through the references provided. The 1400 references date from 1812 to December 1970 and are supplemented with an appendix incorporating later material. It should also be a useful general reference book for postgraduates.

With regard to units, the c.g.s. and SI systems are normally both used together. The SI system is used with interatomic distances (nm) and concentration (mol dm^{-3}) while with temperature °C is employed with K for very low temperatures.

The author would like to express his most grateful thanks to Dr. E. W. Abel for creating the idea and opportunity, and for his encouragement and help during the writing and for reading the completed manuscript. He also thanks the late Professor Sir Ronald Nyholm, Professor A. G. Davies and Turner and Newall Ltd. for encouraging the preliminary work, and Drs. Bird, Clark and Paul for helpful comments. Any errors are the sole responsibility of the author for which apology is now made.

The author thanks all who have granted permission for the use of Figures, Tables *etc.* and acknowledges these in full at the end of the caption or as a superscript reference in the same place, as requested. Grateful thanks are also due to Miss M. Cole for her care in typing the manuscript and to my publishers for their invaluable assistance.

London D. A. A.
1972

Contents

Introduction

Inorganic Chemistry, like other branches of the subject, is expanding at an ever increasing rate. So inevitably, any text on a subject as wide as Inorganic Chemistry must either be very general or, alternatively, become selective. This book was written with these problems in mind, since the material covers many topics only briefly mentioned in the more general texts.

At the outset, two far-reaching limitations were imposed upon the subject matter. Rings including carbon were not to be considered since their inclusion would lay open the vast fields of heterocyclic and chelate chemistry. This would render the subject matter too unwieldy and complex. Indeed, these two subjects are already well-documented in other monographs. However, the incorporation of rings including carbon does occur where this adds to the appreciation of the inorganic ring under consideration, *e.g.*, (i) the organic compounds isoelectronic with various boron–nitrogen rings, (ii) the carboranes, since they are isostructural with many anionic boron hydrides, and (iii) the comparison of the allotropes of carbon with the structural chemistry of the heavier Group IV elements.

Chelates are only included where the donor and acceptor form rings comprising Main Group elements. The decision to exclude transitional elements was based upon the wealth of information in the field of bridged complexes, with the metal in the full range of oxidation states, and of the growing field of metal cluster compounds. Again, this work would not be closely related to the material included and would have meant many additional Chapters. Nevertheless, low-valence transition metal complexes are included where the ligand is cyclic, *e.g.* $Me_6B_3N_3Cr(CO)_3$ and $[EtP_5]Mo(CO)_3$.

These limitations have enabled a reasonably comprehensive and up-to-date survey to be made of the subject which plays a most important part in Inorganic Chemistry, from the theoretical, industrial and geological standpoints, as will be seen from the synopsis below.

The chemistry of these rings is considered through the Group classification of the Periodic Table. The emphasis lies towards determining the structures of these compounds by various physical techniques, and the interpretation of the structural parameters in terms of modern bonding theories. However, other more general aspects naturally fit into the context of this book, *e.g.*, allotropy, and the cyclic catenation of the Main Group elements.

More specifically, the Grignard reagent is surveyed from both a historical and structural standpoint, while the peculiar cyclic chemistry of elemental sulphur is reviewed. The boron hydrides and boron cage compounds illustrate the quite extraordinary nature of this element, and provide the pointers to new approaches for bonding theory incorporating symmetry and topology. The properties of silicate minerals are considered in the light of their structure, while the oxides and sulphides of the Group IV elements are compared with the various forms of silica.

The Rochow process for making chlorosilanes industrially is summarized, since these compounds are the precursors to cyclosiloxanes and the silicones, one of the two major classes of inorganic polymer. The synthesis and bonding in cyclophosphazenes, and their use as forerunners to the second major class of inorganic polymer, provides a contrast between the theoretical picture of bonding and industrial polymer chemistry.

1

The Alkali Metals

Covalent compounds play only a limited role in the chemistry of the alkali metals and, of these, few fall within the category covered by this book. However, sufficient synthetic and structural work has been done with cyclic compounds of these Group I metals which do not contain carbon as a ring member to justify discussing the topic.

ALKOXIDES

The alkali metals all react with alcohols to yield the alkoxide and those of t-butanol have been of particular interest.[1]

$$2M^{I} + 2Bu^{t}OH \longrightarrow 2M^{I}OBu^{t} + H_2$$

These t-butoxides are all white crystalline solids which can be readily sublimed below 200°C under vacuum. They are therefore probably oligomeric and covalent, rather than ionic. This is supported by the steady amassing of structural and spectroscopic data which indicates a close structural similarity with the more widely-studied alkyl-lithium compounds.[2]

t-Butoxylithium is believed to be hexameric from its mass spectrum, which shows the species $Li_6(OBu^t)_5^+$ in high concentration and low appearance potential, compared with the other ions. This indicates inherent stability of the Li_6 cage previously noted for both ethyl- and n-butyl-lithium.[3] The structure is believed to involve an octahedral arrangement of lithium atoms bridged at six of its eight faces by t-butoxy groups (Figure 1.1), giving it a D_{3d} framework.

Trimethylstannoxylithium can be synthesized from methyl-lithium and hexamethylditin oxide in ether.

$$MeLi + (Me_3Sn)_2O \longrightarrow \tfrac{1}{6}[Me_3SnOLi]_6 + Me_4Sn$$

It is a white crystalline solid which dissolves in benzene as a hexamer[4] and could possess a structure similar to t-butoxy-lithium.

While no definitive structural data has been obtained for t-butoxy-sodium, the other alkali metal t-butoxides are all tetrameric. The X-ray structural data for the potassium compound has been sufficiently refined

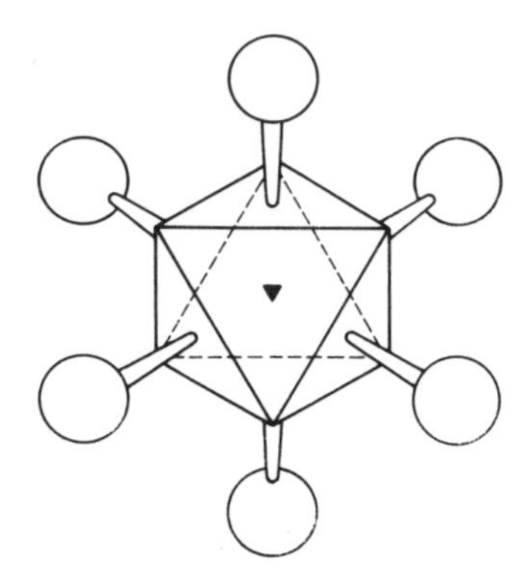

Fig. 1.1. The structure of $[Bu^tOLi]_6$[2]

to show that the four metal atoms are tetrahedrally arranged, with the alkoxide groups oriented above each of the tetrahedral faces. The K—O bond length of 0.256 nm is longer than would be expected for a two-electron dative bond, and so a structure representing the t-butoxide ion as a ligand donating three lone pairs of electrons in a distorted cube (Figure 1.2(a)) is probably inaccurate. The bonding is perhaps better described along the lines applied to the tetrameric alkyl-lithium compounds, where the alkyl anion is considered as a two-electron donor co-ordinating with a three-centre bonding molecular orbital in the triatomic face (Figure 1.2(b)). This could account for the long K—O bond.[1,5]

The trimethylsiloxy derivatives of the alkali metals, which are formed from hexamethyldisiloxane and the amide,

$$2(Me_3Si)_2O + 2M^INH_2 \longrightarrow 2Me_3SiOM^I + (Me_3Si)_2NH + NH_3$$

all possess tetrameric and hexameric structures.[6] The potassium, rubidium and cesium derivatives are tetramers with a structure resembling a cube

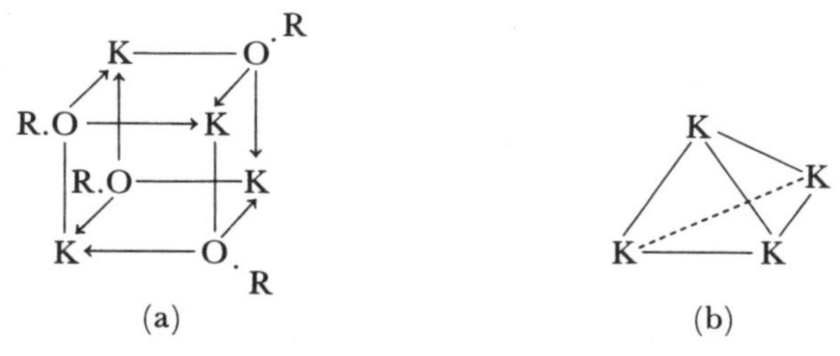

Fig. 1.2. The structure of $[Bu^tOK]_4$

of the metal and oxygen atoms. The short Si—O bond (0.160 nm) and the Si—O—K angle of 123 degrees supports sp^2- hybridized oxygen with electron lone-pair delocalization onto silicon. So the K—O bonding must involve multi-centre orbitals, as there are insufficient electrons for two-electron donor bonds. The mass spectra of the lithium and sodium derivatives give ions consistent with a hexameric structure.

Some alkoxides of elements of Groups II and III are also tetrameric, those most closely resembling the alkali metal t-butoxides being the methoxides of beryllium and monovalent thallium discussed in the next Chapters.[a]

These alkali metal alkoxides are widely used as activating agents and catalysts, and notably, while t-butyl-lithium is stable in cyclohexane heated under reflux, a trace of t-butoxylithium causes ready decomposition.[a]

$$Bu^tLi + Bu^tOLi \longrightarrow Bu^{t-} + Bu^tOLi_2^+$$

This may be initiated by the formation of small concentrations of the t-butyl anion as the oxonium ion $Bu^tOLi_2^+$ appears quite stable as the most abundant peak in the mass spectrum of t-butoxylithium.[3] Indeed, alkyl-lithium compounds act as better nucleophiles if a complexing agent is present. Chelating amines are widely used,[6] but n-butyl-lithium will metallate benzene at 25°C in the presence of t-butoxypotassium.[7]

$$C_4H_9^nLi + C_6H_6 \xrightarrow{C_4H_9^tOK} C_4H_{10} + C_6H_5Li$$

The highly reactive nature of the organo derivatives of the heavier alkali metals, due to the presence of a carbanion, renders these compounds difficult to synthesize. It is, therefore, most convenient that the di-metallation of the t-butoxy group postulated above, and established by mass spectrometry, occurs, since this provides the intermediate for alkali metal exchange between alkyl- and alkoxy-derivatives. Thus, mixing heptane solutions of t-butoxysodium with alkyl-lithium compounds results in the rapid precipitation of the organosodium compound[8] (R = n-butyl, n-octyl or n-dodecyl).

$$Bu^tONa + RLi \xrightarrow{\text{heptane}} RNa + Bu^tOLi$$

n-Butyl-lithium reacts similarly with t-butoxypotassium, to give n-butylpotassium.

Methylpotassium, also synthesized this way, using ether as solvent at −10°C, has been shown to possess a completely ionic structure, unlike the methyl derivatives of the two lighter alkali metals which occur as covalent tetramers.[9]

$$Bu^tOK + MeLi \xrightarrow[70°C]{Et_2O} Bu^tOLi + MeK$$

AMIDES

The oligomeric nature of the alkali metal amides appears to be limited to the derivatives of highly-substituted amines. Thus, while lithium dimethylamide is an insoluble polymer, the lithium derivative of guanidine is crystalline and dimeric in toluene. It results from methyl-lithium, and the imido group of guanidine is thought to be a three-electron donor.[10]

$$2(Me_2N)_2C{=}NH + 2MeLi \longrightarrow (Me_2N)_2C{=}N\underset{Li}{\overset{Li}{\diamond}}N{=}C(NMe_2)_2 \quad (1.1)$$

Both the lithium and sodium salts of hexamethyldisilazane $(Me_3Si)_2NH$ are distillable liquids which crystallize on cooling.[11] They can be synthesized directly from the disilazane,

$$\underset{\text{(dimeric in solution)}}{(Me_3Si)_2NLi} + C_4H_{10} \xleftarrow{Bu^nLi} (Me_3Si)_2NH \xrightarrow[C_6H_6]{NaNH_2} \underset{\text{(dimeric in solution)}}{(Me_3Si)_2NNa} + NH_3$$

and while the lithium derivative dissolves easily in many organic solvents, the more reactive sodium one only dissolves readily in ether and benzene. This increase in ionic character is extended to the potassium, rubidium and caesium compounds, which become progressively less soluble in non-polar solvents and more reactive as the anion character of the $(Me_3Si)_2N^-$ group increases. Thus, the potassium compound dissolves readily in liquid ammonia.[12] This increase in ionization is reflected in the multiple-bond character in the Si—N bond, which arises through p_π–d_π interactions from the nitrogen to the silicon atom. These occur for the dimeric lithium and sodium derivatives, and with the lithium compound a weak Li—N bond indicates a bond order less than one and a N—Li—N three-centre bond characteristic of diborane as discussed in Chapter 3 (p. 45). The structure of the lithium compound in the solid state is known, however. It is a trimer comprising a puckered, 6-membered ring of alternate lithium and nitrogen atoms, with two Me_3Si groups bonded to each nitrogen atom.[13]

As the structural parameters of the amino group support π-bonding from nitrogen on to the two silicon atoms, it is unlikely that the ring bonding involves the co-ordination of two lone pairs to the lithium atoms of the ring, as proposed for the dimeric guanidine compound in equation (*1.1*). The long Li—N bond and the small bond angle at nitrogen supports some form of multi-centre bonding.

The dimeric nature of silylated lithium amides is also evident among more complicated examples. Successive azidosilation of diphenylphosphine, followed by lithiation yields the amide $Ph_2P(NSiMe_3)NLiSiMe_3$.[14]

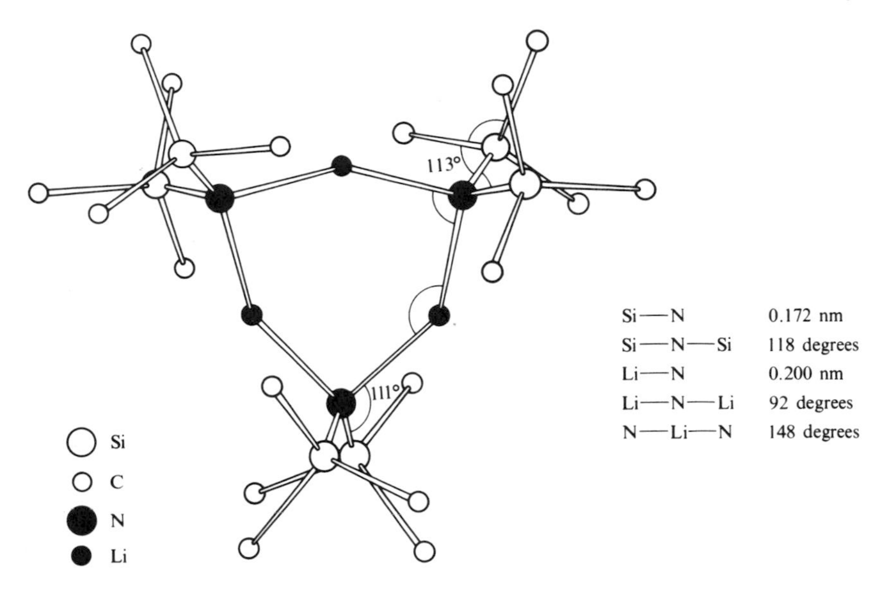

Fig. 1.3. The structure of $[(Me_3Si)_2NLi]_2$[13]

$$Ph_2PH \xrightarrow[-N_2]{Me_3SiN_3} Me_3SiNHPPh_2 \xrightarrow{Me_3SiN_3} Me_3SiNHP(Ph_2){=}NSiMe_3$$

Similarly, t-butyl-dichlorophosphine has been used to prepare $Bu^t(Me_2N)P(NSiMe_3)NLiSiMe_3$ **(1-I)**,[15]

$$Bu^tPCl_2 \longrightarrow Bu^tP(NH_2)_2 \longrightarrow Bu^tP(NHSiMe_3)_2$$

$$\downarrow CCl_4,\ -CHCl_3$$

$$Bu^t(Me_2N)P(NSiMe_3)NLiSiMe_3 \xleftarrow[(2)\ RLi]{(1)\ Me_2NH} Bu^tClP(NSiMe_3)NHSiMe_3$$

and both amides can be sublimed and are dimeric in benzene with **(1-I)** exhibiting equivalent trimethylsilyl groups even at high dilution in the p.m.r. spectrum. This supports a structure involving electron delocalization around the 8-membered ring indicated in Figure 1.4 which probably involves a lithium bridge as with the simple amides.[16]

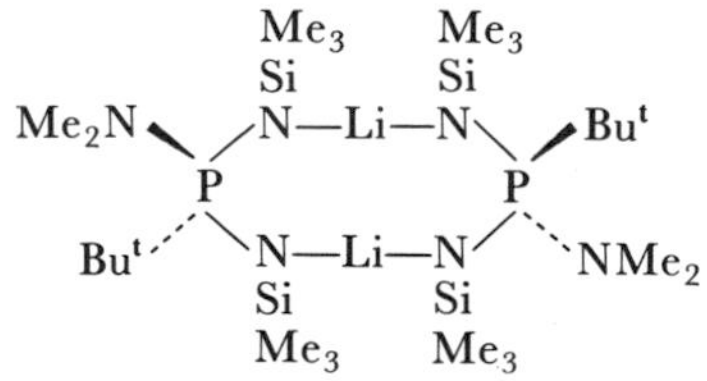

```
               Me3      Me3
               Si       Si
   Me2N\     /N—Li—N\      /But
          P                P
   But/     \N—Li—N/      \NMe2
               Si       Si
               Me3      Me3
```

Fig. 1.4. The proposed structure of compound **(1-I)**

HYDROGEN DERIVATIVES

While the cyclic aspects of hydrogen bonding will not be considered here, as this is a large subject better discussed separately, recent mass spectral and theoretical work has revealed that polyatomic hydrogen cations H_n^+ ($n = 2, 3, 5, 9$) exist. These are produced by bombarding molecular hydrogen with electrons at 3 K.[17] Calculations show that H_3^+ is probably an equilateral triangle with H—H bonds 0.088 nm long. It is thermodynamically stable relative to $H_2 + H^+$ by about 418.4 kJ mol^{-1} (100 kcal mol^{-1}).[18] The cation LiH_2^+ is also predicted to be stable, with a structure involving the three atoms in an isosceles triangle. That such ions should be stable is quite plausible, since overlap of three s orbitals produces a stable bonding orbital (A'_1) to which the two electrons can be ascribed.[19]

Bibliography

General

a G. E. COATES and K. WADE, 'Organometallic Compounds', Methuen and Co. Ltd. London, 3rd Edn., 1967, Vol. 1, 10.

References

1 E. WEISS, H. ALSDORF, H. KÜHR and H-F. GRÜTZMACHER, *Chem. Ber.*, 1968, **101,** 3777.
2 T. L. BROWN, *Adv. Organometallic Chem.*, 1965, **3,** 365. Academic Press, London and New York.
3 G. E. HARTWELL and T. L. BROWN, *Inorg. Chem.*, 1966, **5,** 1257.
4 H. SCHMIDBAUR and H. HUSSEK, *Angew. Chem. Internat. Edn.*, 1963, **2,** 328.
5 E. WEISS, H. ALSDORF and H. KÜHR, *ibid.*, 1967, **6,** 801
6 E. WEISS, K. HOFFMAN and K-F. GRÜTZMACHER, *Chem. Ber.*, 1970, **103,** 1190.
7 M. SCHLOSSER, *J. Organometallic Chem.*, 1967, **8,** 9.
8 L. LOCHMAN, J. POSPISIL and D. LIM, *Tetrahedron Letters*, 1966, 257.
9 E. WEISS and G. SAUERMANN, *Angew. Chem. Internat. Edn.*, 1968, **7,** 133.
10 I. PATTISON, K. WADE and B. K. WYATT, *J. Chem. Soc. (A)*, 1968, 837.
11 *Inorg. Synth.*, 1966, **VIII,** 15, 19.
12 U. WANNAGAT, *Chem. Eng. News*, 1968, **46,** 38.
13 D. MOOTZ, A. ZINNIUS and B. BÖTTCHER, *Angew. Chem. Internat. Edn.*, 1969, **8,** 378.
14 H. SCHMIDBAUR, K. SCHWIRTEN and H-H. PICKEL, *Chem. Ber.*, 1969, **102,** 564.
15 O. J. SCHERER and P. KLUSMANN, *Angew. Chem. Internat. Edn.*, 1968, **7,** 541.
16 O. J. SCHERER, *ibid.*, 1969, **8,** 861.
17 R. CLAMPITT and L. GOWLAND, *Nature*, 1969, **223,** 815.
18 J. EASTERFIELD and J. W. LINNETT, *Chem. Comm.*, 1970, 64.
19 J. N. MURRELL, S. F. A. KETTLE and J. M. TEDDER, 'Valence Theory', Wiley, London, 2nd Edn., 1970, 413.

APPENDIX

Organopotassium compounds can be prepared from organolithium compounds and potassium (—)(IR)-menthoxide in good yields. They are relatively pure and can be used directly to synthesise potassium derivatives of aromatic hydrocarbons.[20]

Concentration and temperature dependent ^{1}H and ^{7}Li n.m.r. spectra show that $LiN(SiMe_3)_2$ undergoes a monomer-dimer equilibrium in tetrahydrofuran and a dimer-tetramer equilibrium in hydrocarbon solvents. Isopiestic molecular weight measurements support this.[21]

References

20 L. LOCHMANN and D. LIM, *J. Organometallic Chem.*, 1971, **28,** 153.
21 B. Y. KIMURA and T. L. BROWN, *ibid*, 1971, **26,** 57.

2

Beryllium and Magnesium

The chemistry of the alkaline earth metals calcium, strontium and barium is dominated by ionic compounds, showing none of the structural characteristics necessary for inclusion in this Chapter. Only beryllium and magnesium fit the conditions of being a member of a "sans carbon" ring. A comparison of the two elements can therefore be conveniently made, commencing with the inorganic salts, the hydride derivatives and finally organometallic compounds.

INORGANIC SALTS

The most striking feature of the chemistry of beryllium is that, despite its electropositive character, the Be^{2+} ion is unknown among compounds of the element. Thus, while magnesium and the alkaline earth metal oxides have a NaCl structure, beryllium oxide has a wurtzite structure characterized by much covalent character to the bonds. Similarly, magnesium difluoride and dichloride both have structures typical of ionic compounds, while the beryllium halides possess ones characterized by covalent compounds. Beryllium difluoride has the β-cristobalite structure of silica with the black circles representing silicon[1] (Figure 2.1).

It is a glassy hygroscopic compound which results from the pyrolysis of ammonium tetrafluoroberyllate, and is a poor conductor. The apparent

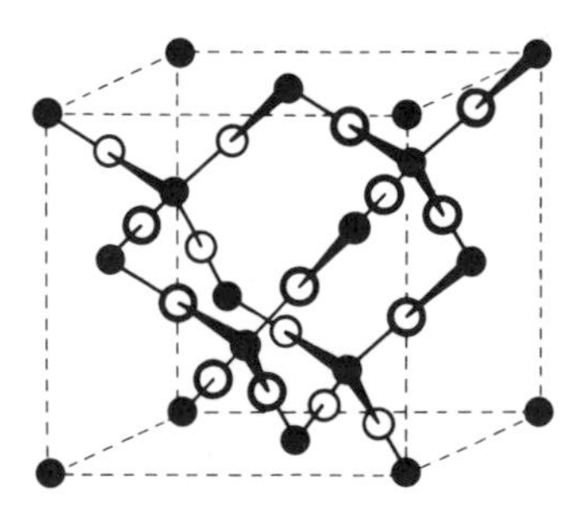

Fig. 2.1. The β-cristobalite structure of beryllium difluoride[1]

covalent character is probably due to the very high polarizing power of the Be^{2+} ion, which will distort the electron cloud of even the smallest and most electronegative of anions.

The chloride results when carbon tetrachloride is passed over heated beryl, and possesses the molecular structure characterized by SiS_2.[1] This involves long chains of $BeCl_4$ tetrahedra sharing opposite edges (Figure 2.2).

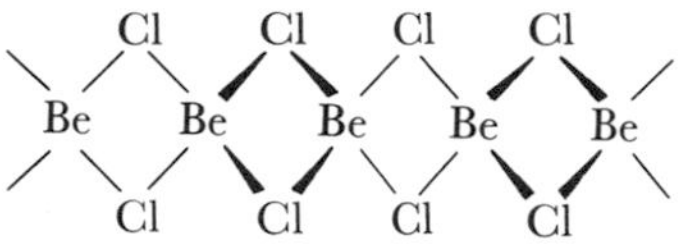

Cl Cl Cl Cl
Be Be Be Be Be
Cl Cl Cl Cl

Fig. 2.2. The structure of beryllium dichloride

Dissolving in water gives the expected hydrated beryllium ions $[Be(H_2O)_4]^{2+}$ which at low concentrations hydrolyze to the oligomeric ion $[Be(H_2O)_2OH]_3^{3+}$.[2] This probably possesses a 6-membered ring involving alternate beryllium and oxygen atoms.

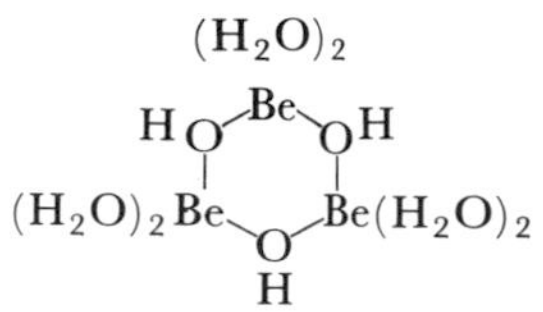

$(H_2O)_2$
HO Be OH
$(H_2O)_2$Be Be$(H_2O)_2$
O
H

Fig. 2.3. Trimeric $[Be(H_2O)_2OH]_3^{3+}$

A variety of inorganic chelating agents have been used to complex with beryllium. These include the phosphinate[3] and azadiphosphinate anions[4] ($R_2PO_2^-$, $[(Ph_2PO)_2N]^-$). The nitrate group also functions as a chelating ligand, as illustrated through a series of reactions starting with beryllium dichloride and dinitrogen tetroxide in ethyl acetate. Solvated covalent beryllium nitrate is formed first and progressively loses N_2O_4 to give colourless crystals of basic beryllium nitrate $Be_4O(NO_3)_6$.[5]

$$BeCl_2 \xrightarrow[\text{EtOCOMe}]{N_2O_4} Be(NO_3)_2{\cdot}N_2O_4 \xrightarrow{50^\circ C}$$

$$Be(NO_3)_2 \xrightarrow[-N_2O_4]{120^\circ C} Be_4O(NO_3)_6$$

This is believed to have a structure involving a tetrahedral array of beryllium atoms edge-bridged by nitrate groups, and having a quadridentate oxygen atom co-ordinating to each beryllium atom from the centre of the tetrahedron (Figure 2.4).

Contaminating a Grignard reagent with oxygen leads to the isolation, in small yields,[6] of a compound formulated as $Mg_4Br_6O{\cdot}4Et_2O$. The

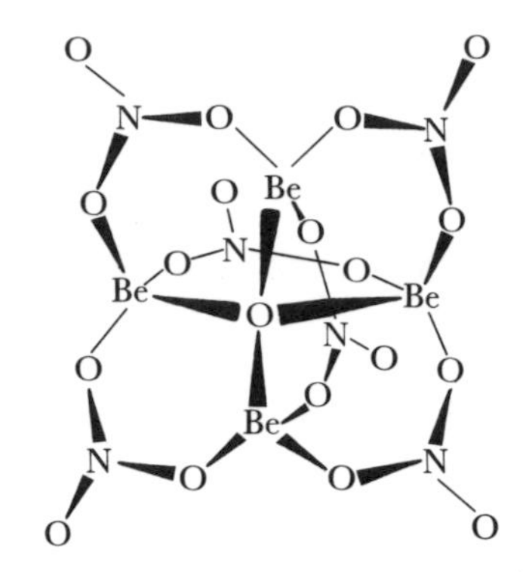

Fig. 2.4. Basic beryllium nitrate[5]

X-ray structure reveals a similar atomic array to that proposed for basic beryllium nitrate, except that the magnesium is 5-co-ordinate, not 4-co-ordinate as in the beryllium compound.

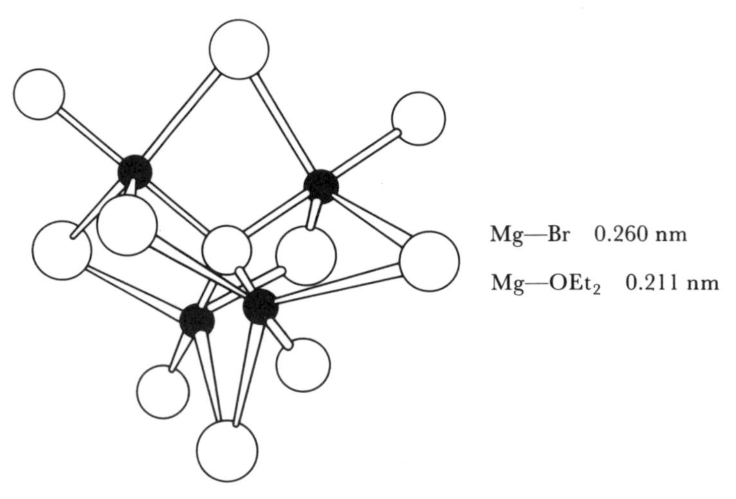

Fig. 2.5. $Mg_4Br_6O(Et_2O)_6$, with spheres representing bromine, oxygen and magnesium in order of decreasing size. (From G. SUCKY and R. E. RUNDLE, *J. Amer. Chem. Soc.* 1964, **86,** 4821).

Thus, the magnesium atoms are bonded to not only the central oxygen atom and three bidentate bromide ions but to ether as well (Figure 2.5).

HYDRIDES

While the alkali metal hydrides are ionic compounds, the high polarizing power of the Be^{2+} ion renders the dihydride covalent, with a structure conveniently represented by that of $BeCl_2$ but with hydride bridges. Disordered cross-linking probably occurs, however, as the X-ray pattern supports an amorphous rather than a crystalline compound.

Beryllium dihydride is conveniently synthesized by pyrolyzing di-t-butylberyllium at 100°C.[7,8]

$$Bu^t_2Be \xrightarrow{100°C} BeH_2 + CH_2{=}CMe_2$$

Isobutene is lost, and the dihydride remaining is stable in air and only slowly hydrolyzed by water, indicating the strength of the hydride-bridge bond, despite its being a three-centre bond. This bridge bond results from the combination of three orbitals, one on each of the two beryllium and hydrogen atoms, and if the local symmetry at the bridge bond is taken as C_{2v}, then the σ atomic orbitals ϕ_1, ϕ_2 and ϕ_3 form the representation Γ_σ shown.

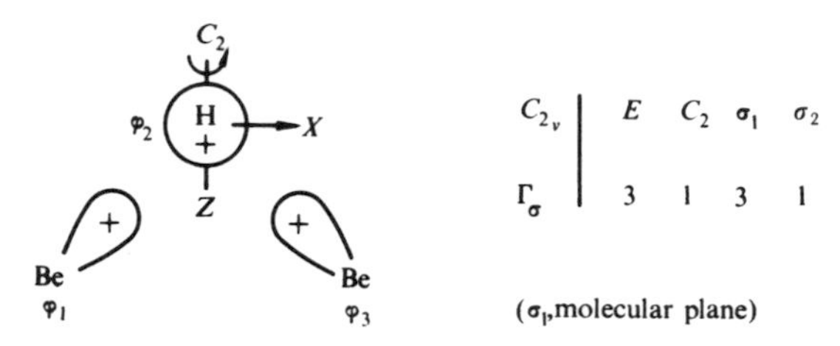

Fig. 2.6. BeH_2 bridge—local symmetry C_{2v}

This comprises two A_1 orbitals and a B_1 orbital. From the character table, the B_1 orbital is obviously

$$\psi(B_1) = a_1\phi_1 - a_1\phi_3$$

The coefficients must be equal, and normalizing gives

$$a_1^2 + a_1^2 = 1; \textit{i.e.}, a_1 = \frac{1}{\sqrt{2}}$$

The two A_1 orbitals can be generally represented as

$$\psi(A_1') = a_1\phi_1 + b_1\phi_2 + c_1\phi_3$$

$$\psi(A_1'') = a_2\phi_1 + b_2\phi_2 + c_2\phi_3$$

For orthogonality, b_1 and b_2 are opposite in sign, *i.e.*, $a_1a_2 + c_1c_2 + b_1b_2 = 0$, and by symmetry $a_1 = c_1$ and $a_2 = c_2$. So $2a_1a_2 = -b_1b_2$. Normalizing, $2a_1^2 + b_1^2 = 2a_2^2 + b_2^2 = 1$. Hence, $a_1 = a_2 = \frac{1}{2}$, and $b_1 = -b_2 = 1/\sqrt{2}$.

The three molecular orbitals are therefore,

$$\psi(B_1) = \frac{1}{\sqrt{2}}(\phi_1 - \phi_3) \qquad \text{(non-bonding)}$$

$$\psi(A_1') = \tfrac{1}{2}(\phi_1 + \phi_3) + \frac{1}{\sqrt{2}}\phi_2 \quad \text{(bonding)}$$

$$\psi(A_1'') = \tfrac{1}{2}(\phi_1 + \phi_3) - \frac{1}{\sqrt{2}}\phi_2 \quad \text{(antibonding).}$$

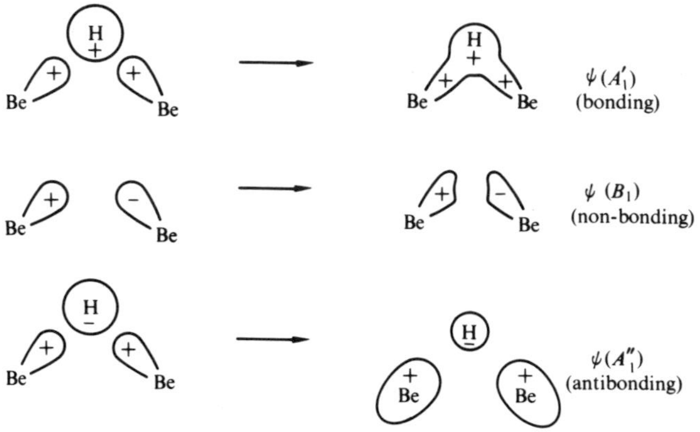

H Be Be → H Be Be $\psi(A_1')$ (bonding)

Be Be → Be Be $\psi(B_1)$ (non-bonding)

H Be Be → H Be Be $\psi(A_1'')$ (antibonding)

Fig. 2.7. Molecular orbitals generated in a BeH_2 bridge

The two electrons available go into the lowest-lying bonding orbital $\psi(A_1')$.

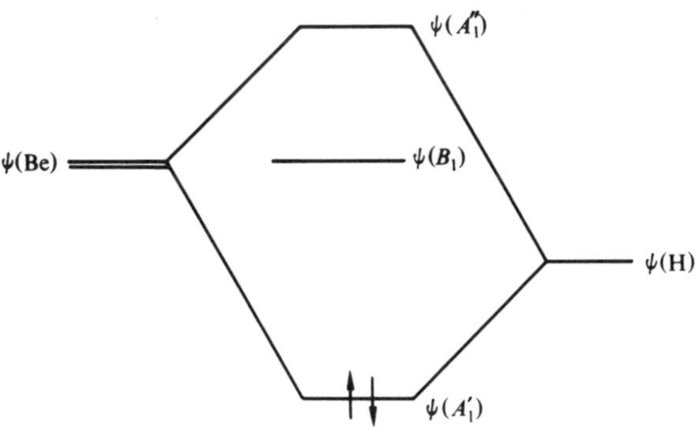

Fig. 2.8. Molecular orbital energy level diagram

This picture can be conveniently used as a definition of electron-deficient compounds, *viz.*, those molecules in which there are insufficient electrons to form two-electron–two-centre bonds, *i.e.*, there are more molecular orbitals than electrons.

Reducing diethylberyllium with diethylaluminium hydride in the presence of tertiary amines yields complexes of dimeric beryllium dihydride.

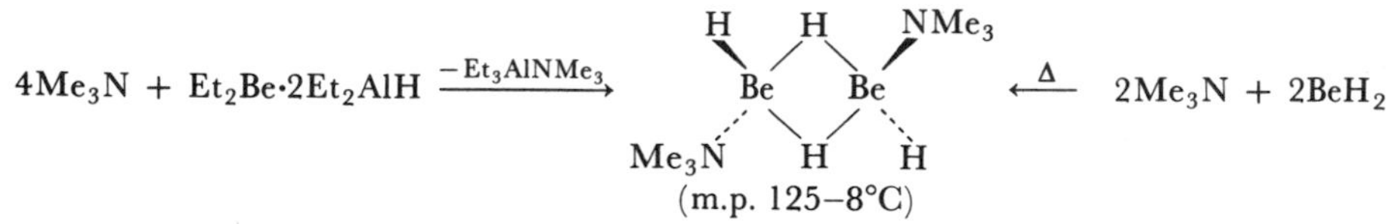

$$4Me_3N + Et_2Be{\cdot}2Et_2AlH \xrightarrow{-Et_3AlNMe_3} [(Me_3N)HBe(\mu\text{-}H)_2BeH(NMe_3)] \xleftarrow{\Delta} 2Me_3N + 2BeH_2$$

(m.p. 125–8°C)

They can also be synthesized directly, and exhibit peaks in the infrared spectrum typical of both bridging and terminal hydride groups.[9] These occur at 1340 cm^{-1} (134×10^3 m^{-1}) and about 1750 cm^{-1} (175×10^3 m^{-1}). With tetramethylethylenediamine (TMED), an insoluble complex is formed which is probably polymeric. This also results from the pyrolysis of the etherate of di-isobutylberyllium in the presence of TMED.[10]

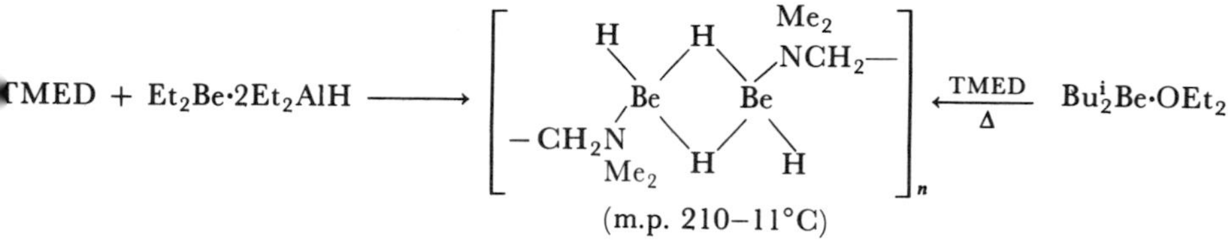

$$TMED + Et_2Be{\cdot}2Et_2AlH \longrightarrow [-CH_2N(Me_2)\text{–}HBe(\mu\text{-}H)_2BeH\text{–}N(Me_2)CH_2-]_n \xleftarrow[\Delta]{TMED} Bu^i_2Be{\cdot}OEt_2$$

(m.p. 210–11°C)

The pyrolysis of other di-alkylberyllium compounds normally removes only one of the alkyl groups. Di-isopropylberyllium evolves propene and leaves the hydride as a viscous polymer, probably involving both propyl and hydride bridges.[11] These compounds can be complexed in the dimeric form. Thus, careful pyrolysis of di-isobutylberyllium in the presence of tetrahydrofuran (THF) or TMED yields the bridged hydride, with a peak in the infrared spectrum at about 1350 cm^{-1}, typical of the BeH_2Be group.[10]

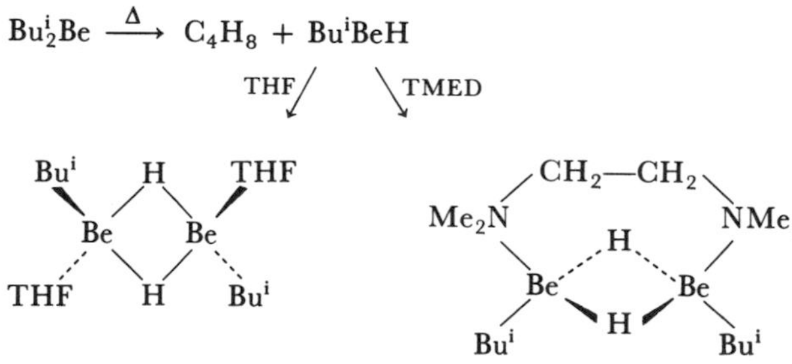

$$Bu^i_2Be \xrightarrow{\Delta} C_4H_8 + Bu^iBeH$$

THF ↙ ↘ TMED

$[Bu^i(THF)Be(\mu\text{-}H)_2Be(THF)Bu^i]$ $\quad$ $[Bu^iBe(\mu\text{-}H)_2BeBu^i(Me_2NCH_2CH_2NMe_2)]$

The dimeric complexes of organoberyllium hydrides are more conveniently synthesized by reducing an equimolar mixture of R_2Be and beryllium dihalide in the presence of a base.[12, 13]

$$Me_2Be + BeBr_2 + 2LiH \xrightarrow{Et_2O} \underset{(oil)}{[MeBeH{\cdot}OEt_2]_2} + 2LiBr$$

$$Ph_2Be + BeCl_2 + 2NaBEt_3H \xrightarrow{Me_3N} [PhBeH{\cdot}NMe_3]_2 + 2Et_3B + 2NaCl$$

$$2Et_2Be + 2Et_3SnH \xrightarrow{Me_3N} \underset{(m.p.\ 90\text{–}1°C)}{[EtBeH{\cdot}NMe_3]_2} + 2Et_4Sn$$

The examples also include the use of triethylstannane which with dimethylberyllium gives the compound $[MeBeH{\cdot}NMe_3]_2$ (m.p. 73°C). At room temperature, the p.m.r. spectrum of this compound in deuterotoluene indicates the presence of two isomers.[14]

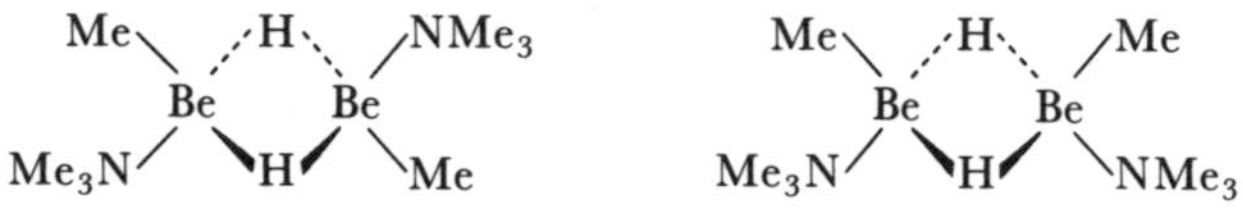

Fig. 2.9. Isomers of $[MeBeH{\cdot}NMe_3]_2$

These are probably *cis* and *trans* forms, of which the *trans* form, with its lower dipole moment, is more stable at higher temperatures. Thus on cooling, the ratio of the areas of the two peaks typical of the Me_3N groups vary so that the *cis* form becomes the predominant isomer.

Beryllium hydride reacts with secondary amines to yield beryllium amides. With dimethylamine the diamide $Be(NMe_2)_2$ results, and is a trimeric crystalline solid, m.p. 94°C.[7]

$$3BeH_2 + 6Me_2NH \xrightarrow{155°C} [Be(NMe_2)_2]_3$$

As the compound has amide bridges, its structure will be considered in the next section.

N-Trimethylethylenediamine displaces only one hydrogen atom from beryllium hydride and the product is trimeric in benzene and thought to have a hydride-bridged structure[15] (Figure 2.10).

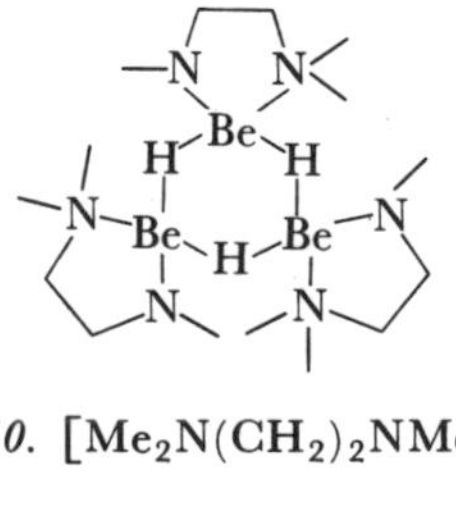

Fig. 2.10. $[Me_2N(CH_2)_2NMeBeH]_3$

While lithium hydride will reduce an organoberyllium halide to the neutral bridging hydride, no alkyl cleavage occurs. Thus, both dimethyl- and diethyl-beryllium yield anionic hydrides in ether.[16]

$$Et_2Be + 2NaH + 2Et_2O \longrightarrow Na_2Et_4Be_2H_2{\cdot}2Et_2O$$

The beryllium hydride residue is again dimeric. A structural determination shows the hydride bridges with Be—H distances of 0.14 nm, a value close to that encountered among bridged boron hydrides (0.133 nm), while the Na—H distance of 0.240 nm is close to that found in sodium hydride. The compound can therefore be formulated as $(Na{\cdot}OEt_2^+)_2Et_4Be_2H_2^{2-}$ with the ether molecules associated with sodium and not beryllium atoms.[17]

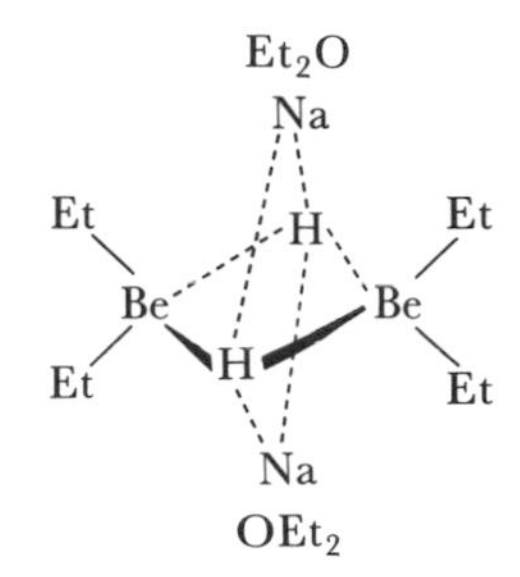

Fig. 2.11. $Na_2Et_4Be_2H_2{\cdot}2Et_2O$

Hydride bridge structures have also been confirmed between beryllium and other elements often associated with electron deficiency. Small yields of beryllium borohydride result from beryllium hydride and diborane,[7] but a better synthetic route is the one first employed for the compound in 1940.[18] This involved the interaction of polymeric dimethylberyllium with diborane.

$$(Me_2Be)_x + xB_2H_6 \longrightarrow 2xCH_4 + xBe(BH_4)_2$$

The borohydride also results from the metathetical exchange between sodium borohydride and beryllium dichloride in the absence of a solvent.

$$2NaBH_4 + BeCl_2 \xrightarrow{125°C} 2NaCl + Be(BH_4)_2$$

Beryllium borohydride is a solid which can be readily sublimed below 100°C. Air and water react with it vigorously at room temperature indicating a weaker hydride bridge than with BeH_2. Careful addition of hydrogen chloride results in the evolution of diborane and hydrogen, thereby providing a convenient way of separating the elements boron and beryllium. The dichloride is left behind on removal of the volatiles from the reaction mixture.

$$Be(BH_4)_2 + 2HCl \longrightarrow BeCl_2 + B_2H_6 + H_2$$

Many amines, phosphines and ethers form 1 : 1 adducts with the borohydride,[19] and the trimethylamine complex is thermally stable up to 140°C. This supports the idea that there is a vacant site for co-ordination in $Be(BH_4)_2$, while the dipole moment of $7.0 \pm 1.7 \times 10^{-30}$ Cm $(2.1 \pm 0.5D)$[20] is inconsistent with the D_{2d} structure (Figure 2.12).

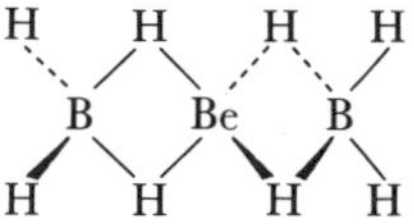

Fig. 2.12. D_{2d} structure of $Be(BH_4)_2$

Indeed the high melting point, (*cf.* propane), suggests a less symmetrical form, though there has been some little dispute as to the exact atomic array. It is agreed to belong to the symmetry point group C_{2v}, and the structure most consistent with molecular parameters encountered elsewhere involves a 6-membered hydrogen-bridged ring[21] (Figure 2.13(a)).

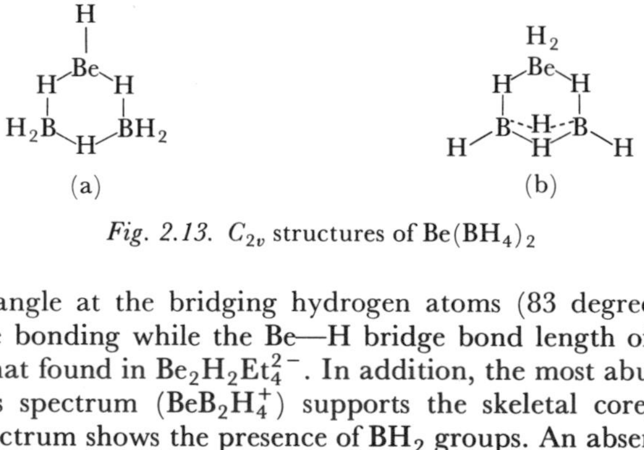

Fig. 2.13. C_{2v} structures of $Be(BH_4)_2$

The small angle at the bridging hydrogen atoms (83 degrees) supports three-centre bonding while the Be—H bridge bond length of 0.14 nm is similar to that found in $Be_2H_2Et_4^{2-}$. In addition, the most abundant peak in the mass spectrum $(BeB_2H_4^+)$ supports the skeletal core, while the infrared spectrum shows the presence of BH_2 groups. An absence of peaks in the region 890–910 cm^{-1} typical of the B—H groups in the isomeric form (Figure 2.13(b)), adds support to the first structure which could readily accept donor molecules at the beryllium atom as it is only 3-co-ordinate.

No magnesium compounds analogous to the beryllium hydride complexes have been isolated, intermediate complexes of the type RMgH·L invariably disproportionating into R_2MgL and MgH_2.[22]

AMIDES OF BERYLLIUM AND MAGNESIUM

The beryllium amide $Be(NMe_2)_2$ has already been briefly mentioned in considering the reactions of beryllium hydride with secondary amines

(p. 14). It is also found when excess dimethylamine reacts with diethyl-[23] or di-isopropylberyllium,[11] and is a trimeric crystalline solid.

$$2R_2Be + 6HNMe_2 \longrightarrow [(Me_2N)_2Be]_3 + 6RH$$

The molecular structure shows a spiran arrangement of atoms involving four bridging amino groups

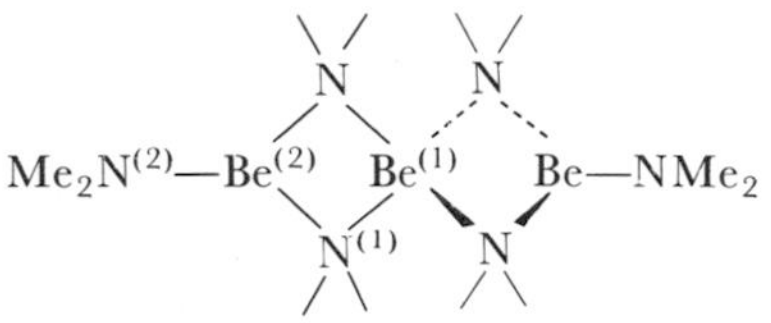

$Me_2N^{(2)}$—$Be^{(2)}$ $Be^{(1)}$ Be—NMe_2; N; N; $N^{(1)}$; N

Fig. 2.14. $[(Me_2N)_2Be]_3$

and two terminal ones.[24] The planar nature of the terminal nitrogen atoms supports the presence of p_π–p_π bonding in the terminal BeN bonds, though a sparsity of data makes the Be—N bond length a poor yardstick for estimating multiple-bond character. Indeed a Be—N bond 0.178 nm long tends to indicate three-centre bonding !. The presence of two absorptions in the n.m.r. spectrum at τ values of 4.28 and 4.90 (intensity ratio 1:2) supports this structure.

Generally speaking, the reactions of secondary amines with R_2Be lead to mono-substituted products.

$$R_2Be + R'_2NH \longrightarrow \frac{1}{n}[RBeNR'_2]_n + RH$$

The oligomeric nature probably results from the inherent electron-deficient nature of the $RBeNR'_2$ monomer, and the strong co-ordinating properties of nitrogen when bonded to an electropositive element—this facet will be encountered later in aluminium chemistry (p. 100).

With dimethylamine and dimethylberyllium an adduct forms which decomposes with methane elimination at its melting point of 44°C.

$$Me_2Be + Me_2NH \longrightarrow Me_2Be{\leftarrow}NHMe_2 \xrightarrow{44°C} \tfrac{1}{3}[MeBeNMe_2]_3 + CH_4$$

The product is trimeric and possesses a 6-membered ring of alternating beryllium and nitrogen atoms.[25]

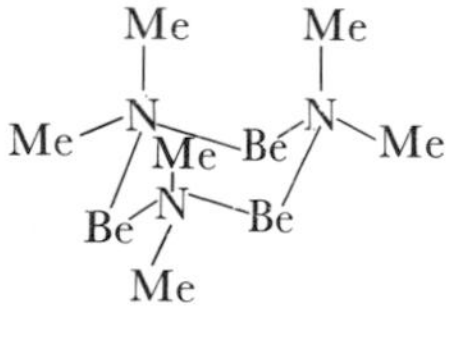

Me Me; Me—N—Be—N—Me; Me; Be—N—Be; Me

Fig. 2.15. The proposed structure of $[Me_2NBeMe]_3$

The more stable chair configuration, nevertheless, has three axial methyl groups and indeed, the synthesis of beryllium amides with larger substituents on the nitrogen yields dimeric oligomers. While this ring contraction produces valence strain in the ring members, it does relieve some of the hindrance that would occur with the trimeric form. Thus, all dimethylamides are trimeric while the dipropyl- and diphenyl-amides are dimeric in benzene.[26]

$$2Et_2Be + 2HNPh_2 \longrightarrow [Ph_2N\text{-}BeEt]_2 + 2C_2H_6$$

Complexing these amides with pyridine produces a similar ring contraction. Making the beryllium atom co-ordinatively saturated will produce axial hindrance at the three beryllium atoms as well as the nitrogen ones. The addition of an excess of pyridine produces the 2:1 monomeric complex,

$$2[Me_2NBeMe]_3 \xrightarrow{6C_5H_5N} 3\,[(C_5H_5N)MeBe\text{-}NMe_2]_2 \xrightarrow{6C_5H_5N} 6(C_5H_5N)_2Be(NMe_2)Me$$

but with $[Pr^n_2NBeMe]_2$ disproportionation results through complexing, Me_2Bepy_2 being precipitated.

$$[MeBeNPr^n_2]_2 + 2py \longrightarrow [(Me(py)BeNPr^n_2)_2] \rightleftharpoons Me_2Bepy_2 + (Pr^n_2N)_2Be$$

N-Trimethylethylenediamine also yields a bridged dimeric compound, and the spare tertiary amino group readily chelates with 3-co-ordinate beryllium. Thus, dimethylberyllium forms a thermally-unstable chelate which readily eliminates methane. Sym-dimethylethylenediamine also forms a similar complex which, while readily subliming at 90°C, polymerizes on further heating.[27]

$$Me_2Be \leftarrow (HN(Me)CH_2CH_2NMe_2) \xrightarrow{18^\circ C} [MeBe\text{-}N(Me)CH_2CH_2NMe_2]_2$$

(m.p. 116–8°C)

$$Me_2Be + (MeNHCH_2)_2 \longrightarrow$$

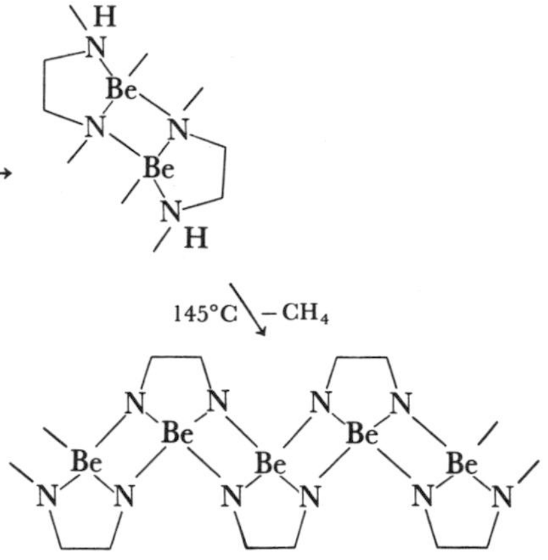

145°C ↘ $-CH_4$

A parallel study on magnesium amides illustrates some differences between the two metals. *N*-trimethylethylenediamine reacts with magnesium dialkyls,[22] in much the same way as dimethylberyllium, giving a chelate-stabilized dimer.

$$R_2Mg + Me_2N(CH_2)_2NHMe \longrightarrow$$

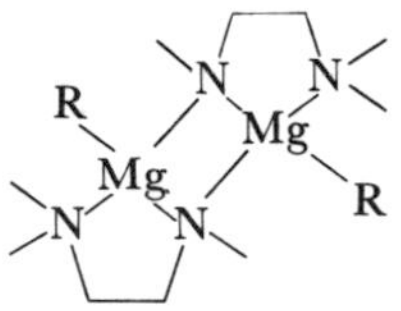

(R = Me, Pr^i, Bu^t)

Only with more hindered amines do the 3-co-ordinatively bound magnesium amide dimers result.[28]

$$Pr^i_2NH + Pr^i_2Mg \longrightarrow [Pr^iMgNPr^i_2]_2$$

$$PhN{=}CHPh + Et_2Mg \longrightarrow$$

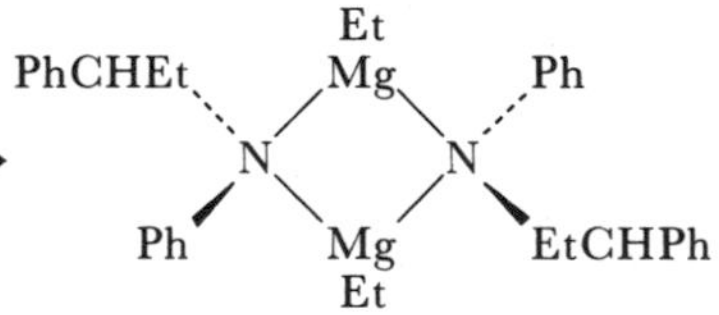

In THF, diethylamine and diethylmagnesium form a solvated dimer which is probably in equilibrium with an amide–ethyl bridged tetramer.

$$Et_2Mg + Et_2NH \xrightarrow{THF}$$

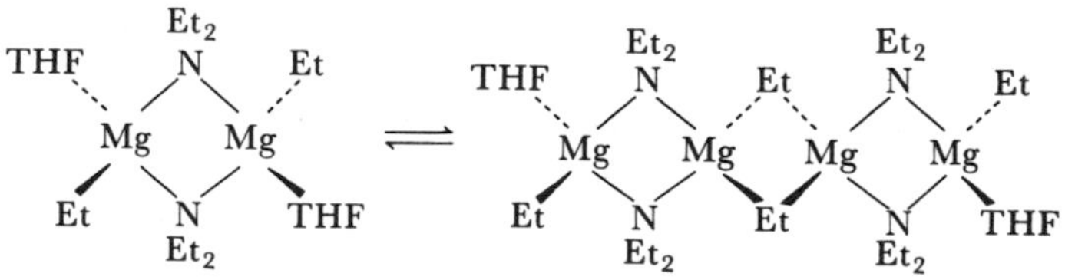

Molecular weight determinations indicate association greater than two, and alkyl bridges are very strong in the magnesium dialkyls.[a]

With less-hindered systems still (*e.g.*, Me_2NH and Et_2Mg), disproportionation into the dialkyl and diamide occurs.

$$Me_2NH + Et_2Mg \longrightarrow [Me_2NMgEt] \longrightarrow [(Me_2N)_2Mg]_x + [Et_2Mg]_n$$

A similar decomposition has already been mentioned for organoberyllium amides in the presence of pyridine or dipyridyl.

ALKOXIDES

Beryllium

Though secondary amine adducts have been prepared with dialkylberyllium compounds (*e.g.*, $Me_2NH{\cdot}BeMe_2$), no such co-ordination compounds are known for alcohols. In this case the alkoxides are formed directly together with the expected alkanes, though this is by no means the only route to organoberyllium alkoxides. The addition of the C—Be bond across a carbonyl group, and the opening of strained small rings and weak bonds, are also useful synthetic routes.[29]

$$[MeBeOPr^n]_4 \xleftarrow{Pr^nOH} Me_2Be \xrightarrow{Pr^iOH} [MeBeOPr^i]_4$$

$$[MeBeOPr^n]_4 \xleftarrow{\text{ethylene oxide}} Me_2Be \xrightarrow{CH_3CHO} [MeBeOPr^i]_4$$

$$Me_2Be + Bu^tOH \longrightarrow [MeBeOBu^t]_4$$

$$Me_2Be + Me_2CO \longrightarrow [MeBeOBu^t]_4 \underset{}{\overset{Me_2CO}{\rightleftharpoons}} [(Bu^tO)_2Be]_3$$

$$Me_2Be + (Bu^tO)_2 \longrightarrow [MeBeOBu^t]_4$$

These are illustrated by the ring opening of ethylene oxide and the cleavage of the peroxide link of di-t-butylperoxide. One point of comparison in passing is that hydrazines, unlike the peroxides, are not cleaved by the C—Be bond. Tetramethylhydrazine and diethylberyllium form a 1 : 2 complex, m.p. 84–7°C,[30] involving co-ordination of the nitrogen lone-pairs.

$$2Et_2Be + Me_4N_2 \longrightarrow Et_2BeNMe_2Me_2NBeEt_2$$

The addition of the one Be—C bond to acetone gives the organo t-butoxide of beryllium, and an excess of acetone adds across the other Be—C bond giving di-t-butoxyberyllium. This also results from dimethylberyllium with an excess of t-butanol and is a sublimable solid, m.p. 112°C. It is trimeric in benzene and the p.m.r. spectrum indicates the presence of two types of proton at τ values 8.62 and 8.78 (ratio 2 : 1).

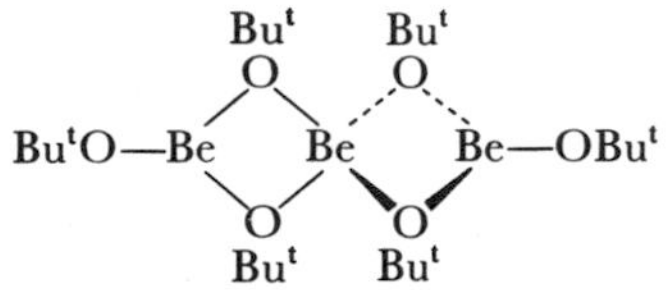

Bu^{t}O—Be(μ-OBu^t)$_2$Be(μ-OBu^t)$_2$Be—OBu^t

Fig. 2.16. $[(Bu^tO)_2Be]_3$

This data supports a structure similar to that found for the dimethylamide of beryllium, though with the alkoxide the more acidic protons appear to be on the bridging rather than terminal groups. As with the amides, increased bulkiness leads to lower oligomers and beryllium triethylmethoxide is no exception. This is dimeric and may possess the simple structure shown.[29]

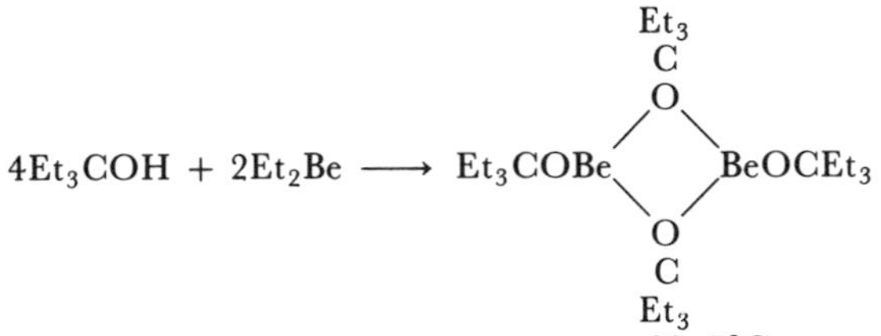

$$4Et_3COH + 2Et_2Be \longrightarrow Et_3COBe(\mu\text{-}OCEt_3)_2BeOCEt_3$$

m.p. 36–7°C

The organoberyllium alkoxides already discussed tend to be tetrameric, especially when the organic substituent on beryllium is small. Large substituents on the beryllium atom give tetrameric alkoxides only with methanol. These include $[MeBeOR]_4$, R = Me, Bu^t, $PhCH_2$; $[R'BeOMe]_4$, R′ = Pr^i, Bu^t, Ph. The structures of these tetramers involve a cube arrangement analogous to that already encountered with potassium t-butoxide, and a structural determination on methyl-trimethylsiloxyberyllium shows an almost perfect cube of Be and O atoms.[31]

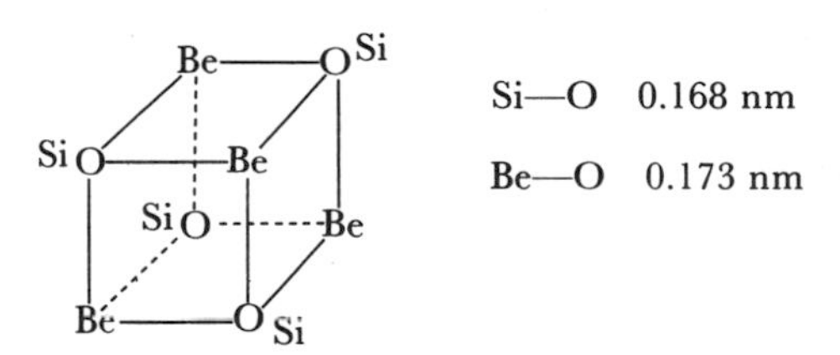

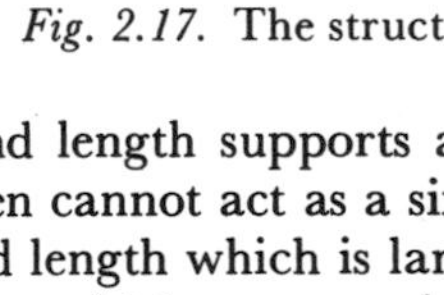

Fig. 2.17. The structure of $[Me_3SiOBeMe]_4$

The Si—O bond length supports a small degree of p_π–d_π bonding and hence the oxygen cannot act as a six-electron donor. This is supported by the Be—O bond length which is larger than in beryl, and the small angle (~90 degrees) would be expected if three- or four-centre bonding were involved.

The melting points of these alkoxides deserve mention, since contrary to what is generally observed, the substitution of a large alkoxy substituent increases the melting point (see Table 2.1). The closeness of the skeleton to a cube structure may well account for the closeness of the melting points

Table 2.1. Melting Points of Organoberyllium Alkoxides

$[MeBeOR]_4$	R = Me	Et	Pr^n	Pr^i	Bu^t	Pr^iBeOMe
m.p.:	23.5°C	28–30°C	38–40°C	134–6°C	93°C	133–5°C

of the two isopropyl compounds. The molecular shape will be very similar for the two compounds.

More hindered alcohols lead to trimeric products with the beryllium atoms 3-co-ordinate. A 6-membered Be_3O_3 ring is believed to be present,

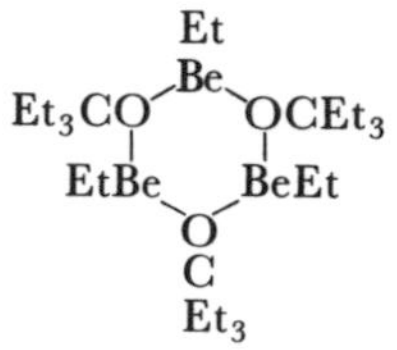

Et
Be
Et_3CO $OCEt_3$
EtBe BeEt
O
C
Et_3

Fig. 2.18. $[Et_3COBeEt]_3$

as has already been reported with beryllium ions in solution ($[BeOH]_3^{3+}$). Triethylmethyl alcohol and diethylberyllium gave such a product, while dimeric alkoxides result from dimethylberyllium and triphenylcarbinol or from t-butanol and di-t-butylberyllium.[29]

$$Et_3COH + Et_2Be \longrightarrow [EtBeOCEt_3]_3$$

$$Bu^tOH + Bu^t_2Be \longrightarrow [Bu^tBeOBu^t]_2$$

$$Ph_3COH + Me_2Be \longrightarrow [MeBeOCPh_3]_2$$

The tetrameric alkoxides appear to dissociate on dilution and with methylberyllium t-butoxide, association falls from 3.8 to 1.97 on diluting from 0.088 to 0.015 moles dm^{-3}.

Recrystallization of some of the more hindered alkoxides from ether and THF yields solvated monomers containing 3-co-ordinate beryllium. Examples include $MeBeOCPh_3{\cdot}OEt_2$ and $MeBeOCHPh_2{\cdot}L$ (L = Et_2O or THF), and generally, the n.m.r. spectra show the absorption position for 3-co-ordinate Me—Be bonds lie between τ 10.8 and 11.5.

An excess of pyridine cleaves the tetrameric cages as with the trimeric and dimeric amides, to give dipyridine complexes.

$$[MeBeOMe]_4 + 8py \longrightarrow 4MeBeOMepy_2$$

$$[MeBeNMe_2]_3 + 6py \longrightarrow 3MeBeNMe_2py_2$$

Unlike the amides, however, which give bridged dimers when complexed with less pyridine, the alkoxides disproportionate into complexed dimethylberyllium and the dialkoxide.

$$2[MeBeNMe_2]_3 + 6py \longrightarrow 3[MeBeNMe_2py]_2$$

$$[MeBeOMe]_4 + 4py \longrightarrow 2Me_2Bepy_2 + 2(MeO)_2Be$$

This has been observed with the more hindered amides.

The n.m.r. spectra of these tetrameric alkoxides show a Me—Be absorption position at τ 10.46–10.77. The absorption positions of the complexes $MeBeOPh{\cdot}OEt_2$ and $MeBeOBu^t{\cdot}py$ indicate a 4-co-ordinate state, and the association between 1.2 and 1.4 support equilibria involving the complexed bridged dimer and the tetramer.

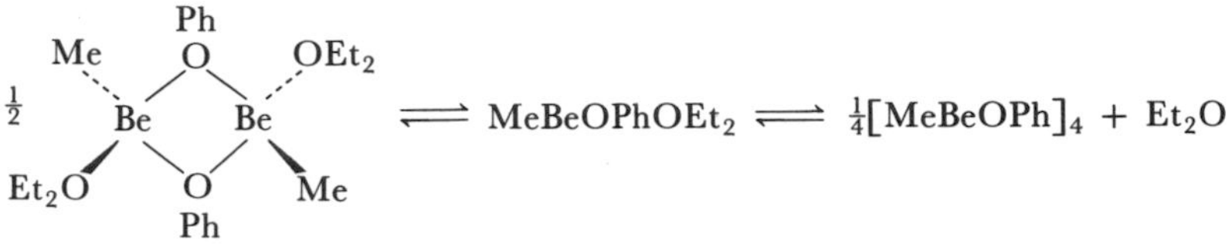

$$\tfrac{1}{2}\,[Me(Et_2O)Be(\mu\text{-}OPh)_2Be(OEt_2)Me] \rightleftharpoons MeBeOPhOEt_2 \rightleftharpoons \tfrac{1}{4}[MeBeOPh]_4 + Et_2O$$

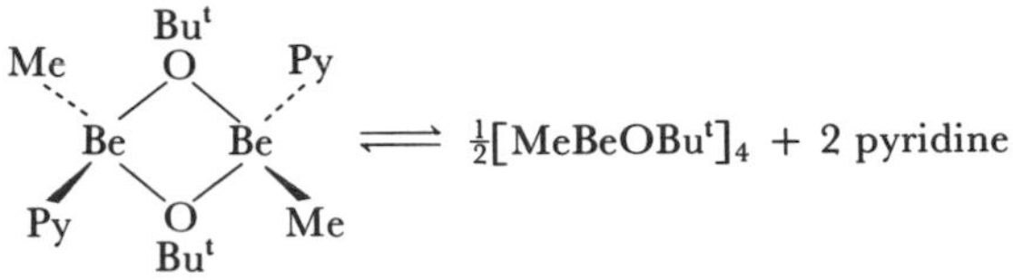

$$[Me(Py)Be(\mu\text{-}OBu^t)_2Be(Py)Me] \rightleftharpoons \tfrac{1}{2}[MeBeOBu^t]_4 + 2\text{ pyridine}$$

The evaporation of solutions of both compounds shows the presence of a third of the ligand in the condensate, in support of these equilibria.

Dimethylberyllium forms the tetrameric alkoxides $[MeBeONCMe_2]_4$, $[MeBeO(CH_2)_2OMe]_4$ and $[MeBeOC_9H_6N]_4$ with acetoxime, 2-methoxyethanol and 8-hydroxyquinoline, respectively.[32] The former two are thought to have the usual cube structure, and the absence of chelation by the methoxyethanol supports the proposal that the donor strength of oxygen bonded to an electropositive metal is greater than that of an ether.[33] This is encountered among the complexes of the Group III metals, especially aluminium and gallium. The yellow quinolinate is believed to comprise non-planar dimeric units bonded in pairs by Be—O bridges (Figure 2.19).

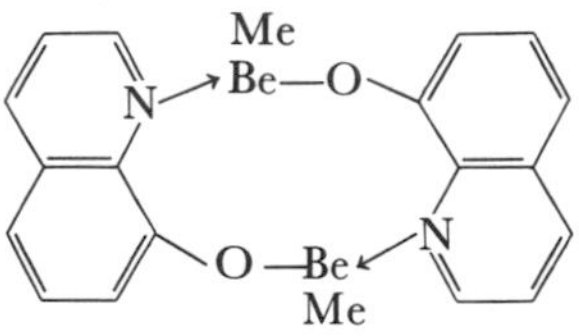

Fig. 2.19. The 8-hydroxyquinoline complex

Magnesium

The most plausible intermediates in the reaction of a Grignard reagent RMgX with carbonyl compounds are magnesium alkoxides, as hydrolysis yields the alcohol.

$$RMgX + R'_2CO \longrightarrow RR'_2COMgX \xrightarrow{H_2O} RR'_2COH + MgX(OH)$$

Consequently, information on the alkoxides of these metals must inevitably help in obtaining an understanding of the chemistry of that most enigmatic of organometallic compounds, the Grignard reagent.

Organomagnesium alkoxides were first prepared by a direct route involving magnesium, the alcohol and alkyl halide.[34] n-Butylmagnesium alkoxides are conveniently synthesized this way, using methylcyclohexane as a solvent,

$$Mg + ROH + Bu^nCl \xrightarrow{100°C} Bu^nMgOR + HCl$$
$$(R = Et, Pr^n, Pr^i \text{ and } Bu^s)$$

but those with a smaller alkyl group on the metal are best synthesized from dialkylmagnesium and 1 mole of the alcohol, as with organoberyllium alkoxides.[35]

$$R_2Mg + R'OH \longrightarrow RMgOR' + RH$$

The alkylmagnesium alkoxides closely resemble the beryllium ones. They occur in dimeric, trimeric and tetrameric states as well as with higher degrees of polymerization. Two dimeric alkoxides are known and both have the alkoxy group substituted at the α- and β-carbon atoms, *viz.*, $EtMgOCEt_3$ and $EtMgOCEt_2Me$.

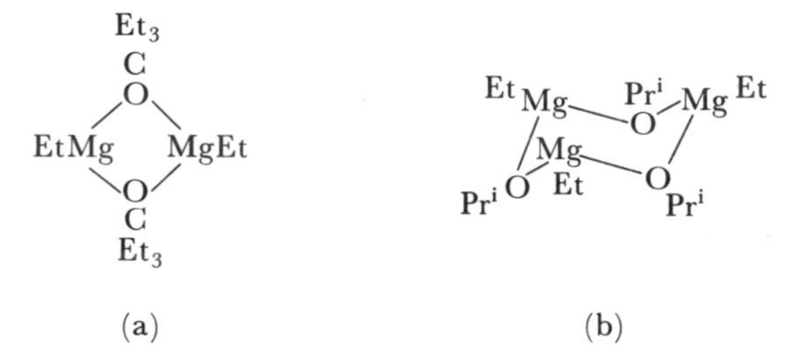

Fig. 2.20. Ethylmagnesium alkoxides: (a) $EtMgOCEt_3$ and (b) $EtMgOPr^i$

n-Butylmagnesium isopropoxide is trimeric and believed to possess a 6-membered Mg_3O_3 ring. The reluctance of the 3-co-ordinate magnesium atoms to complex with ether tends to support the idea of π-bonding from oxygen to the unoccupied 3d orbitals.[36]

In general, however, the polymeric state of the unsolvated alkoxides is higher. Those synthesized from primary alcohols are associated with $n = 8$, while alcohols branched at the α-carbon atom give tetrameric derivatives whose structures probably resemble the beryllium alkoxides already discussed. These include $[EtMgOR]_4$ (R = Pr^i, Bu^t) and $[Pr^iMgOPr^i]_4$.

Solvation results when the more hindered of these alkoxides are recrystallized from ethers. $[EtMgOCEt_3]_2$ forms an etherate with Et_2O, but this readily dissociates. That of THF is more stable, however, and is dimeric in benzene (Figure 2.21), and believed to contain alkoxide bridges.

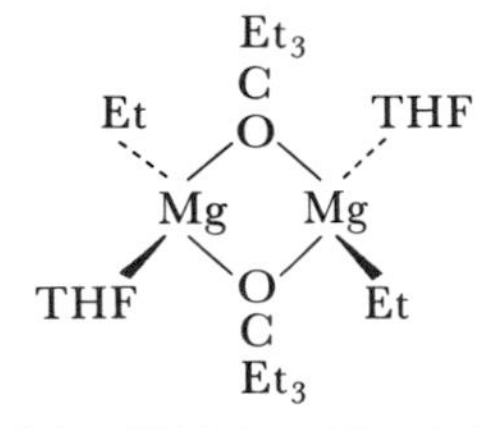

Fig. 2.21. $[EtMg(THF)OCEt_3]_2$

It is of interest to note that increased substitution at the β-carbon atom of the alkoxide bridge leads to the structural change from $[EtMgOBu^t]_4$ to $[EtMgOCEt_3]_2$, and that the tetramers remain so, even in ether.[35]

Products isolated directly from the reactions of carbonyl compounds with a Grignard reagent resemble those already discussed in several ways. They are oligomeric when solvated, and contain alkoxide bridges, but are insoluble, and probably polymeric, when uncomplexed.

Thus, recrystallizing t-butoxymagnesium bromide from ether gives a solvated dimer. This compound can be synthesized in the three ways shown, the diethylether being readily displaced by THF.

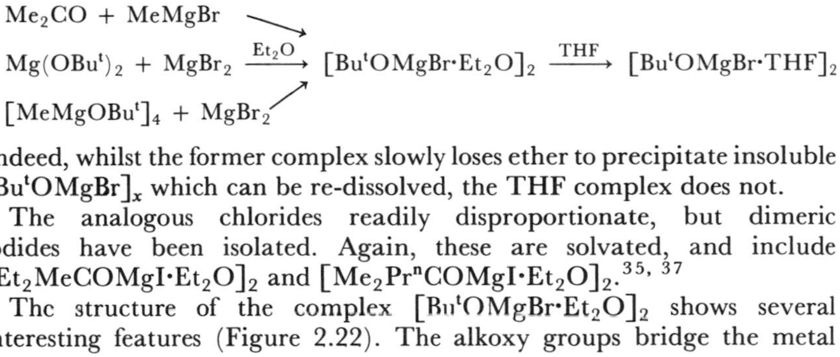

$$Me_2CO + MeMgBr$$

$$Mg(OBu^t)_2 + MgBr_2 \xrightarrow{Et_2O} [Bu^tOMgBr{\cdot}Et_2O]_2 \xrightarrow{THF} [Bu^tOMgBr{\cdot}THF]_2$$

$$[MeMgOBu^t]_4 + MgBr_2$$

Indeed, whilst the former complex slowly loses ether to precipitate insoluble $[Bu^tOMgBr]_x$ which can be re-dissolved, the THF complex does not.

The analogous chlorides readily disproportionate, but dimeric iodides have been isolated. Again, these are solvated, and include $[Et_2MeCOMgI{\cdot}Et_2O]_2$ and $[Me_2Pr^nCOMgI{\cdot}Et_2O]_2$.[35, 37]

The structure of the complex $[Bu^tOMgBr{\cdot}Et_2O]_2$ shows several interesting features (Figure 2.22). The alkoxy groups bridge the metal

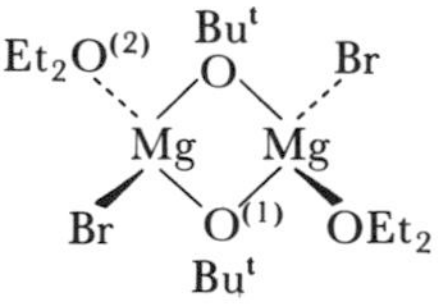

Mg—$O_{(2)}$ 0.201 nm
Mg—$O_{(1)}$ 0.191 nm
Mg—$O_{(1)}$—Mg 97 degrees

Fig. 2.22. $[Et_2OMg(OBu^t)Br]_2$

atoms, and the groups around both kinds of oxygen atom are in its plane.[38] This, and the short Mg—$O_{(1)}$ distance of 0.191 nm, may well support the p_π–d_π bonding already proposed to explain why $[Bu^nMgOPr^i]_3$ fails to complex with ether. The Mg—$O_{(2)}$ distance is close to that encountered among etherated Grignard reagents[39] ($PhMgBr{\cdot}2Et_2O$, 0.201, 0.206; $EtMgBr{\cdot}2Et_2O$, 0.203, 0.206 nm).

A comparatively interesting set of molecular parameters has been obtained from the X-ray structural determination of $Mg[(MeO)_2AlMe_2]_2{\cdot}C_4H_8O_2$.[40] This compound is synthesized by the partial oxidation of magnesium tetramethylaluminate in dioxan.

$$Mg(AlMe_4)_2 + O_2 \xrightarrow{\text{dioxan}} Mg[(MeO)_2AlMe_2]_2{\cdot}C_4H_8O_2$$

It is polymeric, through bridging dioxan molecules. The structure consists of planar $Al(OMe)_2Mg(OMe)_2Al$ residues cross-linked by dioxan through the 6-co-ordinate magnesium atoms.

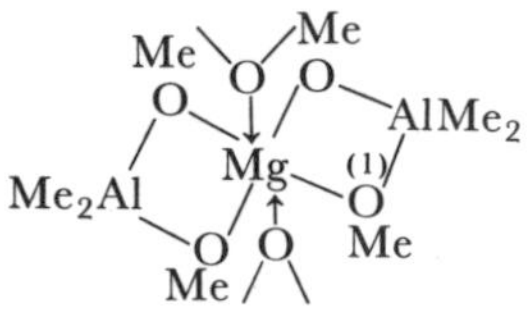

Mg—$O_{(1)}$ 0.205, 0.206 nm
Al—O 0.180, 0.186 nm
Mg—O 0.224 nm
Al—C 0.199 nm

Fig. 2.23. Co-ordination around the magnesium atom in $Mg[(MeO)_2AlMe_2]_2{\cdot}C_4H_8O_2$ showing alkoxy and dioxan oxygen atoms.

The planarity of the bridging oxygen atoms indicates the possibility of p_π–d_π bonding, but the Mg—O distances are both larger than those in $[Bu^tOMgBr{\cdot}Et_2O]_2$, though in each case the ether–magnesium bond length was the longer. In addition, if it is justifiable to compare the Mg—C and Mg—O bond lengths using covalent radii and the electronegativity correction, the Mg—O bond should be the shorter by about 0.012 nm. The structure of ethylmagnesium bromide (solvated with ether) supports this, with C—Mg 0.216 nm and Mg—O 0.203 and 0.206 nm. The alkoxide–magnesium distance of 0.191 nm in $[Bu^tOMgBr{\cdot}OEt_2]_2$, again with 4-coordinate magnesium, may therefore support p_π–d_π bonding. With 6-co-ordinate magnesium, in the aluminate, again two distinctly different Mg—O distances were measured. To propose π bonding to explain this would be premature with the limited data available to date, though the bridged Mg—O distance is shorter than in magnesium oxide, hydroxide or hydrate.[a]

THIO DERIVATIVES OF BERYLLIUM AND MAGNESIUM

The reactions of dialkyl derivatives of these metals with thiols[41] follow similar lines to those observed with alcohols. Thio-bridged dimers, trimers and tetramers have been obtained, but there appears to be a greater tendency for the alkylthio derivative to disproportionate than with alkoxides. This may well be due to the weaker donating ability of sulphur to "a"-type metals such as beryllium and magnesium, compared with oxygen.[42]

Considering the beryllium tetramers first, these have been synthesized by three routes from diethylberyllium in ether, and resemble those used for beryllium alkoxides.

$$Et_2Be \text{ in } Et_2O \xrightarrow{Be(SEt)_2,\ EtSH,\ EtSSEt} [EtBeSEt]_4$$

Only four tetramers have been synthesized, $[EtBeSR]_4$ (R = Et, Pr^i, Bu^t) and $[MeBeSBu^t]_4$, and, apart from $[EtBeSEt]_4$, they are best synthesized in a non-donor solvent. The less hindered of these tetramers, $[EtBeSR]_4$ (R = Et, Pr^i), yield no complexes with ether or pyridine and are stable to disproportionation, though $[EtBeSPr^i]_4$ is probably dimeric in ether at high dilution. Isomeric $[Pr^iBeSEt]$, however, is stabilized as dimeric solvates, the etherate reacting with pyridine to give the pyridine complex.

$$Pr^i_2Be + EtSH \xrightarrow{Et_2O} [Pr^i(Et_2O)Be(\mu\text{-}SEt)_2Be(OEt_2)Pr^i] \xrightarrow{py} [Pr^i(py)Be(\mu\text{-}SEt)_2Be(py)Pr^i]$$

Further hindrance using propane-2-thiol gives a monomer as the only complex, while increasing steric hindrance further with 2-methylpropane-2-thiol causes disproportionation to occur.

$$Pr^i_2Be + Pr^iSH \xrightarrow{py} Pr^iBeSPr^i \cdot py_2$$

$$2Pr^i_2Be + 2Bu^tSH \xrightarrow{Et_2O} 2[Pr^iBeSBu^t] \longrightarrow Pr^i_2Be + Be[SBu^t]_2$$

Hence, the reaction has reversed in going from two ethyl groups to a Pr^i/Bu^t combination, illustrating a highly labile system.

Both $[MeBeSBu^t]_4$ and $[EtBeSBu^t]_4$ slowly disproportionate, and the sulphur bridges are readily broken by ethers and amines. Molecular weight determinations establish the equilibria (equation (*2.1*))

$$\tfrac{1}{4}[EtBeSBu^t]_4 + L \rightleftharpoons EtBeSBu^t \cdot L \rightleftharpoons \tfrac{1}{2}[EtBeSBu^t \cdot L]_2 \qquad (2.1)$$

which with an excess of THF lie well over to the left. Thus, the molecularity of $EtBeSBu^tTHF$ is greater than unity when measured in benzene, and THF is detectable in the evaporated solvent. With pyridine (1 mole per Be atom), the complex formed analyzes for $EtBeSBu^t \cdot py$ with a degree of association of two. Hence, the equilibrium is predominantly to the right.

So while the weaker base (ether) leads to oligomerization and desolvation, the stronger amines not only complex but in excess will break the sulphur bridge.

$$[EtBeSBu^t{\cdot}py]_2 \xrightarrow{2py} \underset{\text{yellow}}{EtBeSBu^t{\cdot}py_2}$$

The sulphides $Be(SR)_2$ ($R = Et, Bu^t$) are insoluble in hydrocarbons and ether, and are probably polymeric, unlike $[Be(OBu^t)_2]_3$. The sulphur bridges are readily broken by pyridine and 2,2′-bipyridyl to yield 4-co-ordinate monomers, $Be(SBu^t)_2py_2$ and $Be(SBu^t)_2bipy$, the former dissociating in benzene without precipitation implying bridged-dimer formation.

$$\frac{2}{x}[Be(SBu^t)_2]_x \xrightarrow{4py} 2Be(SBu^t)_2py_2 \rightleftarrows [(Bu^tS)(py)Be(\mu\text{-}SBu^t)_2Be(py)(SBu^t)] + 2py$$

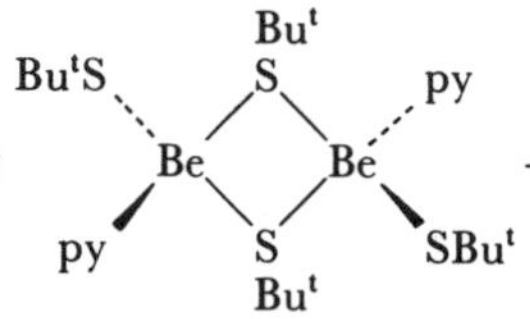

Thio- and seleno-phenol give undissociated dimeric ether complexes $[MeBeSPh{\cdot}Et_2O]_2$ and $[EtBeSePh{\cdot}Et_2O]$, in contrast to $[MeBeOPh{\cdot}Et_2O]_2$ which is in equilibrium with ether, and probably the tetramer, in solution.

The only trimeric organoberyllium thio derivative is chelate-stabilized and contains four fused rings.[32] It is almost insoluble in ether, as would be expected for 4-co-ordinate beryllium, and results from 2-dimethylaminoethane thiol and dimethylberyllium in ether.

$$3Me_2NCHCH_2SH + 3Me_2Be \longrightarrow [MeBeSCH_2CH_2NMe_2]_3 + 3CH_4$$

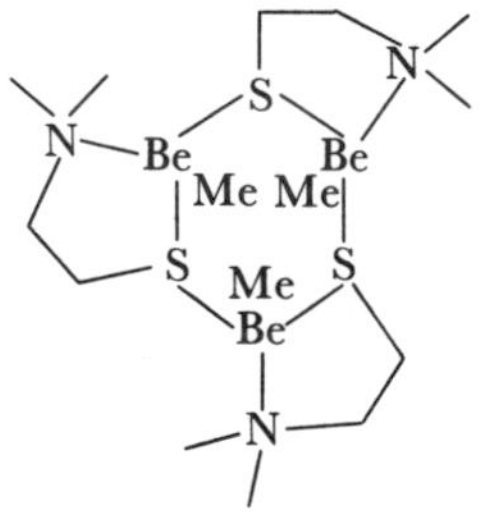

(m.p. 194°C dec.")

Fig. 2.24. The structure proposed for $[MeBeS(CH_2)_2NMe_2]_3$

The only tetrameric organomagnesium analogue of the thioberyllium compounds discussed is $[EtMgSBu^t]_4$, and this has to be synthesized in an ether-free system.[43]

$$4Et_2Mg + 4Bu^tSH \xrightarrow{\text{toluene}} [EtMgSBu^t]_4 + 4EtH$$

As with the tetrameric alkoxides, this compound is believed to have a structure involving a Mg_4S_4 cube.

2-Dimethylaminoethane thiol forms a chelated bridged-dimer complex with di-t-butylmagnesium.

$$2Me_2NCH_2CH_2SH + 2Bu^t_2Mg \longrightarrow [Bu^tMgSCH_2CH_2NMe_2]_2 + 2Bu^tH$$

With dimethylmagnesium, this thiol yields a tetramer. The structure probably involves the expected Mg_4S_4 cube with an amino group weakly co-ordinating with each magnesium atom, rendering them 5-co-ordinate, as found in $(Et_2O)_4Mg_4OBr_6$ (this Chapter, p. 10).

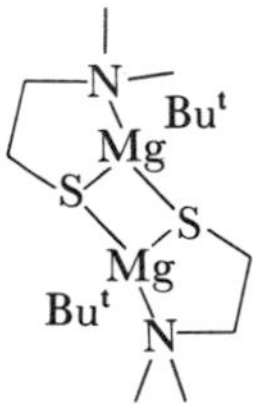

Fig. 2.25. $[Bu^tMgS(CH_2)_2NMe_2]_2$

Generally speaking, the compounds RMgSR′ can only be obtained in an oligomeric state if complexed. Thus, while $EtMgSBu^t$ can be obtained as a tetramer in an ether-free system, the presence of THF gives a dimeric complex, (m.p. 72–4°C).

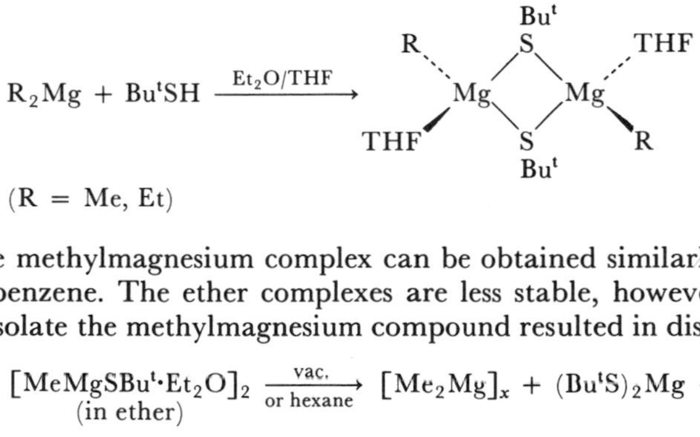

$$R_2Mg + Bu^tSH \xrightarrow{Et_2O/THF} [R(THF)Mg(\mu\text{-}SBu^t)_2Mg(THF)R]$$

(R = Me, Et)

The methylmagnesium complex can be obtained similarly and is dimeric in benzene. The ether complexes are less stable, however, and attempts to isolate the methylmagnesium compound resulted in disproportionation.

$$\underset{\text{(in ether)}}{[MeMgSBu^t{\cdot}Et_2O]_2} \xrightarrow[\text{or hexane}]{\text{vac.}} [Me_2Mg]_x + (Bu^tS)_2Mg + 2Et_2O$$

This is probable related to the stability of the bridging $MgMe_2Mg$ group, compared with other alkylmagnesium systems.[44] Thus, while polymeric dimethylmagnesium is partly dissociated in diethylether, it can be recrystallized ether-free. Diethylmagnesium is less associated in ether, and recrystallizes as a solvate.[45]

Since the t-butyl group would be expected to have even weaker bridging properties, it is understandable that di-t-butylmagnesium and propane-2-thiol form an etherate.

$$2Bu^t_2Mg + 2Pr^iSH \xrightarrow{2Et_2O} [Bu^tMgSPr^i{\cdot}Et_2O]_2 + 2Bu^tH$$

The reaction of di-t-butylmagnesium with 2-methyl-propane-2-thiol might be expected to give a similar compound in ether, but unexpectedly, only disproportionation products result.

$$2Bu^t_2Mg + 2Bu^tSH \xrightarrow[\text{or THF}]{Et_2O} 2[Bu^tMgSBu^t] \longrightarrow Bu^t_2Mg + Mg(SBu^t)_2$$

This may be due to the combined effect of hindrance in the bridged dimer, and to the ready solvation of di-t-butylmagnesium.[22]

ORGANOBERYLLIUM AND ORGANOMAGNESIUM HALIDES

Organomagnesium halides

The most versatile and widely used of organometallic compounds is the Grignard reagent. It has been used to alkylate Main Group elements such as beryllium and arsenic, the transition metals iron, chromium, rhodium and others, as well as performing its wide variety of functions in organic chemistry.

This reagent was discovered in 1900 in the laboratory of Professor Barbier at the University of Lyon. Grignard, then a graduate student, had been given a problem to optimize conditions for what is now known as the Barbier reaction. The system he considered involved (4-methylbut-3-enyl)methyl ketone, which when hydrolysed in the presence of methyl iodide and magnesium gave the alcohol.

$$Me_2C{=}CH(CH_2)_2COMe + MeI + Mg \xrightarrow{H_2O} Me_2C{=}CH(CH_2)_2CMe_2OH$$

Grignard found that alkyl halides and magnesium react readily in ether, and yields of the alcohol formed after addition of the carbonyl compound, followed by hydrolysis, were higher than Barbier had reported.[b] He formulated his magnesium intermediate as RMgX, and its versatility was rapidly recognized. This led to Grignard being awarded the Nobel Prize in 1912.[c]

One of the paradoxes of chemistry is that despite its almost limitless use as a synthetic intermediate, the structure of the species called "the Grignard reagent" is, as yet, far from being understood. In the past ten years, however, the advent of new physical techniques, and improved methods of handling air-sensitive compounds, has led to both a deeper understanding and an appreciation of the complexity of this most enigmatic of reagents.

The molecular structure of PhMgBr recrystallized from THF, and

EtMgBr recrystallized from ether, show 4-co-ordinate magnesium σ-bonded to the halogen and organic group.[39] These compounds are also solvated and monomeric at high dilution, supporting the formulation RMgX proposed by Grignard.[d]

However, the species in solution and in the crystalline state are not necessarily the same, and the addition of dioxan to ether solutions of a Grignard reagent precipitates the complexed dihalide, suggesting that what is now known as the Schlenk equilibrium holds.[46]

$$2RMgX \rightleftharpoons R_2Mg + MgX_2$$

In addition, solutions obtained by mixing equimolar amounts of $MgBr_2$ (labelled with radioactive ^{25}Mg) and diethylmagnesium in ether were shown to statistically exchange magnesium, as expected for the above equilibrium.[47]

The monomeric RMgX is abundant at low concentrations in ether. Solutions of diethylmagnesium and magnesium halides in THF are also separately monomeric at high dilution. As concentrations increase, polymerization begins to occur. This is rapid for the dihalides, and complexes analogous to $BeCl_2$ or solvated rings are probably formed.[48]

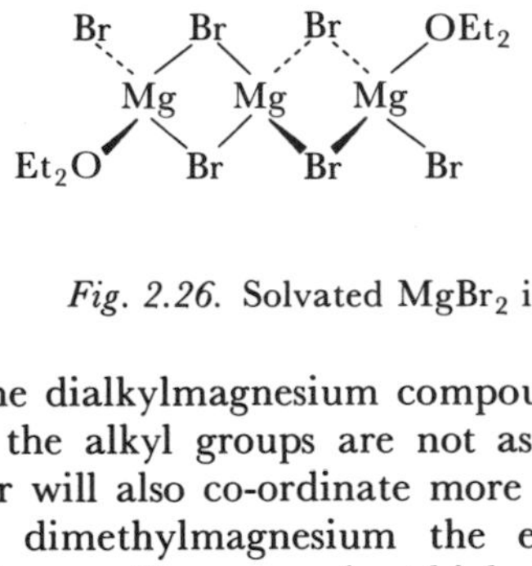

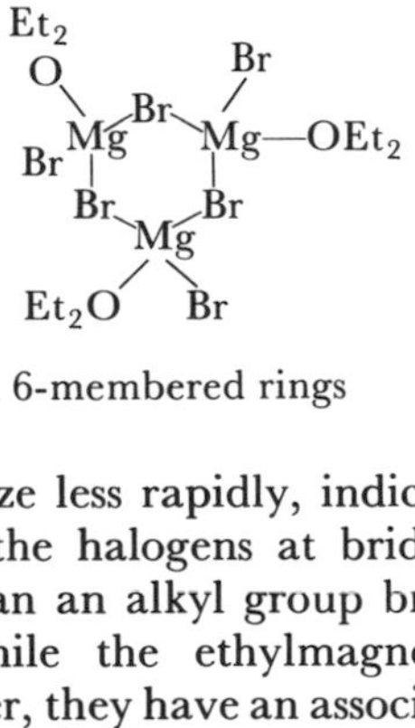

Fig. 2.26. Solvated $MgBr_2$ involving 4- and 6-membered rings

The dialkylmagnesium compounds polymerize less rapidly, indicating that the alkyl groups are not as effective as the halogens at bridging. Ether will also co-ordinate more effectively than an alkyl group bridge, with dimethylmagnesium the exception. While the ethylmagnesium halides are all monomeric at high dilution in ether, they have an association factor $i = 2.5$ at concentrations of 2–3 moles dm^{-3}.[49] t-Butylmagnesium chloride is dimeric over a wide concentration range, however, and the poor bridging character of the t-butyl group, already encountered with the alkylthio derivatives, supports a halogen-bridged structure. The monoetherate is also dimeric.

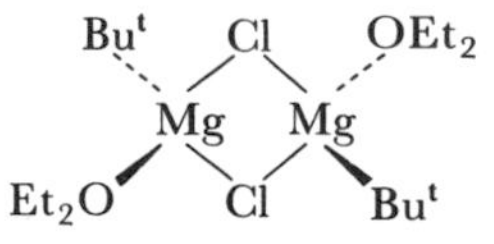

Fig. 2.27. $[Bu^tMgCl \cdot Et_2O]_2$

In THF, the stronger donating power of the solvent over ether renders solutions of the Grignard reagents monomeric over a wide concentration range, in contrast to the ether solutions. The reagents considered were alkylmagnesium chlorides and bromides and phenylmagnesium bromide.

Triethylamine will also readily complex with the Grignard reagent, the amine complex precipitating readily from an ether solution of ethylmagnesium bromide.[50] Under reduced pressure, this readily loses half of complexed amine, possibly due to the bulkiness of triethylamine compared with ether and THF.

$$\text{EtMgBr in ether} \xrightarrow{2Et_3N} (Et_3N)_2EtMgBr \xrightarrow[-Et_3N]{\text{vac.}} [Et_3NEtMgBr]_2$$

Indeed Bu^tMgCl is dimeric as an etherate. The dimeric character of the amine complex has been confirmed by molecular weight and X-ray structural determinations.[51]

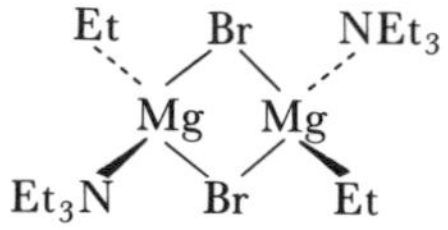

Et Br NEt3
Mg Mg
Et3N Br Et

Fig. 2.28. $[EtMgBr{\cdot}Et_3N]_2$

The Mg–Br distance compares closely with that found in $(Et_2O)_4$-Mg_4OBr_6, and is significantly larger than the Mg–Br distance in both phenyl and ethylmagnesium bromides (0.246 nm).

So it is quite apparent that the species present in the solution of a Grignard reagent are numerous, and that their concentrations depend on the conditions encountered, *viz.*, dilution, solvent, substituents. This is best summarized through the Schlenk equilibrium, to include the bridged dimer.[52]

$$R_2Mg + MgX_2 \rightleftharpoons 2RMgX \rightleftharpoons RMg\underset{X}{\overset{X}{\diamond}}MgR$$

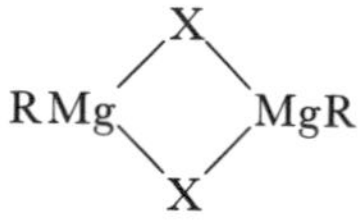

Further evidence for the presence of a halogen-bridged structure has been found from the low-temperature n.m.r. spectral data obtained for mixtures of dimethylmagnesium and methylmagnesium bromide in ether.[53] The data obtained are summarized in Figure 2.29. At 30°C, only one peak is observed for dimethylmagnesium, indicating rapid exchange of the methyl group. Cooling slows this process, and at −105°C three peaks are observed, as shown by curve A. These have been tentatively assigned to bridging (τ 11.32), terminal (in Me_2Mg oligomer, τ 11.69) and terminal sites (in Me_2Mg monomer, τ 11.74) on the magnesium.

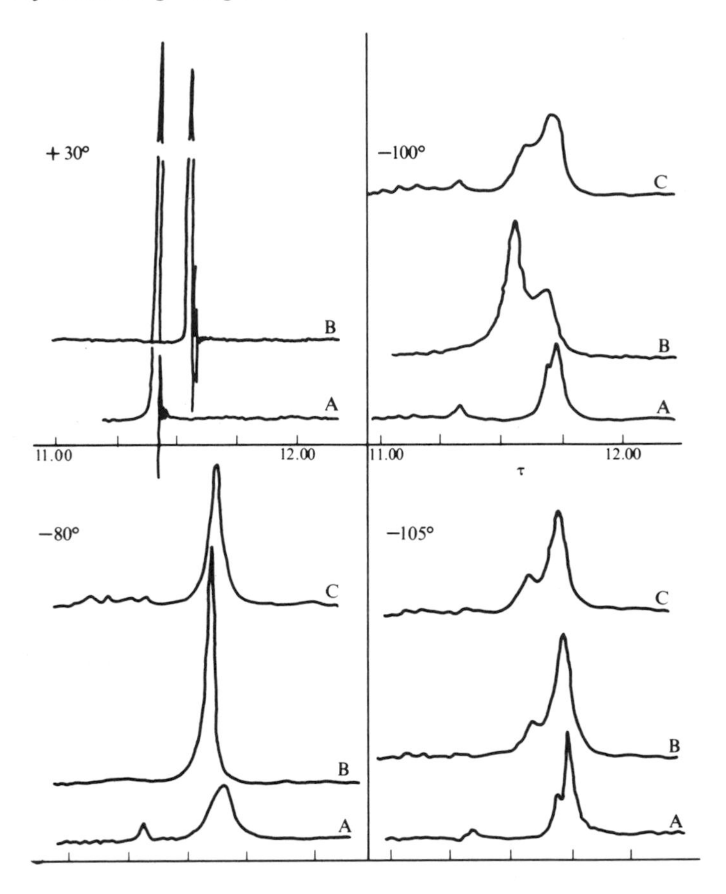

Fig. 2.29. Low-temperature 100 MHz n.m.r. spectra of A, Me_2Mg; B, MeMgBr; and C, a 1 : 1 mixture of A and B[53]

Curve B for MeMgBr does not resolve until −100°C, and changes drastically with temperature and time. This is consistent with the disproportionation of the Grignard reagent into Me_2Mg and $MgBr_2$, the latter being precipitated.

$$2MeMgBr \rightleftharpoons Me_2Mg + MgBr_2\downarrow$$

So the peaks observed at −100°C can be assigned to the Grignard reagent (lower field) and Me_2Mg, the intensity of the latter increasing with time and temperature fall to −105°C. In addition, a doublet appears at τ 11.10 and 11.16 which also occurs when Me_2Mg is rapidly cooled. This doublet is thought to be caused by particular methyl bridging groups.

All peaks characteristic of bridging groups are absent in the p.m.r. spectrum of the Grignard reagent at $-100°C$ (before significant disproportionation has occurred). This is consistent with the halogen being the predominant bridging group in ether solutions of MeMgBr, as supported by molecular weight data.

Curve C represents the 1 : 1 mixture of MeMgBr : Me_2Mg. Again, the three peaks ascribed to bridging methyl groups are observed along with an increase in the intensity of the monomeric Me_2Mg peak at τ 11.74.

This data is consistent with the extended Schlenk equilibrium incorporating an intermediate involving the formation of an alkyl group bridge, *e.g.*, $MgMe_2MgX_2$.

$$R_2Mg + MgX_2 \rightleftharpoons [X_2MgR_2Mg] \rightleftharpoons RMg\langle X_2 \rangle MgR \rightleftharpoons 2RMgX$$

The position of the equilibrium can be adjusted by changing the functional groups, the temperature and the donating ability of the solvent employed.

The part played by the solvent has already been summarized, but only in the cases of polar compounds. It is often implied that they are necessary in the preparation of a Grignard reagent, but under suitable conditions these compounds can be prepared in high yield in hydrocarbons and other non-donor media.[c, d] However, the absence of the donor solvent influences the chemical behaviour of the RMgX compound. This is, therefore, better referred to as an unsolvated organomagnesium halide, rather than a Grignard reagent, when synthesized in non-donor media.

These compounds have been synthesized from the halide and freshly-powdered magnesium in high boiling hydrocarbon solvents such as isopropylbenzene and tetrahydronaphthalene. Dried and re-distilled "Pink Paraffin" is also suitable!

$$RX + Mg \xrightarrow[N_2]{\text{hydrocarbon}} RMgX$$

$(R = Bu^n;\ X = Cl, Br, I;\ R = Ph,\ X = Cl, Br)$

The formation of the organomagnesium halide is promoted by light metal alkoxides,[54, 55] and the products are best formulated as polymers of "R_3Mg_2X" since they have organic substituents in excess of RMgX.

The structure proposed involves a linear-bridged arrangement of alkyl and halogen groups, rather similar to that of beryllium dichloride.

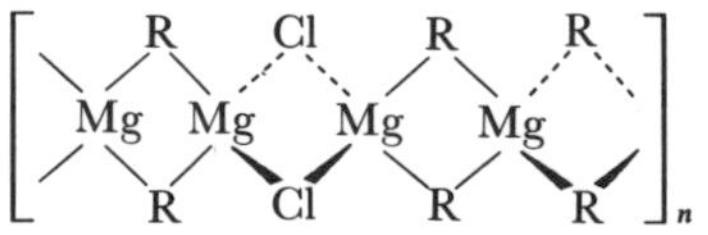

Fig. 2.30. Polymeric $[R_3Mg_2X]_n$

Indeed, hydrogen chloride attacks all Mg—C bonds to give an active form of $MgCl_2$, believed to have the covalent $BeCl_2$-type structure. This work supports the presence of alkyl-bridged species in the Schlenk equilibrium.

The presence of ions in the solution of a Grignard reagent is suggested by conductivity measurements. Low conductance indicates their presence in low concentrations, and ionization giving carbanions seems unlikely. Solutions of carbanions (*e.g.*, Bu^nLi in tetramethylethylenediamine) readily deprotonate ethers, tertiary amines, organic sulphides *etc.* Electrolysis deposits magnesium at both electrodes, indicating the presence of magnesium in cations and anions.[56] Dissociation may well proceed according to equation (*2.2*),

$$R\text{—}Mg\langle{}^{X}_{X}\rangle Mg\text{—}R \rightleftharpoons RMg^+ + RMgX_2^- \qquad (2.2)$$

and the formation of $RMgX_2^-$ would be consistent with the manner in which magnesium catalyzes a Friedel–Crafts reaction between n-butyl chloride and benzene. Whilst n-butyl iodide and magnesium form Bu^nMgI in high yields in benzene, the chloride forms sec-butylbenzene.[57]

From the work discussed, it will be appreciated that the structure of the Grignard reagent is very complex, and, in addition to the variations considered to date, the purity of the magnesium plays an important role in determining the reactivity of this reagent. Thus, isopropylmagnesium bromide can be isomerized to the n-propyl Grignard reagent in the presence of small quantities of titanium.[57a]

Indeed, it is probably safe to say that if reagents were obtained pure enough to make accurate kinetic studies, this would be atypical of the conditions employed by the majority of chemists using this reagent.

Organoberyllium halides

The work devoted to these compounds is sparse and spans 50 years. Nevertheless, some interesting comparisons with analogous magnesium compounds have emerged.

Organoberyllium iodides are conveniently synthesized directly from the metal. Heating the iodide with a catalyst in ether,[58] or without either,[59] is successful, and the presence of a Be—C bond is confirmed by the alkylation of carbonyl compounds.

$$Be + RI \xrightarrow[HgCl_2]{Et_2O/100^\circ} RBeI$$

$$(R = Me, Et, Bu^n, Ph)$$

$$Be + Bu^nI \xrightarrow[12\ h]{130^\circ C} Bu^nBeI$$

The chlorides and bromides are more conveniently made by the coproportionation of the dihalide with a diorganoberyllium compound.[60, 61]

The equilibrium resembles the Schlenk one proposed for the magnesium system, but a careful study reveals distinct differences. In ether, beryllium dichloride and dibromide rapidly equilibrate with R_2Be (R = Me, Et, Ph), and the addition of a stronger base will precipitate the complexed organoberyllium halide.

$$R_2Be + BeX_2 \rightleftharpoons 2RBeX$$

This is contrary to the Grignard system, where dioxan precipitated the complexed dihalide.

$$RBeX + L \longrightarrow RBeX{\cdot}L$$
(L = dioxan or bipyridyl)

Molecular weight determinations in solution show the compounds RBeX to be monomeric over a wide concentration range. P.m.r. spectra of mixtures of Me_2Be and $BeCl_2$ or $BeBr_2$ in ether in various ratios shows conclusively that the Schlenk equilibrium for this system is predominantly to the right, supporting the precipitation work. The p.m.r. spectral data are summarized below for the dichloride.

Table 2.2. P.m.r. Data for $Me_2Be/BeCl_2$ Mixtures

Solution \ Temperature	35°C	0°C	−45°C	−85°C
Me_2Be	70	70.5	74.5	86
$Me_2Be:BeCl_2$ (1:1)	75	76	79	82
2:1	72.5	74.5	76	81 and 86.5 (1:1)
3:1	72	73.5	75.5	81.6 and 86.5 (1:2)

(measurements recorded in cps upfield from TMS)

A large shift is observed for dimethylberyllium, but the peak remains as a singlet indicating only terminal MeBe groups in contrast to dimethylmagnesium. A smaller shift occurs for the 1:1 mixture of Me_2Be and $BeCl_2$, and again only one peak is observed, indicating but one kind of Me—Be group. This is probably solvated MeBeCl, and the absence of a peak at 86 cps establishes that the Schlenk equilibrium applied to MeBeCl is predominantly to the right. Solutions containing 2:1 and 3:1 mixtures of Me_2Be and $BeCl_2$ show the presence of both Me—Be species at −85°C, but methyl exchange between Me_2Be and MeBeCl is so rapid above −85°C that only one peak is observed. Unlike the case of the Grignard reagents, $BeCl_2$ is only precipitated if present in excess of the Me_2Be.

This data supports the Schlenk equilibrium with the co-proportionation product as the major participant, unlike the situation with the Grignard reagent.

$$R_2Be + BeX_2 \rightleftharpoons 2RBeX$$

However, both t-butylberyllium chloride and bromide etherates are dimeric[62] in benzene as is $Bu^tMgCl{\cdot}Et_2O$, but the chloride is monomeric in ether. This indicates that while the monomeric complex tends to dimerize in benzene, probably because of steric hindrance, an excess of ether will break the $BeCl_2Be$ bridge.

$$[Bu^t(Et_2O)Be(\mu\text{-}Cl)_2Be(OEt_2)Bu^t] \longrightarrow Bu^tBeCl(Et_2O)_2$$

(in benzene)

So less-hindered alkylberyllium halides might be expected to be monomeric. This also indicates that the Schlenk equilibrium might be complicated for beryllium as well.

Although a rigorous survey of cyclic zinc, cadmium and mercury compounds will not be included in this book, organic derivatives of these elements do show a close similarity to those of the parent Group metals. The organohalogeno compounds are well known, while many amino-, alkoxy- and alkylthio- derivatives have structural analogues among the compounds already considered, *e.g.*, $[EtZnNPh_2]_2$, $[MeCdSBu^t]_4$ and $[PhHgOMe]_3$.

Bibliography

General

a G. E. COATES and K. WADE, 'Organometallic Compounds', Methuen and Co. Ltd., London, 3rd Edn., Vol. I, 1967.
b V. GRIGNARD, *Compt. rend.*, 1900, **130,** 1322.
c E. C. ASHBY, *Quart Rev.*, 1967, **21,** 259.
d B. J. WAKEFIELD, *Organometallic Chem. Rev.*, 1966, **1,** 131.

References

1 A. F. WELLS, 'Structural Inorganic Chemistry', Oxford University Press, 3rd Edn., 1962.
2 H. HAKIHANA and L. G. SILLÉN, *Acta Chem. Scand.*, 1956, **10,** 985; B. CARELL and A. OLIN, *ibid.*, 1961, **15,** 1875.
3 G. E. COATES and D. S. GOLIGHTLY, *J. Chem. Soc.*, 1962, 2523.
4 K. L. PACIORCK and R. H. KRATZER, *Inorg. Chem.*, 1966, **5,** 538.
5 C. C. ADDISON and A. WALKER, *Proc. Chem. Soc.*, 1961, 242.
6 G. STUCKY and R. E. RUNDLE, *J. Amer. Chem. Soc.*, 1964, **86,** 4821.
7 G. E. COATES and F. GLOCKLING, *J. Chem. Soc.*, 1954, 2526.
8 E. L. HEAD, C. E. HOLLEY, JR., and S. W. RABIDEAU, *J. Amer. Chem. Soc.*, 1957, **79,** 3687.
9 L. H. SHEPHERD, G. L. TER HAAR and E. M. MARTLETT, *Inorg. Chem.*, 1969, **8,** 976.
10 G. E. COATES and P. D. ROBERTS, *J. Chem. Soc. (A)*, 1969, 1008.
11 G. E. COATES and F. GLOCKLING, *J. Chem. Soc.*, 1954, 22.
12 N. A. BELL and G. E. COATES, *J. Chem. Soc. (A)*, 1966, 1069.
13 *Idem, ibid.*, 1965, 692.
14 N. A. BELL, G. E. COATES and J. W. EMSLEY, *ibid.*, 1966, 1360.

15 N. A. BELL and G. E. COATES, *ibid.*, 1968, 823.
16 G. E. COATES and G. F. COX, *Chem. and Ind.*, 1962, 269.
17 G. W. ADAMSON and H. M. M. SHEARER, *Chem. Comm.*, 1965, 240.
18 A. B. BURG and H. I. SCHLESINGER, *J. Amer. Chem. Soc.*, 1940, **62,** 3425.
19 L. BANFORD and G. E. COATES, *J. Chem. Soc.* (*A*), 1966, 274.
20 J. W. NIBLER and J. MCNABB, *Chem. Comm.*, 1969, 134.
21 T. H. COOK and G. L. MORGAN, *J. Amer. Chem. Soc.*, 1969, **91,** 774.
22 G. E. COATES and J. A. HESLOP, *J. Chem. Soc.* (*A*), 1968, 514.
23 F. M. PETERS and N. R. FETTER, *J. Organometallic Chem.*, 1965, **4,** 181.
24 J. L. ATWOOD and G. D. STUCKY, *Chem. Comm.*, 1967, 1169; *J. Amer. Chem. Soc.*, 1969, **91,** 4426.
25 G. E. COATES, F. GLOCKLING and N. D. HUCK, *J. Chem. Soc.*, 1952, 4512.
26 G. E. COATES and A. H. FISHWICK, *J. Chem. Soc.* (*A*), 1967, 1199; G. E. COATES and M. TRANAH, *ibid.*, 236.
27 G. E. COATES and S. I. E. GREEN, *J. Chem. Soc.*, 1962, 3340.
28 G. E. COATES and D. RIDLEY, *J. Chem. Soc.* (*A*), 1967, 57.
29 G. E. COATES and A. H. FISHWICK, *J. Chem. Soc.* (*A*), 1968, 477.
30 N. R. FETTER, *Canad. J. Chem.*, 1964, **42,** 861.
31 D. MOOTZ, A. ZINNIUS and B. BÖTTCHER, *Angew. Chem. Internat. Edn.*, 1969, **8,** 378.
32 G. E. COATES and A. H. FISHWICK, *J. Chem. Soc.* (*A*), 1968, 640.
33 E. G. HOFFMAN, *Annalen*, 1960, **629,** 104.
34 D. BRYCE-SMITH and B. J. WAKEFIELD, *Proc. Chem. Soc.*, 1963, 376.
35 G. E. COATES, J. A. HESLOP, M. E. REDWOOD and D. RIDLEY, *J. Chem. Soc.* (*A*), 1968, 1118.
36 D. BRYCE-SMITH and I. F. GRAHAM, *Chem. Comm.*, 1966, 559.
37 G. E. COATES and D. RIDLEY, *ibid.*, 1966, 560.
38 P. T. MOSELEY and H. M. M. SCHEARER, *ibid.*, 1968, 279.
39 G. D. STUCKY and R. E. RUNDLE, *J. Amer. Chem. Soc.*, 1964, **86,** 4825; L. J. GUGGENBERGER and R. E. RUNDLE, *ibid.*, 5344.
40 J. C. ATWOOD and G. D. STUCKY, *J. Organometallic Chem.*, 1968, **13,** 53.
41 G. E. COATES and A. H. FISHWICK, *J. Chem. Soc.* (*A*), 1968, 635.
42 R. S. AHRLAND, J. CHATT and N. R. DAVIES, *Quart. Rev.*, 1958, **12,** 265.
43 G. E. COATES and J. A. HESLOP, *J. Chem. Soc.* (*A*), 1968, 631.
44 E. C. ASHBY and F. WALKER, *J. Organometallic Chem.*, 1967, **7,** P17.
45 W. SCHLENK, *Ber.*, 1931, **64,** 734.
46 W. SCHLENK and W. SCHLENK, JR., *ibid.*, 1929, **62,** 920.
47 R. E. DESSY and co-workers, *J. Amer. Chem. Soc.*, 1957, **79,** 3476; 1958, **80,** 5824.
48 E. C. ASHBY and F. WALKER, *J. Organometallic Chem.*, 1967, **7,** P17.
49 *Idem*, *J. Amer. Chem. Soc.*, 1969, **91,** 3845.
50 E. C. ASHBY, *ibid.*, 1965, **87,** 2509.
51 J. TONEY and G. D. STUCKY, *Chem. Comm.*, 1967, 1168.
52 J. G. ASTON and S. A. BERNHARD, *Nature*, 1950, **165,** 485.
53 E. C. ASHBY, G. PARRIS and F. WALKER, *Chem. Comm.*, 1969, 1464.
54 D. BRYCE-SMITH and G. F. COX, *J. Chem. Soc.*, 1961, 1175.
55 E. T. BLUES and D. BRYCE-SMITH, *Chem. and Ind.*, 1960, 1533.
56 W. V. EVANS and co-workers, *J. Amer. Chem. Soc.*, 1933, **55,** 1474; 1942, **64,** 2865; R. E. DESSY and R. M. JONES, *J. Org. Chem.*, 1959, **24,** 1685.
57 D. BRYCE-SMITH and W. J. OWEN, *J. Chem. Soc.*, 1960, 3319.
57a C. W. BIRD, 'Transition Metal Intermediates in Organic Synthesis', Logos/Academic, 1967, p. 78.
58 H. GILMAN and F. SCHULZE, *J. Amer. Chem. Soc.*, 1927, **49,** 2904.
59 L. I. ZAKHARKIN, O. Y. OKHLOBYSTIN and B. N. STRUNIN, *Izvest. Akad. Nauk S.S.S.R.*, 1961, 2254; *Chem. Abs.*, 1962, **57,** 13785.
60 N. A. BELL, *J. Organometallic Chem.*, 1968, **13,** 513.

61 J. R. SANDERS, JR., E. C. ASHBY and J. H. CARTER, *J. Amer. Chem. Soc.*, 1968, **90,** 6385.
62 G. E. COATES and P. D. ROBERTS, *J. Chem. Soc. (A)*, 1968, 2651.

APPENDIX

The skeleton encountered for $[Be(OBu^t)_2]_3$ and $[Be(NMe_2)_2]_3$ occurs with ketimine complexes and the t-butoxy compound obtained from $(PhC{\equiv}C)_2Be{\cdot}2L$ and Bu^tOH.[63]

Monomer-dimer equilibrium has been established for $MeBeBH_4$ while further studies on $Be(BH_4)_2$ fail to completely resolve the structure of the compound.[64] Magnesium hydride has been quantitatively obtained from sodium hydride and magnesium bromide or iodide in ether solvents.[65] The chemistry of HMgBr is surveyed,[66] while reducing alkylmagnesium amides yield aminomagnesium hydrides R_2NMgH. These are associated, probably through H and amino bridges. The analogous alkoxy magnesium hydrides readily disproportionate at room temperature.[67] Reducing

$$2ROMgH \longrightarrow (RO)_2Mg + MgH_2$$

diethylmagnesium with AlH_3 produces various substituted magnesium hydrides.[68]

A tetrameric Grignard reagent from EtCl/Mg in tetrahydrofuran analyses for $[EtMg_2Cl_3(C_4H_8O)_3]_2$.[69] It possesses a peculiar structure with 2 three-co-ordinated bridging chlorine atoms and magnesium in both 5- and 6-co-ordinate sites. Six-co-ordination is also encountered in polymeric $Mg(O_2PCl_2)_2(OPCl_3)_2$.[70] Both magnesium bromide and alkylmagnesium bromides are ionic in hexamethyl-phosphoramide, supporting further solvent controlled equilibria in the Schlenk equation.[71]

All attempts to prepare organomagnesium fluorides prior to 1970 were unsuccessful. Two routes are now known involving direct synthesis and exchange. All are dimeric in polar solvents over a wide concentration range, showing they do not behave according to the Schlenk equilibrium (compare alkoxytin halides). The stability is probably due to strong fluoride bridges.[72]

$$RF + Mg \longrightarrow RMgF \qquad (R = Me, Et, Hex).$$
$$R_2Mg + BF_3 \longrightarrow RMgF \qquad (R = Me, Et, Ph).$$

The mechanism of the reaction of MeMgBr and ketones has been closely examined,[73] while the p.m.r. spectra of alkyl and aryl Grignard reagents show alkyl and aryl exchange to be slow at room temperature and slower in co-ordinating solvents.[74]

Mossbauer spectroscopy shows $Ph_3SnMgBr$ to be a 6 membered ring dimer with tin incorporated.[75] The addition of dicyclopentadienyl tin(II) to PhMgBr gives a bromide bridged dimer with a pseudo-carbene

cp_2Sn having inserted into the C—Mg bond. A trialkylsilylmagnesium compound is also suspected as an intermediate[76] in another system.

The reaction of peroxides and disulphides[77] with Grignard reagents was investigated along with a study of alkyl group isomerisation[78] in the reagent, and the catalytic effect of Fe^{III} and Ni^{II} on reactions.[79]

References

63 C. SUMMERFORD, K. WADE and B. K. WYATT, *J. Chem. Soc.* (*A*), 1970, 2016, G. E. COATES and B. R. FRANCIS, *ibid*, 1971, 160.
64 T. H. COOK and G. L. MORGAN, *J. Amer. Chem. Soc.*, 1970, **92,** 6487; J. W. NIBLER and T. DYKE, *ibid*, 2920.
65 E. C. ASHBY and R. D. SCHWARTZ, *Inorg. Chem.*, 1971, **10,** 355.
66 K. D. BERLIN, T. E. SINDER and O. C. DERMER, *Tetr. Letters*, 1970, 3991.
67 R. G. BEACH and E. C. ASHBY, *Inorg. Chem.*, 1971, **10,** 906.
68 S. C. SRIVASTAVA and E. C. ASHBY, *ibid*, 186.
69 J. TOVEY and G. D. STUCKY, *J. Organometallic Chem.*, 1971, **28,** 5.
70 J. NYBORG and J. DANIELSON, *Acta Chemica Scandanavica*, 1970, **24,** 59.
71 J. DUCOM and B. DENISE, *J. Organometallic Chem.*, 1971, **26,** 305.
72 E. C. ASHBY, S. H. YU and R. G. BEACH, *J. Amer. Chem. Soc.*, 1970, **92,** 433; E. C. ASHBY and J. A. NACKASHI, *J. Organometallic Chem.*, 1970, **24,** C17; E. C. ASHBY and S. H. YU, *ibid*, 1971, **29,** 339.
73 E. C. ASHBY, J. LAEMULE and H. M. NEUMANN, *J. Amer. Chem. Soc.*, 1971, **93,** 4601; T. HOLM, *J. Organometallic Chem.*, 1971, **29,** C45.
74 D. F. EVANS and G. V. FAZAKERLEY, *J. Chem. Soc.* (*A*), 1971, 184.
75 P. G. HARRISON, J. J. ZUCKERMAN and J. G. NOLTES, *J. Organometallic Chem.*, 1971, **31,** C23.
76 K. TAMAO, M. KUMADA and A. NORO, *ibid.*, 169.
77 A. W. P. JARVIE and D. SKELTON, *ibid.*, 1971, **30,** 145.
78 A. G. DAVIES, B. P. ROBERTS and R. TUDOR, *ibid.*, 1971, **31,** 137.
79. L. FARADAY and L. MARKO, *ibid.*, 1971, **28,** 159; M. TAMURA and J. KOCH, *ibid.*, 1971, **31,** 289.

3

The Group III Elements

Boron is second to no other element in the diversity of compounds it will form. Among the cyclic ones, these range from 4-membered rings, which also typify the heavier elements of the Group, to cage compounds, *e.g.*, $B_{12}H_{12}^{2-}$, resembling the low valency transition metal halide clusters. Explaining the bonding, and hence the structures, of these compounds is complex, but greatly assisted by the use of symmetry arguments. The main feature of these boron cages is their electron-deficient nature, and the necessity to form polyatomic delocalized molecular orbitals to bind together. Multiple bonding involving p_π orbitals is also widely recognized in boron chemistry. The borazoles, or borazines, undergo many reactions analogous to benzene, while boron halides are stable as monomers, unlike the post-boron trihalides of Group III. Borate mineral chemistry is unique among the elements of this Group, and illustrates as do the hydrides, the similarity to silicon.

Consequently, our attention will be focussed on boron, while the other Group III elements will be discussed in a comparative fashion where appropriate. The nomenclature of boron compounds is complicated, and for this reason the proposals of the Boron sub-Committee, which were generally approved,[a] are followed in this Chapter.

ELEMENTAL BORON AND ITS DERIVATIVES

Of the five Group III elements being considered, boron is the only one to exhibit, as an element, structural properties involving cyclic units. It is best obtained by pyrolyzing diborane at 700°C, but cruder samples result electrolytically or by reducing a boron halide with hydrogen in a hot tube or on a hot tantalum filament. Boron is best purified by sublimation, and is structurally unique among the elements with its three valence electrons too localized for metallic character and too few for simple covalent structures. It is nearly as hard as diamond (9.3 on the Mohs scale), and the three structural forms already fully characterized contain the B_{12} icosahedral array of boron atoms.[b]

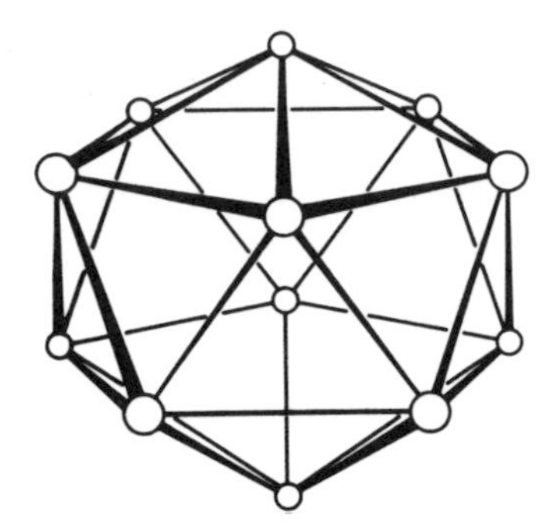

Fig. 3.1. The icosahedral B_{12} cage

The reactivity of boron increases with temperature, making its isolation difficult. It is hardly affected by acids or oxidizing agents, but fluorine reacts at room temperature while the heavier halogens all oxidize boron on heating. It will reduce metal fluorides to the metal (cadmium, zinc and lead), while silver fluoride and boron detonate on contact. Carbon tetrachloride is also readily reduced by boron.

$$4B + 3CCl_4 \xrightarrow{200-250°C} 3C + 4BCl_3$$

Most of the non-metallic elements of Groups IV to VI react with boron above 1100°C to give compounds varying in stoichiometry, but based on the B_{12} framework, *i.e.*, $B_{12}X_n$ ($X_n = C_3$, $Si_{\sim 4}$, S, O_2, $P_{1.8}$, $As_{1.8}$). The structures involve a three-dimensional array of B_{12} icosahedra with interstices suitable for the non-metal, which is often found occupying the position of a boron atom in the B_{12} cage.

Metal borides are usually synthesized by reducing the metal oxide with carbon and boron carbide, or directly from the elements, in an inert atmosphere.[b,c] They vary in structure from isolated boride ions to three-dimensional frameworks occluding metal ions. Thus, the highest fully-characterized boride UB_{12} comprises an octahedral array of distorted icosahedral B_{12} units linked to give a three-dimensional array and containing uranium ions in the octahedral holes. Dodecaborides also exist for zirconium and the post-gadolinium rare earths.

All the rare earth and alkaline earth metals form hexaborides, as does uranium. While the dodecaborides are built on the Na^+Cl^- framework of 6-co-ordination, the hexaborides, with their Cs^+Cl^- structure, comprise eight B_6 octahedra at the corners of a cube, enclosing the divalent metal ions in the centre.

The B_6^{2-} anions can be conveniently considered as bonded together by inter-octahedral B—B links, through sp-hybridized orbitals.[1,2] This leaves one sp orbital and two 2p orbitals per boron atom for the intra-octahedral bonding and 14 electrons. The representation shows (a) that the six sp

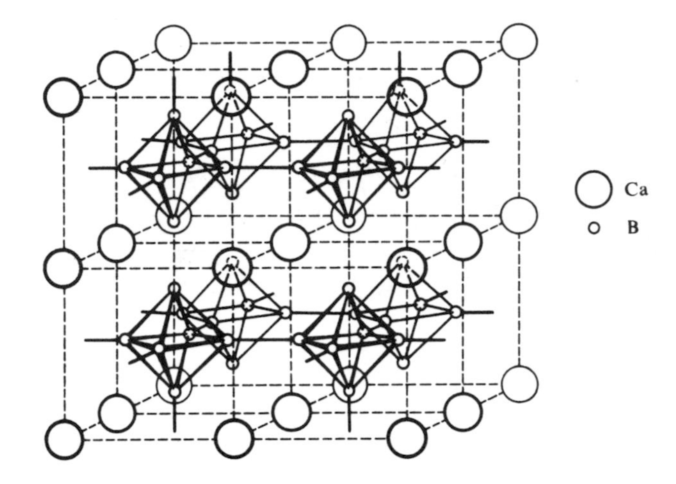

Fig. 3.2. The structure of CaB_6[c]

hybrids behave as $\mathbf{a}_{1g} + t_{1u} + e_g$, and (b) that the twelve p orbitals behave as $t_{1g} + \mathbf{t}_{2g} + \mathbf{t}_{1u} + t_{2u}$.

Orbitals	E	$8C_3$	$3C_2$	$6C_2'$	$6C_4$	i	$8S_6$	$3\sigma_h$	$6\sigma_d$	$6S_4$
Six sp hybrids	6	0	2	0	2	0	0	4	2	0
Twelve p orbitals	12	0	−4	0	0	0	0	0	0	0

The strongly-bonding representations are in bold type with the t_{1u} of (a) probably interacting with the t_{1u} bonding representation to make it even lower in energy. The 14 electrons can thus be readily incorporated into these seven orbitals.

Metallic diborides are among the best electrically conducting, highest melting, hardest and most refractory of all the borides. They are known for 24 metals including magnesium, aluminum, silver, gold, many transition elements of all three rows, rare earths and uranium. The structural characteristic of these compounds is the hexagonal laminar array of boron atoms, rather similar to graphite, with a similar metallic array above and below the boron sheet.

The tetragonal tetraborides are known for the rare earth metals, calcium and yttrium, molybdenum, tungsten and uranium. They are structural hybrids of the di- and hexa-borides. The planar arrangement of B_6 octahedra linked through chains of two boron atoms allows the metal ions to occupy the eight co-ordination positions around the B_6 species and the trigonal-prism arrangement around the single boron atoms.

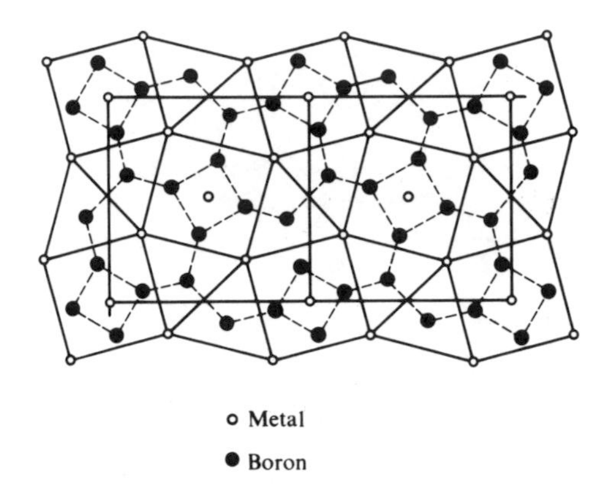

Fig. 3.3. The projected MB_4 lattice[c]

The only other boride structure incorporating a cyclic arrangement is that of M_3B_4 (M is Cr, Mn, Ta and Nb). This comprises a double-chain arrangement with interchain bonding the stronger. It is thus intermediate between the chain array in TaB (B—B distance 0.172 to 0.191 nm) and the

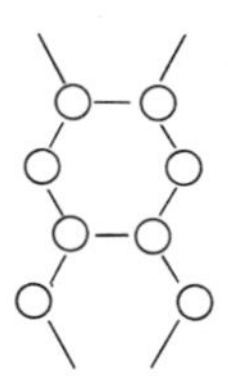

Fig. 3.4. The structure of Ta_3B_4, showing the boron skeleton[c]

hexagonal sheet array of TaB_2 (B—B distance 0.172–0.179 nm). In Ta_3B_4 the interchain bonding seems much stronger than that of the chains themselves.

DIBORANE(6)

The acid hydrolysis of magnesium boride, carefully prepared from B_2O_3 and magnesium, led Stock to isolate tetraborane(10).[d] Pyrolysis at 95°C gave diborane(6) (b.p. −90°C),[3] but in low yields. The reduction of boron halides with $LiAlH_4$ or BH_4^- leads to better yields, while iodine and BH_4^- gives diborane(6) in 98% yield.[4] However, the purest samples result if potassium borohydride is slowly added to conc. phosphoric acid.[5]

$$2BH_4^- + 2H^+ \longrightarrow B_2H_6 + H_2$$

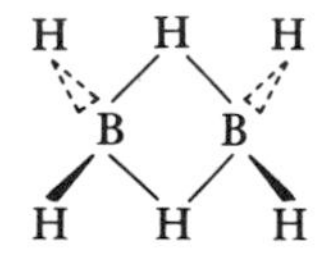

Fig. 3.5. The structure of diborane(6)

The bridge structure has been established by infrared spectroscopy and electron diffraction,[6] and subsequently confirmed by n.m.r. spectroscopy.[7] The bridging protons resonate at a higher field than the terminal ones, while Lewis bases (*e.g.*, CO, Me_3N, H^-) cleave this bridge to give monomeric adducts, and halogenation yields the trihalides. Partial halogenation with the hydrogen halides is catalyzed by AlX_3, while hydrolysis is rapid and oxidation explosive. Pyrolysis products will be discussed later in the section dealing with polyboranes (p. 50). Partial bridge cleavage occurs with BH_4^-, giving the anion $B_2H_7^-$. This possesses a bridging hydrogen bond which, like that of HF_2^-, is very strong and probably linear.[8]

Trialkyl- and triaryl-boranes will undergo rapid exchange reactions with B_2H_6, giving alkyl- and aryl-diboranes.[9] Penta-alkyldiboranes are, as yet, unknown, while in the absence of excess Me_3B trimethyldiborane disproportionates. Sym-tetramethyldiborane, so formed, possesses a hydride bridge.[10]

$$Me_3B_2H_3 \longrightarrow 3Me_4B_2H_2 + B_2H_6$$

B,*B*′-dialkyldiboranes are normally absent from the products of this coproportionation but can be made by the disproportionation of monoalkyldiborane in the presence of dimethylether.[11]

$$2MeB_2H_5 + 2Me_2O \longrightarrow 2Me_2O{\cdot}BH_3 + [MeBH_2]_2$$

Surprisingly, triphenylborane gives only the *B*,*B*′-diphenyldiborane.[12]

The addition of diborane(6) to olefines, acetylenes and buta-1,3-dienes has been used widely in the synthesis of sterically-hindered diboranes which are used as selective reducing agents.[13] Tri-substituted olefines yield the tetra-alkyldiboranes, while tetra-substituted ones give dialkyldiboranes.[e] Subsequent oxidation gives the alcohol.

$$RCH{=}CH_2 + {>}B{-}H \longrightarrow RCH_2CH_2B \xrightarrow{O_2} RCH_2CH_2OH$$

Butadiene forms the boracyclopentane (borolanes), which functions as an effective exchange catalyst whilst monomeric but its efficiency is lost on dimerization to the diborolane.[14]

$$2C_4H_6 + B_2H_6 \longrightarrow 2\,\text{BH} \rightleftharpoons \text{B}\overset{H}{\cdots}\text{B (H-bridged)}$$

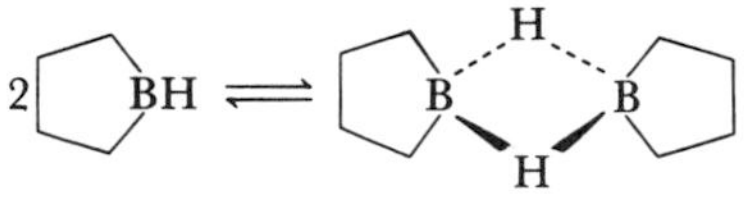

Borohydride will reduce trialkylboranes in the presence of HCl, or vinyl and allyl bromides,[15] to the alkyl diboranes, while lithium aluminium hydride reduces *B*-chloroborolane to bis-(borolane).

$$(6-n)LiBH_4 + (6-n)HCl + nBR_3 \longrightarrow 3B_2H_{6-n}R_n + (6-n)(LiCl + H_2)$$

$$CH_2CHCH_2Br + NaBH_4 \longrightarrow \tfrac{1}{2}(Pr^n_2B_2H_4) + NaBr$$

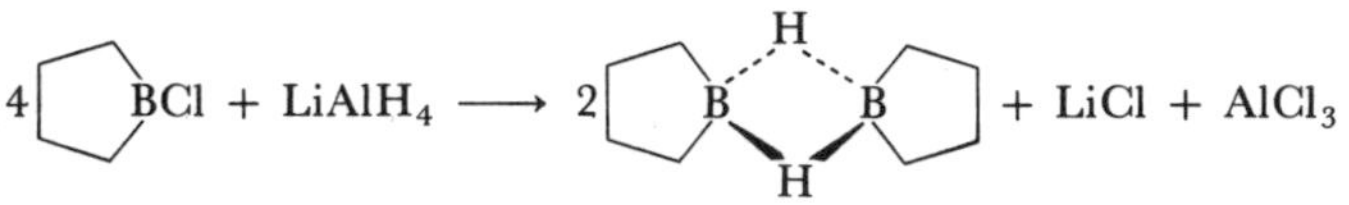

With diborane at 120°C, diborolane yields compound **(3-I)**,

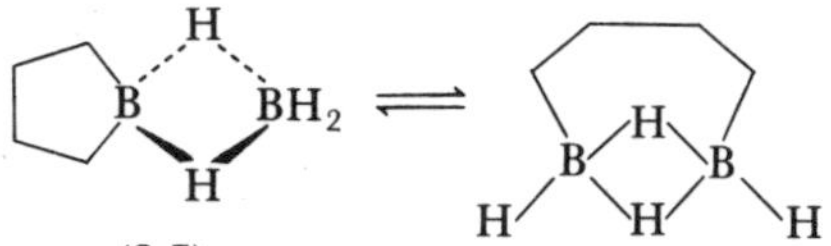

(**3-I**)

while at 100°C, diborane and butadiene form a polymer and the two isomers of (**3-I**). These equilibrate on warming.[16]

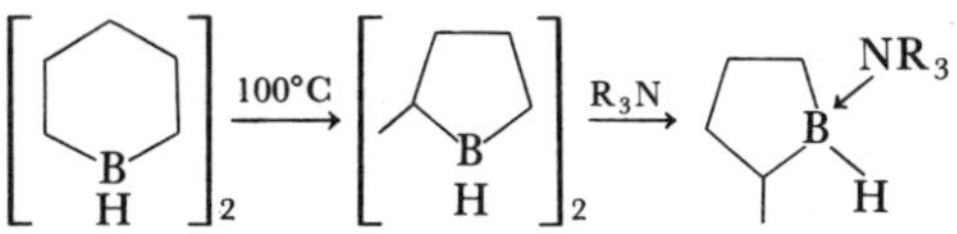

Diborolanes are more stable than the open chain borane/olefine adducts,[f] resisting hydrolysis, oxidation and olefine addition up to 100°C. The borocyclohexane dimer isomerizes at 100°C, and base will break the hydride bridge.[g]

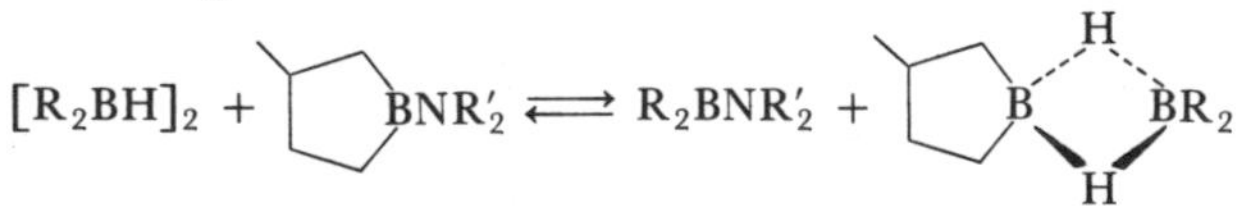

Consequently, aminoboracyclopentanes and diboranes readily exchange.

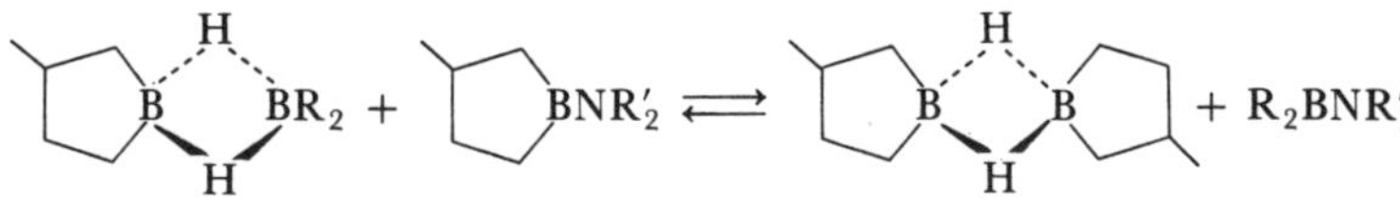

While no group congeners of diborane are known, aluminium borohydride will give the AlH_2B bridge with dimeric triethylaluminium. This also results from ethylene and $Al(BH_4)_3$.[17]

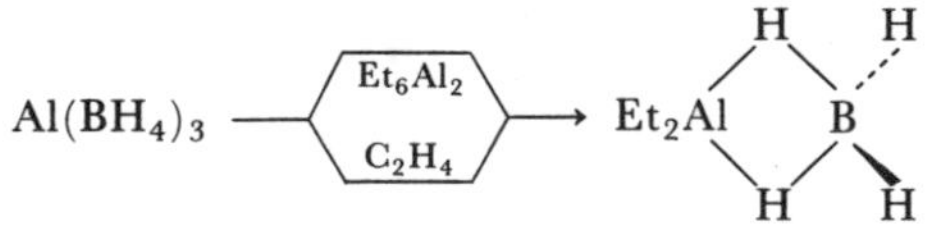

A 2:3 mixture of trimethylgallium and diborane gives the gallium analogue, which slowly decomposes at room temperature.

$$2Me_3Ga + 3B_2H_6 \longrightarrow \underset{\text{(m.p. 1.5°C)}}{2Me_2GaBH_4} + 2MeB_2H_5$$

Dialkylalanes are usually prepared by reducing a halide, though dimethylalane is conveniently prepared from trimethylborane and $LiAlH_4$.[18] It is trimeric in benzene and dimeric in the gaseous phase between 83°C and 167°C.

$$3LiAlH_4 + 3Me_3B \longrightarrow 3LiBH_3Me + [AlHMe_2]_3$$

These polymeric alanes[19] react with Lewis bases to form monomeric adducts,

$$Me_4Al_2H_2 \xrightarrow{2Me_3N} Me_3NAlMe_3 + MeAlH_2{\cdot}NMe_3$$

$$[Et_2AlH]_3 + 3EtLi \longrightarrow 3Li^+[Et_3AlH]^-$$

while the hydride bridge in trimers is easily cleaved by EtLi and Et_2M (M = Mg, Zn, Cd, Hg), giving excellent yields of triethylalane.[20] As with boranes, addition of olefines to the Al—H and Al—C bonds has proved of commercial importance in polymer synthesis.[h]

POLYBORON HYDRIDES

Remarkably, the first new polyboron hydride synthesized since Stock's classical work 40 years ago was in 1958 when B_9H_{15} was isolated.[21] Since then, renewed interest through improved techniques and new methods of analysis has resulted in the isolation of more new hydrides than Stock ever made. Most of these compounds are air-sensitive, and while the formula can be readily determined, the structure of the boron cage and the position of the hydrogen atoms at bridging or terminal sites is often difficult to predict with any degree of certainty. Indeed, the unambiguous structure of some polyboranes is still not known, and X-ray analysis is often the only way to fully characterize them.

An infrared analysis, followed by deuteration, helps to characterize the hydrogen atoms, since exchange occurs most readily at the bridging positions where the stretching frequency is significantly lower. Both boron isotopes (^{11}B; 81% nuclear spin $\frac{3}{2}$, ^{10}B; spin 3), as well as H and D (spin 1), render n.m.r. spectroscopy probably the most useful analytical tool for boranes.[i]

The acidolysis of magnesium boride led to the isolation of the six polyboranes B_4H_{10}, B_5H_9, B_5H_{11}, B_6H_{10}, B_6H_{12} and $B_{10}H_{14}$, of which all but B_6H_{12} have now been well characterized.

Tetraborane(10), B_4H_{10}, (b.p. 16°C) was also prepared by Stock through a Wurtz-type reaction on iododiborane.[22] Yields are poor, however, and tetraborane(10) is now synthesized by allowing diborane(6) to disproportionate under pressure, or under the influence of an electric discharge or temperature gradient. It also results from the BH_4^- reduction of diboron tetrachloride and the hydrolysis of B_5H_{11} or a $B_3H_8^-/B_2H_6$ mixture.[23]

$$4BH_4^- + B_2Cl_4 \longrightarrow B_4H_{10} + B_2H_6 + 4Cl^-$$

The structure of B_4H_{10} is given in Figure 3.6.

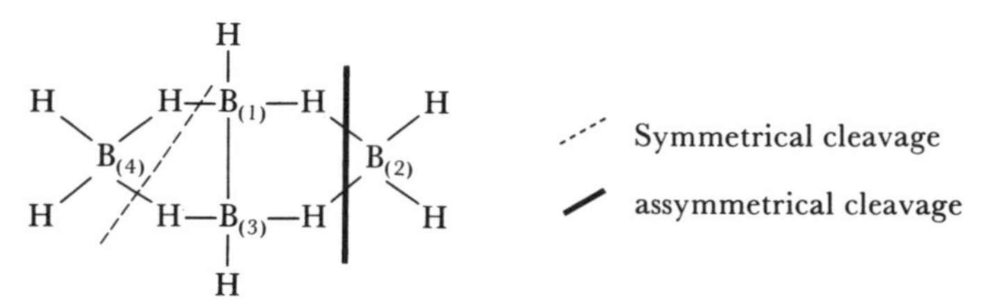

Fig. 3.6. Tetraborane(10)

Both terminal and bridging hydrogen atoms are present as well as a B—B bond in the puckered B_4 rhombus. The bonding can be conveniently considered on the basis of two- and three-centre bonds, the latter accounting for the four hydride bridges. Six terminal B—H bonds and the B—B bond account for the available 22 electrons.

Pyrolysis[24] and oxidation[25] are both thought to involve cleavage of B_4H_{10} into BH_3 and B_3H_7. The former process leads to B_5H_{11} (pentaborane(11)), while oxidation yields boric oxide via BH_3O. However, degradation into B_4H_8 and H_2 is also reasonable.[26] While acid and neutral hydrolysis yields boric acid,[d] base gives $B_3H_8^-$ (to be considered later, (p. 66), BH_4^- and $B(OH)_4^-$ anions.[d]

In general, nucleophiles will degrade the B_4H_{10} cage but to varying extents. Ethers, thioethers and tertiary amines all form adducts of triborane(7), $L{\cdot}B_3H_7$. The dimethylether adduct reacts with $[MeBH_2]_2$ to give 2-MeB_4H_9, together with dimethyl isomers. Better yields of 2-MeB_4H_9 result from the Me_2Hg and B_4H_{10}. The trimethylamine complex is further degraded by triphenylphosphine to a complex of diborane(4).[27]

$$THF{\cdot}B_3H_7 + B_2H_6 \xleftarrow{THF} B_4H_{10} \xrightarrow{Me_3N} Me_3NBH_3 + Me_3NB_3H_7$$

$$THF{\cdot}B_3H_7 \xrightarrow{Ph_3P} Ph_3PB_3H_7 + THF \quad (\text{m.p. } 161°C)$$

$$Me_3NB_3H_7 \xrightarrow[C_6H_6]{60°C,\ Ph_3P} Ph_3PBH_2BH_2PPh_3 \ (\text{m.p. } 185°C) + Ph_3PBH_3 \ (\text{m.p. } 187–8°C) + Me_3N$$

The nature of these triborane(7) complexes has been deduced from their n.m.r. spectra. With the ether complex, all boron atoms are equivalent and coupled equivalently to all seven protons. This can be accounted for on the basis of both hydrogen and ligand tautomerism. The more basic Me_3N gave a complex whose spectrum showed two octets, the upfield one being half the intensity of the other. So base exchange is much slower but hydrogen tautomerism is observed.[28]

Ethylene forms the complex $B_4H_8C_2H_4$ with tetraborane(10) by adding across B_2 and B_4. The structure resembles a basket with the C—C bond as handle, and results from the initial degradation of B_4H_{10} and not by addition of B—H across the double bond.[29]

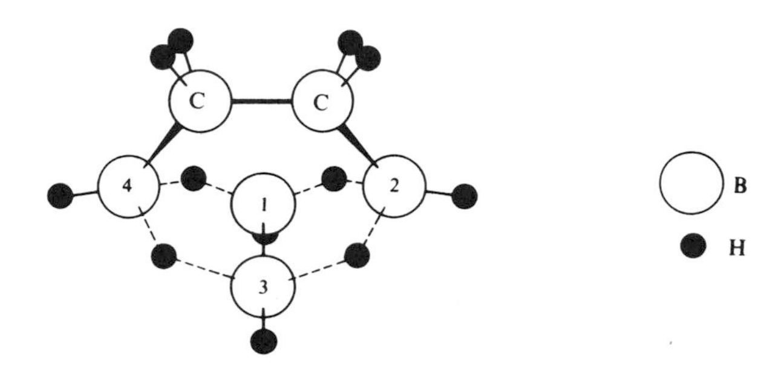

Fig. 3.7. The structure of μ-ethylenetetraborane(10)[9]

This has been established using deuteroethylene, as mass spectrometry shows the tetradeutero-compound to be the only product.

$$B_4H_{10} + C_2D_4 \longrightarrow B_4H_9CD_2CD_2H \begin{cases} \xrightarrow{-HD} B_4H_8CD_2CDH \text{ (2 parts)} \\ \xrightarrow{-H_2} B_4H_8CD_2CD_2 \text{ (1 part)} \end{cases}$$

$$B_4H_{10} + C_2D_4 \xrightarrow{-H_2} B_4H_8 + C_2D_4 \longrightarrow B_4H_8C_2D_4$$

Carbon monoxide will also form a complex, $B_4H_8{\cdot}CO$ from either B_4H_{10} or B_5H_{11}. The CO is readily displaced by fluorophosphines (PF_3, Me_2NPF_2 and HPF_2).[30]

$$B_4H_{10} + CO \longrightarrow B_4H_8CO \xrightarrow{PF_2X} B_4H_8PF_2X$$

Its lability implies weak σ-bonding at position 1 in the borane structure, rather than a B—C—O—B link resembling the BC_2B bridge of $C_4H_8C_2H_4$. The structure of the aminophosphine complex shows co-ordination at $B_{(1)}$ while the P—N bond (0.1593 nm) and P—F bonds (0.1584 nm) are shorter than in Me_2NPF_2 (0.1628, 0.1610 nm).

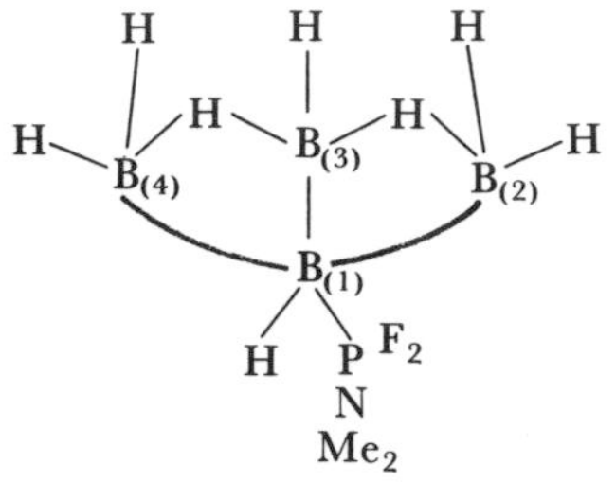

H H H
H—$B_{(4)}$ H $B_{(3)}$ H $B_{(2)}$—H
$B_{(1)}$
H PF_2
N
Me_2

Fig. 3.8. The structure of $Me_2NPF_2B_4H_8$

This implies a weak interaction between the cage and the π-orbitals of the phosphine. Even deuterium[31] will displace the CO giving the μ,1-dideutero product $B_4D_2H_8$ ($\nu_{(B-H)bridge}$ 2150; $\nu_{(B-D)terminal}$ 1945; $\nu_{(B-D)bridge}$ 1605 cm^{-1}).

Pentaborane(9) (B_5H_9, m.p. $-46°C$, b.p. $58°C$) has the square-pyramidal structure indicated,

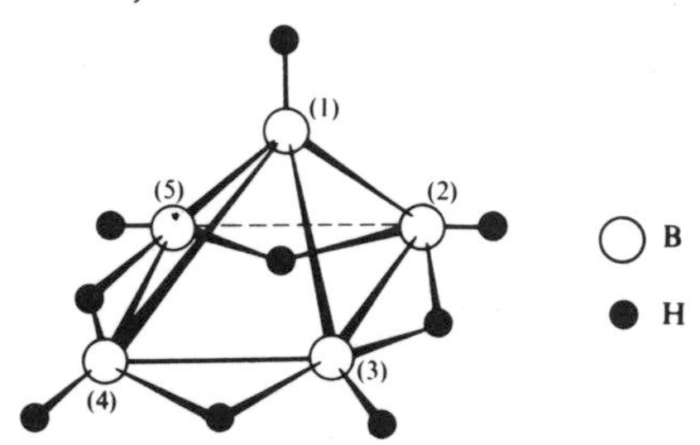

Fig. 3.9. The structure of pentaborane(9)

and is prepared by pyrolyzing diborane and excess hydrogen at 250°C for 3 s.[32] It also results from the base–triborane catalyzed conversion of B_4H_{10}, and the interaction of B_4H_{10} and B_5H_{11}. Hexaborane(12) also dissociates into B_5H_9.[b]

$$2B_4H_{10} \xrightarrow{R_2O\cdot B_3H_7} B_5H_9 + 3B_2H_6$$

$$B_6H_{12} \xrightarrow{Me_2O} B_5H_9 + \tfrac{1}{2}B_2H_6$$

The structure of the cage cannot readily be explained solely in terms of two- and three-centre bonds discussed previously, and has to be considered along the same lines as B_6^{2-}. With the four base boron atoms sp^3-hybridized, only one orbital per boron is available for cage-bonding. The other three are used in terminal and bridge B—H links. These four sp^3 hybrid orbitals from the four basal boron atoms, along with the sp hybrid of B_1 and its two p orbitals, transform as shown below under C_{4v} point group symmetry. The a_1 and e representations of the sp^3 hybrids will give three bonding and three antibonding molecular orbitals with the representations of the apical boron. These orbitals will conveniently house the six electrons available.[2]

Orbitals	E	C_2	$2C_4$	$2\sigma_v$	$2\sigma_v'$
Four sp³ hybrids from basal B, Γ_{sp^3}	4	0	0	2	0
Apical B sp hybrid, Γ_{sp}	1	1	1	1	1
Two p orbitals, Γ_p	2	−2	0	0	0

$\Gamma_{sp^3} = a_1 + e + b_1$; $\Gamma_{sp} = a_1$; $\Gamma_p = e$

The n.m.r. spectrum of pentaborane(9) shows the proton of the apical boron to be at a higher field than the basal terminal protons (τ 9.47 and 7.51). Bridging (μ) protons are higher still (τ 12.28). The ^{11}B n.m.r. spectrum gives a high-field doublet due to the apical boron atom. Consequently, electrophilic attack readily occurs at the apical position, and chlorination with $AlCl_3$ (catalyst) gives 1-chloropentaborane(9) in 90% yield.[33] DCl will exchange at the same position.[34]

$$DCl + B_5H_9 \xrightarrow{AlCl_3} 1\text{-}DB_5H_8 + HCl$$

Halogenation with bromine and iodine gives apical substitution, but isomerization occurs in the presence of a basic catalyst, probably through a B_5-cage rearrangement rather than a group migration. This electrophilic substitution, followed by rearrangement using hexamethylenetetramine (HMTA), is now well-documented in the synthesis of a variety of polymethyl halogenopentaboranes.[35]

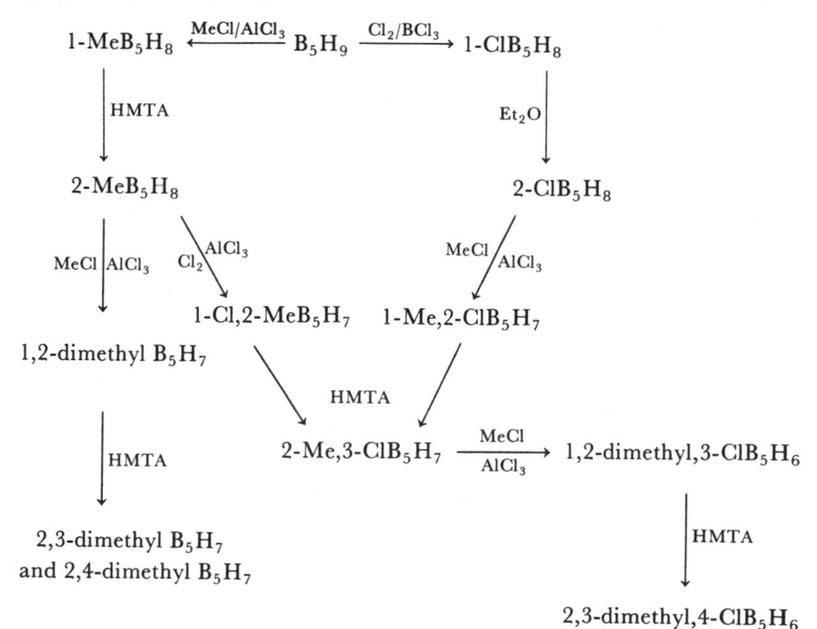

Hydrolysis and alcoholysis are slow, but yield the expected hydrogen and borates, while oxidation to boric oxide is thought to proceed through the unstable intermediate μ-oxo-bis-diborane $B_4H_{10}O$.

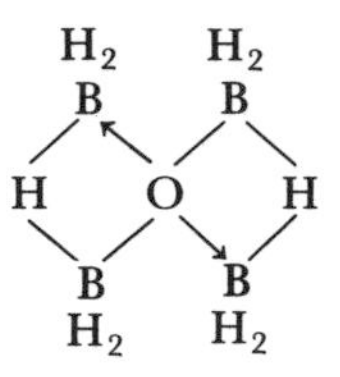

Fig. 3.10. The structure proposed for $B_4H_{10}O$

n-Butyl-lithium gives the anion $B_5H_8^-$, probably through the loss of the more acidic bridging proton.[36] Deuteration with DCl gives exclusively μ-DB_5H_8.

Remarkably, halogenotrialkyl derivatives of the Group IV elements also give μ-$R_3MB_5H_8$ compounds, which means that these elements are also stable in three-centre bonds.[37] Carbon, however, does not.

$$R_3MX + B_5H_8^- \longrightarrow \mu\text{-}R_3MB_5H_8 + X^-$$

(M = Si, Ge, Sn, Pb)

The trimethylsilyl compound, m.p. 16–17°C brominates at position 1 in the pentaborane structure without cleaving the silyl bridge, while hydrolysis gives boric acid, siloxane and H_2.

$$\mu\text{-}Me_3SiB_5H_8 \xrightarrow{Br_2} 1\text{-}Br,\mu\text{-}Me_3SiB_5H_7$$

$$2\mu\text{-}Me_3SiB_5H_8 \xrightarrow{31H_2O} (Me_3Si)_2O + 10B(OH)_3 + 24H_2$$

Dimethylether catalyzes the isomerization to the 2-substituted compounds which are more volatile.

$$e.g., \mu\text{-}Me_3GeB_5H_8 \xrightarrow{Me_2O} 2\text{-}Me_3GeB_5H_8$$

μ-$Me_3GeB_5H_8$ (m.p. 11.5°C; v.p. not detectable at room temp.)

2-$Me_3GeB_5H_8$ (liquid; v.p. ~1 mm at 26°C)

This is thought to proceed through bridge opening followed by hydride closing.

H Si Me_3 H $\xrightarrow{Me_2O}$ H Me_2O H $SiMe_3$ $\longrightarrow$ H H $SiMe_3$

In addition to its catalytic activity, Me_2O is cleaved by 1-iodopentaborane(9), giving 2-$MeOB_5H_8$.[38] Indeed, it is as well to remember that both bromo- and iodo-boranes will readily cleave ethers.

$$1\text{-}IB_5H_8 \xrightarrow{Me_2O} 2\text{-}MeOB_5H_8 + MeI$$

1-Bromopentaborane(9) reacts with the anions of manganese and rhenium carbonyls in ether solvents to give the 2-substituted borametal carbonyl. No 1-substituted product was isolated, presumably due to base-catalyzed isomerization.

$$1\text{-}BrB_5H_8 + M(CO)_5^- \longrightarrow 2\text{-}(CO)_5MB_5H_8$$

The complexes are examples of an ever-widening class of σ-bonded transition metal boron compounds.

Though weak bases do not react with B_5H_9, amines produce a complex series of products, depending on the amount and nature of the amine used. Secondary amines yield adducts which decompose to aminoboranes and diboranes.[b, i]

$$B_5H_9(NHR_2)_3 \longrightarrow 2B_2H_5NR_2 + BH_2NR_2 \xrightarrow{-H_2} B_5H_9(NHR_2)_4$$

$$BH_2NR_2 \xleftarrow{-2H_2} B_5H_9(NHR_2)_5$$

Pentaborane(11), B_5H_{11}, (m.p. −123°C, b.p. 63°C), is the initial pyrolysis product of diborane(6) at 100°C, and has a more open structure than B_5H_9. It resembles an icosahedral fragment, with the $B_{(2)}$—$B_{(5)}$ bridge of B_5H_9 open.

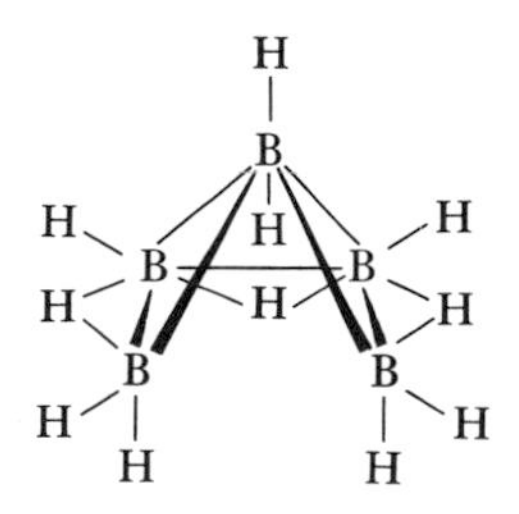

Fig. 3.11. The structure of pentaborane(11)

Hence it has a lower melting point. The ^{11}B n.m.r. spectrum supports this structure, with a high field doublet for $B_{(1)}$ and a low field triplet for $B_{(2)}$ and $B_{(5)}$. So bridging hydrogens cause no splitting. This is one of the lesser stable hydrides, yielding diborane(6) and a little decaborane(14) slowly at 60°C. Varying the heating time and the excess of hydrogen at 100°C can give di-, tetra-, deca- and penta-borane(9).[39]

Ethyldiborane reversible ethylates B_5H_{11},[40] while ethylene yields ethylpentaborane(11) and μ-ethylenetetraborane(10).

$$[EtBH_2]_2 + 2B_5H_{11} \rightleftharpoons B_2H_6 + EtB_5H_{10}$$

$$B_5H_{11} + C_2H_4 \longrightarrow EtB_5H_{10} + C_2H_4B_4H_8$$

Phosphorus trifluoride and carbon monoxide cleave B_5H_{11} similarly,[41]

$$B_5H_{11} + 2PF_3 \longrightarrow BH_3{\cdot}PF_3 + B_4H_8PF_3$$

while ethers catalyze the disproportionation of B_5H_{11} by the two routes shown.

$$B_5H_{11} \xrightarrow{R_2O} \begin{cases} B_4H_8 + \tfrac{1}{2}B_2H_6 \\ \tfrac{1}{2}B_6H_{10} + B_2H_6 \end{cases}$$

The formation of hexaborane(10) B_6H_{10} (m.p. −63°C, b.p. 108°C) from diborane(6) occurs almost quantitatively with dimethylether or diglyme.[42] Its structure resembles a pentagonal pyramid, with non-equivalent basal boron atoms bridged by four hydrogen atoms.

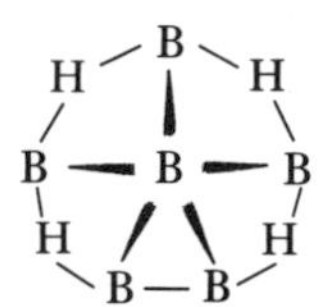

Fig. 3.12. The boron cage of hexaborane(10)

However, the n.m.r. spectrum of this compound shows them to be equivalent, indicating some valence tautomerism.[43]

As with B_5H_9, nucleophilic attack by n-BuLi occurs exclusively at the bridging hydrogen. Deuteration of the $B_6H_9^-$ gives μ-DB_6H_9, while deuterated hexaborane(10) with basal terminal deuterium gives only methane with MeLi. The acidity of these bridging hydrogen atoms seems to increase with cage size, since larger ones can be lithiated with LiB_5H_8.[44]

$$B_5H_8^- + B_6H_{10} \longrightarrow B_6H_9^- \xrightarrow{B_{10}H_{14}} B_{10}H_{13}^-$$

Hexaborane(12), B_6H_{12}, (m.p. −83°C, b.p. 80–90°C), is produced in 4% yield from metaphosphoric acid and $B_3H_8^-$.[45] It decomposes in dimethylether to B_5H_9 while B_4H_{10} results with water. It is less stable than B_6H_{10} and has a more open structure.

Among the lesser important boranes are octaborane(12) and octaborane(18). The former results from the electric discharge of B_2H_6 and B_5H_9, and from the low pressure decomposition of B_9H_{15}, while B_8H_{18} is formed on acidifying $B_3H_8^-$ with polyphosphoric acid. The preferred structure involves two B_4H_9 residues linked through a B—B bond. Indeed, hydrogen gives some B_4H_{10} while CO gives B_4H_8CO and B_4H_{10} in good yields.[46]

n-Nonaborane(15), (m.p. 26°C, b.p. 28°C/0.8 mm), was the first new boron hydride to be synthesized and characterized[21] since the work of Stock, and resulted in small amounts from the silent electric discharge of diborane(6). Higher yields result when B_5H_{11} decomposes on the surface of HMTA or in the presence of diborane(6). The structure can be readily accounted for on the basis of BBB and BHB three-centre bonds, along with two-centre ones.[47]

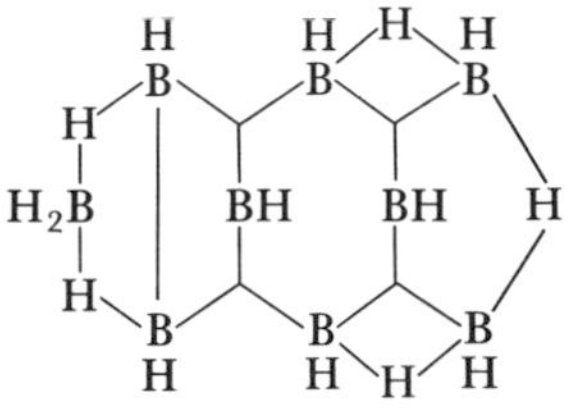

Fig. 3.13. The topological structure of n-nonaborane(15)

It readily decomposes to B_8H_{12} which re-forms B_9H_{15} with diborane.

Decaborane(14), (m.p. 99°C, b.p. 213°C), can be degraded with OH^- to the $B_9H_{14}^-$ anion by loss of the extreme boron atoms (6 or 9).

$$B_{10}H_{14} \xrightarrow{OH^-} B_{10}H_{13}^- \longrightarrow B_{10}H_{13}OH^{2-} \xrightarrow[H_2O]{H_3O^+} B_9H_{14}^- + B(OH)_3 + H_2$$

Bases form neutral adducts of B_9H_{13},

$$B_{10}H_{14} \xrightarrow{2Et_2NH} B_{10}H_{13}NHEt_2^- \, Et_2NH_2^+ \xrightarrow[2H_2O]{H_3O^+} B_9H_{13}NHEt_2 + B(OH)_3 + H_2$$

$$B_{10}H_{12}(SMe_2)_2 \xrightarrow{3EtOH} B_9H_{13}SMe_2 + Me_2S + B(OEt)_3 + H_2$$

while acidification gives iso-B_9H_{15}, isomeric with n-B_9H_{15}. In the presence of ether, $B_9H_{13}OEt_2$ is formed which readily undergoes ligand displacement.[48]

$$B_9H_{14}^- \xrightarrow[-80°C]{HCl} \text{iso-}B_9H_{15}$$

$$B_9H_{14}^- \xrightarrow[Et_2O]{HCl} B_9H_{13}OEt_2 \xrightarrow{L} B_9H_{13}\cdot L + Et_2O$$

The structure resembles part of the $B_{10}H_{14}$ cage, with a triangle of boron atoms removed from the B_{12} icosahedron.

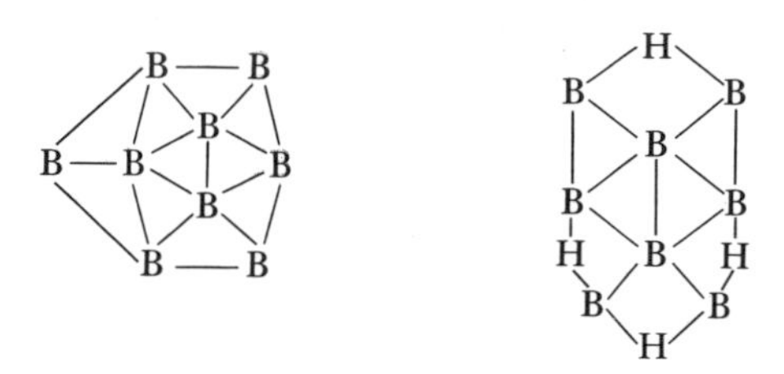

Fig. 3.14. The skeletons of iso-nonaborane(15) and octaborane(12)

Hydrogen is easily lost, and the B_9H_{13} so formed readily disproportionates and dimerizes.

$$B_9H_{15} \xrightarrow{-H_2} B_9H_{13} \begin{cases} \longrightarrow B_8H_{12} + B_{10}H_{14} \\ \longrightarrow \text{n-}B_{18}H_{22} \end{cases}$$

The B_8H_{12} structure is based on the B_9H_{15} cage.

Decaborane(14) is a pyrolysis product of diborane(6) produced through B_5H_9.[49] Yields are highest in the presence of Me_2O, while mixtures of B_4H_{10} and B_5H_9, B_2H_6 and B_4H_{10}, or B_4H_{10} alone all give decaborane(14) when warmed under pressure. The structure shows the presence of four bridging protons, all of which are labile and readily exchange in acid.

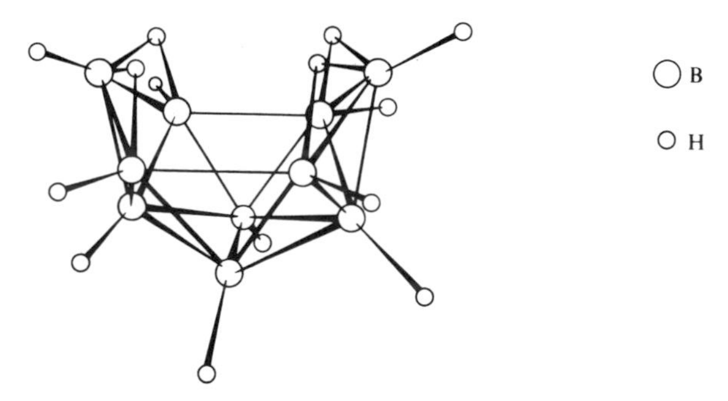

Fig. 3.15. The structure of decaborane(14)[c]

In D_2O, the bridge protons exchange first, followed by terminal ones in the order 6 and 9, then 5, 7, 8, 10 and finally the other positions.[50] Exchange is more rapid in basic ethers, and if bridge-deuterated $B_{10}D_4H_{10}$ "ages" in dioxan H—D exchange occurs only at positions 5, 7, 8 and 10. The mechanism is thought to involve base attack at $B_{(6)}$ and bridge opening. Base removal will give exchange.

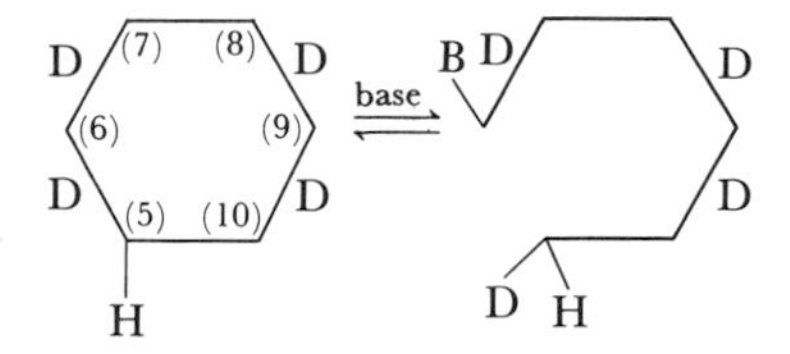

Bases will readily abstract a bridging proton—indeed the hydride is acidic in water,[51] while alkali eventually causes cage contraction.

$$B_{10}H_{14} + H_2O \rightleftharpoons B_{10}H_{13}^- + H_3O^+$$

Sodium hydride will abstract two protons, the product $B_{10}H_{12}^{2-}$ possessing no bridging hydrogen atoms (ν_{B-H} 2300–2400 cm^{-1}, 1700–2000 cm^{-1} blank).

$$B_{10}H_{14} + H^- \longrightarrow B_{10}H_{13}^- \xrightarrow{H^-} B_{10}H_{12}^{2-}$$

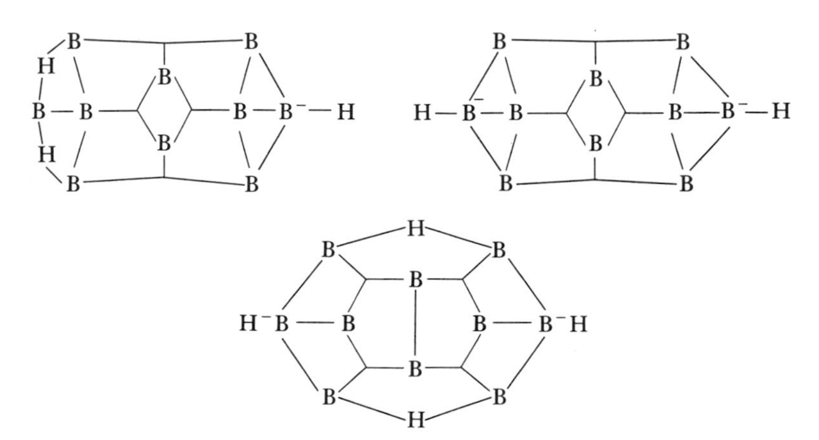

Fig. 3.16. The topological structures of $B_{10}H_{13}^-$, $B_{10}H_{12}^{2-}$ and $B_{10}H_{14}^{2-}$ (one terminal proton omitted per boron atom)

Reducing with Na/NH_3(liq.) or BH_4^- produces the anion $B_{10}H_{14}^{2-}$ which possesses bridging hydrogen atoms. Many derivatives are now known in which the two extra hydrogen atoms on $B_{(6)}$ and $B_{(9)}$ are replaced by a bridging metal or metalloid, or by two molecules of a base.

Me_3NAlH_3 gives the AlH_2-derivative with two bridging hydrogen atoms across $B_{(7)-(8)}$ and $B_{(5)-(10)}$. Diorgano-zinc and -cadmium give similar derivatives while the decaborane Grignard reagent $B_{10}H_{13}MgI$ gives the borane–mercury 4-co-ordinate complex with methylmercury halides.[52]

All possess bridging hydrogen atoms, and direct boron–metal bonds.

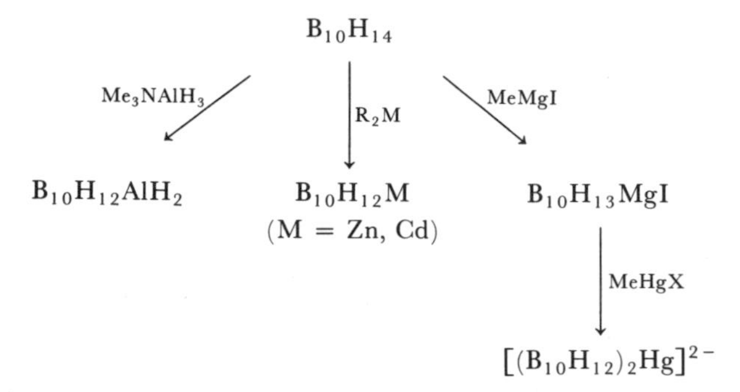

Acetonitrile will substitute $B_{10}H_{14}$ in the 6 and 9 positions, and a wide variety of bases (*e.g.*, amides, tertiary amines, sulphoxides, phosphines and their oxides) will displace MeCN from the complex.[53]

$$B_{10}H_{14} + 2MeCN \longrightarrow B_{10}H_{12}(MeCN)_2 \xrightarrow{2L} B_{10}H_{12}L_2$$

All possess two bridging hydrogen atoms and have the same topological structure as $B_{10}H_{14}^{2-}$. With alkyl isocyanides similar compounds form, while a 1 : 1 adduct formed directly is believed to be a

$$B_{10}H_{12}(SEt_2)_2 + 2RNC \longrightarrow B_{10}H_{12}(CNR)_2$$

$$B_{10}H_{14} + EtNC \longrightarrow EtNH_2CB_{10}H_{12}$$

zwitterion with a CB_{10} cage isoelectronic with $B_9C_2H_{13}$ and $B_{11}H_{13}^{2-}$.

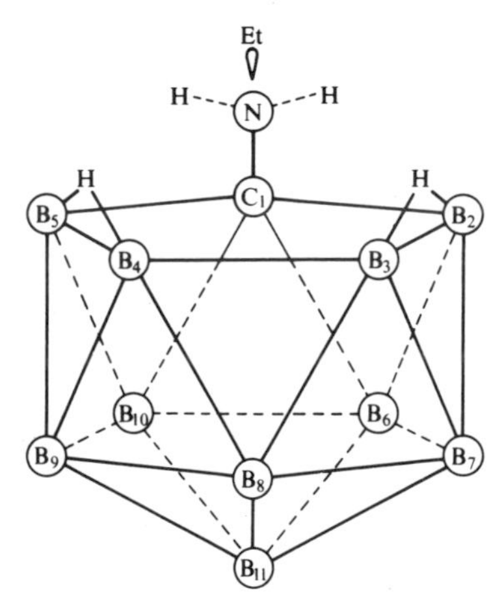

Fig. 3.17. A proposed structure for $EtNH_2CB_{10}H_{12}$[53]

Soft bases form adducts with decaborane(14) rather than the ionic products of $B_{10}H_{10}^{2-}$ typical of the harder amines. Consequently, $B_{10}H_{10}^{2-}$ can be converted into the decaborane(14) adduct with diethyl sulphide, substitution occurring at the 6 and 9 positions, thereby reversing the normal reaction.

$$B_{10}H_{10}^{2-} + 2Et_2S + 2H^+ \longrightarrow B_{10}H_{12}(Et_2S)_2$$

Just as aluminium can be put into the decaborane framework, so boron will insert using $Et_3N{\cdot}BH_3$. The $B_{11}H_{14}^-$ anion is formed under mild conditions;

$$(Et_3NH^+)_2B_{12}H_{12}^{2-} \xleftarrow{2Et_3N\cdot BH_3} B_{10}H_{14} \xrightarrow{Et_3N\cdot BH_3} Et_3NH^+B_{11}H_{14}^-$$

this probably has an icosahedral structure with one site vacant and the lost BH replaced by H_3^+ (both two-electron donors) centred on the five-fold axis.[54]

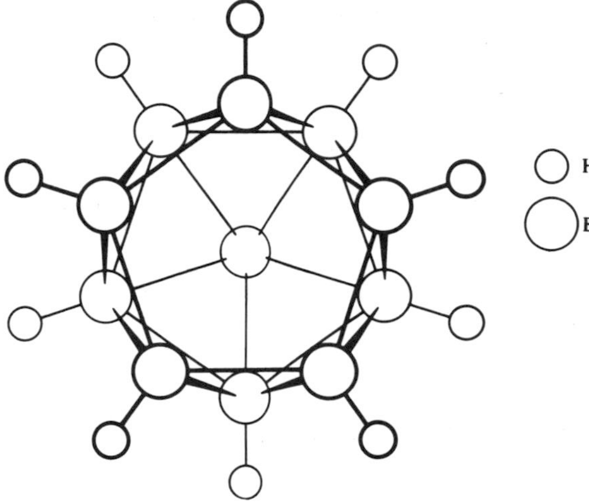

Fig. 3.18. A view of the proposed structure of $B_{11}H_{14}^-$ down the five-fold axis. The three additional protons are thought to be bonded in the open pentagonal face[54]

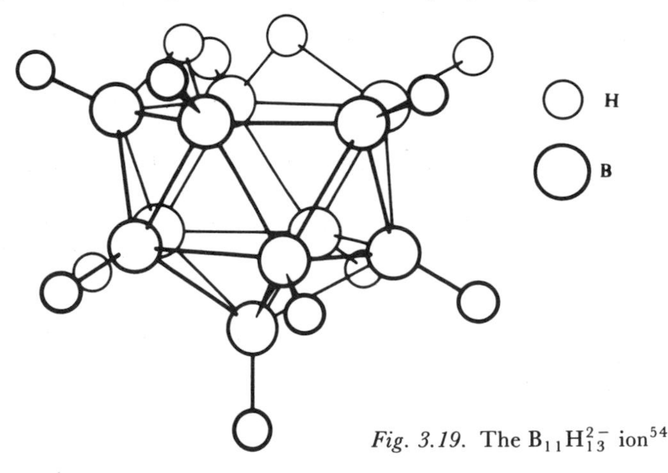

Fig. 3.19. The $B_{11}H_{13}^{2-}$ ion[54]

There may well be some interaction between these three hydrogen atoms as H_3^+ has been characterized at low temperatures, while $B_{11}H_{13}^{2-}$, formed by base attack, has the two bridging protons within 0.17 nm. So the strongly bonding character of H_3^+ proposed in Chapter 1 is indirectly supported here. The $B_{12}H_{12}^{2-}$ ion, which also results by this reaction, was first predicted to be stable in 1955 and to possess 13 cage-bonding molecular orbitals (A_g, T_{1u}, G_g, H_{1u}). These are conveniently filled by the 26 electrons available. This was isolated in 1960 as a minor product from the reaction of Et_3N and 2-iododecaborane(14).[55]

$$Et_3N + B_{10}H_{13}I \xrightarrow{C_6H_6} Et_3NH^+ + B_{10}H_{10}^{2-} + B_{12}H_{12}^{2-}$$

Both $B_{11}H_{14}^-$ and $B_{12}H_{12}^{2-}$ can be synthesized from decaborane(14) and BH_4^-.[56]

$$B_{12}H_{12}^{2-} + 5H_2 \xleftarrow[\text{diglyme}]{100°C} B_{10}H_{14} \xrightarrow[\text{glyme}]{80°C} B_{11}H_{14}^- + 2H_2$$

Electron density calculations on the $B_{10}H_{14}$ cage indicate preferred electrophilic attack at $B_{(1)-(4)}$. This is supported by the low field ^{11}B n.m.r. absorption position for $B_{(1)}$ and $B_{(3)}$.[54] Friedel–Crafts substitution occurs in these four positions[57] readily, while iodination in the presence, or absence, of a catalyst occurs at similar sites.[58] Substitution is limited, except in the presence of methanol, when complete cage breakdown occurs.

$$B_{10}H_{14} + 20I_2 + 30MeOH \longrightarrow 10B(OMe)_3 + 40HI + 2H_2$$

THE ANIONIC BORON HYDRIDES $B_nH_n^{2-}$

$B_{10}H_{10}^{2-}$

Though triethylamine reacts with 2-iododecaborane to give a trace of $B_{12}H_{12}^{2-}$ and $B_{10}H_{10}^{2-}$, the latter is better obtained by refluxing decaborane or its acetonitrile adduct with triethylamine in benzene.[53]

$$\begin{matrix}(CH_3CN)_2B_{10}H_{12}\\ \text{or } B_{10}H_{14}\end{matrix} \xrightarrow[C_6H_6]{2Et_3N} (Et_3N)_2B_{10}H_{12} \longrightarrow [Et_3NH]_2^{2+}B_{10}H_{10}^{2-}$$

This synthesis again illustrates the ligand displacement reactions operational for such compounds. The mechanism of formation of the $B_{10}H_{10}^{2-}$ ion involves "zipping-up" the mouth of decaborane in the presence of base to yield a closed-cage *closo*-like structure, which resembles two inverted skew square-pyramids (D_{4d}).

```
            H
         /     \
       B         B(5)
      /(10)       \
 Et3NB             BNEt3  ---->  B----B----B 2-          + 2Et3NH+
      \(9)        /(6)          /(9)\ /(10)\ /(5)
   (8)B           B(7)         B----B----B
         \     /              (8)  (7)  (6)
            H
```

The chemistry of both this cage and its derivatives has been widely studied so will only be briefly summarized here. The ^{11}B n.m.r. spectrum shows a low-field doublet one quarter of the intensity of the high-field one, supporting the proposed structure,[59] and with the low-field borons having lesser ligancy.

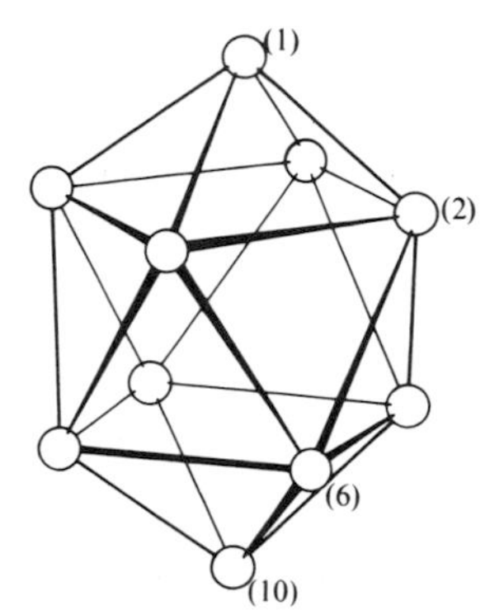

Fig. 3.20. The structure of the $B_{10}H_{10}^{2-}$ cage showing the two kinds of boron atoms

Large cations will stabilize this anion in crystalline lattices, while the heavy coinage metals form weak bonds along the $B_{(1)-(2)}$ edge (*i.e.*, covalent salts) without reduction to the free metal.[60] Whilst it is difficult to fragment the $B_{10}H_{10}^{2-}$ cage, it can be oxidized by inorganic cations or electrochemically to $B_{20}H_{18}^{2-}$ and its derivatives.

$$2B_{10}H_{10}^{2-} + 4Fe^{3+} \longrightarrow 2H^+ + B_{20}H_{18}^{2-} + 4Fe^{2+}$$

$$2B_{10}H_9SMe_2^- + 4Ce^{4+} \longrightarrow 2H^+ + B_{20}H_{16}(SMe_2)_2 + 4Ce^{3+}$$

These compounds are thought to involve two B_{10} cages linked through their $B_{(1)-(2)}$ edges by two three-centre bonds. $B_{20}H_{18}^{2-}$ can be readily isomerized using u.v. light in acetonitrile. The structure of this "photo-isomer" is thought to involve a $B_{(2)-(2')}$ and a $B_{(6)-(9')}$ hydrogen bridge between the two B_{10} cages. Reduction with sodium/liq. NH_3 yields $B_{20}H_{18}^{4-}$, which comprises two $B_{10}H_9^{2-}$ residues linked through a B—B bond. The three isomers comprising apical–apical, equatorial–equatorial and apical–equatorial linking have all been characterized, with the former the most stable.[61] Partial oxidation gives $B_{20}H_{19}^{3-}$, with two $B_{10}H_9^{2-}$ residues bridged by a proton in some way. The low ν_{BHB} at 1850 cm^{-1} supports the facile hydrolysis

$$(Et_3NH^+)_3B_{20}H_{19}^{3-} \xrightarrow{OH^-} B_{20}H_{18}^{4-} + H_2O$$

(m.p. 163°C)

Halogenation of the $B_{10}H_{10}^{2-}$ cage produces poly- and fully-substituted anions which are singularly stable both thermally and to strong nucleophiles. Olefins give alkylated derivatives, while the S—S bond of disulphides is readily cleaved. Dimethylsulphoxide is reduced to the dimethylsulphide adduct.

$$B_{10}H_9C_2H_4R^{2-} \xleftarrow{RCHCH_2} (H_3O^+)_2B_{10}H_{10}^{2-} \xrightarrow[X_2]{H_2O} B_{10}X_{10}^{2-} \quad (X = Cl, Br, I)$$

$$(H_3O^+)_2B_{10}H_{10}^{2-} \xrightarrow{R_2S_2} B_{10}H_8(SR)_2^{2-} + H_2$$

$$(H_3O^+)_2B_{10}H_{10}^{2-} \xrightarrow{Me_2SO} B_{10}H_8(SMe_2)_2$$

The most useful intermediate for cage substitution appears to be the neutral diazonium compound $B_{10}H_8(N_2)_2$. Nitrogen is readily displaced by many ligands and stereochemical integrity is maintained.

$$B_{10}H_{10}^{2-} \xrightarrow[soln.]{HNO_2} \text{intermediate} \xrightarrow[MeOH]{BH_4^-} 1,10\text{-}B_{10}H_8(N_2)_2$$

$$1,10\text{-}B_{10}H_8L_2 \xleftarrow{2L} 1,10\text{-}B_{10}H_8(N_2)_2 \xrightarrow{2CO} 1,10\text{-}B_{10}H_8(CO)_2$$

The dicarbonyl readily forms a perhalogeno derivative, which is more stable than the parent compound, and undergoes similar reactions.[62]

$$B_{10}H_8(CO)_2 \xrightarrow{Cl_2} B_{10}Cl_8(CO)_2$$

$$B_{10}H_8(CO)_2 \xrightarrow{MeOH} B_{10}H_8(COOMe)_2^{2-}$$

$$B_{10}H_8(CO)_2 \xrightarrow{N_3^-} B_{10}H_8(NCO)_2$$

$$B_{10}H_8(CO)_2 \xrightarrow{NH_3} B_{10}H_8(CONH_2)_2^{2-}$$

$$B_{10}H_8(CO)_2 \xrightarrow{H_2O,\ H_2NOSO_3H} B_{10}H_8(NH_3)_2$$

$$B_{10}H_8(CO)_2 \xrightarrow{LiAlH_4} B_{10}H_8Me_2$$

$B_{12}H_{12}^{2-}$

This anion, the first simple compound with the icosahedral framework (I_h point group) is conveniently synthesized from $B_{10}H_{14}$ and either Et_3NBH_3 or BH_4^-. The n.m.r. spectrum shows all 12 boron atoms to be equivalent.

The chemistry of this anion closely resembles that of $B_{10}H_{10}^{2-}$. Halogenation occurs readily, but in some cage substitution reactions differences have been observed, with $B_{10}H_{10}^{2-}$ functioning as a more organic substituent.[63] Thus, benzoylation gives a ketonic residue substituted equatorially, as expected for electrophilic attack. This can be readily protonated and forms a carbazone. With $B_{12}H_{12}^{2-}$, compounds containing the B—O bond result.

$$B_{10}H_{10}^{2-} \xrightarrow[H^+]{PhCOCl} B_{10}H_9COPh^{2-} \xrightarrow{Cl_2} B_{10}Cl_9COPh^{2-}$$

$$B_{10}H_9COPh^{2-} \xrightarrow{H_2NNHCONH_2} B_{10}H_9C(NNHCONH_2)Ph^{2-}$$

However, alkylation with olefins and amination with hydroxylamine-*O*-sulphuric acid are directly similar, while carbonylation occurs directly giving 1,7- and 1,12-isomers.

$$B_{12}H_{12}^{2-} + 2CO \longrightarrow B_{12}H_{10}(CO)_2$$

Unlike the waterstable $B_{10}H_8(CO)_2$, the carbonylated B_{12} isomers are hygroscopic and ν_{CO} disappears as the acid derivatives form.

$$B_{12}H_{10}(CO)_2 + 2H_2O \longrightarrow B_{12}H_{10}(CO_2H)_2^{2-} + 2H^+$$

Attempts to oxidize $B_{12}H_{12}^{2-}$ chemically to coupled cage anions only yielded BO_3^{3-}, but electrochemical oxidation in acetonitrile yields the ion $B_{24}H_{23}^{3-}$. This ion is thought to have a structure like $B_{20}H_{19}^{3-}$ involving a B—H—B linkage between two $B_{12}H_{11}$ cages. The high stretching frequency of this linkage (2250 cm^{-1}) supports the hydrolytic stability. Reduction with Na/liq. NH_3 gives $B_{12}H_{12}^{2-}$, in contrast to the $B_{20}H_{18}^{2-}$ anion which is not bisected. Halogenation and deuterium exchange occur readily.[64]

$$B_{12}H_{12}^{2-} \underset{\text{Na/liq. }NH_3}{\overset{\text{Oxidation}}{\rightleftarrows}} B_{24}H_{23}^{3-} \xrightarrow{\text{DCl}} B_{24}D_{23}^{3-}$$

$$B_{24}H_{23}^{3-} \longrightarrow B_{24}H_{23-n}X_n^{3-} \text{ and } B_{24}H_{22-n}X_n^{4-}$$

$B_nH_n^{2-}$ (n = 6–9, 11)

The anion $B_6H_6^{2-}$ has an octahedral structure, and the bonding of the cage can be explained in the same way as the B_6^{2-} polymer. This hexahydrohexaborate anion results when diborane(6) and BH_4^- are heated under reflux in diglyme, and crystals can be obtained by extracting with Me_4N^+. It is also formed when $Na_2B_9H_9 \cdot H_2O$ is heated in THF, along with $B_7H_7^{2-}$ and $B_8H_8^{2-}$.

Of the anions $B_nH_n^{2-}$ ($n = 6-12$), $B_7H_7^{2-}$ is the least stable, and may well have a D_{5h} structure as the ^{11}B n.m.r. spectrum shows two peaks (area ratio 5:2). It is the least stable of the $B_nH_n^{2-}$ anions in the presence of water.[65]

Though $B_8H_8^{2-}$ can be readily brominated without disrupting the cage, acid hydrolysis gives boric acid and hydrogen. This ion has the D_{2d} structure which is also found in B_8Cl_8 and $B_6C_2H_6Me_2$.

Salts of $B_9H_9^{2-}$ can be readily extracted from the thermal decomposition products of $B_3H_8^-$. The caesium salt is stable to 600°C, whilst that of the $B_{11}H_{11}^{2-}$ ion, formed from $Cs_2B_{11}H_{13}$ by loss of H_2 at 250°C, disproportionates at 400°C.

$$Cs_2B_{11}H_{13} \xrightarrow[-H_2]{250°C} Cs_2B_{11}H_{11} \xrightarrow{400°C} Cs_2B_{10}H_{10} + Cs_2B_{12}H_{12}$$

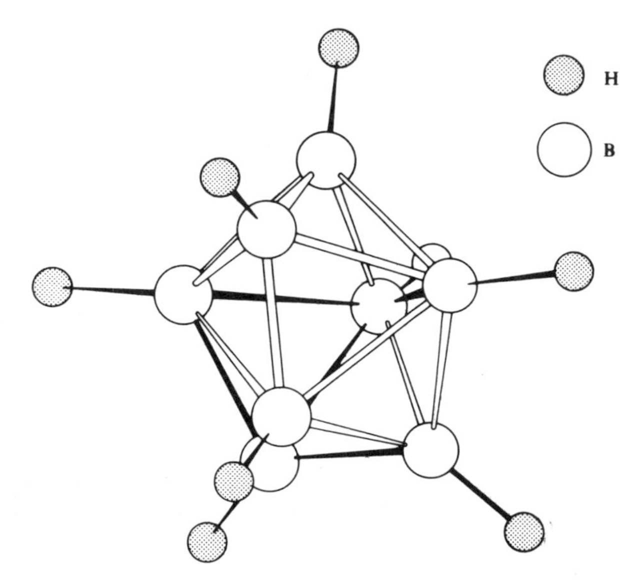

Fig. 3.21. The molecular structure of $B_8H_8^{2-}$ [65]

Whilst $B_9H_9^{2-}$ has the structure of a tricapped triangular prism, that of $B_{11}H_{11}^{2-}$ is thought to involve the $B_{10}H_{14}$ framework bridged by **BH** across the mouth.

So all the ions $B_nH_n^{2-}$ ($n = 6$–12) have been synthesized. All have neutral analogues $B_{n-2}C_2H_n$, to be discussed later (p. 72), and have hydrolytic stabilities decreasing in the order for n of $12 > 10 \gg 11 > 9 \sim 8 \sim 6 > 7$.

BOROHYDRIDES

The borohydride derivatives of the elements show a wide range of properties, varying from ionic compounds, *e.g.*, $K^+BH_4^-$, which are air- and water stable, and soluble in polar solvents, to the covalent hydrogen-bridged derivatives of aluminium. This is a low boiling liquid (estimated b.p. 45°C), which is spontaneously inflammable. It adopts a hydrogen-bridge structure, as do most borohydrides at low temperatures. At room temperature, the majority are unstable, especially the post-Group III ones. So consideration will be given only to those of the first three Groups.

An interesting structural variation occurs with lithium borohydride.[b] At room temperature it is not isomorphous with the other alkali metal borohydrides, but comprises a laminar array of boron and lithium atoms which are hydrogen-bridged. At 110°C, it becomes isomorphous with the other Group I borohydrides. The best yields are obtained from lithium hydride and diborane(6) or boron trifluoride, while the metathetical reaction between lithium chloride and sodium borohydride proves the best method economically.

$$4LiH + 4BF_3 \longrightarrow LiBH_4 + 3LiBF_4$$

Beryllium borohydride has already been discussed in the last Chapter (p. 15), and, in support of the diagonal relationship between first and second period elements, aluminium also forms a volatile borohydride. It is the only stable Group III borohydride, and with its low boiling point (estimated as 45°C) and high heat of combustion it has received much attention as a potential fuel. It was first prepared by Schlesinger in 1939 from diborane(6) and trimethylaluminium below 80°C.[66] Better yields result using aluminium halides and the borohydride ion.

$$[Me_3Al]_2 + 4B_2H_6 \longrightarrow 2Al(BH_4)_3 + 2Me_3B$$

$$AlX_3 + 3MBH_4 \longrightarrow Al(BH_4)_3 + 3MX \quad (80\text{–}95\% \text{ yield})$$

Aluminium borohydride is a very reactive compound which slowly loses hydrogen at room temperature. It spontaneously enflames in air and hydrolyzes to give diborane(6) as the first product. Alkenes add to give alkylaluminium compounds, while ethyl orthoformate and tetraethyl silicate are reduced to the ether and alkoxysilane.

Bases readily form 1 : 1 adducts, amines and phosphines then reacting further until four moles of ligand are used.[67]

$$Al(BH_4)_3 + R_3M \xrightarrow[(M\,=\,N,\,P)]{} R_3MAl(BH_4)_3 \longrightarrow 3R_3M{\cdot}BH_3 + 3R_3MAlH_3$$

The 1 : 1 adduct with trimethylamine has a structure approximating to a pentagonal-bipyramid, with the Me_3N group and one hydrogen atom *trans*, and five hydrogen atoms in the five-fold plane. So the BH_4 groups still double-bridge.

Ammonia also complexes with $Al(BH_4)_3$, but the final products are ionic and not the simple borane/alane adducts formed with Me_3N. The n.m.r. spectrum indicates the presence of complexed boron and aluminium cations in addition to BH_4^- anions.

$$Al(BH_4)_3 + 6NH_3 \longrightarrow AlH_2(NH_3)_4^+{\cdot}BH_2(NH_3)_2^+{\cdot}(BH_4^-)_2$$

On the basis of electron diffraction data, $Al(BH_4)_3$ is believed to have the D_{3h} structure associated with three BH_4 residues, rendering the aluminium atom 6-co-ordinate. The p.m.r. spectrum at high temperatures (>50°C) supports a structure in which internal rotation of the BH_4 residue occurs through the breaking and re-forming of bridge bonds. Consequently, the spectrum resembles the quartet (and superimposed septet) of the borohydride anion. Indeed, the Al—H distance is long

(0.21 nm). The fine structure is lost on cooling, consistent with rotation slowing below the p.m.r. time-scale, and spin–spin coupling with ^{27}Al nuclei and their large electric quadrupole moment.[68]

Alane, AlH_3, is polymeric, unlike diborane(6). Each aluminium atom is surrounded octahedrally by six hydrogen atoms at the same distance of 0.172 nm.[69] Each bridges two aluminium atoms, and this distance lies between the Al—H bond distances of $Al(BH_4)_3$ (long) and $LiAlH_4$ (short), supporting the three-centre–two-electron bonding of the bridge.

Gallium and indium borohydrides[b] are unstable, but dimethylgallium borohydride is more stable, as described above in the discussion of diborane(6) (p. 47). Similarly, thallic chloride is reduced to thallous borohydride, which is isomorphous with the ionic alkali metal borohydrides.

THE OCTAHYDROTRIBORATE ANION $B_3H_8^-$

This ion is usually prepared by reducing BF_3, or by base-cleavage of tetraborane(10). However, the most convenient route involves refluxing $NaBH_4$ with diborane(6) in diglyme at 100°C.[70]

$$4BF_3 + 5BH_4^- \longrightarrow 3BF_4^- + 2B_3H_8^- + 2H_2$$

$$2R_2O + B_4H_{10} \longrightarrow B_3H_8^- + (R_2O)_2BH_2^+$$

$$NaBH_4 + B_2H_6 \longrightarrow NaB_3H_8 + H_2$$

The elucidation of the structure has proved to be an interesting problem. The ^{11}B n.m.r. spectrum shows all protons to be equivalent in compounds that are essentially ionic. With $(Ph_3P)_2CuB_3H_8$, two peaks are observed.[71]

The Group VI metal carbonyls readily complex with $B_3H_8^-$. The salts formed exhibit ν_{B-H} at 2400–2500 cm^{-1} and ν_{B-H-B} at about 2100 cm^{-1} in their infrared spectra.[72]

$$B_3H_8^- + M(CO)_6 \longrightarrow [M(CO)_4B_3H_8]^- + 2CO$$

The structures of these, and of $(Ph_3P)_2CuB_3H_8$, show that, in general, the metal hydride bridges with the B—B bond. The metal–carbon bonds *trans* to these hydride bridges are generally shorter than normally encountered in *cis*-di-substituted octahedral complexes.[72] This may well be due to increased π-bonding between the metal and these equatorial carbonyl groups, indicating little d-orbital interaction with the B_3H_8 group.

The ion slowly decomposes in refluxing diglyme to yield $B_{12}H_{12}^{2-}$, while metaphosphoric acid gives B_4H_{10} and B_6H_{12}.

$$5B_3H_8^- \longrightarrow B_{12}H_{12}^{2-} + BH_4^- + 8H_2$$

Alkylammonium halides yield aminetriborane(7) derivatives.[73]

$$B_3H_8^- + R_3NH^+ \longrightarrow H_2 + R_3N{\cdot}B_3H_7$$

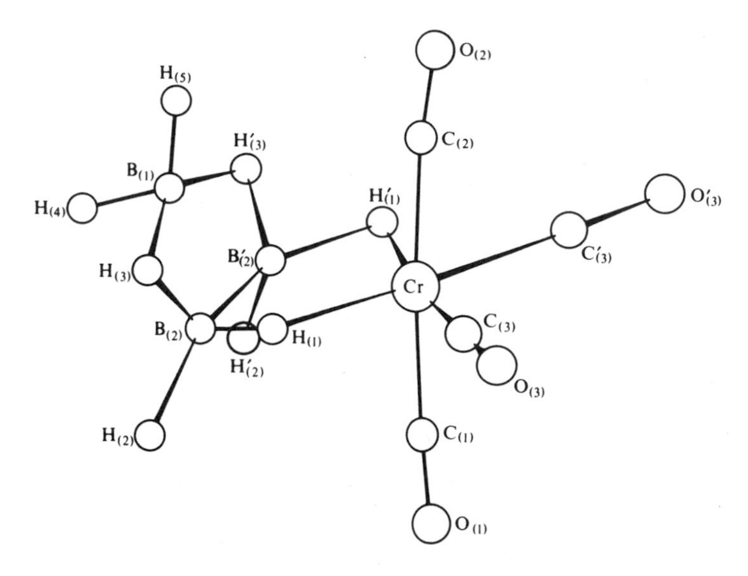

Fig. 3.22. The molecular structure of $(CO)_4CrB_3H_8^-$ [72]

THE CARBORANES

These are compounds containing carbon as an integral part of the boron cage, and include *nido*-carboranes which possess an open-cage structure in which some framework members are linked by hydrogen bridges. The *closo*-carboranes have closed-cage structures in which hydrogen bridges are absent, and are structurally analogous to the $B_nH_n^{2-}$ anions already discussed, with B^- replaced by isoelectronic C.

It will be appreciated from the structure adopted by benzene and $B_6H_6^{2-}$ that only four electrons are necessary to change the closed cage $B_6H_6^{2-}$ structure into the planar one adopted by benzene. Changing C for B in a closed cage will therefore lead to a more open, flatter structure. Thus, heating *cis*-diborylalkenes or attacking diethylchloroborane with the $(RC{\equiv}C{\cdot}BEt_3)^-$ adduct leads to the substituted pentagonal-pyramidal cage $C_4B_2H_6$.[74] Heating 1,2-tetramethylenediborane(6) at 550°C gives low yields of the parent carborane $C_4B_2H_6$, the fifth member of the series in Figure 3.25.

$$[Et_3B{\cdot}C{\equiv}CR]^- + Et_2BCl \longrightarrow (Et)(Et_2B)C{=}C(R)(BEt_2) \xrightarrow{150^\circ C} Et_6C_4B_2 \text{ (pentagonal-pyramidal cage)}$$

Et Et
C—C
Et
B
Et B — C Et
C
Et

Open-Cage- or *Nido*-carboranes

These compounds have structures based on some of the boron hydrides but incorporating one, two or three carbon atoms, and having at least one bridging hydrogen atom in the skeleton. One peculiar feature common to all carboranes is that to date no compound has been synthesized with either carbon bridging two boron atoms in a three-centre–two-electron bond or acting as one end of a hydride bridge.

Small nido-carboranes

The only example isoelectronic with B_5H_9 is 1,2-dicarba-*nido*-pentaborane(7), $C_2B_3H_7$.[75]

$$B_4H_{10} + \underset{\text{(10-times excess)}}{C_2H_2} \xrightarrow{50°C} C_2B_3H_7 \text{ (3–4\% yield)}$$

Unlike pentaborane(9), however, it has only two bridging hydrogen atoms, as the two carbon atoms provide the extra two electrons, with the bridging B—H bond replaced by C.

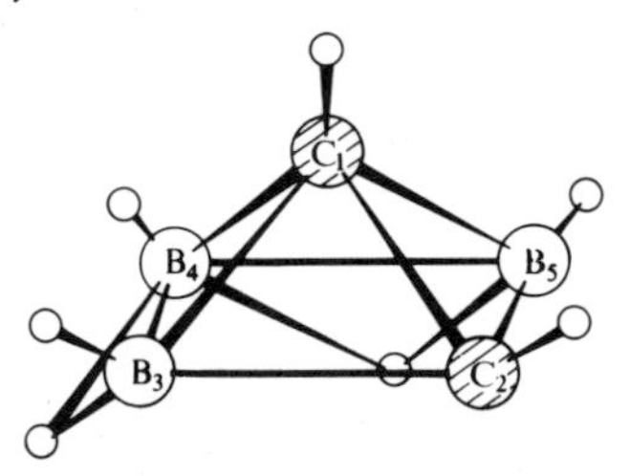

Fig. 3.23. The proposed structure of 1,2-$C_2B_3H_7$[75]

Like most carboranes, the structure is conveniently determined by infrared and n.m.r. spectroscopy.

Monocarba-*nido*-hexaborane(7) CB_5H_7 (v.p. 503 mm/26°C) results from the silent electric discharge of 1-methylpentaborane(9). It has an octahedral structure, with a hydride bridge,[76] probably between two boron atoms *cis* to carbon, and is the conjugate acid of $CB_5H_6^-$ (*cf.* $B_6H_6^{2-}$).

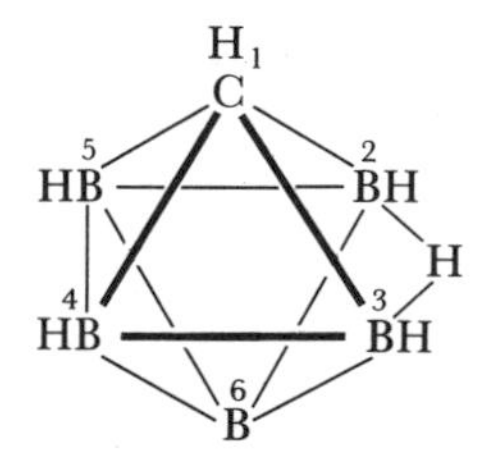

Fig. 3.24. The structure of CB_5H_7[76]

Several carboranes have been synthesized based on the cage structure of B_6H_{10}.

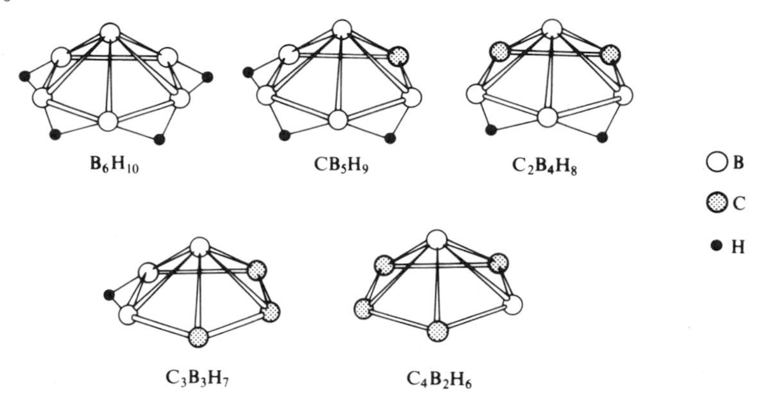

Fig. 3.25. The isosteric compounds $C_xB_{6-x}H_{10-x}$ (x = 0–4)[78]

Although the parent compound 2-carba-*nido*-hexaborane(9) is unknown, three of its monomethyl isomers,[77] together with dicarba-*nido*-hexaborane(8), result when acetylene is heated with pentaborane(9). They are thought to be formed by the addition of the B—H bond to acetylene, thereby forming a vinylborane which rearranges through intramolecular hydroboration. This monocarba-*nido*-hexaborane(9) cage is also formed from ethyldifluoroborane and lithium as the *B*-penta-ethyl, *C*-methyl homologue (b.p. 84°C/10^{-3} mm), while others are formed from dialkylhalogenoboranes, R_3B and $BF_3{\cdot}Et_2O$ with an alkali metal.[78] The organo-substituted homologues are only slowly oxidized at room temperature, but pyrolyze to dicarbo-*closo*-pentaborane and *closo*-heptaborane.

Dialkylhalogenoboranes and sodium, or the pyrolysis of acetylenes with ethylboranes, yields derivatives of the tetrahedral cage B_3C.[79] The fully-alkylated homologues are generally fairly stable, but they will slowly oxidize to other carboranes.

2,3-Dicarba-*nido*-hexaborane(8), and its homologues, provide another example in the series of compounds $C_xB_{6-x}H_{10-x}$ (x = 0–4). They are conveniently synthesized by the base catalysis or pyrolysis of alkynes with pentaborane(9), and isomers result by varying these starting materials.[80]

$$RC{\equiv}CR' + B_5H_9 \xrightarrow[\text{or base}]{\text{heat}} RR'C_2B_4H_6$$

The structure closely resembles that of B_6H_{10} with two bridging hydrogen atoms missing and a C—C bond distance of 0.143 nm, close to that of benzene. In *closo*-carboranes it is normally longer, as would be expected for these more electron-deficient carboranes.

The reaction of $C_2B_4H_8$ by H^- in diglyme leads to the loss of a bridging proton, which would be expected from the pentaborane(9) and hexaborane(10) work. Direct deuteration leads to exchange on the B—H bonds but not the C—H bonds.[81] Acid-catalyzed halogenation occurs at $B_{(4)}$,[82] while with trimethylgallium the compound $MeGaC_2B_4H_6$ (m.p. 34°C) forms with the two acidic bridging protons forming methane.

$$Me_3Ga + C_2B_4H_8 \longrightarrow MeGaC_2B_4H_6 + 2CH_4$$

This *closo*-carborane is isoelectronic with $B_7H_7^{2-}$ and $C_2B_5H_7$, and has the MeGa and BH groups in apex positions of the pentagonal-bipyramid.[83] Many hetero-substituted carboranes have now been synthesized.

The three-carbon carborane, tricarba-*nido*-hexaborane(7), the fourth member of the series shown in Figure 3.25, is formed from tetraborane(10) and acetylene.[84] The parent compound 2,3,4-tricarba-*nido*-hexaborane(7) has been isolated along with methyl-substituted homologues, but only the latter are stable at room temperature. The *C,C'*-dimethyl derivative reacts with sodium hydride.

$$2,4\text{-}Me_2C_3B_3H_5 + H^- \longrightarrow Me_2C_3B_3H_4^- + H_2$$

The sole bridging hydrogen is lost, and deuteration with DCl gives the bridged deutero product $Me_2C_3B_3H_4D$.

Large nido-carboranes

Dicarbo-*nido*-undecaborane(13), $C_2B_9H_{13}$, is the second member of the class of *nido*-carborane $C_2B_nH_{n+4}$ (n = 4 or 9). The parent carborane and its *C*-substituted derivatives can be prepared by the base degradation of *ortho*-carborane (1,2-dicarba-*closo*-dodecaborane $C_2B_{10}H_{12}$).[85]

$$1,2\text{-}C_2B_{10}H_{12} \xrightarrow{MeO^-} C_2B_9H_{12}^- \xrightarrow{H^+} C_2B_9H_{13} \xrightarrow{-H_2} C_2B_9H_{11}$$

It is isoelectronic with $B_{11}H_{13}^{2-}$, in which the two extra protons bridge around the open pentagonal ring. The p.m.r. spectrum and facile dehydrogenation of $C_2B_9H_{13}$ at 100°C support this weak bonding. Indeed, it can be titrated as a monoprotic acid, while the anion $C_2B_9H_{12}^-$ can be readily reduced.

$$R_2C_2B_9H_{10}^- \xrightarrow{\text{Na or NaH}} R_2C_2B_9H_9^{2-}$$

This anion possesses no bridging hydrogen so is named (3),1,2-dicarba-*closo*-undecaborane(11) dianion with boron missing from the 3 position of the *ortho*-carborane framework. This name is abbreviated to (3),1,2-dicarbollide (Spanish "olla", kettle or jar). The ion resembles a deep jar, and the open pentagonal face has three filled molecular orbitals of symmetry suitable for π-bonding to a transition metal, just as $C_5H_5^-$ does. These, and other such complexes, will be discussed in a later section.

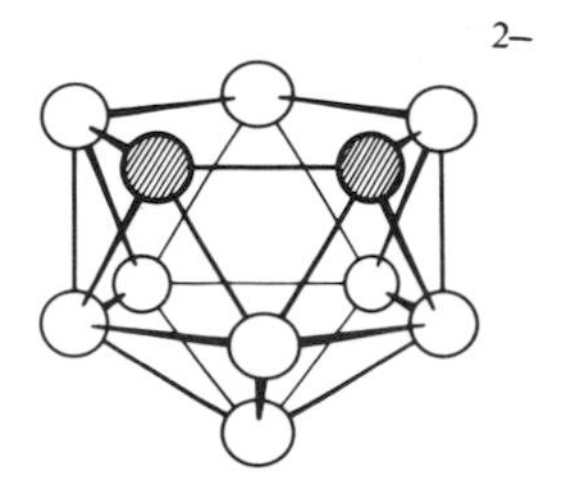

Fig. 3.26. The dicarbollide anion $C_2B_9H_{11}^{2-}$ with the open π-bonding pentagonal face

The *closo*-undecaborane(11) cage results when $C_2B_9H_{13}$ is heated. This can be oxidized by dichromate to derivatives of dicarba-*nido*-nonaborane(13), which also result when (3),1,7-$C_2B_9H_{12}^-$ is hydrolyzed.[86]

$$R_2C_2B_9H_9 \xrightarrow[MeCO_2H]{Cr_2O_7^{2-}/0°C} R_2C_2B_7H_{11}$$

$$\underset{meta\text{-carborane}}{1,7\text{-}C_2B_{10}H_{12}} \xrightarrow[KOH]{EtOH} (3),1,7\text{-}C_2B_9H_{12}^- \xrightarrow{Cr_2O_7^{2-}} C_2B_7H_{13}$$

The structure closely resembles part of the decaborane(14) cage. It is represented topologically in Figure 3.27, and contains two methylene groups bound into the polyborane framework.[87]

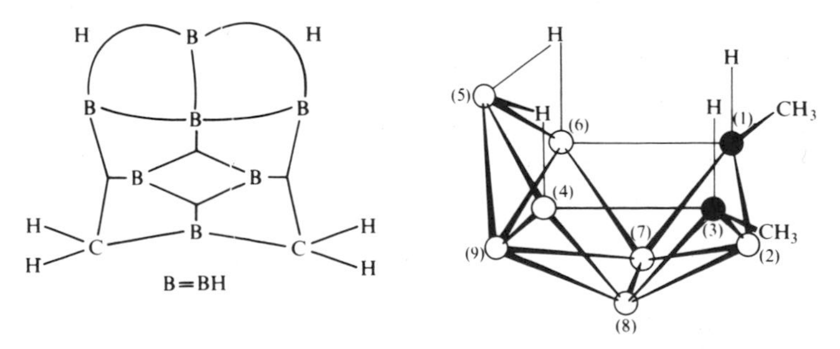

Fig. 3.27. The topological representation and the structure of $B_7C_2H_{11}(CH_3)_2$ where ● = carbon and ○ = B—H[86]

1,6-Dicarba-*closo*-decaborane loses one boron atom with base, again giving the conjugate base of $C_2B_7H_{13}$ namely 1,3-$C_2B_7H_{12}^-$. At 200°C this decomposes to the nonacarborane 1,7-$C_2B_7H_9$, while other carboranes also form when 1,3-$C_2B_7H_{13}$ is heated.[88]

$$\underbrace{C_2B_7H_9 + C_2B_6H_8 + 1,6\text{-}C_2B_8H_{10}}$$

$$\uparrow 215°C \quad 200°C \uparrow \quad \downarrow \text{base}$$

$$1,3\text{-}C_2B_7H_{13} \xrightarrow{NaH} 1,3\text{-}C_2B_7H_{12}^-$$

$$\downarrow B_2H_6$$

$$\text{mainly } 1,6\text{-}C_2B_8H_{10}$$

Closed-cage- or *Closo*-carboranes

These are compounds with the general formula $C_2B_nH_{n+2}$ (n = 3–10) in which the substituents are only terminal. No bridging hydrogen atoms are present in the C_2B_n skeleton, and the structures are analogous to the isoelectronic $B_nH_n^{2-}$ cages. They will be considered in three groups: (a) small, n = 3–5, (b) large n = 6–9, and (c) dicarba-*closo*-dodecaborane.

The Small Carboranes

The effect of a silent electric discharge on a mixture of acetylene and either diborane(6) or pentaborane(9) was first used to produce the small *closo*-carboranes, *i.e.*, 1,5-dicarbo-*closo*-pentaborane(5), the 1,2- and 1,6-hexaboranes(6) and the 2,4-heptaborane(7) (1,5-$C_2B_3H_5$, 1,2- and 1,6-$C_2B_4H_6$ and 2,4-$C_2B_5H_7$). Yields are only small, but can be improved by heating 2,3-dicarba-*nido*-hexaborane(8) at 450°C/10 mm for 1–3 s.[89] The ethyl homologues of these small carboranes result when the 2-carba-*nido*-hexaborane(9), or $Et_4B_2H_2$ and acetylene, are heated.

The *closo*-carboranes are relatively stable compared with other carboranes, and though 1,5-dicarba-*closo*-pentaborane(5), like the two hexaborane(6) compounds, loses hydrogen at 150°C, it is stable at room temperature to water, air, amines and acetone. Deuterium exchange with B_2D_6 is faster with small carboranes and occurs at the B—H bond, while bromination takes place at the alkyl side chains and cleaves no B—C bonds. Indeed, the *B*-alkyl groups resist exchange with diborane(6), unlike the alkyldiboranes.

Though the 1,5-*closo*-pentaborane(5) was predicted to be the most stable isomer, and was isolated first, recyclization of an acetylene/diborane(6) mixture through a silent electric discharge yields *C*,3-dimethyl-1,2-dicarba-*closo*-pentaborane(5), the methyl groups appearing to stabilize the cage to rearrangement,[90] for though 1,5-dicarba-*closo*-pentaborane(5) compounds are present, there was no trace of the parent 1,2-isomer.

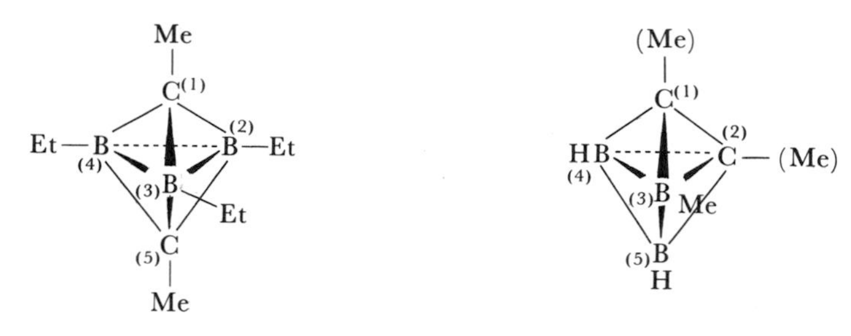

(m.p. −84°C; b.p. 58°C/11 mm)

Fig. 3.28. Derivatives of the isomeric *closo*-pentaboranes

The two isomeric *closo*-hexaboranes mentioned earlier can be considered as *cis* and *trans* isomers of the octahedral cage, the *trans* compound being the more thermally stable.[91] The two isomers are isostructural and isoelectronic with the $B_6H_6^{2-}$ ion.

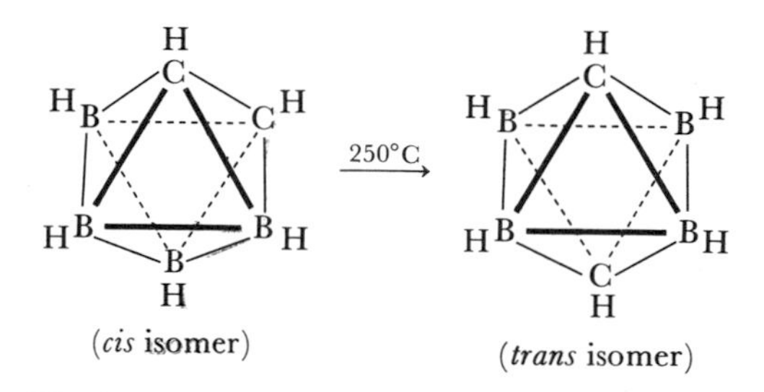

(*cis* isomer) (*trans* isomer)

Fig. 3.29. The *closo*-hexaborane isomers $C_2B_4H_6$

Of the synthetic pathways used to synthesize the *closo*-heptaboranes, only the pyrolysis of ethyldiboranes and acetylene gives an isomer other than the 2,4 one, *i.e.*, the 1,7 (*cf.* two C_2B_3 and two C_2B_4 isomers known).[92] 2,4-Dicarba-*closo*-heptaborane(7) was predicted to be the most stable isomer, with the carbon atoms not directly bonded.

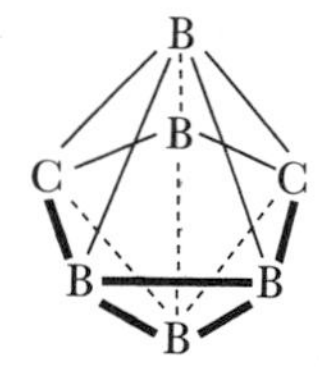

Fig. 3.30. Dicarba-*closo*-heptaborane(7)

A microwave spectrum shows the apical–equatorial B—B and B—C distances (0.182 and 0.171 nm) to be greater than their equatorial counterparts (0.165 and 0.156 nm), supporting the idea of a weak σ- and π-overlap between apical groups and the molecular orbitals formed from the orbitals of the equatorial plane. One noteworthy feature of the infrared spectrum of 2,4-$C_2B_5H_7$ is the absence of the C—H stretching band.

All of these small *closo*-carboranes have n.m.r. spectra supporting the proposed structures. 1,5-Dicarba-*closo*-pentaborane(5) shows a low-field proton singlet and high-field quartet (ratio 2 : 3) along with a sharp ^{11}B doublet. The *C*,3-dimethyl-1,2-dicarba isomer gives an ^{11}B n.m.r. spectrum with two doublets and a singlet (ratio 1 : 1 : 1). This supports methyl substitution at $B_{(3)}$ or $B_{(4)}$, and presents the possibility of optical as well as geometrical isomerism among these cage compounds.

The Large Carboranes

The chemistry of the compounds $C_2B_nH_{n+2}$ (n = 6–9) is so closely knit that they have to be considered together. The thermolysis of 1,3-$C_2B_7H_{13}$ and 1,3-$C_2B_7H_{12}^-$ has recently been mentioned (p. 72), and leads to the first three members, while 1,6-$C_2B_8H_{10}$ readily isomerizes at 300°C into 1,10-$C_2B_8H_{10}$.[88] The *closo*-undecaborane cage results when $C_2B_9H_{12}^-$ is acidified and heated.[85]

$$1,2\text{-}R,R'C_2B_{10}H_{12} \xrightarrow{MeO^-} 1,2\text{-}R,R'C_2B_9H_{10}^- \xrightarrow[100^\circ C]{\text{Polyphosphoric acid}} R,R'C_2B_9H_9$$

$$1,2\text{-}R,R'C_2B_9H_{10}^- \xrightarrow{HCl} R,R'C_2B_9H_{11} \xrightarrow{130^\circ C} R,R'C_2B_9H_9$$

Better yields of the octaborane and its organic homologues result if the homologues of $C_2B_7H_{13}$ are gently warmed.[93] The structure of the 1,7-dimethyl compound, (m.p. −40°C, b.p. 62°C/134 mm), which can also be made from hexaborane(10) and dimethylacetylene, is based on the bi-capped triangular prism.[94] The carbon atoms are found one in the prism and the other above the face opposite.

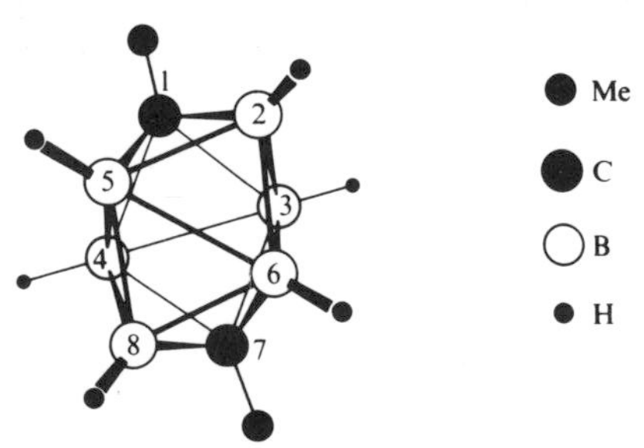

Fig. 3.31. The molecular structure of $Me_2C_2B_6H_6$[94]

M.O. predictions, based on the co-ordination number and relative electronegativities, lead to the conclusion that electrophilic attack at boron would occur in the order $B_{(2),(8)}$, $B_{(5),(6)}$, $B_{(3),(4)}$. It possesses two 5-co-ordinate carbon atoms, as does $C_2B_7H_9$ and its derivatives.

The ^{11}B n.m.r. spectrum of this carba-*closo*-nonaborane(9) shows three doublets which, starting from low field, are in the ratio 1:2:4. The structure of the *C,C'*-dimethyl homologue, which results when $Me_2C_2B_7H_{11}$ is heated to 200°C in diphenylether, supports this, with two faces of a triangular prism capped by MeC groups, the third by BH.[95]

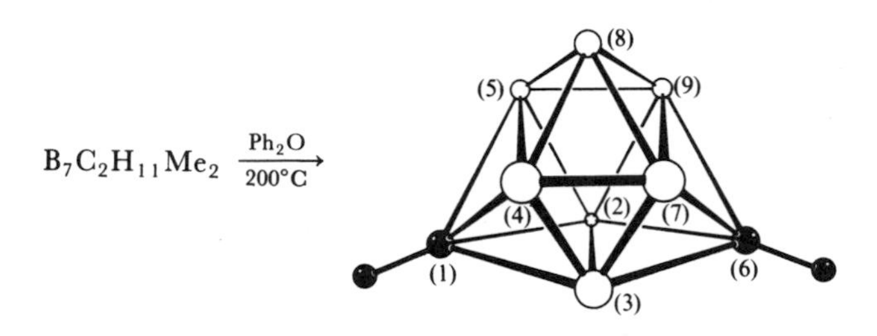

This gives it C_{2v} symmetry, and M.O. calculations indicate that electrophilic substitution should occur at $B_{(8)}$ first, followed by $B_{(4),(5),(7),(9)}$, as is found from acid-catalyzed alkylation and bromination.

$$Me_2C_2B_7H_{11} \xrightarrow[B_2H_6]{Ph_2O/200°C} \underset{\text{(m.p. 1°C, b.p. 73°C/32 mm)}}{1,6\text{-}Me_2C_2B_8H_8 \ (41\%)} \xrightarrow[10\ h]{350°C} \underset{\text{(m.p. 27°C)}}{1,10\text{-}Me_2C_2B_8H_8}$$

The *C,C'*-dimethyldicarba-*closo*-decaboranes(10) both have structures comprising two inverted skew square-pyramids,[96] and are isoelectronic and isostructural with $B_{10}H_{10}^{2-}$. (See Figure 3.32 on top of next page.)

Though the precursors to the formation of dicarba-*closo*-undecaborane can be two isomers with different cage structure, the subsequent pyrolysis of these precursors gives one and the same compound.[97] This is demonstrated in the reaction sequence on the next page, the degradation of the undecaborane demonstrating how the two carbon atoms have been separated from being nearest neighbours in (3)-1,2-$C_2B_9H_{11}Ph^-$ to next nearest ones in (3)-1,7-$C_2B_9H_{11}Ph^-$.

The *closo*-undecaborane has a structure closely resembling the alane

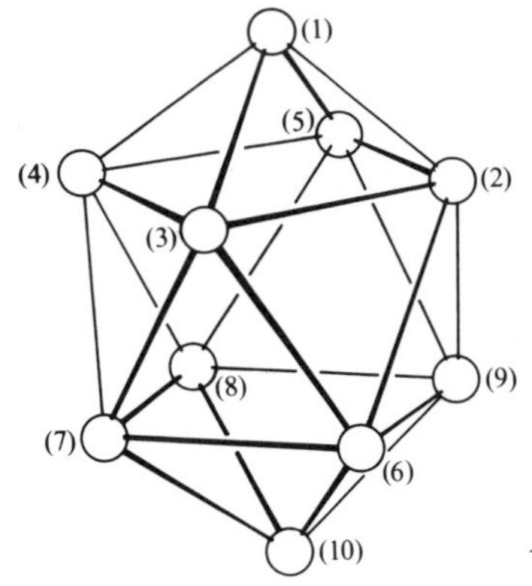

Fig. 3.32. The dicarba-*closo*-decaborane(10) cage

$1,2\text{-}C_2B_{10}H_{11}Ph \xrightarrow{410°C} 1,7\text{-}C_2B_{10}H_{11}Ph$ (m.p. 55°C)

$1,2\text{-}C_2B_{10}H_{11}Ph \xrightarrow{Pd/H_2} PhC_2H_5$

$1,2\text{-}C_2B_{10}H_{11}Ph \xrightarrow{MeO^-} (3)\text{-}1,2\text{-}C_2B_9H_{11}Ph^-$

$1,7\text{-}C_2B_{10}H_{11}Ph \xrightarrow{EtO^-} (3)\text{-}1,7\text{-}C_2B_9H_{11}Ph^-$

$(3)\text{-}1,2\text{-}C_2B_9H_{11}Ph^- \xrightarrow{300°C} (3)\text{-}1,7\text{-}C_2B_9H_{11}Ph^-$

$(3)\text{-}1,2\text{-}C_2B_9H_{11}Ph^- \underset{}{\overset{H^+}{\rightleftharpoons}} (3)\text{-}1,2\text{-}C_2B_9H_{12}Ph$

$(3)\text{-}1,7\text{-}C_2B_9H_{11}Ph^- \underset{}{\overset{H^+}{\rightleftharpoons}} (3)\text{-}1,7\text{-}C_2B_9H_{12}Ph$

$(3)\text{-}1,2\text{-}C_2B_9H_{12}Ph \xrightarrow{100°C} C_2B_9H_{10}Ph \xleftarrow{75°C} (3)\text{-}1,7\text{-}C_2B_9H_{12}Ph$

$C_2B_9H_{10}Ph \xrightarrow{Pd/H_2}$ toluene

and metal adducts of decaborane(14), with the hetero-boron atom bridging the decaborane basket. The extra boron atom bridges the two carbon atoms in the 6,9 positions of the decaborane(14) cage.[98]

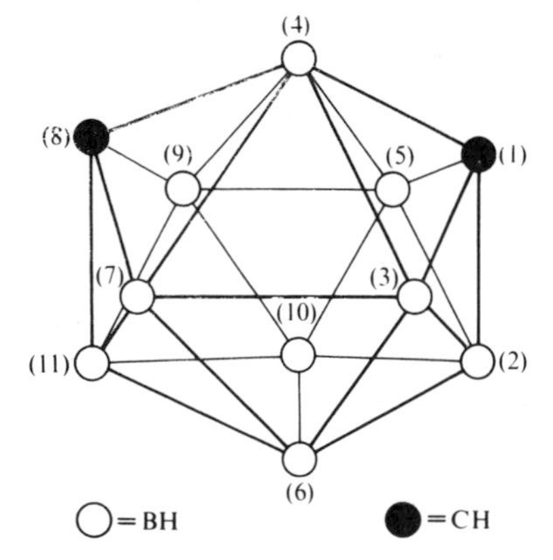

Fig. 3.33. The structure of 1,8-dicarba-*closo*-undecaborane(11)[98]

Adducts of this cage reversibly form with phosphines, amines and isocyanides.[97]

$$1,8\text{-}C_2B_9H_{11} + L \rightleftharpoons (3)\text{-}1,7\text{-}C_2B_9H_{11}^-L^+$$

These complexes probably involve partial cage opening to a $C_2B_9H_{12}^-$ structure, especially as with BH_4^- (3)-1,7-$C_3B_9H_{12}^-$ is obtained in high yield. Methyl-lithium gives a similar complex (3)-1,7-$C_2B_9H_{11}Me^-$, with the methyl group σ-bonded to an atom of the open face and the displaced hydrogen bonding in the delocalized π-electrons of this face. The anions of *C*-methyl-*o*-carborane and *C*-phenyl-1,10-dicarba-*closo*-decaborane(10) also form similar complexes involving two carborane cages.[99]

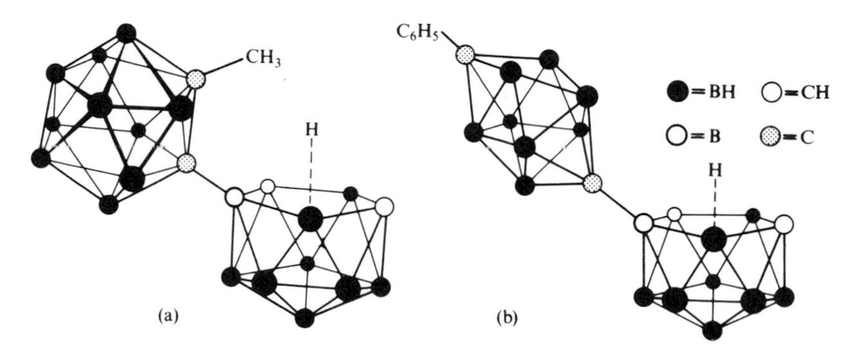

Fig. 3.34. The proposed structure of (a) the $B_9C_2H_{11}B_{10}C_2H_{10}CH_3^-$ ion and (b) the $B_9C_2H_{11}B_8C_2H_8C_6H_5^-$ ion[99]

Dicarba-closo-dodecaboranes (o-, m-, and p-"carborane")

That plastics be serviceable at even higher temperatures is a major demand of modern technology. Carbon polymers are now replaced by silicones (Chapter 4) and polyphosphazenes (Chapter 5) in many plastics, and it is the desire to stabilize this polymer that has led to the synthesis of ball and chain backbones to polymers using the thermally-stable electron-deficient carborane cage as the ball. This reduces mobility, and provides the delocalized electron sink for thermal and electronic stability. Such polymers are highly stable, and it is because of this that boron cage research has expanded so in the past decade.[100]

Unlike the dicarba-*closo*-decaboranes, where two of the structural isomers have been characterized, all three dicarba-*closo*-dodecaboranes are known. These are the 1,2-, 1,7- and 1,12-cage substituted carboranes, conveniently known as *o*-, *m*- and *p*-carborane. It is this latter isomer, linked through Me_2SiO, that has been widely studied as a high-temperature elastomer.

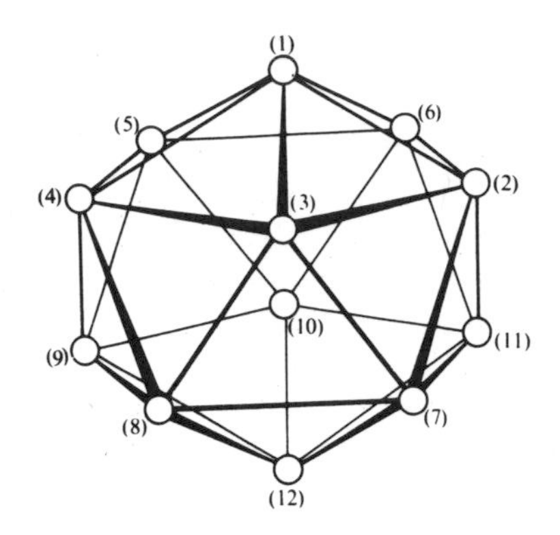

Fig. 3.35. The carborane cage, with numbering

o-Carborane is the only isomer that can be synthesized directly, and results from the base-catalyzed reaction of acetylenes with decaborane(14) or via $B_{10}H_{12}L_2$.

$$B_{10}H_{14} + 2L \xrightarrow{-H_2} B_{10}H_{12}L_2 \xrightarrow{R_2C_2} R_2C_2B_{10}H_{10} + H_2 + 2L$$

$$2Et_3N + B_{10}H_{14} \longrightarrow (Et_3NH^+)_2B_{10}H_{10}^{2-} + H_2$$

No carborane results in the absence of base, and amines give the decahydrodecaborate anion which is stable to acetylenes.

The C—C bond distance of 0.168 nm in *o*-carborane is larger than in the less electron-deficient $Me_2C_2B_4H_6$ where it is 0.143 nm. Isomerization to the *meta* and *para* isomers occurs on heating *o*-carborane strongly.[101]

$$o\text{-carborane} \underset{\text{Na/liq. NH}_3}{\overset{470°\text{C}}{\rightleftarrows}} m\text{-carborane} \xrightarrow{615°\text{C}} p\text{-carborane}$$

As carborane cage substitution has become a vast subject, only an outline will be considered, followed by a comparison of the three cages.

The most acidic protons of the cage are those on carbon. Lithiation readily occurs with n-BuLi, so providing the opening into *C*-substituted carboranes. The strongly electron-withdrawing power of the C_2B_{10} cage renders *o*-carborane-acetic acid $B_{10}H_{10}CHC{\cdot}CH_2CO_2H$ as strong an acid as chloro- and fluoro-acetic acid. Similarly, electrophilic addition to an olefinic double bond in the carbon side-chain only occurs if the double bond is at the end of a four-carbon chain. A double bond nearer to the cage remains unoxidized.[i]

Electrophilic substitution on the cage normally occurs on the boron atoms farthest from carbon. Thus with *o*-carborane, both catalyzed chlorination and bromination occur first at $B_{(9)}$ and $B_{(12)}$, and last at $B_{(3)}$ and $B_{(6)}$. Progressive chlorination renders the C—H protons more acidic and readily ionized.[102]

$$o\text{-}C_2H_2B_{10}H_{10} \xrightarrow[\text{u.v.}]{Cl_2} o\text{-}C_2H_2B_{10}Cl_{10} \xrightarrow{2Et_3N} (Et_3NH)_2^+C_2B_{10}Cl_{10}^{2-}$$
$$\xrightarrow{Br_2} Br_2C_2B_{10}Cl_{10}$$

All isomers resist attack by acid, consistent with the reduced hydridic character of the B—H bonds. Base attack of *o*- and *m*-carborane readily occurs,[85,103] however, to give the 1,2- and 1,7-isomers of $C_2B_9H_{12}^-$. Reacting these ions with n-BuLi, followed by organoboron dihalides, re-forms the carba-*closo*-dodecarborane cage.[104]

$$C_2B_9H_{12}^- \xrightarrow{\text{n-BuLi}} C_2B_9H_{11}^{2-} \xrightarrow{RBX_2} C_2B_{10}H_{11}R$$

Degradation of the 1,2-isomer is exothermic at room temperature in piperidine, while the 1,7-isomer requires weeks under reflux. This may be due to the higher electron-attracting nature of the C_2 group as opposed to isolated carbon atoms. *p*-Carborane has not been degraded by base.

By comparison, 1,6-$B_8H_8(CMe)_2$ is attacked by piperidine giving $B_7C_2H_{11}Me_2$, whereas the 1,10-isomer remains intact.

The very proximity of the two carbon atoms in *o*-carborane renders much of the *C*-substitution chemistry unique. Thus, it readily yields an anhydride, will form a chelating ligand and give small exocyclic rings, which are generally more stable than the parent carborane.[105] Benzo-carborane, formed as indicated in the reaction sequence, is very stable, even to conc. H_2SO_4 and bromine in CCl_4 (see p. 80).

Large Carboranes containing a single Carbon Atom

C-Trialkylamine carba-*nido*-undecaborane(12) derivatives are formed as indicated, and with an excess of NaH[106] give the anion $HCB_{10}H_{12}^-$ which is probably isostructural with $B_{11}H_{13}^{2-}$.

$$B_{10}H_{14} + RNC \longrightarrow RNH_2CB_{10}H_{12} \xrightarrow[Me_2SO_4]{2NaH}$$
$$RNMe_2CB_{10}H_{12} \xrightarrow{NaH} HCB_{10}H_{12}^-$$

This is also formed using sodium via the intermediate $Na_3B_{10}H_{10}CH\text{-}(THF)_{1.8}$ which hydrolyzes to $HCB_{10}H_{12}^-$ in 80% yield. Oxidation of this intermediate with iodine yields $B_{10}H_{10}CH^-$ and compares with the oxidation of $B_{10}H_{14}^{2-}$ to decaborane.[107] The anion is isoelectronic with $C_2B_9H_{11}$ and $B_{11}H_{11}^{2-}$ and probably has a similar structure.

The monocarbon analogues of $B_{10}H_{10}^{2-}$ and $B_{12}H_{12}^{2-}$ result when $HCB_{10}H_{12}^-$ decomposes at 300°C, a reaction which is similar to the disproportionation of $B_{11}H_{11}^{2-}$ already mentioned.

$$2CsB_{10}H_{12}CH \xrightarrow{300°C} Cs\text{-}1\text{-}B_9H_9CH + CsB_{11}H_{11}CH + 2H_2$$

The larger cage can also be made through cage expansion using the complex $Et_3N{\cdot}BH_3$.[108]

$$CsB_{10}H_{12}CH + Et_3N{\cdot}BH_3 \xrightarrow{180°C} CsB_{11}H_{11}CH + Et_3N + 2H_2$$

$$LiO_2CCCLiB_{10}H_{10} + A \xrightarrow[H_2O]{CO_2} (HO_2C)_2B_{10}H_{10} \xrightarrow{PCl_5} (OC)_2O\,C_2B_{10}H_{10}$$

$$\underset{(A)}{H_2C_2B_{10}H_{10}} \xrightarrow{BuLi} \underset{(B)}{LiHC_2B_{10}H_{10}} \xrightarrow{CO_2} LiO_2CCCHB_{10}H_{10} \xrightarrow{B} LiO_2CCCLiB_{10}H_{10} + A$$

$$H_2C_2B_{10}H_{10} \xrightarrow{2BuLi} Li_2C_2B_{10}H_{10} \xrightarrow[NiCl_2]{2Ph_2PCl} H_{10}B_{10}C_2(PPh_2)_2NiCl_2$$

$$Li_2C_2B_{10}H_{10} \xrightarrow{S,\ H_2O} (HS)_2C_2B_{10}H_{10} \xrightarrow{PhPCl_2} H_{10}B_{10}C_2S_2PPh$$

$$(HS)_2C_2B_{10}H_{10} \xrightarrow{(Ph_3P)_2NiCl_2} H_{10}B_{10}C_2S_2Ni(PPh_3)_2$$

$$Li_2C_2B_{10}H_{10} \xrightarrow{cis(ClCH_2)_2C_2H_2} (CH_2CH{=}CHCH_2)C_2B_{10}H_{10} \xrightarrow{NBS\ |\ AIBN} (CH{=}CHCH{=}CH)C_2B_{10}H_{10}$$

(NBS—*N*-bromosuccinimide;
AIBN—azo-iso-butyronitrile)

That these two new anions are isostructural with the boron cages is supported by the n.m.r. spectra. With $B_9H_9CH^-$, the three peaks in the ^{11}B spectrum, area ratio 1:4:4, are in accordance with the $B_{10}H_{10}^{2-}$ polyhedron with $B_{(1)}$ replaced by carbon. That of $CsB_{11}H_{11}CH$ gives three peaks of area ratio 1:5:5.

Carboranes containing a Heteroatom

During the past three years there has been a steady increase in the variety of hetero atoms introduced into a carborane cage. These now include Be, Ge, Sn, P, As and Sb.

The compound (3)-1,2-$C_2B_9H_{13}$, with its two protons in the pentagonal face, reacts with Me_2Be in ether to give white crystalline $B_9BeC_2H_{11}OEt_2$.[109] This is water- and oxygen-sensitive, but gives a more stable Me_3N complex (m.p. 221–3°C).

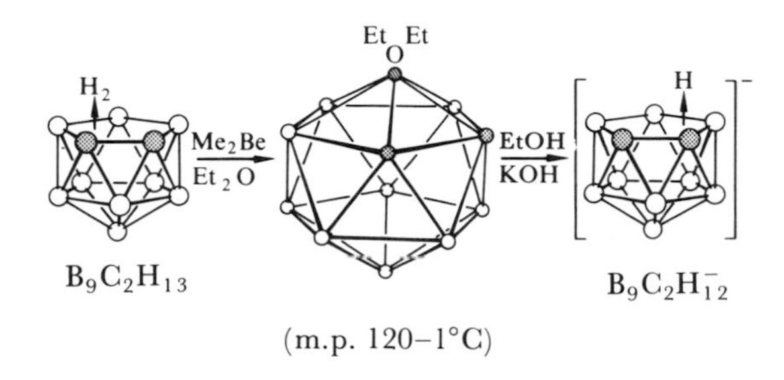

(◎ carbon, ◎ heteroatom, ○ boron; terminal protons omitted)

The incorporation of transition metals into the carba-*closo*-dodecaborane(12) framework will be mentioned in the next Section, using the anion $C_2B_9H_{11}^{2-}$ (dicarbollide) as an example. Stannous chloride or Me_2SnCl_2 react similarly to give the heterocarborane $B_9C_2SnH_{11}$.[110] Base degradation again yields $B_9C_2H_{12}^-$.

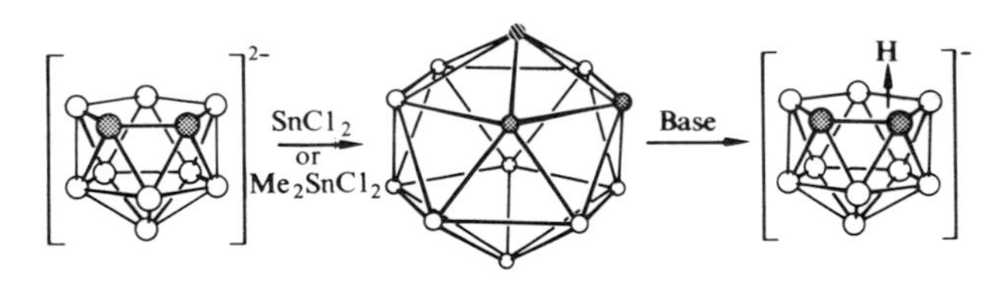

(◎ carbon, ◎ heteroatom, ○ boron; terminal protons omitted)

The monocarba-*closo*-carborane anion of $Na_3B_{10}H_{10}CH(THF)_2$, mentioned in the last Section, reacts with $MeGeCl_3$ to give the germanacarborane.[111] Piperidine readily demethylates it, but with methyl iodide this process is reversed.

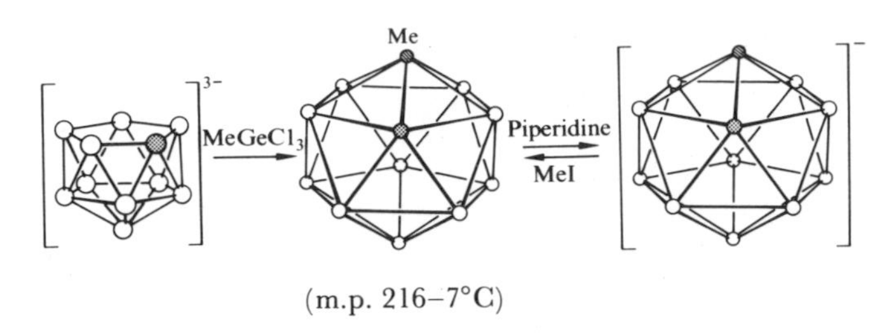

(m.p. 216–7°C)

(⊛ carbon, ⊘ heteroatom, ○ boron; terminal protons omitted)

Phosphorus, arsenic and antimony can all be inserted into the icosahedral cage using this reaction. The initial product is the 1,2-compound, but this isomerizes on heating to the *meta*- and then *para*-phosphacarboranes.[111,112] It is interesting that reaction with piperidine removes a boron atom, not the hetero one, unlike the cases of tin and beryllium.

$$B_{10}H_{10}CH^{3-} + MX_3 \longrightarrow 1,2\text{-}B_{10}H_{10}CHM \xrightarrow{\text{base}} [B_9H_{10}CHM]^-$$

The phosphacarborane anion can be readily methylated on the phosphorus atom, and the 1,7-isomer gives the phosphacarbollide anion with sodium hydride.

$$B_9H_9CHP^{2-} \xleftarrow{\text{NaH}} (3)\text{-}1,7\text{-}B_9H_{10}CHP^- \xrightarrow{\text{MeI}} (3)\text{-}1,7\text{-}B_9H_{10}CHPMe$$
(m.p. 112–3°C)

This readily bonds to transition metals like a π-cyclopentadienyl group, and can be readily methylated at phosphorus while on or off the metal.

Boron hydride cages incorporating a Heteroatom

Sulphur, nitrogen and phosphorus have been successfully incorporated into boron cages. The icosahedral cage $B_{11}H_{11}PPh$, (m.p. 157°C) results from $B_{11}H_{13}^{2-}$ and $PhPCl_2$.[112] The azaborane cage $B_9H_{12}NH^-$ is isostructural and isoelectronic with the more stable $B_9H_{12}S^-$, which is thought to have the $B_{10}H_{14}^{2-}$ structure with S and NH in position 6. It can be synthesized from decaborane(14) using thionitrosodimethylamine Me_2NNS. Two products result, and reduction of the diazo compound gives the azaborane cage.[113]

$$B_{10}H_{14} + Me_2NNS \longrightarrow B_{10}H_{11}S^- + B_9H_{12}NNMe_2^- \xrightarrow[\text{THF}]{\text{Na}} B_9H_{12}NH^-$$

It can be recrystallized from aqueous acetic acid as the Me_4N^+ salt.

The thiaborane $B_9H_{12}S^-$ cage results from $B_{10}H_{14}$ and polysulphide, and also forms $B_{10}H_{11}S^-$ indirectly. This is used widely in the synthesis of various other cage and transition metal compounds, as indicated.

$$B_{10}H_{14} \xrightarrow[4H_2O]{S^{2-}} B_9H_{12}S^- + B(OH)_4^- + 3H_2$$

$$B_9H_{12}S^- \xrightarrow{200°C} B_{10}H_{11}S^- \qquad B_9H_{12}S^- \xrightarrow{\text{oxid}^n,\ L} LB_9H_{11}S \qquad B_9H_{12}S^- \xrightarrow{H^+} B_9H_{11}S$$

$$B_{10}H_{11}S^- \xleftarrow[200°C]{Et_3NBH_3} LB_9H_{11}S \xleftarrow{\text{Ligand}} B_9H_{11}S$$

$$B_{10}H_{11}S^- \underset{\text{base (aq)}}{\overset{H_3O^+}{\rightleftarrows}} B_{10}H_{12}S \qquad B_{10}H_{11}S^- \xrightarrow{Bu^nLi} B_{10}H_{10}S^{2-}$$

$$B_{10}H_{10}S^{2-} \xrightarrow[C_5H_5^-]{CoCl_2} B_{10}H_{10}SCoC_5H_5 \text{ (orange m.p. 268°C)} + cp_2Co^+(B_{10}H_{10}S)_2Co^- \text{ (m.p. 287°C)}$$

$$B_{10}H_{12}S \longrightarrow B_{10}H_{10}S^{2-} \xrightarrow{-S} B_{10}H_{10}^{2-}$$

$$B_{10}H_{10}S^{2-} \xrightarrow{Re(CO)_5Cl} [B_{10}H_{10}SRe(CO)_3]^-$$

$$B_{10}H_{10}S^{2-} \xrightarrow{Fe^{2+}} [(B_{10}H_{10}S)_2Fe]^{2-}$$

$$B_{10}H_{10}S^{2-} \xrightarrow{PhBCl_2} PhB_{11}H_{10}S \text{ (icosahedral)}$$

$$PhB_{11}H_{10}S \xrightarrow{\text{Base}} PhB_{10}H_{10}S^-$$

$$PhB_{10}H_{10}S^- \xrightarrow[PhBCl_2]{Bu^nLi} Ph_2B_{11}H_9S \text{ (icosahedral)}$$

$$PhB_{10}H_{10}S^- \xrightarrow{BuLi} PhB_{10}H_9S^{2-} \xrightarrow{Fe^{2+}} [(PhB_{10}H_9S)_2Fe]^{2-}$$

The transition metal complexes mentioned above are typical of the many now known between metals and these six π-electron boron cages.

TRANSITION METAL COMPLEXES OF BORON CAGES

A recent review[114] lists the total number of such complexes as about a hundred. So a brief summary of the ligands and synthetic methods will be made along with the relevant structures and bonding picture where possible.

The Spanish noun "olla", meaning water jar, was used to give the trivial name "ollide" to the hypothetical anion $B_{11}H_{11}^{4-}$. The isoelectronic hetero-atom ions $B_{10}CH_{11}^{3-}$, $B_9C_2H_{11}^{2-}$ and $B_9CPH_{10}^{2-}$ are called "carbollide", "dicarbollide" and "phosphacarbollide". All can be synthesized from the icosahedral carborane, and with transition metal halides, metal carbonyls and their derivatives form carbollide transition metal complexes in which the 11-atom cage formally bonds to the metal in a manner reminiscent of the π-cyclopentadienyl ring.

Many complexes of the dicarbollide ligand have been prepared which formally resemble the metallocenes, but include examples as yet unknown in this series.[115] Thus, 1,2-$B_9C_2H_{11}^{2-}$ in aqueous base will react with copper sulphate giving the blue cupricene analogue $(1,2\text{-}B_9C_2H_{11})_2Cu^{2-}$, which oxidizes in air to red $(1,2\text{-}B_9C_2H_{11})_2Cu^{-}$. Similar Au^{II}, Au^{III}, Pd^{II} and Pd^{IV} complexes are known which have no π-cyclopentadienyl analogues. The 1,7-dicarbollide isomer also forms such complexes, as do $B_{10}CH_{11}^{3-}$ and the 1,2- and 1,7-isomers of $B_9H_9CHP^{2-}$ and $B_9H_9CHAs^{2-}$.

Examples of the complexes synthesized are formulated below.

$$1,2\text{-}B_9C_2H_{11}^{2-} + Mo(CO)_6 \xrightarrow{THF} (1,2\text{-}B_9C_2H_{11})Mo(CO)_3^{2-}$$

$$(1,2\text{-}B_9C_2H_{11})Mo(CO)_3^{2-} \xrightarrow{W(CO)_6} (1,2\text{-}B_9C_2H_{11})Mo(CO)_3W(CO)_5^{2-}$$

$$1,2\text{-}B_9C_2H_{11}^{2-} + C_5H_5^{-} \xrightarrow[air]{FeCl_2} \pi\text{-}C_5H_5Fe(1,2\text{-}B_9C_2H_{11})$$
(deep red; m.p. 181–2°C)

$$1,7\text{-}B_9H_9CHP^{2-} + FeCl_2 \longrightarrow (1,7\text{-}B_9CHP)_2Fe^{2-} \text{ (red-violet)}$$

$$Na_3B_{10}H_{10}CH(THF)_2 + Mn(CO)_5Br \longrightarrow (B_{10}H_{10}CH)Mn(CO_3^{2-}$$

$$1,2\text{-}B_9C_2H_{11}^{2-} \xrightarrow{Pd(acac)_2} (1,2\text{-}B_9C_2H_{11})Pd^{2-} \xrightarrow{I_2} (1,2\text{-}B_9C_2H_{11})_2Pd$$

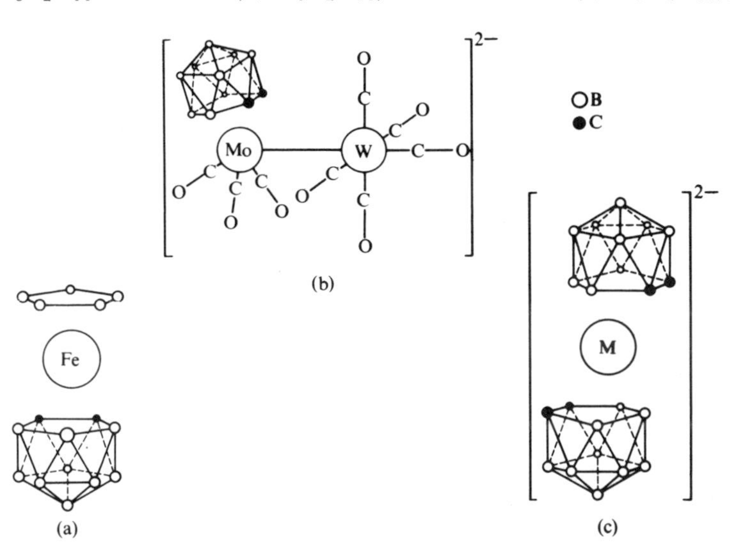

Fig. 3.36. Examples of the structural nature of carbollide complexes:[114] (a) structure of $(\pi\text{-}C_5H_5)Fe(1,2\text{-}B_9C_2H_{11})$, (b) probable structure of the $[(1,2\text{-}B_9C_2H_{11})Mo(CO)_3W(CO)_5]^{2-}$ ion, and (c) general "π-slipped" structure of $(1,2\text{-}B_9C_2H_{11})_2M$ where M = Ni^{II}, Cu^{II}, Cu^{III} and Au^{III}

Bis-dicarbollides have now been synthesized for most of the oxidation states of the metals of the first transition series ($d^5 \rightarrow d^9$, $Fe^{III} \rightarrow Cu^{II}$). The rings appear to bond in two distinctly different ways. Symmetrical

π-sandwich complexes occur with metals of formal configuration d^7 or less, while with the d^8 and d^9 metals the dicarbollide groups bond in a manner similar to a π-allyl system, with six boron atoms from the two rings formally interacting with the metal ion. This concept is supported by structural data and has an interesting parallel in metallocene chemistry. While cp_2Fe, cp_2Co and cp_2Ni (cp = cyclopentadiene) are all isomorphous and have the same melting point, nevertheless the reactions of nickelocene show that it reacts as if one ring functions as a three-electron donor, leaving the double bond for addition reactions *etc.*[j]

Table 3.1. Structural Parameters of Carbollide Transition Metal Complexes

Compound	Distance in nm for M-cage			
	⊥*	M—C	M—B	C—C
$C_5H_5Fe^{III}C_2B_9H_{11}$ (deep red)	0.147	0.204	0.209	0.158
$(CO)_3Re^{I}C_2B_9H_{11}^-$ (pale yellow)		0.231	0.234	0.161
$Co^{III}(C_2B_9H_{11})_2^-$ (yellow) disorder	0.147	(0.207)		
$Cu^{II}(C_2B_9H_{11})_2^{2-}$ (blue)	0.179	0.257	0.220	0.153
$Cu^{III}(C_2B_9H_{11})_2^-$ (red)	0.170	0.253	0.211	0.149

* Shortest distance from the metal ion to the cage.

The colours and magnetic data support the formal oxidation states proposed for the metals in the above complexes, rendering the first three (d^5 and d^6) as π-symmetrical and the two copper ones (d^9 and d^8) as π-allyl. There is no significant structural differences between these two copper complexes.[116]

In the ferricinium-like complex, the M—C and M—B bond lengths are similar in both rings, and closely agree with the values found in the Co^{III} compound. With the Re^{I} compound, the lower oxidation state expands the orbital size, making the M—C and M—B distances greater, though again, these agree within experimental error.

With the two copper complexes, the copper atoms are closer to the ligand boron atoms than to the carbon ones. The C—C distance is correspondingly shorter (0.15 nm) than in complexes where the M—C interaction is important.[117]

Smaller carboranes will also form π-complexes with transition metals. Just as base cleavage of a carborane cage followed by BuLi gives a six π-electron donor, so with $B_7C_2H_{13}$ two moles of NaH gives a cage capable of bonding similarly. The bis-carborane cobalt complex isomerizes on heating with migration of the carbon atoms in the cage.[118]

$$B_7C_2H_{13} \xrightarrow{2NaH} B_7C_2H_{11}^{2-} \xrightarrow[-H_2]{CoCl_2} (B_7C_2H_9)_2Co^-$$

$$B_7C_2H_{11}^{2-} \xrightarrow{CoCl_2,\ C_5H_5} C_5H_5CoB_7C_2H_9$$

Attempts to make simple metal carbonyl derivatives resulted in cage contraction.

$$B_7C_2H_{11}^{2-} + Mn(CO)_5Br \xrightarrow{THF} B_6C_2H_8Mn(CO)_3^-$$

The ^{11}B n.m.r. spectrum gave 4 doublets (ratio 1:2:2:1) and both CH groups equivalent. The structure proposed is thought to involve a B_5Mn triangular prism with the two carbon atoms above faces adjacent to the Mn atom. The BH group is above the face opposite.[119]

Just as $C_2B_4H_8$ lost the two bridging protons with Me_3Ga and gave a pentagonal-bipyramidal compound, so 2-$MeC_3B_3H_5$ loses its one bridging proton with $Mn_2(CO)_{10}$ and gives an analogous compound, with $Mn(CO)_3$ bonding to the pentagonal face of the carborane.[120]

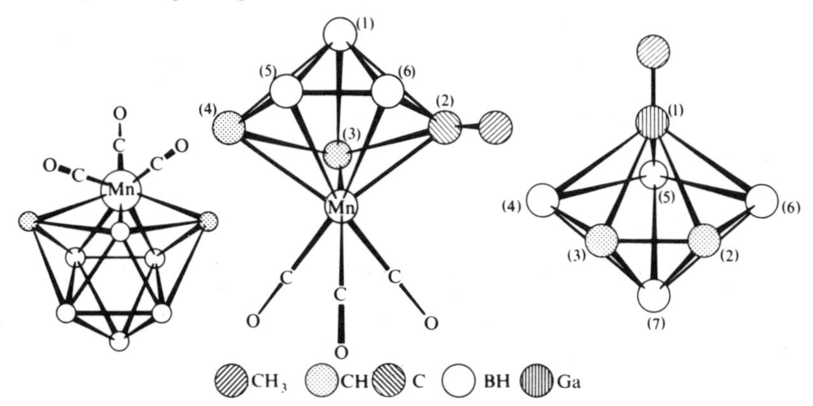

Fig. 3.37. Proposed structures of the $B_6C_2H_8Mn(CO)_3^-$ ion,[119] of $(CH_3C_3B_3H_5)$-$Mn(CO)_3$[120] and of $CH_3GaC_2B_4H_6$[83]

The cobalt complex $Cs_2(C_6H_{32}B_{26}Co_2)H_2O$ results as a by-product in the synthesis of $(1,2\text{-}C_2B_9H_{11})_2Co^-$ in the presence of base. It has the structure indicated, with the formulation $[(B_9C_2H_{11})Co(B_8C_2H_{10})Co$-$(B_9C_2H_{11})]^{2-}$, and may result through base attack on $(C_2B_9H_{11})_2Co^-$ followed by the capture of $(C_2B_9H_{11})Co$, as the carbollide unit is mobile in base.

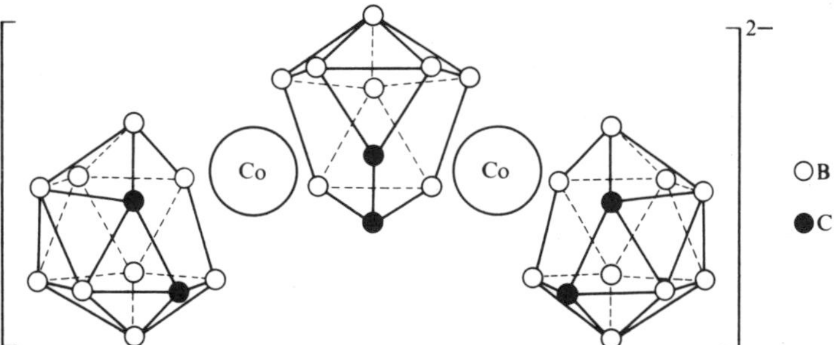

Fig. 3.38. The structure of the $[(B_9C_2H_{11})Co(B_8C_2H_{10})Co(B_9C_2H_{11})]^{2-}$ ion[114]

Nickel(II) slowly exchanges with $(C_2B_9H_{11})_2Co^-$ under these conditions.[121] This has been extended to include four cages and three cobalt atoms.

Among the more interesting reactions of decaborane(14) derivatives with transition metal compounds is that of $B_{10}H_{13}^-$ with the Group VI metal carbonyls.[122]

$$B_{10}H_{13}^- + M(CO)_6 \xrightarrow[350\ nm]{h\nu} B_{10}H_{10}COHM(CO)_4^- \xrightarrow{H^-} B_{10}H_{10}COMCO(CO)_3^{2-}$$

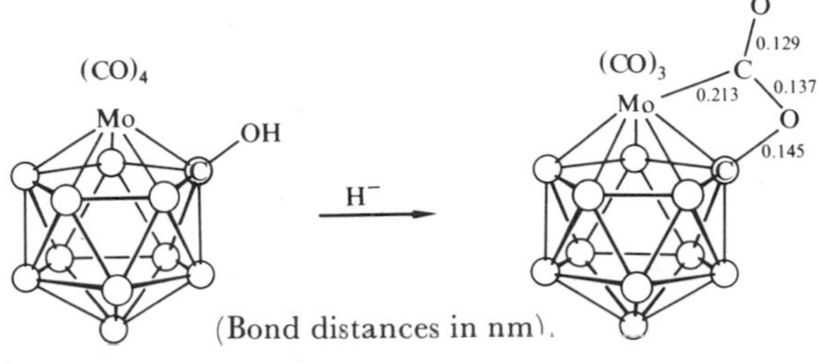

(Bond distances in nm).

The metal and carbon are incorporated to make up the icosahedral cage, as the structure of the molybdenum compound indicated.

Deprotonation of the hydroxy compound probably induces nucleophilic attack at a Mo—CO group. Such a reaction renders this CO group a co-ordinated carbene, and the Mo—CO_2 distance is longer than that of the terminal carbonyl groups, and close to the values observed in other transition metal carbene complexes.

The $MoCO_2C$ distances are close to those found in organic esters, and base readily cleaves the carborane C—O bond and displaces carbon not boron from the cage, to yield the *nido*-metalloboranes $B_{10}H_{12}M(CO)_4^{2-}$. These contain only terminal carbonyl groups, and are probably similar to the complexes formed between $B_{10}H_{12}^{2-}$ and the late and post-transition

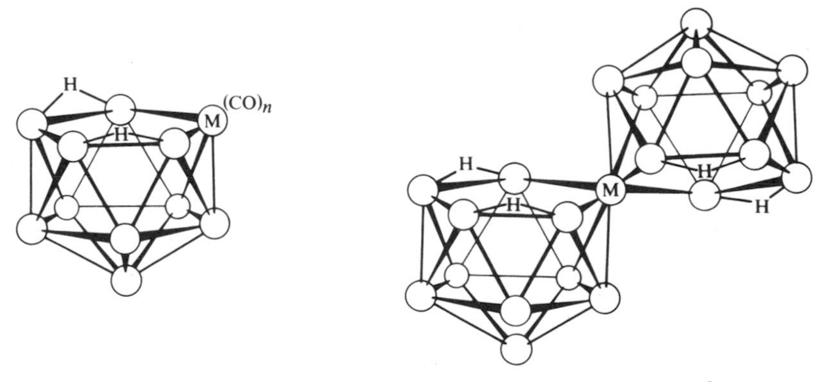

Fig. 3.39. The complexes $B_{10}H_{12}M(CO)_n$ and $M(B_{10}H_{12})_2^{2-}$

metals. They probably contain the metal as part of an 11-atom icosahedral fragment like the dicarbollide anion.

$$B_{10}H_{13}^- + \text{metal complex} \longrightarrow M(B_{10}H_{12})_2^{2-}$$
$$(M = Co, Ni, Pd, Pt, Zn)$$

Low-valency complexes can also be made, *e.g.*, $B_{10}H_{12}M(CO)_3^-$ (M is Co, Rh, Ir), with the general structures indicated.[123] (See Figure 3.39.)

HALOGENATED BORON CAGE COMPOUNDS

Fully-halogenated cage structures are becoming increasingly common in boron chemistry. They result from the decomposition of B_2Cl_4, while the derivatives $B_{10}X_{10}^{2-}$ and $B_{12}X_{12}^{2-}$ result from the halogenation of the corresponding hydride ions. Complete substitution of these latter ions can be accomplished, but the decrease in the reactivity of the halogen with its increase in atomic weight often necessitates the use of both ultraviolet light and *N*-halogenosuccinimides to effect complete replacement. All these ions possess high thermal stability, and do not react with air or aqueous media, unlike the neutral lower chlorides B_nCl_n.

Diboron tetrachloride spontaneously decomposes at room temperature. Three chlorinated boron cage compounds were isolated.[124] The first and most volatile, B_4Cl_4, is a pale yellow crystalline solid, (v.p. 34 mm/68°C, m.p. 95°C). The tetrahedral B_4 cage, with its chlorine atoms attached terminally to each boron, has B—Cl bond lengths shorter than expected for single bonds, and the π-bonding indicated here probably stabilizes the molecule. This may well partly account for the non-existence to date of B_4H_4, and for the limited methylation to MeB_4Cl_3.

Hydrolysis yields three moles of H_2, as would be expected for a molecule with six B—B bonds,

$$B_4Cl_4 + 6H_2O \longrightarrow 3H_2 + 4HCl + [B_4O_4(OH)_2]_n$$

while alkyl chlorides result with both methanol and diethylether. A peculiar reaction is proposed with dimethylamine involving a cage rearrangement. The infrared absorptions typical of the B_4 cage are absent in $[Me_2NB]_4$ supporting a square structure rather than a tetrahedral one. B_4Cl_4 ignites in dry air, forming BCl_3 and polymers.[125]

Octaboron octachloride B_8Cl_8 was isolated as a red solid from the decomposition of B_2Cl_4 by recrystallization from BCl_3. A complete structural analysis has been made, but a study of the chemistry was halted by the inability to repeat the synthesis. With approximate D_{4d} symmetry and B—Cl bonds longer than in B_4Cl_4, B_8Cl_8 can be conveniently considered as an Archimedian antiprism folded along the top and bottom diagonals.[126]

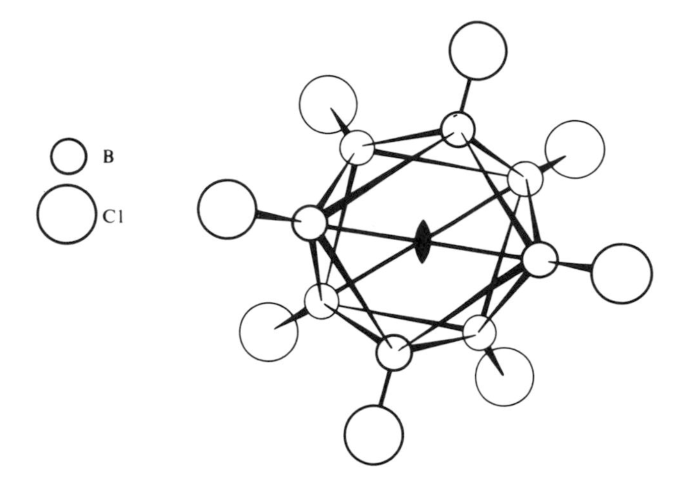

Fig. 3.40. The structure of B_8Cl_8[126]

The most thermally stable of the perchloropolyboranes appears to be B_9Cl_9. Again, the thermolysis of B_2Cl_4 yields this compound which forms thin yellow crystalline plates, deepening in colour as their thickness increases. The structure is not yet known, but the ^{11}B n.m.r. spectrum shows eight equivalent boron atoms.[127]

Heating the hydrated acid of $B_{10}Cl_{10}^{2-}$ to 200°C gives the sublimable B_9Cl_8H as red crystals, melting point 224°C. It is readily hydrolyzed to boric acid via the anion $B_9Cl_8H^{2-}$, and is thought to have a structure resembling a monocapped square-antiprism with the BH group on the uncapped face.[128]

CYCLIC NITROGEN DERIVATIVES OF THE GROUP III ELEMENTS

These derivatives are many and varied in number, especially those of boron. Variations in ring size and co-ordination number occur, but generally only boron forms a variety of oligomeric 3-co-ordinate compounds, since its ability to form strong $p\pi-p\pi$ bonds with donor atoms renders further oligomerization more difficult. Monomeric 3-co-ordinate B—N compounds, like amino derivatives of the heavier Group III elements, will oligomerize, *e.g.*, $[BH_2NH_2]_3$, $[Et_2AlNMe_2]_2$, but the borazines (borazoles) have no aluminium congeners. For this reason, these compounds will be considered along these lines with 4-, 6- and 8-membered rings involving 3- and 4-co-ordination.

1,3,2,4-Diazadiboretidines $[RBNR']_2$

This class of compound has been known since 1963 when the highly-hindered t-butylamino derivative was isolated.[129]

$$(Bu^tNH)_2BCl \xrightarrow[-HCl]{Et_3N} (Bu^tNH)_2BN(Bu^t)B(Cl)NHBu^t \xrightarrow[-HCl]{Bu^tNH_2} [(Bu^tNH)_2B]_2NBu^t \xrightarrow{Heat} [Bu^tNHBNBu^t]_2$$

(b.p. 72–4°C/0.005 mm)

The interest in these compounds stems from their formal analogy with cyclobutadiene. However, the latter is unstable and predicted to be paramagnetic, while boretidines are diamagnetic and stable in an inert atmosphere.

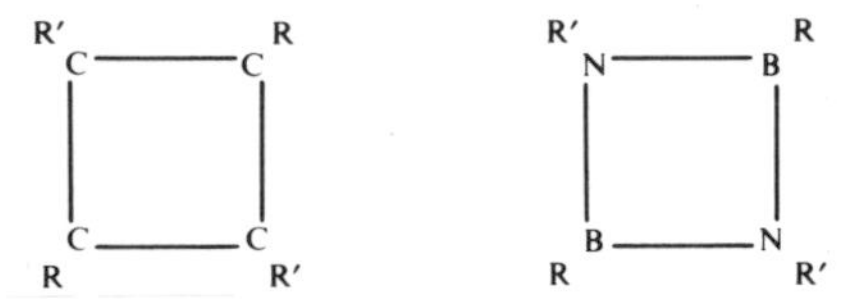

Fig. 3.41. Isostructural cyclobutadienes and boretidines

The subsequent isolation of other derivatives has shown that the substituents all possess the inherent ability to function as exocyclic π-bonding groups. The fully-silylated analogue $[(Me_3Si)_2NBNSiMe_3]_2$ has been prepared by reacting the sodium salt of hexamethyldisilazane with $[(Me_3Si)_2N]_2BCl$, or heating the fluoride.[130]

$$2[(Me_3Si)_2N]_2BF \xrightarrow[-2Me_3SiF]{200°C} [(Me_3Si)_2NBNSiMe_3]_2 \quad (m.p.\ 212°C)$$

$$2[(Me_3Si)_2N]_2BCl + 2(Me_3Si)_2NNa \xrightarrow{-2(Me_3Si)_3N,\ -2NaCl} [(Me_3Si)_2NBNSiMe_3]_2$$

The p.m.r. spectrum shows those protons of the Me_3Si—N group to be at a higher field than $(Me_3Si)_2N$—B, implying that exocyclic π-bonding may be stronger. This idea is not supported by the B—N bond lengths (exocyclic 0.144 nm, endocyclic 0.145 nm).[131]

While alkyl-substituted boretidines are unknown, the aryl ones have been synthesized by heating diphenylboron azide[132] or an arylboron dihalide with an aromatic amine in benzene.[133]

$$2Ph_2BN_3 \xrightarrow{heat} [PhBNPh]_2 + 2N_2$$

$$2C_6F_5BCl_2 + 2p\text{-}MeOC_6H_4NH_2 \longrightarrow [C_6F_5BNC_6H_4OMe\text{-}p]_2 + 4HCl$$

The decomposition of Ph_2BN_3 is accompanied by an aryl group migration rather like a Curtius rearrangement.[134] However, this is preceded by loss of N_2, since the unstable "nitrene" Ph_2BN will insert into the NH bond of Me_2NH.

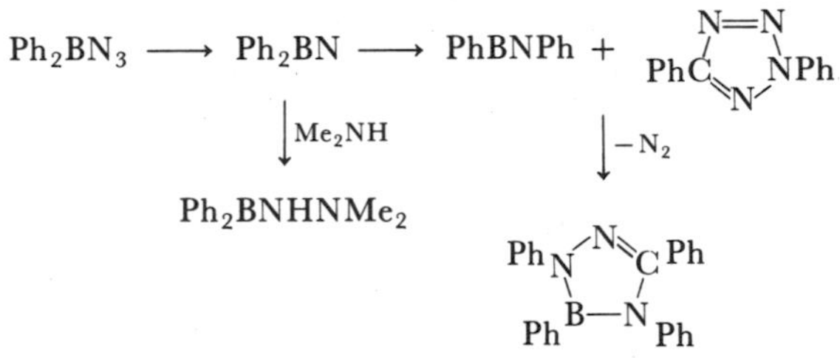

$$Ph_2BN_3 \longrightarrow Ph_2BN \longrightarrow PhBNPh + PhC\overset{N=N}{\underset{N}{}}NPh$$

$Ph_2BN \xrightarrow{Me_2NH} Ph_2BNHNMe_2$; $\xrightarrow{-N_2}$ Ph N–N=C Ph, Ph B—N Ph (five-membered ring)

The monomeric borazyne intermediate is thought to precede dimerization, and can be trapped as shown.[132]

The stability of these monomers ArBNAr′ seems to depend on the relative electron-withdrawing power of the substituents.[133] The B—N bond is polarized to nitrogen, so this will be offset by an electronegative group on boron. Thus, $C_6F_5BNC_6H_4OMe$-*p* and C_6F_5BN–mesityl both occur as monomers.

$$C_6F_5BCl_2 + ArNH_2 \longrightarrow \underset{(80\%\ \text{yield})}{C_6F_5B{=}NAr}$$

The B—N stretch is observed in the Raman at about 1700 cm^{-1}, but not in the infrared, supporting the non-polar nature of the B—N bond. The perfluorophenyl group will also strengthen the tendency of nitrogen to π-bond with boron. Rather surprisingly, aniline and phenylboron dichloride give a similar monomer with ν_{B-N} 1658 cm^{-1}.[135]

Borazines $[RBNR']_3$

These compounds have attracted much attention partly through being isoelectronic with benzene.[i,k,l,m]

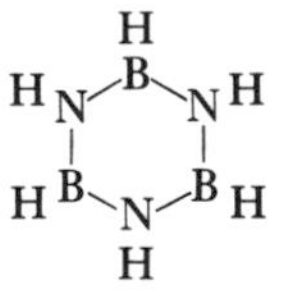

(b.p. 55°)

Fig. 3.42. Borazine

The parent compound $B_3N_3H_6$ superficially resembles benzene in its physical properties, with the ratio (b.p. borazine derivative)/(b.p. benzene derivative) = 0.93 ± 0.01. Though density, Trouton constant and surface tension agree more closely, the name "inorganic benzene" becomes inappropriate when considering the nature of the B—N bond, which is polarized to nitrogen through σ-bonding and to boron through π-bonding. This will be considered in more detail later (p. 94).

Although borazine itself was first isolated in 1926 by the pyrolysis of a diborane/ammonia mixture,[136] it was not until the 1950's that it could be prepared under mild conditions, involving the reaction of ammonium chloride or trichloroborazine with BH_4^- in ethers.

$$3BH_4^- + 3NH_4Cl \longrightarrow B_3N_3H_6 + 9H_2 + 3Cl^-$$

Substituted borazines result from amine hydrohalides, or by pyrolyzing primary and tertiary amine/borane adducts, while *B*-substituted compounds result with alkyl-lithium compounds.

$$\rangle B{-}H + RLi \longrightarrow \rangle B{-}R + LiH$$

The simplest method, however, involves reacting halogenoboranes with amines, and though not recognized as trimeric, the first borazine was isolated in 1889 from boron trichloride and aniline.

$$3BCl_3 + 6PhNH_2 \longrightarrow [ClBNPh]_3 + 3PhNH_2{\cdot}HCl$$

B-trichloroborazine (m.p. 84°C) is best prepared by passing boron trichloride over ammonium chloride on glass beads at 200°C in a hot tube. Excellent yields result in the presence of metal catalysts.

Though *B*-tribromoborazines result from boron tribromide and ammonia, fluoro- and iodo-borazines[137] have to be prepared indirectly.

$$Cl_3B_3N_3H_3 + TiF_4 \longrightarrow \underset{\text{(m.p. 122°C)}}{F_3B_3N_3H_3}$$

$$3BI_3 + (Me_3Si)_2NH \longrightarrow \underset{\text{(m.p. 134°C)}}{H_3N_3B_3I_3} + 6Me_3SiI$$

Various condensation reactions have been shown to yield borazines, notably those involving bis-(alkylamino)-substituted boranes, or by HX elimination, *e.g.*,

$$3XB(NHR)_2 \longrightarrow [XBNR]_3 + 3RNH_2$$
$$(X = Ph, R_2N, RS \text{ or alkyl})$$

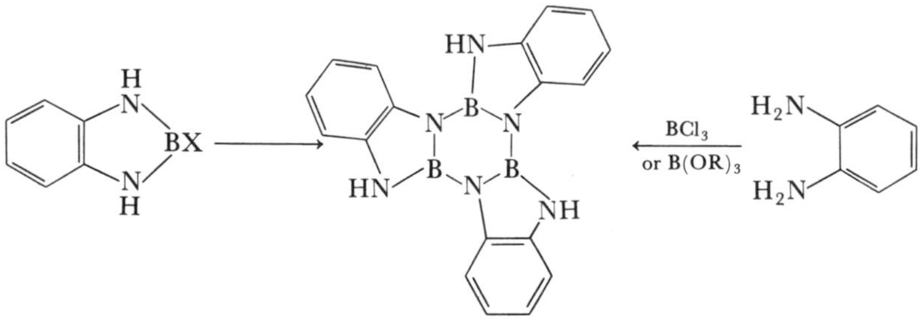

H N N H BX ⟶ HN B N N B B N NH HN ⟵ BCl_3 or $B(OR)_3$ H_2N H_2N

B-trichloroborazine is a most useful precursor for other borazines, Grignard reagents and amines affording direct substitution, while diazomethane and ethylene oxide insert into the B—Cl bond. Metathetical reactions also occur with alkoxides and pseudo-halide salts.

Both of the main synthetic routes to borazines, namely amine/borane[l] and amine/boron halide,[138] involve various intermediates, many of which can be isolated if the right substituent is present.

$$RNH_2 + H_2BR' \rightleftharpoons RH_2N \rightarrow BH_2R' \xrightarrow{-H_2} RHNBHR' \xrightarrow{-H_2} RNBR'$$

$$RHNBHR' \rightleftharpoons [RHNBHR']_3 \xrightarrow{-3H_2} [RNBR']_3$$

$$Bu^tNH_3^+ + PhBCl_3^- \longrightarrow Bu^tNH_3^+PhBCl_3^- \xrightarrow{R_3N} [Bu^tNBPh]_3 + [Bu^tNBPh]_4$$

$$Bu^tNH_3^+BPh_4^- \longrightarrow Bu^tNH_2BPh_3 \longrightarrow Bu^tNHBPh_2$$

$$Pr^nNH_3^+PhBCl_3^- \xrightarrow{-C_6H_6} Pr^nNH_2BCl_3 \longrightarrow [Pr^nNBCl]_3$$

Perhaps the most peculiar of the borazines is the fully-chlorinated one.[139] This melts at 176°C, and is formed from a BCl_3/NCl_3 mixture, from the thermolysis of trimeric dichloroboron azide $[Cl_2BN_3]_3$ or by cleaving the Si—N bonds of *N*-chlorohexamethyldisilazane with BCl_3.

$$3NCl_3 + 3BCl_3 \longrightarrow B_3N_3Cl_6$$

$$3(Me_3Si)_2NCl + 3BCl_3 \longrightarrow B_3N_3Cl_6$$

$$B_3N_3Cl_6 \xleftarrow[200^\circ C]{-3N_2} [Cl_2BN_3]_3$$

The borazine ring is relatively stable to hydrolysis.[l] Thus, $B_3N_3H_6$ forms a trihydrate with ice-water which evolves H_2 at 100°C. Methyl-substituted derivatives are hydrolyzed at 150°C or with boiling aqueous acid. At 100°C, *N*-methylborazine gives the hydroxyborazine, while the *B*-methyl isomer yields the boroxine (boroxole).

$$Me_3N_3B_3(OH)_3 \xleftarrow[H_2O]{100^\circ C} Me_3N_3B_3H_3 \xrightarrow{150^\circ C} 3B(OH)_3 + 3MeNH_2 + 3H_2$$

$$Me_3B_3O_3 \xleftarrow[H_2O]{100^\circ C} Me_3B_3N_3H_3 \xrightarrow{150^\circ C} 3MeB(OH)_2 + 3NH_3$$

Generally speaking, borazines are thermally stable unless substituents are present which will readily condense.

The polar nature of the B—N bond in borazines is supported by the addition reactions involving the hydrogen halides. With borazine itself, HCl forms an adduct which dissociates to *B*-trichloroborazine at 100°C.

$$[HBNH]_3 \xrightarrow[25^\circ C]{3HCl} [ClHBNH_2]_3 \xrightarrow{100^\circ C} [ClBNH]_3 + 3H_2$$

$$[RBNR]_3 + 3HX \xrightarrow[100-130^\circ C]{20-90^\circ C} [RXBNHR]_3$$

Methyl-substituted borazines form similar adducts which decompose reversibly.

Ever since the isolation and full characterization of benzene and cyclopentadienyl complexes of low valency transition metals, which showed that the multi-centre π-orbitals of the rings bonded to the vacant metal d orbitals, it has been the desire of inorganic chemists to bond an inorganic ring in the same way. The formal analogy between benzene and borazine makes the latter a strong contestant. The hexamethylborazine complex of chromium carbonyl was the first to be isolated,[140] and subsequently a series of hexa-alkylborazines have been shown to give these π-complexes by heating with *cis*-tris-(acetonitrile)chromium tricarbonyl in dioxan.[141] The complexes also form if the borazine and $Cr(CO)_6$ are irradiated with ultraviolet light.

$$(MeCN)_3Cr(CO_3 \underset{Me_6B_3N_3}{\overset{MeCN}{\rightleftarrows}} Me_6B_3N_3Cr(CO)_3 \xrightarrow{PH_3} (H_3P)_3Cr(CO)_3$$

$$(MeCN)_3Cr(CO)_3 \underset{C_4H_8O_2}{\overset{MeCN}{\rightleftarrows}} (CO)_3Cr\begin{matrix} OC_4H_8O \\ OC_4H_8O \\ OC_4H_8O \end{matrix}Cr(CO)_3 \underset{C_4H_8O_2}{\overset{Me_6B_3N_3}{\rightleftarrows}} Me_6B_3N_3Cr(CO)_3$$

The products are air-stable sublimable solids with infrared spectra supporting π-bonding from the borazine ring. These were similar to the spectrum of $Me_6C_6Cr(CO)_3$, while ν_{B-N} had fallen about 30 cm^{-1} through complexing, a drop similar to that observed with hexa-ethylbenzene and its complex. These spectra differ markedly from those of tris-amino metal tricarbonyl complexes.

Similar π-complexes are reported between hexamethylborazine and hexamethylbenzene with tetracyanoethylene, and the similarity of their ultraviolet spectra, supports π-delocalization in the borazine.[142] This is also upheld by the planar nature of the ring, the B—N bonds being a little shorter than expected for a single bond, and all being equal.[143]

Fused and Polycyclic Borazines

The "inorganic benzene" analogy of borazine has been further extended to diphenyl and naphthalene.[m] The analogues of these organic compounds were first suspected as by-products in the synthesis of borazine,[d] and have now been isolated from the thermal decomposition of borazine between 340°C and 440°C.[144]

```
                                                                N-B-N
      H     H            H H     H H                            |   |
HB-N-B-N-BH               N-B     N-B                        N-B-N-B-N
|    |    |           HB     N-B     NH                      |   |   |
HN-B-N-B-NH               N-B     N-B                        B-N-B-N-B
   H    H                H H     H H
```

(m.p. 27–30°) (m.p. 59–60°)

Fig. 3.43. Polycyclic B—N rings

The borazanaphthalene adds five moles of hydrogen halides reversibly, while the less-volatile components are thought to include amino derivatives of borazanaphthalene and borazadiphenyl together with $B_6N_7H_9$, the fused tricyclic compound. Such products could only result from extensive ring cleavage, unlike the case of benzene which undergoes intermolecular dehydrogenation yielding polyphenyls only. Phosphorus–nitrogen rings form similar fused compounds.

Borazadiphenyl derivatives have resulted from lithium halide elimination and deamination, the products possessing a B—N ring bridge.[145] Further deamination leads to a two-dimensional graphite-like polymer. Though borazadiphenyl derivatives with a N—N ring bridge are unknown, potassium will condense *B*-monochloroborazines to B—B-linked borazadiphenyl compounds.

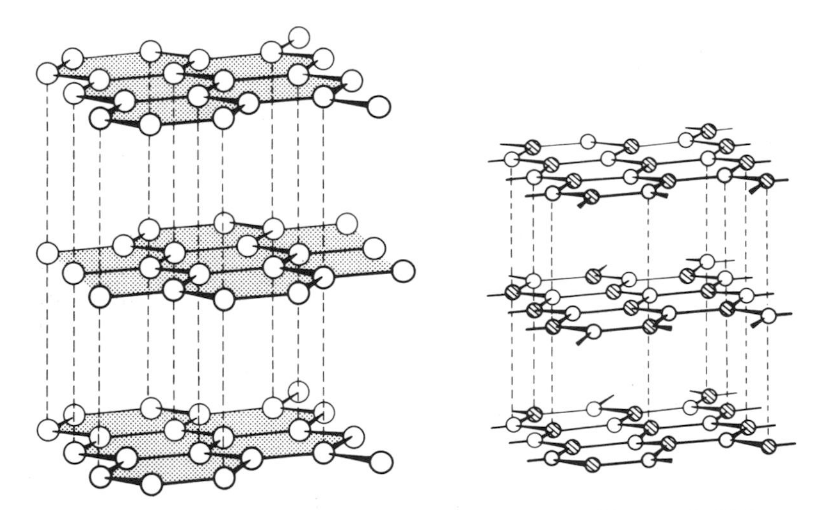

Fig. 3.44. The crystal structures of graphite and boron nitride[c]

Tetrameric borazynes $[XBNR]_4$

Although a few borazines have been found to be resistant to hydrolysis, so providing possible precursors to useful heat- and water-resistant materials, the search has generally proved fruitless. Resistance to hydrolysis generally results through steric crowding, which, in turn, leads to a different structure formation. So the dehydrohalogenation of the adduct $Bu^tNH_2 \cdot BCl_3$ yields the tetrameric borazyne $[ClBNBu^t]_4$,[146] which has been shown to possess a boat structure with localized π-bonding.[m]

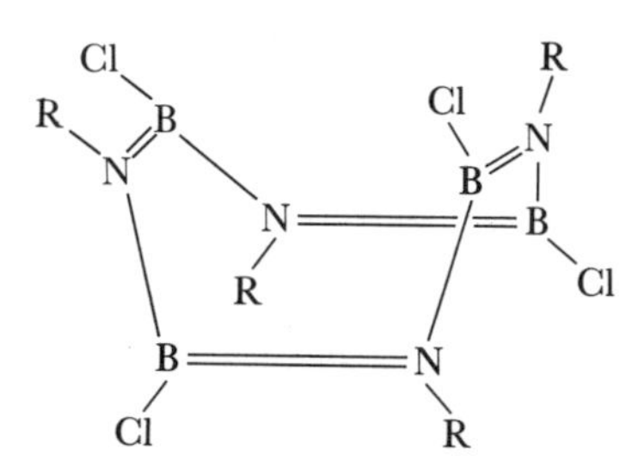

Fig. 3.45. The structure of the tetrameric borazyne

The stability of this compound to hydrolysis and metallo-organic reagents has been attributed to steric hindrance. However, metathetical exchange reactions are successful, the tetrachloroborazyne reacting with alkali metal pseudo-halides in an organic solvent.

$$[ClBNBu^t]_4 + 4MX \longrightarrow [XBNBu^t]_4 + 4MCl$$
$$(X = NCO, NCS, NCSe, N_3)$$

The structure of the isothiocyanato compound shows a tub-shaped B_4N_4 skeleton with localized π-bonds between alternate pairs of atoms.[147] The larger B—N bonds of the ring (0.1456 nm) are close to the length estimated for a single bond (0.145 nm), while the shorter ones (0.1402 nm) exhibit considerable multiple-bond character. This is close to the value reported for trichloroborazine (0.1415 nm).[148]

A conversion from tetramer to trimer occurs with the *B*-phenyl, *N*-isobutyl rings. Both result from the dehydrohalogenation of the adduct $(Bu^iNH_2 \cdot PhBCl_2)$ or salt $(Bu^iNH_3^+ PhBCl_3^-)$ with triethylamine, and heating the tetramer just above its melting point yields the lower melting trimer.[149]

$$\underset{\text{(m.p. 232°C)}}{3[Bu^iNBPh]_4} \xrightarrow{250°C} \underset{\text{(m.p. 134°C)}}{4[Bu^iNBPh]_3}$$

Cyclic 3-co-ordinate aluminium–nitrogen compounds have so far proved elusive: even tris-aminoalanes are dimeric through nitrogen-bridging, while polymers result from the reaction of triethylalane and methylamine hydrochloride. However, the elimination of two moles of benzene from the arylamine/triphenylalane adduct gave the tetrameric product $[PhAlNAr]_4$.

$$4Ph_3Al + 4ArNH_2 \longrightarrow [PhAlNAr]_4 + 8C_6H_6$$

This was thought to have a boat structure similar to the borazynes but was later shown to possess a cube structure.[150]

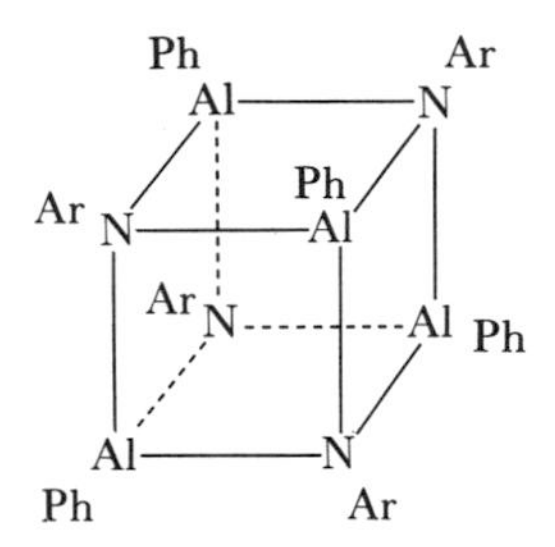

Fig. 3.46. The structure of $[PhAlNAr]_4$

These cage compounds possess high thermal stability, $[EtAlNPh]_4$ subliming at 250°C/0.001 mm. Hydrogen chloride adds across the Al—N bond before cleaving the Al—C one.[151]

$$Et_3Al + PhNH_2 \longrightarrow \underset{\text{(m.p. 58°C)}}{[Et_2AlNHPh]_2} \xrightarrow[-C_2H_6]{150°C} [EtAlNPh]_4$$

$$Et_2AlCl + PhNH_2 \longrightarrow \underset{\text{(m.p. 58–61°C)}}{[EtClAlNHPh]_2} \xrightarrow{162°C} \underset{\text{(sublimes 300°C/0.001 mm)}}{[ClAlNPh]_4}$$

$$[EtAlNPh]_4 \xrightarrow{4HCl} 2[ClEtAlNHPh]_2 \xrightarrow{4HCl} 4Cl_2EtAlNH_2Ph \xrightarrow{HCl} C_2H_6$$

Four Co-ordinate Compounds

It is apparent from all the ring systems so far discussed that any tendency in boron to form 4-co-ordinate compounds is offset by internal π-bonding, which can only be quenched by highly polar compounds such as the hydrogen halides. The situation is considerably different with boron and the heavier Group V elements, and with nitrogen and the heavier Group III ones. Internal compensation for electron deficiency is only weak, so that no Group III analogues of the B—N compounds already described at length exist, while oligomerization of the monomers $R_2MNR'_2$ is common.

Oligomers of $R_2BNR'_2$

The first self-association of an aminoborane was reported in 1933. The low reactivity of dimethylaminodichloroborane led to the proposed bridged-dimer structure. This is supported by the low B—N stretching frequency (900 cm^{-1}) and the low dipole moment,[152] as well as by the molecular weight in benzene. This dimerization can be reversed by heating, the monomer distilling as a fuming liquid that deposits the solid dimer on standing. While Me_2NBMe_2 shows no such tendency to dimerize, the mixed $Me_2NBMeCl$ slowly deposits a dimeric solid[153] which can be washed with water, unlike the hydrolytically unstable monomers.

$$[Me_2NBCl_2]_2 + 2Me_2NBMe_2 \xrightarrow{-25°C} \begin{array}{c} Cl_2 \qquad Me_2 \\ B\text{—}N \\ | \quad\; | \\ N\text{—}B \\ Me_2 \qquad Me_2 \end{array}$$

(m.p. 63–5°C)

$$\xrightarrow{170°C} \begin{array}{c} Me \\ Cl\;B\text{—}N\;Me_2 \\ | \quad\; | \\ Me_2\;N\text{—}B\;Me \\ Cl \end{array}$$

(m.p. 100–102°C)

The other dimethylaminoboron dihalides form by a similar route and are dimeric in the solid phase, except the di-iodide which showed no tendency to dimerize on standing.[154]

$$2(Me_2N)_3B + 4BX_3 \longrightarrow 3[Me_2NBX_2]_2 \qquad (X = F, Cl, Br)$$

The difluoride possesses a planar structure with the $[BN]_2$ ring almost square.[155] The B—N bonds are larger than would be expected for single bonds (0.160 nm), which may well indicate the presence of multi-centre rather than two-centre bonding (*cf.* $[LiN(SiMe_3)_2]_3$ and $[(Me_2N)_2Be]_3$).

Increasing the size of the substituents on nitrogen again decreases the tendency to dimerize. Thus, Et_2NBF_2 is dimeric as a solid and in solution, but exhibits a monomer–dimer equilibrium as a liquid, and is monomeric in the vapour phase. The analogous chloride and bromide show no dimer character as a liquid.[156]

Dimerization is also enhanced by strongly electron-withdrawing groups on nitrogen. The lithium salt of diphenylketimine gives dimeric amino-boron dihalides with BX_3 (X = Cl, Br, I).[157] The chloride and bromide are both stable in moist air at 20°C.

$$2Ph_2C{=}NLi + 2BX_3 \longrightarrow [Ph_2C{=}NBX_2]_2 + 2LiX$$

Unlike Me_2NBMe_2, which is exclusively monomeric, $Ph_2C{=}NBMe_2$ is dimeric ($\nu_{C=N}$ 1662 cm^{-1}), while $Ph_2C{=}NBPh_2$ is a monomer ($\nu_{C=N}$ 1786 cm^{-1}).

The oligomeric state of $RR'NBH_2$ (R and R' alkyl or hydrogen) appears to depend on the size of the functional group, as was observed with organo-beryllium amides (p. 17). These compounds can be synthesized by pyrolyzing primary amine–borane adducts, or reacting these adducts with dimethylaminoborane.[158]

$$RNH_2{\cdot}BH_3 \longrightarrow [RNHBH_2]_3 \ (R = Me, Et, Pr^n, Bu^n, Bu^t)$$

$$RNH_2{\cdot}BH_3 \longrightarrow [RNHBH_2]_2 \ (R = Pr^i, Bu^t)$$

$$Me_2NBH_2 + RNH_2{\cdot}BH_3 \longrightarrow Me_2N\langle H_2B{-}NRH,\ H_2B{-}NRH\rangle BH_2 \ (R = Pr^n, Pr^i)$$

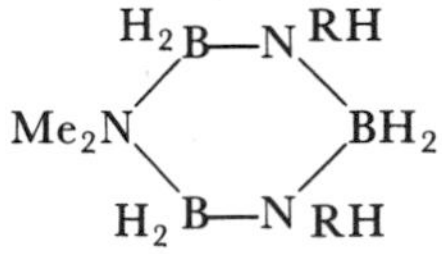

The symmetrical products are thought to result from intermediates such as $H_2B(NHMe)_2^+BH_4^-$ and $MeH_2NBH_2NHMeBH_2NH_2Me^+BH_4^-$ by dehydrogenation. This proceeds further in the presence of Me_2NBH_2, giving borazines.

$$Me_2NBH_2 + [RNHBH_2]_3 \longrightarrow Me_2NH{\cdot}BH_3 + [RNBH]_3 \ (R = Et, Pr^n)$$

The p.m.r. spectra of $[RNHBH_2]_3$ show the presence of isomers, which are probably the *cis* and *trans* forms of the substituted chair conformer.

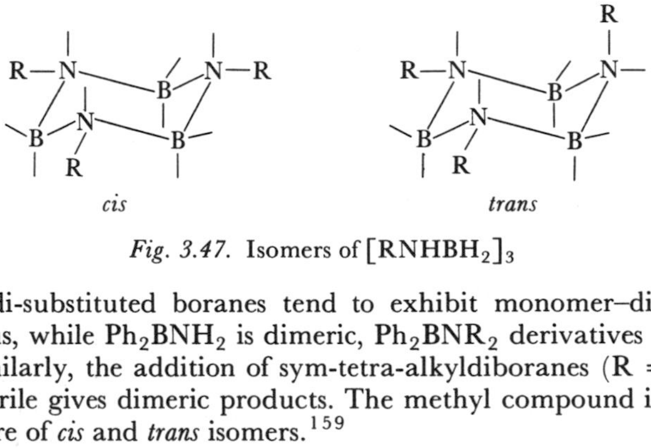

cis *trans*

Fig. 3.47. Isomers of $[RNHBH_2]_3$

Amino-di-substituted boranes tend to exhibit monomer–dimer equilibria. Thus, while Ph_2BNH_2 is dimeric, Ph_2BNR_2 derivatives are monomeric. Similarly, the addition of sym-tetra-alkyldiboranes (R = Me, Et) to acetonitrile gives dimeric products. The methyl compound is obtained as a mixture of *cis* and *trans* isomers.[159]

$$[Me_2BH]_2 + 2MeCN \longrightarrow [MeCH{=}NBMe_2]_2$$

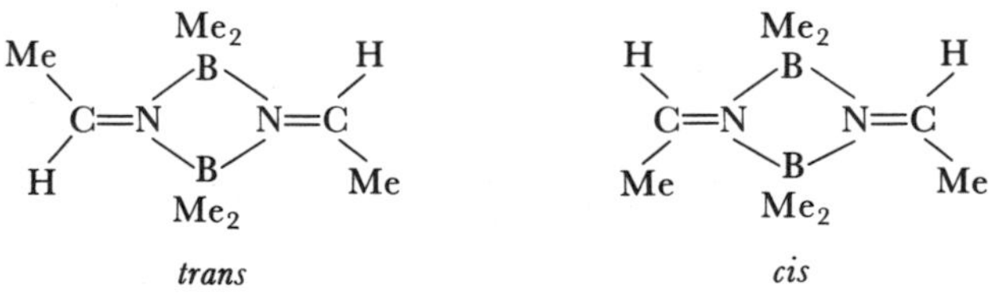

trans *cis*

Fig. 3.48. Isomers of $[MeCHNBMe_2]_2$

Heating tri-n-butylborane with t-butylcyanide gives the aminoborane which is monomeric in the vapour and has a boiling point close to that of monomeric $Bu^n_2BNEt_2$ (77°C/0.3 mm).[160]

$$Bu^n_3B + Bu^tCN \xrightarrow{160^\circ C} Bu^n_2B{-}N{=}CHBu^t \xrightarrow{\text{cool}} [Bu^n_2B{-}N{=}CHBu^t]_2 \quad (\text{m.p. } 74\text{–}6^\circ C)$$

For the dimer, $\nu_{(B-N-C)}$ occurs at 1670 cm^{-1}, but rises to 1850 cm^{-1} in the monomer, supporting the allene structure with strong B—N π-bonding.

Oligomers of aminoborane $[H_2BNH_2]_n$ result when the diammonate of diborane reacts with sodamide in liquid ammonia, where $n = 2, 3, 4$ or 5.

$$2BH_2(NH_3)_2^+BH_4^- \xrightarrow{NH_2^-} \frac{1}{n}[H_2BNH_2]_n + BH_4^- + NH_3$$

The pentamer is the major product, and is only hydrolyzed by boiling water. The dimer readily sublimes and ring-expands to the trimer in the presence of bases. This is better prepared by reducing the adduct of borazine and hydrogen chloride, and is stable to ice-water in the absence of ether.[161]

$$B_3N_3H_6 \xrightarrow{3HCl} Cl_3H_3B_3N_3H_6 \xrightarrow{3BH_4^-} B_3N_3H_{12} + \tfrac{3}{2}B_2H_6 + 3Cl^-$$

Trimeric aminoborane is isostructural and isoelectronic with cyclohexane. It is a solid with a dipole moment of 3.2 *D* (1.067×10^{-29} Cm) acting along the three-fold axis, which may well account for its being a solid and having a heat of sublimation about 3 times as great as C_6H_{12}.

The crystal structure of $[Me_2NBH_2]_3$ supports the above proposals, and the B—N bond length (0.159 nm) is close to that found in the difluoride dimer.[162]

Oligomers of $R_2MNR'_2$ (M = Al, Ga, In, Tl)

Most of the compounds of this kind which have been prepared are dimeric. The simplest synthetic route is analogous to that widely used for beryllium and magnesium compounds, and involves the elimination of a hydrocarbon from an amine adduct.

$$R_2NH + R'_3M \longrightarrow R_2NH{\cdot}MR'_3 \xrightarrow{-R'H} \frac{1}{n}[R_2NMR'_2]_n$$

This is generally the case with large functional groups, but trimers and higher oligomers do result with smaller substituents (*e.g.*, Me, H). Thus, Me_2NAlH_2 is dimeric in ether, but trimeric in benzene or in the vapour phase. Likewise, Me_2NGaMe_2 and Me_2NTlMe_2 are both dimeric in benzene and crystalline, while the aluminium and indium compounds are oligomeric glasses.[163]

The Al—N compounds formed thus from tin amines and AlR_3, and by using diphenylketenimine or nitriles, give dimeric products.[164]

$$2Me_3SnNMe_2 + 2AlEt_3 \longrightarrow 2Me_3SnEt + [Et_2AlNMe_2]_2$$

$$Ph_2C{=}NH \xrightarrow{Me_3Al} Ph_2C{=}NH{\cdot}AlMe_3 \ (\text{m.p. } 76°C) \xrightarrow{100°C} [Ph_2C{=}NMMe_2]_2$$

$$Ph_2C{=}NH \xrightarrow{Me_3Ga} Ph_2C{=}NH{\cdot}GaMe_3 \xrightarrow{100°C} [Ph_2C{=}NMMe_2]_2$$

Hexamethyldisilazane, or its *N*-bromo derivative, reacts with $AlCl_3$ to give the silylaminoaluminium dimer,[165] while acetoxime forms dimers with Me_3M[166] (M = B, Al, Ga, In, Tl). All are crystalline solids with 6-membered MONMON rings.

$$2(Me_3Si)_2NX + 2AlCl_3 \xrightarrow{-2Me_3SiCl} [Me_3SiNXAlCl_2]_2 \ (X = H, \text{m.p. } 163°C)$$
$$(X = H \text{ or } Br)$$

$$2Me_2C{=}NOH + 2Me_3M \longrightarrow Me_2C{=}N\overset{O-MMe_2}{\underset{Me_2M-O}{\ }}N{=}CMe_2$$

In the presence of $(Bu^nNH)_3Al$, dichloroalane and n-butylamine the aminoalane $(Cl_2AlNBu^n)_3Al$ is formed.[167] This can be reduced and complexes with amines, ethers, *etc.* The structure of the product, $Base{\cdot}Al(NBu^nAlH_2)_3$ is thought to involve a 6-membered $[Al—N]_3$ ring bridged by a complexed Al atom.

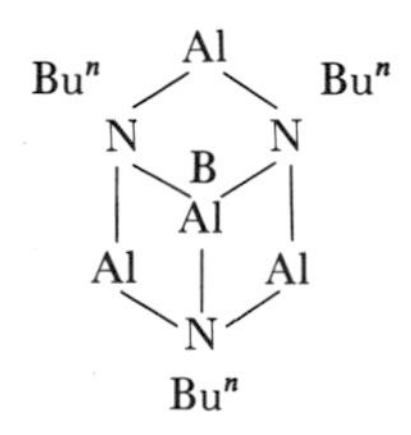

Fig. 3.49. The structure proposed for $Base{\cdot}Al(NBu^nAlH_2)_3$

The Al—N bond lengths in trimeric ethyleniminodimethylaluminium (0.191 nm) are shorter than those found in $Me_5Al_2NPh_2$ (0.201 nm), while the Al—N—Al bond angle (120 degrees) suggests sp^2-hybridization and multi-centre bonding in the 3-membered ethylenimine ring.[168]

Based on the ideas of the last Chapter, it is puzzling that hindrance should be greater than in dimeric $[Me_2AlNMe_2]_2$, especially as the less-hindered $[Me_2AlPMe_2]_3$ is trimeric. However, the electronic requirements inevitably differ in ethylenimino compounds, and it has been suggested that ring valence strain plays only a limited role in determining the structures of compounds of elements of the second and third periods. This accounts for the $[Me_2AlSMe]_2$ (vapour) and $[Me_2AlOMe]_3$ situation though the former is polymeric as a solid, but not the $[Me_2AlNMe_2]_2$, $[Me_2AlPMe_2]_3$ anomaly.

Like the azidodihalogenoboranes, dialkylaluminium and gallium azides are trimeric. They result from the mixed halogen XN_3 and either BX_3 or R_3M, and have a structure involving a 6-membered ring $[X_2MN_3]_3$ in accordance with the vibrational spectrum.[169]

$$3XN_3 + 3BX_3 \longrightarrow 3X_2 + [X_2BN_3]_3$$

$$3R_3Al + 3XN_3 \longrightarrow [R_2AlN_3]_3 + 3RX$$

Diethylindium azide is dimeric and covalent, while the thallium compound $Et_2Tl^+N_3^-$ is ionic.

Nitrides

The condensation polymerization already considered for $Ph_3Al{\cdot}PhNH_2$ has been extended to give polymeric aluminium nitride. This has the wurtzite structure comprising sheets of chair 6-membered [Al—N]$_3$ units cross-linked with Al—N bonds, the one sheet being the mirror image of the next, but with aluminium and nitrogen atoms interchanged. Gallium nitride, obtained from gallium and ammonia at 1200°C, and indium nitride resulting from the pyrolytic decomposition of $(NH_4)_2InF_6$, have similar structures.

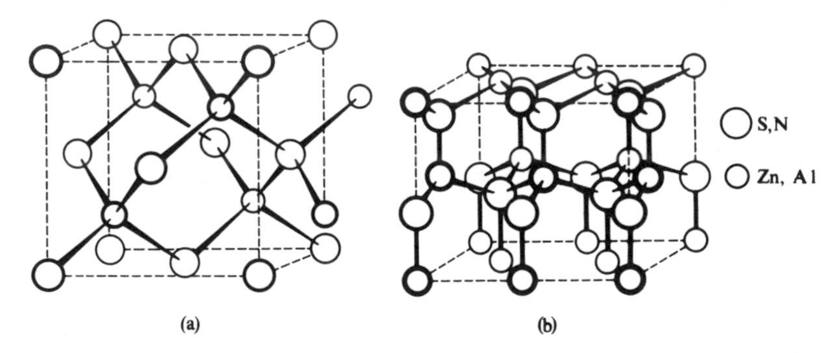

Fig. 3.50. The crystal structures of (a) zinc blende and (b) wurtzite, adopted by the nitrides of the Group III elements[c]

Boron nitride, however, has the polymeric adamantane cage structure characteristic of diamond or zinc blende.

Boron nitride is one of the oldest-known boron–nitrogen compounds, and closely resembles diamond in its properties.[m] Many synthetic methods have been tried, and involve considerable technical difficulty. Reactions between boron or its oxide, and nitrogen compounds such as oxides, ammonia, cyanides, urea and amides have been utilized. However, the best laboratory process involves the fusion of borax and ammonium chloride, or ammonia and boron tribromide. Industrially, the fusion of urea and boric acid gives a turbostratic structure with poorly-orientated hexagonal layers.

$$2(HO)_3B + CO(NH_2)_2 \longrightarrow 2BN + 5H_2O + CO_2$$

At 1800°C, however, the hexagonal form resembling graphite results, but formally resembles the wurtzite structure with its hexagonal rings flattened (see Figures 3.44 and 3.50). The cubic form of boron nitride can be obtained from the hexagonal form at 1800°C and 85 000 atm in the presence of alkali metals,

$$\underset{\text{(hexagonal)}}{BN} \underset{\text{50 000 atm/2500°C}}{\overset{\text{1800°C/85 000 atm, alkali metal}}{\rightleftarrows}} \underset{\text{(cubic (zinc blende))}}{BN}$$

and, strangely, possesses a B—N bond distance of 0.1446 nm (borazine 0.144 nm), indicating partial double-bonding. It is a good insulator and lubricant, while the cubic form is harder than diamond and less liable to oxidize.

Boron nitride is attacked by HF and fluorine, and no trace of fluoroborazines has been found, since H_3NBF_3 decomposes irreversibly into BN and $NH_4^+BF_4^-$.

$$2BN + 3F_2 \longrightarrow 2BF_3 + N_2$$

$$4BN + 12HF \longrightarrow 4[H_3NBF_3] \longrightarrow 3NH_4^+BF_4^- + BN$$

Hydrazine derivatives

Hydrazines yield derivatives of varying ring size. Tetramethylhydrazine forms an adduct with triethylalane, the p.m.r. spectrum indicating that the monomer has equivalent methyl groups.[170]

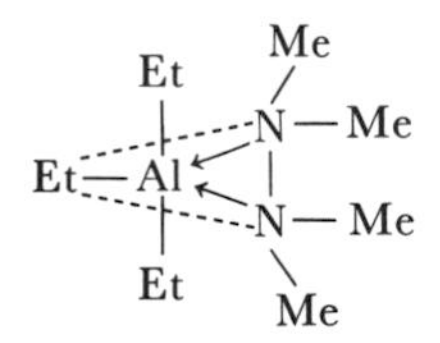

Fig. 3.51. The tetramethylhydrazine adduct of triethylaluminium

A bridged dimer results from 1,1-dimethylhydrazine and diethylaluminium chloride, which probably possesses a 4-membered nitrogen-bridged ring, as the nitrogen atom bonded to the electropositive aluminium is the stronger donor.

$$2Me_2NNH_2 + 2Et_2AlCl \longrightarrow 2C_2H_6 + [EtClAlNHNMe_2]_2$$

Also Al—C bond cleavage occurs in preference to Al—Cl cleavage.

Symmetrical 6-membered rings involving boron in the 1 and 4 positions can be synthesized by various routes involving trans-amination, halide elimination and the addition of diborane to azobenzene.[171]

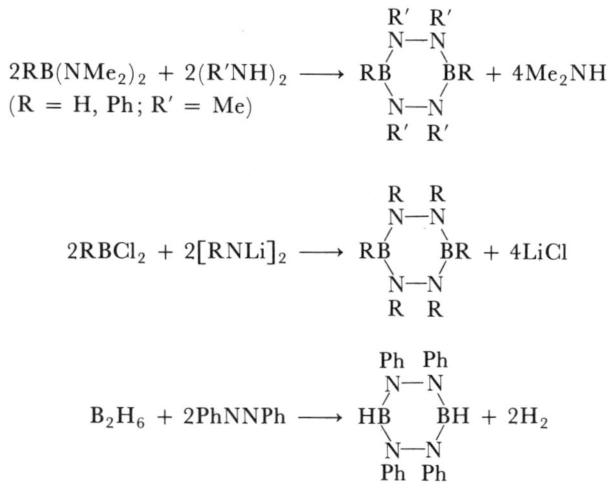

```
                                         R' R'
                                         N—N
                                        /    \
2RB(NMe2)2 + 2(R'NH)2  ——→  RB      BR + 4Me2NH
(R = H, Ph; R' = Me)                    \    /
                                         N—N
                                         R' R'
```

```
                                 R  R
                                 N—N
                                /    \
2RBCl2 + 2[RNLi]2  ——→  RB      BR + 4LiCl
                                \    /
                                 N—N
                                 R  R
```

```
                              Ph Ph
                               N—N
                              /    \
B2H6 + 2PhNNPh  ——→  HB      BH + 2H2
                              \    /
                               N—N
                              Ph Ph
```

As with the borazines, HCl will readily form adducts, but two kinds result due to the presence of more nitrogen than boron atoms.[172]

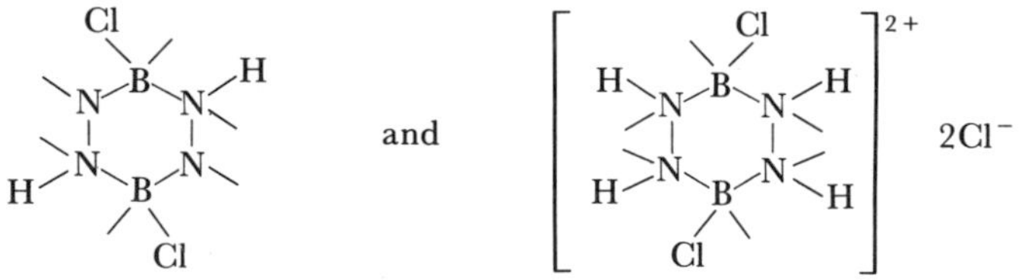

```
       Cl                              Cl          2+
        \ /                            \ /
     \N—B—N/H                      H\N—B—N/H
      |    |\          and          /|    |\            2Cl⁻
     /N—B—N\                        \N—B—N/
    H    / \                       H    / \  H
          Cl                            Cl
```

Fig. 3.52. Hydrogen chloride adducts of cyclic hydrazinoboranes

The product resulting from the dehydrogenation of the adduct $Bu^tBH_2{\cdot}N_2H_4$ has a singular structure. It analyzes for tetrameric $C_4H_{11}BN_2$ and possesses no B—H bonds.[173]

$$\mathrm{Bu^tBH_2{\cdot}N_2H_4} \xrightarrow{-\mathrm{H_2}} \left[\begin{array}{c} \mathrm{H\ \ H_2} \\ \mathrm{Bu^t\ N{-}N\ H} \\ \mathrm{B\qquad B} \\ \mathrm{H\ \ N{-}N\ Bu^t} \\ \mathrm{H_2\ \ H} \end{array}\right] \xrightarrow[-\mathrm{H_2}]{0^\circ\mathrm{C}} [\mathrm{Bu^tBN_2H_2}]_4$$

(m.p. 93°C)

It comprises two 6-membered B_2N_4 rings cross-linked through N ⟶ B co-ordinate bonds. The $B_{(1)}$—$N_{(1)}$ bond (0.166 nm) is slightly larger than expected for the single bond length of $B_{(1)}$—$N_{(3)}$ (0.160 nm), while $B_{(1)}$—$N_{(2)}$ (0.149 nm) shows some multiple-bond character. Two of the six valence angles at boron are significantly less than the tetrahedral angle to allow the incorporation of boron into the 5-membered ring ($N_{(2)}\hat{B}_{(1)}N_{(3)}$ and $N_{(1)}\hat{B}_{(1)}N_{(3)}$).

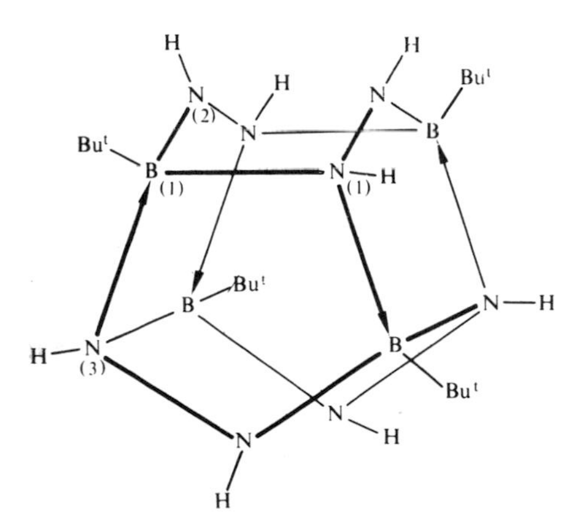

Fig. 3.53. The structure of $[\mathrm{Bu^tBN_2H_2}]_4$[173]

Miscellaneous cyclic compounds

Five- and 7-membered rings containing boron and nitrogen have been synthesized by the methods illustrated below, which resemble those employed for the hydrazine compounds.[174]

$$\mathrm{HN(BPhNMe_2)_2 + (MeNH)_2} \longrightarrow \mathrm{HN}\begin{array}{l} \mathrm{B(Ph){-}NMe} \\ \qquad\ \ | \\ \mathrm{B(Ph){-}NMe} \end{array} \xrightarrow{\mathrm{H_2O}} \mathrm{O}\begin{array}{l} \mathrm{B(Ph){-}NMe} \\ \qquad\ \ | \\ \mathrm{B(Ph){-}NMe} \end{array}$$

(m.p. 106–8°C) (m.p. 202–4°C)

These compounds possess high thermal stability, but are readily hydrolyzed, and the triazadiborolane (TADB) is readily lithiated and reacts with ferrous chloride to give a diamagnetic brown complex resembling ferrocene.

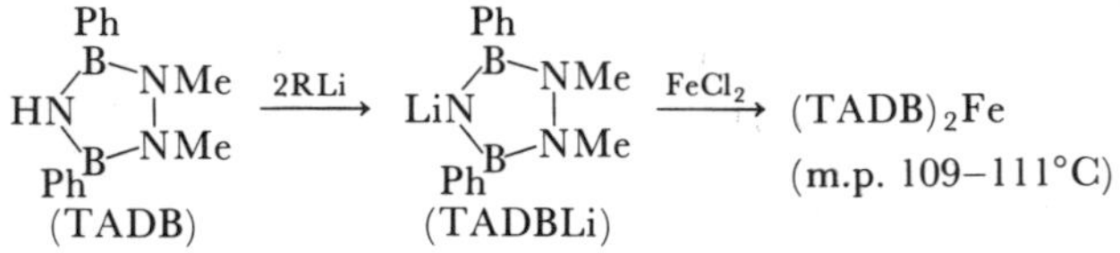

$$\text{HN}\langle\text{B(Ph)–NMe–NMe–B(Ph)}\rangle\ (\text{TADB}) \xrightarrow{2\text{RLi}} \text{LiN}\langle\text{B(Ph)–NMe–NMe–B(Ph)}\rangle\ (\text{TADBLi}) \xrightarrow{\text{FeCl}_2} (\text{TADB})_2\text{Fe}$$

(m.p. 109–111°C)

Mercuric chloride yields a σ-bonded mercury amide.

Five-membered rings incorporating Si—Si and B—B bonds can be synthesized in a similar way,[175]

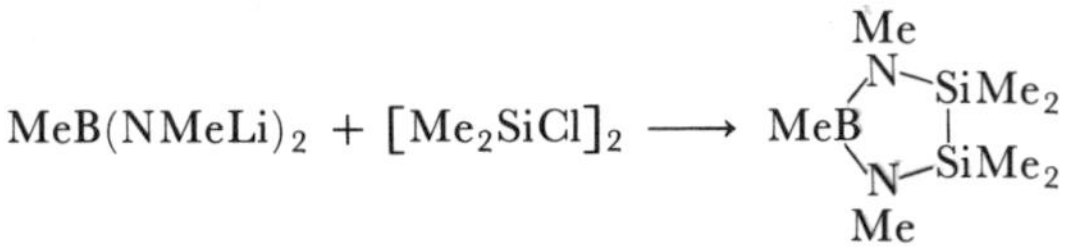

$$\text{MeB(NMeLi)}_2 + [\text{Me}_2\text{SiCl}]_2 \longrightarrow \text{MeB}\langle\text{N(Me)–SiMe}_2\text{–SiMe}_2\text{–N(Me)}\rangle$$

(b.p. 76°C/10 mm)

$$\text{MeB(NMeLi)}_2 + [\text{MeNBCl}]_2 \longrightarrow \text{MeB}\langle\text{N(Me)–BNMe}_2\text{–BNMe}_2\text{–N(Me)}\rangle$$

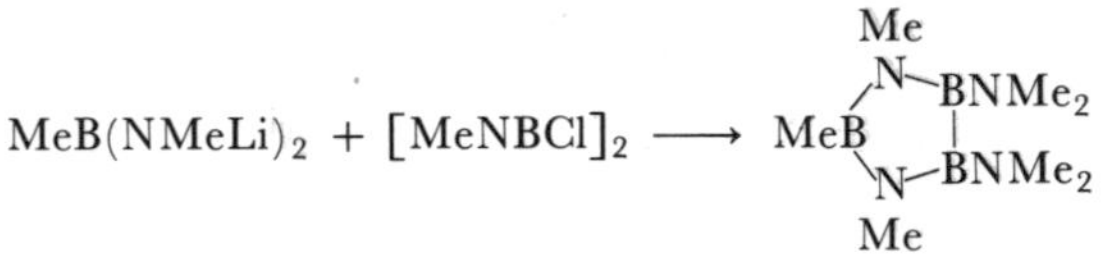

(b.p. 54°C/high vac.)

while tetrazaborolane derivatives result from the addition of borazynes to phenyl azide,[176] or methyl azide to primary amine borane adducts.[177]

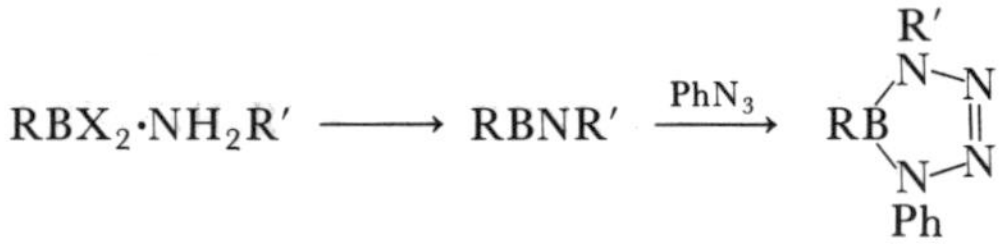

$$\text{RBX}_2\cdot\text{NH}_2\text{R}' \longrightarrow \text{RBNR}' \xrightarrow{\text{PhN}_3} \text{RB}\langle\text{N(R}')\text{–N=N–N(Ph)}\rangle$$

(R = Bu^n, C_6F_5, *p*-$MeOC_6H_4$; R′ = Me, C_6F_5, Ph)

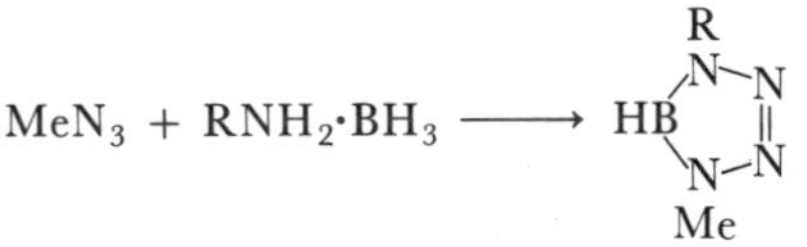

$$\text{MeN}_3 + \text{RNH}_2\cdot\text{BH}_3 \longrightarrow \text{HB}\langle\text{N(R)–N=N–N(Me)}\rangle$$

(R = Me; m.p. 11°C; R = Ph; b.p. 80°C/0.01 mm)

Spectroscopic properties of the disilane ring suggest it is planar, with substantial B—N π-bonding. Such bonding is also evident in the tetrazaborolane ring, where the B—N bond length is 0.1413 nm.[178]

Unsymmetrical 4-membered rings

Both boron and aluminium form compounds of the type indicated.

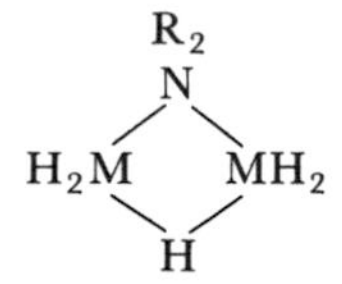

Fig. 3.54. μ-Dialkylamino compounds

μ-Dimethylaminodiborane is best prepared from diborane(6) and the anion $Me_2NBH_3^-$.

$$Me_2NH{\cdot}BH_3 \xrightarrow[\text{glyme}]{\text{NaH}} Me_2NBH_3^- \xrightarrow{B_2H_6} H_2B(\mu\text{-}NMe_2)(\mu\text{-}H)BH_2$$

(v.p. 101 mm/0°C)

The aluminium homologue is formed when dimethylaminochloroalane is reduced by AlH_4^-, and is a solid which decomposed at 130°C before melting; it is the first example of a derivative of the unknown dialane Al_2H_6.[179]

$$AlH_4^- + Me_2NAlH(Cl) \rightarrow\rightarrow H_2Al(\mu\text{-}NMe_2)(\mu\text{-}H)AlH_2 + Cl^-$$

Trimethylphosphinimine Me_3PNH, and its phenyl analogue, have been widely used to prepare small linear and cyclic hetero compounds. They readily form crystalline adducts with trimethyl-aluminium, -gallium and -indium which eliminate methane on heating, leaving the thermally-stable 4-membered heterocycle,[180] *e.g.*,

$$Me_3PNH \xrightarrow{Me_3Al} Me_3PNH{\cdot}AlMe_3 \xrightarrow{\text{heat}} [Me_2Al(\mu\text{-}N{=}PMe_3)]_2$$

Me_3PNH (m.p. 59–60°C, b.p. 70°C/1 mm); $Me_3PNH{\cdot}AlMe_3$ (m.p. 44–6°C, (decomposed)); $Me_3P{=}N$–$AlMe_2$ ring dimer (m.p. 129–131°C)

This particular example also results by the longer process involving the *N*-silylphosphimine and an aluminium halide.

$$Me_3PNH \xrightarrow[2.\ Me_3SiCl]{1.\ RLi} Me_3PNSiMe_3 \xrightarrow{AlX_3} Me_3PNSiMe_3\cdot AlX_3$$

$$\xrightarrow[heat]{-Me_3SiX} Me_3P{=}N\langle AlX_2 \rangle_2 N{=}PMe_3 \xrightarrow{4MeLi} Me_3P{=}N\langle AlMe_2 \rangle_2 N{=}PMe_3$$

Diazidodimethylsilane reacts step-wise with trimethylphosphine. The de-nitrogenated adduct complexes with Me_3M (M = Al, Ga, In), and the p.m.r. spectra of these compounds show the presence of rapid interchange between the two co-ordination positions.[181]

$$Me_2Si(N_3)_2 \xrightarrow{Me_3P} Me_2Si(N{=}PMe_3)_2 \xrightarrow{Me_3M} Me_3P{=}N\langle SiMe_2,\ MMe_3 \rangle N{=}PMe_3$$

The three compounds are crystalline solids (m.p. 93–5°C, (Al); 80–2°C, (Ga); 74–6°C, (In)) which show singlets in the p.m.r. spectra for all but the P—Me groups (doublet-split by ^{31}P) at 30°C. With the first two compounds, this peak again splits at −60°C into two doublets. This has been taken to show that the metal is not in a 5-co-ordinate environment but in a rapidly changing 4-co-ordinate one, which is too slow at −60°C to cause coalescence of the peaks representing the different Me_3P groups. The indium compound shows no splitting, supporting 5-co-ordination or very rapid interchange.

An excess of the acceptor does not give a double complex but causes ionization,[182]

$$Me_2Si(NPMe_3)_2 \xrightarrow{2Me_3M} [Me_3P{=}N\langle SiMe_2,\ MMe_2 \rangle N{=}PMe_3]^+\ MMe_4^- \qquad (M = Al, Ga)$$

while the sequence below shows aluminium to be the better acceptor as it is more electropositive.

$$\begin{array}{c} PMe_3 \\ \| \\ N \\ Me_2Si \quad AlMe_3 \\ N \\ \| \\ PMe_3 \end{array} \xrightarrow{GaMe_3} \begin{array}{c} Me_3 \\ P \\ \| \\ N \\ Me_2Si \; (+) \; AlMe_2 \\ N \\ \| \\ P \\ Me_3 \end{array} \; GaMe_4^- \xleftarrow{AlMe_3} \begin{array}{c} PMe_3 \\ \| \\ N \\ Me_3Ga \quad SiMe_2 \\ N \\ \| \\ PMe_3 \end{array}$$

All are crystalline solids which decompose on melting at about 136°C.

With excess Me_3In, the di-adduct (m.p. 84–6°C) readily forms, while with the 1 : 1 Me_3Ga complex the mixed di-adduct m.p. 82–4°C results.

$$\begin{array}{c} Me_3 \\ P \\ \| \\ N \\ Me_2Si \quad InMe_3 \\ N \\ \| \\ P \\ Me_3 \end{array} \xrightarrow{Me_3Ga} \begin{array}{c} Me_2 \\ Si \\ Me_3P{=}N \quad N{=}PMe_3 \\ \downarrow \qquad \downarrow \\ In \qquad Ga \\ Me_3 \quad Me_3 \end{array} \xleftarrow{Me_3In} \begin{array}{c} Me_3 \\ P \\ \| \\ N \\ Me_3Ga \quad SiMe_2 \\ N \\ \| \\ P \\ Me_3 \end{array}$$

These structures are supported by the p.m.r. spectra, which show the ionic compounds to have a MeSi absorption at a lower field than with the di-adduct. Similarly, the Me_2M group absorbs at a lower field than Me_4M^-.

CYCLIC PHOSPHORUS AND ARSENIC DERIVATIVES OF THE GROUP III ELEMENTS

From the discussion on cyclic nitrogen derivatives of the Group III elements, it appears that p_π–p_π bonding is sufficient to stabilize boron in certain 3-co-ordinate systems. Such bonding is weaker in bonds between the heavier elements of both Groups III and V, so the classes of compound encountered are generally 4- or 5-co-ordinate. Thus, there are no boron–phosphorus analogues of the borazines.

The three main methods of synthesizing these compounds involve elimination of HX from $R_2PH{\cdot}BX_3$,[183] R′H from MR_3-adducts of phosphines and arsines,[184] and hydrogen from $R_2PH{\cdot}BH_3$ or $R_2AsH{\cdot}BH_3$.[185]

Hydrogen halide elimination from diphenylphosphine–boron trihalide adducts yields dimeric phosphinoboranes.

$$2Ph_2PH{\cdot}BX_3 + 2Et_3N \longrightarrow [Ph_2PBX_2]_2 + Et_3NHX\ (X = \text{Br or I})$$

The iodide is hydrolytically stable, and does not react with alcohols, amines, methyl iodide, N_2O_4 or a phenyl Grignard reagent. The bromide, however, yields tris-aminoboranes with amines, and also slowly hydrolyzes.

Though no chloride was isolated by this method, diethylphosphinoboron dichloride dimer results from the reaction of boron trichloride on phosphinosilanes[186] or Et_2PLi.[187] Aluminium chloride reacts similarly.

$$\underset{\text{(b.p. 106°C/10 mm)}}{2Me_3SiPBu^n_2} + 2BBr_3 \longrightarrow [Bu^n_2PBBr_2]_2 + 2Me_3SiBr$$

$$[Et_2PAlCl_2]_3 \xleftarrow{AlCl_3} Et_2PLi \xrightarrow{BCl_3} \underset{\text{(m.p. 136–8°C)}}{[Et_2PBCl_2]_2}$$

The oligomerization of these phosphinoboranes is unnecessary if the boron is bonded to strong π-donors as well. Thus, Et_2PLi reacts with both $(Me_2N)_2BCl$ and Et_2NBCl_2 to give monomeric aminated phosphinoboranes. Unlike the oligomers, these compounds readily form phosphonium salts, and are cleaved at the P—B bond by alcohols and amines.[187]

$$(Me_2N)_2BPMeEt_2^+I^- \xleftarrow{MeI} \underset{\text{(b.p. 134°C/53 mm)}}{(Me_2N)_2BPEt_2} \xleftarrow{(Me_2N)_2BCl} Et_2PLi \xrightarrow{Et_2NBCl_2} \underset{\text{(b.p. 79°C/0.001 mm)}}{Et_2NB(PEt_2)_2} \xrightarrow{HCl} Et_2NB(Cl)PEt_2$$

$$(Me_2N)_2BPEt_2 \xrightarrow{ROH\ \text{or}\ R_2NH} Et_2PH$$

Trimethylalane, -gallane and -indane form adducts with diphenylamine, -phosphine and -arsine. All the alane and gallane adducts are stable at room temperature, but decompose on warming to yield methane and bridged dimers, $[Me_2M^{III}M^{V}Ph_2]_2$, *e.g.*,

$$2Me_3Ga{\cdot}HAsPh_2 \xrightarrow{30°C} [Me_2GaAsPh_2]_2 + 2CH_4$$

Triethylalane and diphenylphosphine, lithium diethylphosphide and chlorodiethylalane form similar dimers, while it is reported that the tetramer results from chlorodimethylalane and sodium diphenylphosphide.

This tendency to form dimers has been attributed to the entropy of translation being larger for the smaller rings, so tending to overcome the valence angle strain present in the small rings. Such strain is less with heavier atoms, while steric hindrance is also reduced in the small ring oligomers.

All the dimers are sensitive to moisture, the aluminium ones enflaming in air and forming adducts with trimethylamine.

$$[Me_2AlAsPh_2]_2 + 2Me_3N \underset{70°C\ vac.}{\rightleftarrows} 2Me_2AlAsPh_2{\cdot}NMe_3$$

This contrasts with the inertness of the boron compounds, where co-ordinative saturation prevented easy nucleophilic attack.

Phosphinoboranes have attracted much attention through their inherent inertness and high thermal stability. Such polymers were first synthesized by pyrolyzing the borane adducts of dimethylphosphine. Hydrogen bromide elimination has also been used. Borane reduces Me_3SiPEt_2 to the trimer, while halides give the less-hindered dimers, or linear polymers.

$$Me_2PH{\cdot}BH_3 \xrightarrow{150°C} 3H_2 + [Me_2PBH_2]_3$$

$$3Me_2PH{\cdot}Me_2BBr \xrightarrow{3Et_3N} [Me_2PBMe_2]_3 + 3Et_3NHBr$$

$$3Me_3SiPEt_2{\cdot}BH_3 \longrightarrow 3Me_3SiH + [Et_2PBH_2]_3$$

The borano–dimethylarsine adduct loses hydrogen steadily at 100°C to yield mainly trimeric dimethylarsinoborane. A small amount of polymer also forms, which decomposes into trimer and tetramer, the latter slowly ring-contracting to the trimer at 180°C. Both are less stable than their phosphorus analogues but nevertheless, can be recrystallized on the open bench.

$$3Me_2AsH{\cdot}BH_3 \xrightarrow{100°C} [Me_2AsBH_2]_3$$

The inherent stability of these phosphinoboranes has led to wide industrial interest through their potential as precursors to thermoplastic materials." This has led to ingenious large-scale synthetic routes not involving air-sensitive phosphines and diborane. Good yields of the trimer result from the BH_4^- reduction of both dimethylphosphonyl chloride and tetramethyldiphosphine disulphide.[188]

$$Me_2POCl \xrightarrow{BH_4^-} [Me_2PBH_2]_3 \xleftarrow{BH_4^-} Me_2P(S)P(S)Me_2$$

Small yields of thermally-stable polymer result on pyrolysis of these cyclic oligomers, but better yields result by pyrolyzing the tetramethyl-diphosphine–borane adduct.

The stability of these compounds has provided an interesting problem in bonding, for while the majority of monomeric phosphine–borane adducts are reactive, these cyclic oligomers resist air, acids, bases and water, hydrolyzing only slowly at 300°C, in the presence of HCl. So they provide the most stable B—H bonds, and are not cleaved by excess diborane, unlike Me_2NBH_2, which gives the aminodiborane.[189]

This has been accounted for by considering the hydrogen atoms of the BH_2 group to be less hydridic than usual, due to hyperconjugation with the residual electrons on boron forming a π-bond with the 3d/4s orbitals of phosphorus. This idea is supported by the structures of $[Me_2PBH_2]_{3\text{ and }4}$ which show wide angles at the boron and phosphorus atoms, in support of a higher contribution of s character to the ring σ-bonds from boron.[190] The B—P bonds (0.193 nm) are shorter than in the cyclic dimer $[Ph_2PBI_2]_2$ (0.200 nm).[191]

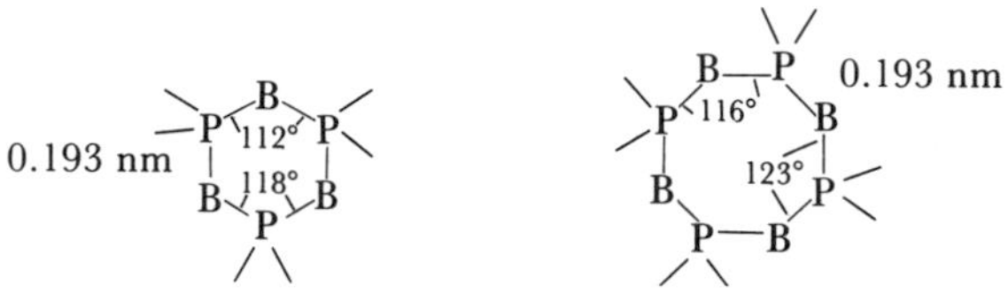

Fig. 3.55. The structure of cyclic phosphinoboranes

Accordingly, the absence of this π-bonding should nullify any tendency to oligomerize in $(CF_3)_2PBH_2$ and $(CF_3)_2AsBH_2$, since trifluoromethyl phosphines and arsines are very weak donors. However, the electronegative CF_3 groups should contract the lowest unoccupied orbitals of phosphorus and arsenic, and hence assist π-bonding.

$$(CF_3)_2PH + B_2H_6 \longrightarrow \underset{\text{(m.p. 30.5°C)}}{[(CF_3)_2PBH_2]_3} \text{ and } \underset{\text{(m.p. 116°C)}}{[(CF_3)_2PBH_2]_4}$$

$$(CF_3)_2AsH + B_2H_6 \longrightarrow \underset{\text{(m.p. 3°C, decomposes above 40°C)}}{[(CF_3)_2AsBH_2]_3}$$

These compounds exist for both phosphorus and arsenic, but decompose on heating into BF_3 products, though not until 200°C with the phosphorus compounds. This supports the expectation that the 4d orbitals of arsenic would be less effective in π-bonding than the 3d orbitals of phosphorus.

In examining possible organic reactions applicable to the phosphinoborane trimers, attempts to alkylate the boron by a Friedel–Crafts reaction in an analogous way to B_5H_9 and $B_{10}H_{14}$ resulted in halogenation.

$$[RR'PBH_2]_3 + 6R''Cl \xrightarrow{AlCl_3} [RR'PBCl_2]_3 + 6R''H$$

(R'' = Me, $HCCl_2$, CCl_3, or Bu^n)

Step-wise iodination can be accomplished as indicated.

$$[Me_2PBH_2]_3 + 3CH_2I_2 \longrightarrow [Me_2PBHI]_3 + 3MeI \longrightarrow [Me_2PBI_2]_3 + 3CH_4$$

N-Chloro- and *N*-bromosuccinimides also halogenate the ring, giving smaller yields, while HF is required for the fluorinated trimer.[188, 192]

$$[Me_2PBH_2]_3 + 6HF \longrightarrow [Me_2PBF_2]_3 + 3H_2$$

Aluminium bromide catalyzed trans-halogenations readily take place, chloroform quantitatively converting the hexabromocycloboraphane into its chloro analogue.

$$[Me_2PBBr_2]_3 + 6CHCl_3 \xrightarrow{AlBr_3} [Me_2PBCl_2]_3 + 6CHCl_2Br$$

Again, the logical extension of these compounds incorporates the phosphides, arsenides and antimonides of the Group III elements (boron antimonide is not known). All adopt the zinc blende structure based on the adamantane framework, and like cubic boron nitride.

CYCLIC GROUP III OXYGEN COMPOUNDS

Borates

On a planet enveloped by an atmosphere containing oxygen, many non-metals exist naturally as salts of their oxy-anions. This is particularly so with the elements boron, aluminium and silicon, which form many condensed cyclic anionic polymers comprising 4- and 6-membered rings.[c]

Oxy-anions of boron are many and varied in number, providing an interesting structural facet of Geochemistry. The metaborate ion represents the simplest structure, and is present in the crystalline form of metaboric acid, the $B_3O_6^{3-}$ ions being hydrogen-bonded into layer structures.

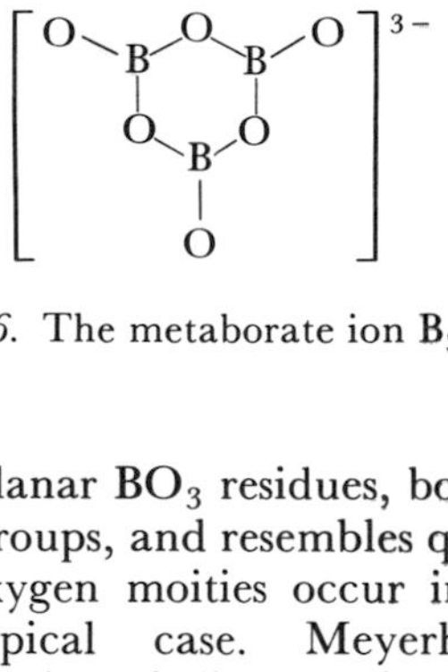

Fig. 3.56. The metaborate ion $B_3O_6^{3-}$

While this ion comprises planar BO_3 residues, boron trioxide contains an array of tetrahedral BO_4 groups, and resembles quartz.

Both of these boron–oxygen moities occur in complex anions, with $KH_4B_5O_{10}{\cdot}2H_2O$ a typical case. Meyerhofferite and inyoite, $CaB_3O_3(OH)_5{\cdot}xH_2O$ ($x = 1$ and 4), contain two BO_4 and one BO_3 residues, while borax, though represented as a $B_4O_7^{2-}$ derivative, is better formulated as $Na_2[B_4O_5(OH)_4]{\cdot}8H_2O$, with the anion containing two BO_4 and two BO_3 residues. (See Figure 3.57.) The $B_4O_7^{2-}$ ion is unknown in boron chemistry. Bond lengths of 0.136 nm and 0.148 nm have been obtained for the B—O bonds of the planar BO_3 and tetrahedral BO_4 moities, indicating strong π-bonding in the former.

Though cyclic aluminate chemistry is virtually non-existent, many

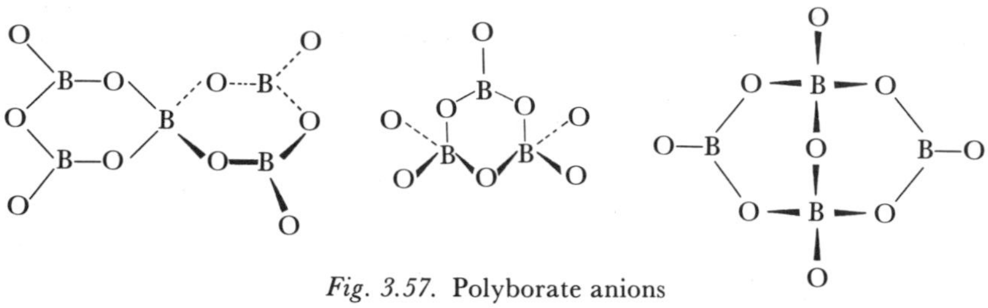

Fig. 3.57. Polyborate anions

complex ions are known in which B^- and Al^- ions replace Si in a polysilicate framework, and this will be discussed in the next Chapter.

As with the Group V elements just considered, the tendency of the Group III elements to form strong $p_\pi - p_\pi$ bonds is limited to boron and oxygen, so 4-, 5- and 6- co-ordination again is characteristic of the oxygen derivatives of the heavier elements, while analogues of boroxine $[XBO]_3$ are unknown.

Boroxine and its derivatives

Boroxines are formulated as $[XBO]_3$, and possess a 6-membered ring. The parent boroxine $H_3B_3O_3$ can be prepared by the high-temperature hydrolysis of boron, and by exploding oxygen with B_5H_9 or diborane(6).[193]

$$\underset{}{B_2H_6} + O_2 \longrightarrow \underset{0.32}{H_3B_3O_3} + \underset{2.8}{H_2} + \underset{0.4\ (\text{moles})}{B_2O_3}$$

Although this compound is thermodynamically unstable at room temperature with respect to B_2H_6 and boric oxide, nevertheless it can be characterized,

$$H_3B_3O_3 + \tfrac{1}{2}O_2 \longrightarrow H_2B_2O_3 + \tfrac{1}{6}B_2H_6 + \tfrac{1}{3}B_2O_3$$

and the molecular structure shows it to be planar, with ring angles of 120 degrees and the B—O bond length 0.1375 nm, indicating π-bonding.[194]

Lewis bases such as PF_3 or CO give boric oxide and the borane adduct,

$$H_3B_3O_3 + \text{base} \longrightarrow H_3B{\cdot}\text{base} + B_2O_3$$

while in the presence of BH_4^- *B*-trimethylboroxine forms in high yield.[195]

$$3B_2H_6 + 6CO \xrightarrow{BH_4^-} \underset{(\text{b.p. } 80°\text{C})}{2[MeBO]_3}$$

These boroxines are usually prepared by dehydrating the corresponding boronic acids, either azeotropically with benzene, or at low pressure.[o]

$$RBX_2 \longrightarrow RB(OH)_2 \xrightarrow{-H_2O} [RBO]_3$$

boron and phosphorus halides

They also result from the oxidative hydrolysis and dehydration of trialkylboranes. With the mesityl compound, however, hindrance appears so large that only the 4-membered ring size was isolated. The pyrolysis of alkylborinic anhydrides $(R_2B)_2O$, boronic esters $RB(OR')_2$, acyl boronates or trialkylborane/boric oxide mixtures all yield symmetrical boroxines. Assymmetrical ones result from the condensation of acyl boronic anhydrides and boronic esters but the products disproportionate on heating into the symmetrical boroxines.

$$[MeCOB(Ph)]_2O + Pr^nB(OBu^n)_2 \xrightarrow{-Bu^nOAc} \text{(PhB)}_2\text{(Pr}^n\text{B)O}_3 \text{ ring} \xrightarrow{\text{heat}} [PhBO]_3 + [Pr^nBO]_3.$$

Hydrolysis and alcoholysis occur readily, while alkyl group migration with aluminium alkyls yields trialkyl boranes and oxidation gives the alkoxyboroxines.

$$RB(OR')_2 \xleftarrow{R'OH} [RBO]_3 \xrightarrow{R_3Al} R_3B + Al_2O_3$$

$$[RBO]_3 \xrightarrow{O_2} [ROBO]_3 \qquad [RBO]_3 \xrightarrow{\text{B or P halides}} RBX_2 + B_2O_3 \text{ or P oxides.}$$

Alkoxyboroxines are generally prepared, however, by azeotropically refluxing equimolar quantities of an alcohol with boric and metaboric acids, or from the alcohol and boric oxide. Both this latter method, and heating triethyl borate with boric oxide, were employed over 100 years ago.

$$(EtO)_3B + B_2O_3 \longrightarrow [EtOBO]_3$$

Hydrolysis of trimethylborate yields the methoxyboroxine, while the re-distribution reaction of a borate/boron tribromide mixture, followed by disproportionation gives the alkoxyboroxine.

$$9(MeO)_3B + 3H_2O \longrightarrow [MeOBO]_3 + 6(MeO)_3B\cdot MeOH$$

$$2(MeO)_3B + BBr_3 \longrightarrow 3[(MeO)_2BBr] \longrightarrow [MeOBO]_3 + MeBr$$

Hindered metaborates can be prepared by trans-esterification, through the removal of the low boiling alcohol.

$$\text{R-C}_6\text{H}_2\text{(Bu}^t)_2\text{-OH} + [Bu^nOBO]_3 \xrightarrow{\text{toluene}} [\text{R-C}_6\text{H}_2\text{(Bu}^t)_2\text{-OBO}]_3 + 3Bu^nOH.$$

Again, hydrolysis and alcoholysis yield borates and boric acid, while pyrolysis forms not only orthoborate and boric oxide but olefin also. Unless the orthoborate is removed from the reaction zone, however, no decomposition occurs; trialkylsilylmetaborates survive at 250°C for 50 h in a sealed tube.

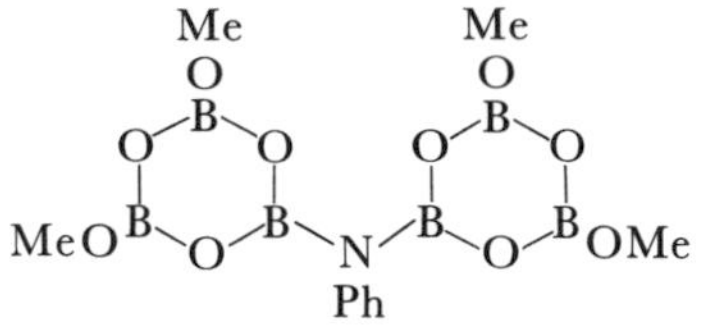

Me Me
O O
B B
O O O O
MeO B O B N B O B OMe
Ph

Fig. 3.58. Condensed boroxines

Aniline displaces two moles of methanol and bridges two boroxine rings, while carboxylic acids, anhydrides and acid chlorides produce organic esters. Hydrogen and covalent halides (boron, phosphorus or sulphur) all give the alkyl halide, while BF_3 and BCl_3 yield alkoxyboron dihalides.

Alkylation occurs with aluminium-, zinc- and magnesium-alkyls, while ammonia produces a polyborate complex which decomposes to the turbostratic boron nitride mentioned earlier.[196]

Halogenoboroxines and alkoxyboron halides

The oxides of aluminium, silicon, titanium and boron all react with BF_3 at 450°C to yield cyclic fluoroboroxine.

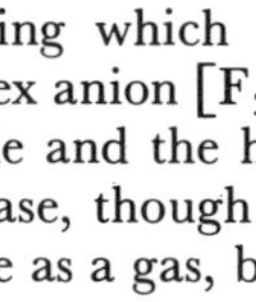

F B O B F
O O
B
F

Fig. 3.59. Fluoroboroxine

It comprises a 6-membered ring which readily decomposes, and with fluoride ion, it forms the complex anion $[F_2BO]_3^{3-}$. A similar chloride and bromide result from boric oxide and the halide. These are also unstable, especially in the condensed phase, though trichloroboroxine prepared by a photo-ionized process is stable as a gas, but decomposes on glass.

$$BCl_3 \xrightarrow{h\nu} BCl \xrightarrow{O_2} BClO_2 + BCl \longrightarrow BOCl \longrightarrow [BClO]_3$$

The aminoboroxine resulting from B_2O_3 and tris-dimethylaminoborane at 260°C is thermally stable, but with primary amines gives alkylaminoboroxines which readily de-aminate.

$$(Me_2N)_3B + B_2O_3 \longrightarrow [Me_2NBO]_3 \xrightarrow{RNH_2} [RNHBO]_3 \xrightarrow[-RNH_2]{heat} [RNB_2O_2]_x$$

As with B—N compounds, multiple-bonding contributes to the electronic saturation of boron in many boron–oxygen rings. Cases of oligomerization are known, however, especially among the halogeno derivatives.

Alkoxyboron difluorides result from alcohols or orthoborates as indicated.[o]

$$ROH \xrightarrow{BF_3} [ROBF_2]_n \xleftarrow{BF_3} (RO)_3B$$

They can also be conveniently synthesized from tin alkoxides and boron halides.[197]

$$R'_2Sn(OR)_2 + 6BX_3 \xrightarrow{X = Cl} 6ROBCl_2 + 3R'_2SnX_2$$

$$R'_2Sn(OR)_2 + 6BX_3 \xrightarrow[X = F]{} 2[ROBF_2]_3 + 3R'_2SnX_2$$

The chlorides are monomeric, while the fluorides are trimers,

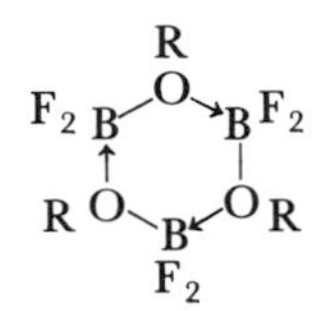

Fig. 3.60. Trimeric $ROBF_2$

and both readily complex with pyridine.

$$[ROBF_2]_3 + 3\text{pyridine} \longrightarrow 3ROBF_2\cdot\text{pyridine}$$

Bis-alkoxyfluoroboranes are monomeric and thermally unstable, while 2-chloro-1,3,2-dioxaborolane (b.p. 74°C/1 mm) is dimeric though oxygen bridging has not been established.

Among the notable reactions of these compounds are their ready hydrolysis to boric acid, thermal decomposition and disproportionation, the latter predominating at low temperatures.

$$3ROBCl_2 \longrightarrow (RO)_2BCl \longrightarrow (RO)_3B$$

$$3ROBCl_2 \longrightarrow 3RCl + 3ClBO \longrightarrow B_2O_3 + BCl_3$$

The tendency to form oligomers is more marked with the heavier elements. Not only does this occur by the oligomerization processes encountered for boron, but through the formation of the cube structures already encountered among Al—N compounds.

Oxygen derivatives of the post-boron elements

Trimethyl derivatives of aluminium, gallium and indium react with alcohols, yielding dimeric and trimeric alkoxides. As with the Group V derivatives of these elements, steric hindrance encourages smaller ring formation, and with aluminium only the methyl-substituted alanes are trimers.[p]

$$ROH + R'_3Al \longrightarrow \frac{1}{n}[R'_2AlOR]_n + R'H$$

(n = 3 when R or R′ is methyl, except when R is Bu^t)

Phenol initially gives dimeric $[PhOAlMe_2]_2$ in toluene, but the molecular weight and p.m.r. spectrum shows this to be unstable relative to the trimer, which slowly forms on standing.[198]

An interesting situation arises with the derivative of 2-ethoxyethanol. No chelation occurs, but an oxygen-bridged dimer is formed.

$$2Et_3Al + 2EtO(CH_2)_2OH \longrightarrow EtO(CH_2)_2O\!\!<\!\!\begin{matrix}AlEt_2\\AlEt_2\end{matrix}\!\!>\!\!O(CH_2)_2OEt$$

and not $Et_2Al\!<\!\!\begin{matrix}O\!-\!CH_2\\ O(Et)\!-\!CH_2\end{matrix}$ (five-membered chelate ring)

This has been attributed to the electropositive nature of aluminium enhancing the donor ability of the oxygen bonded directly to it. 2-Dimethylamino-ethanol forms an oxygen-bridged dimer with trimethylgallane. Though alkoxy-bridged dimers of aluminium and gallium are stable to trimethylamine, aryloxy, and acyl, sulphur- and selenium-bridged ones are all cleaved, some reversibly, *e.g.*,

$$\underset{\text{(m.p. 132°C)}}{[PhOGaMe_2]_2} \underset{\text{vac}}{\overset{2Me_3N}{\rightleftharpoons}} \underset{\text{(m.p. 39°C)}}{Me_3N\cdot GaMe_2OPh}$$

The gallium compounds slowly hydrolyze to the hydroxide $[Me_2GaOH]_4$. This is tetrameric in the solid state but trimeric in solution.

While indium behaves very similarly, the existence of two stable valency states for thallium adds interest to the structural aspects of the chemistry. The thallic alkoxides are dimeric, the fully-methylated compound having a remarkably high melting point (179°C), since the ethoxy compound is a liquid, (b.p. 110–120°C/15 mm) (similar features occur with the thallous and aluminium compounds).

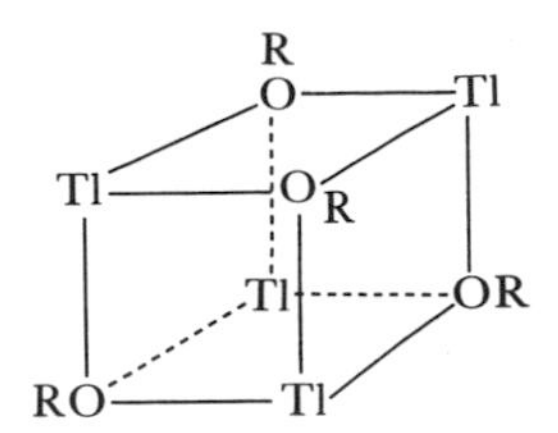

Fig. 3.61. Thallous alkoxides

They are synthesized from thallous alkylates and dimethylthallic bromide,

$$2Me_2TlBr + 2TlOR \longrightarrow [Me_2TlOR]_2 + 2TlBr$$

and are readily hydrolyzed, the thallium content being conveniently titrable.

Thallous alkoxides result from air-oxidized alcoholysis of thallium metal. They are tetrameric, with a distorted cube structure analogous to that of $[Me_3SiOZnMe]_4$ and $[Me_3SiOCdMe]_4$, and hydrolyze reversibly.

A wide variety of siloxy and germanoxy derivatives of these elements have been studied and all are dimeric with a 4-membered ring.

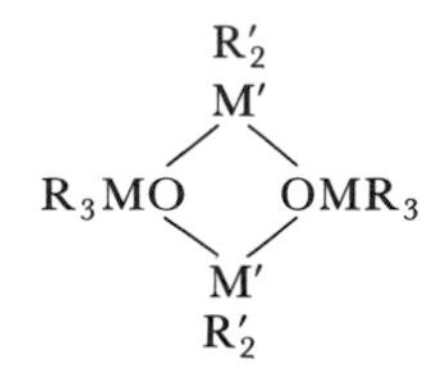

Fig. 3.62. Organo-oxy derivatives of the Group III elements

Those of aluminium, gallium and indium are known, the methyl derivatives being low-melting sublimable solids and the phenyl compounds high-melting solids.

These compounds are usually synthesized from the silanol, germanol or an alkali metal derivative, and either complexed R_3M' or $R_2M'X$,[199] *e.g.*,

$$2Me_3SiOH + 2Et_2O{\cdot}InMe_3 \xrightarrow[-2CH_4]{-2Et_2O} [Me_2InOSiMe_3]_2 \quad \text{(m.p. 16°C, b.p. 57°C/3 mm)}$$

$$2Me_3SiONa + 2Me_2TlCl \xrightarrow{-2NaCl} [Me_2TlOSiMe_3]_2 \quad \text{(m.p. 98–100°C)}$$

$$2Me_3GeOLi + 2Me_2GaCl \longrightarrow [Me_2GaOGeMe_3]_2 \quad \text{(m.p. 42–4°C, b.p. 84°C/1 mm)}$$

$$2Ph_3GeOH + 2Ph_3Ga \xrightarrow{-2C_6H_6} [Ph_2GaOGePh_3]_2 \quad \text{(m.p. 291–3°C)}$$

The analogous dihalogeno compounds result using the Group III trihalide which can be reduced or alkylated readily.[200]

$$2R_3SiONa + 2AlX_3 \longrightarrow [R_3SiOAlX_2]_2 \xrightarrow{4RLi} [R_3SiOAlR_2]_2$$

The structure of $[Me_3SiOAlBr_2]_2$ shows that π-bonding between from oxygen to silicon is unlikely as the bond length is close to that of a single bond.[201]

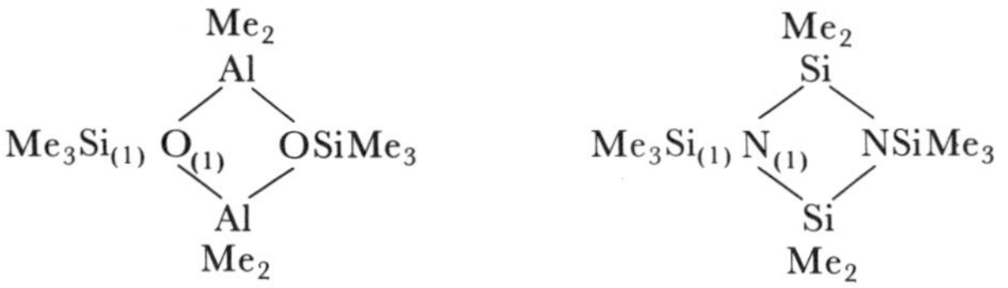

Me_2 Al; $Me_3Si_{(1)}$ $O_{(1)}$ $OSiMe_3$; Al Me_2 — Me_2 Si; $Me_3Si_{(1)}$ $N_{(1)}$ $NSiMe_3$; Si Me_2

Fig. 3.63. Isostructural $[Me_3SiOAlMe_2]_2$ and $[Me_3SiNSiMe_2]_2$

This contrasts markedly with isostructural $[Me_3SiNSiMe_2]_2$ where π-bonding probably occurs in the extra-annular Si—N bond with the angle C—$Si_{(1)}$—$N_{(1)}$ greater (112 degrees) and C—$Si_{(1)}$—$O_{(1)}$ less (106 degrees) than the tetrahedral angle.

Calcium indate $CaIn_2O_4$ is prepared by heating $CaCO_3$ and indium trioxide. This has the calcium ferrite structure AB_2O_4 typified by $CaFe_2O_4$, along with Eu_3O_4 and MSc_2O_4 (M is Ca, Mg).[202] The structure comprises infinite chains of edge-shared octahedra. Consequently, they comprise 4-membered rings, and are thought to exist for most of the rare earth metals. The In—O bridge distances are a little larger than the Sc—O ones (0.21–0.22 nm).

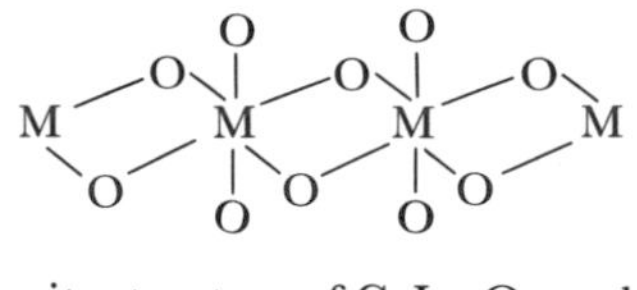

Fig. 3.64. The ferrite structure of $CaIn_2O_4$ and rare earth oxides

Boron, aluminium and gallium can be completely silanolated using an excess of Me_3SiONa.[203]

$$MCl_3 + 3Me_3SiONa \longrightarrow (Me_3SiO)_3M + 3NaCl$$
$$(M = B, Al, Ga)$$

While the latter two are crystalline solids with high melting and boiling points (m.p. 208°C for the gallium compound, b.p. 155°C/1 mm for the aluminium) the boron compound is a liquid (b.p. 34°C/1 mm). Molecular weight determinations show the solids to be dimeric, and the p.m.r. spectra show two types of Me_3Si group. The boron compound is probably stabilized as a monomer through π-bonding. This is supported by its reluctance compared with the other two compounds to react with excess Me_3SiO^-.

$$[(Me_3SiO)_3Al]_2 + 2Me_3SiO^- \longrightarrow 2[(Me_3SiO)_4Al]^-$$

These siloxy compounds are fairly stable thermally, but will decompose above 250°C into the oxide and hexamethyldisiloxane.

$$[(Me_3SiO)_3Al]_2 \xrightarrow{280°C} 3(Me_3Si)_2O + Al_2O_3$$

While trialkylborates are volatile monomers, the analogues of the heavier Group III elements are oligomeric,[q] either as dimers, trimers or tetramers. The isoatomic carbon counterparts of the siloxy compounds just considered, namely $(Bu^tO)_3M$, (M = Al, Ga), are also dimeric, but decreasing the hindrance of the substituent helps to stabilize higher oligomers. Thus, in non-polar solvents the alkoxides in the following list have the molecular state indicated in parentheses:[204] $Al(OPr^i)_3$ (4) $Ga(OPr^i)_3$ (2) $Al(OCH_2Ph)_3$ (4) $Al(OCH_2CCl_3)_3$ (2) $Ga(OR)_3$ (4) (R = Et, Pr^n, Bu^n), $Ga(OMe)_3$ (polymer).

The higher degree of association of the isopropoxyaluminium compound in comparison with its gallium analogue is thought to be due to the more polar nature of the Al—O bond rendering association easier.

In polar solvents, however, solvent co-ordination occurs with a reduction in association. Thus, all alkoxy-alanes are less associated in donor solvents, though the only monomers in such solvents are $(Cl_3CCH_2O)_3Al$ (dioxan) and $(PhCH_2O)_3Al$ (pyridine).[205] The facile solvation of the 2,2,2-trichloroethyl compound reflects the strongly electrophilic nature of aluminium, and this compound, unlike the non-chlorinated alkoxy-aluminium compounds, will insert CH_2 from diazomethane into the Al—O bond.[206]

$$CH_2N_2 \xrightarrow{Al—OPr^i} \frac{1}{n}(CH_2)_n + N_2 \text{ only}$$

$$Cl_3CCH_2OAl + CH_2N_2 \longrightarrow Cl_3CCH_2OCH_2Al + N_2 + (CH_2)_n$$

The Meerwein–Ponndorf method for reducing aldehydes and ketones uses aluminium isopropoxide,[207] while secondary alcohols can be conveniently oxidized to the ketone with the t-butoxide (Oppenauer method).

$$RR'CO + Al(OPr^i)_3 \rightleftharpoons Me_2CO + (RR'CHO)_3Al \xrightarrow{H^+} RR'CHOH$$

$$\underset{(excess)}{RR'CHOH + Me_2CO} \xrightleftharpoons{Al(OBu^t)_3} RR'CO + Me_2CHOH$$

The benzyl compounds thermally decompose at 300°C into the dibenzyl ether and further dissociation products.[208]

$$[(PhCH_2O)_3Al]_4 \xrightarrow{300°C} 2(PhCH_2)_2O + 2[(PhCH_2O)_2Al]_2O \longrightarrow$$

These compounds are tetrameric and the p.m.r. spectra show two methylene peaks of equal intensity. This supports a structure involving aluminium in two different co-ordination states as shown.

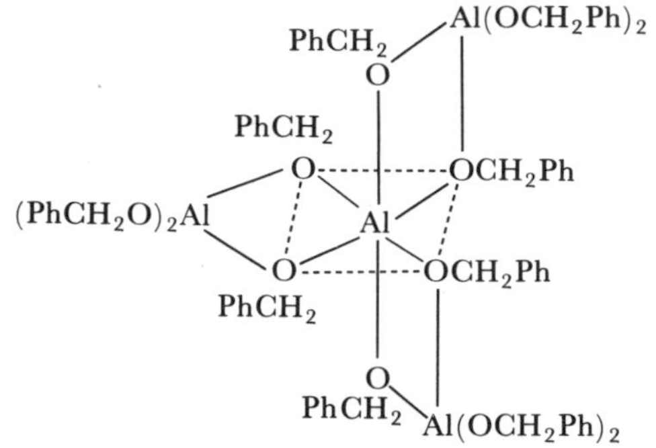

Fig. 3.65. The structure of $[(PhCH_2O)_3Al]_4$

The low-field peak is split into an AB pattern typical of non-equivalent protons on the group bonded to the 6-co-ordinate aluminium. Using a 220 MHz, n.m.r. spectrometer resolves the AB structure of the other methylene groups, showing that each proton of the methylene group is in a different magnetic environment.[209]

Non-equivalent methyl groups occur in $[(Pr^iO)_3Al]_4$, while $[(Cl_3CCH_2O)_3Al]_2$ shows two peaks as expected for the two kinds of methylene group. No splitting could be observed due to proton non-equivalence.

The isopropanol complexes of lanthanum and praseodmium both react with potassium aluminium isopropoxide. The products $M[Al(OPr^i)_4]_3$ melt at about 100°C and can be vacuum-distilled at about 200°C. Both probably possess a structure analogous to the benzyl compound just discussed, with the rare earth metal in the 6-co-ordinate site.[210]

Both boron and aluminium trihalides cleave the Si—O bond in linear siloxanes, to give R_3SiOMX_2.[r]

$$(R_3Si)_2O + MX_3 \longrightarrow R_3SiOMX_2 + R_3SiX$$
$$(X = Cl, Br, I)$$

The products are all monomeric with boron, but the aluminium compounds again are dimeric,[211] and possess the usual 4-membered ring.[212] The gallium analogues cannot be made the same way since $GaCl_3$ cleaves the Si—C bond. Thus, with hexamethyldisiloxane, polymeric dimethylsiloxane forms.

$$(Me_3Si)_2O + GaCl_3 \longrightarrow MeGaCl_2 + Me_3SiCl + \frac{1}{n}[Me_2SiO]_n$$

Consequently a process involving a lower activation energy has to be employed, namely the silanolate.[213]

$$2Me_3SiO^- + 2GaCl_3 \xrightarrow{-2Cl^-} [Me_3SiOGaCl_2]_2$$

Dialkylpoly- and -cyclosiloxanes react with the boron halides to give a mixture of products including $(R_2XSiO)_3B$, B_2O_3, $R_2XSiOBX_2$, $(R_2XSi)_2O$ *etc.*[214] While gallium trichloride yields polymeric methylchlorosiloxane with $[Me_2SiO]_4$,

$$[Me_2SiO]_4 + 4GaCl_3 \longrightarrow 4MeGaCl_2 + \frac{4}{n}[MeClSiO]_n$$

the aluminium halides give a product analyzing for $(R_2Si)_4O_6Al_3X_5$ (X = Cl, Br). This also results from the polysiloxane and $[Me_3SiOAlCl_2]_2$.[215] The chloro compound readily sublimes and melts at 147–150°C.

$$3Me_3SiOAlCl_2 + [Me_2SiO]_4 \longrightarrow (Me_2Si)_4O_6Al_3Cl_5 + Me_3SiCl + (Me_3Si)_2O$$

The ethyl compound can be made similarly, and structural determinations show the presence of 4-membered Al_2O_2 rings and 6-membered $AlSi_2O_3$ ones, with aluminium 4- and 5-co-ordinate in this case, the latter state being rare for the element. The five groups around the aluminium atom could be represented by a distorted trigonal-bipyramid or square-pyramid.[216]

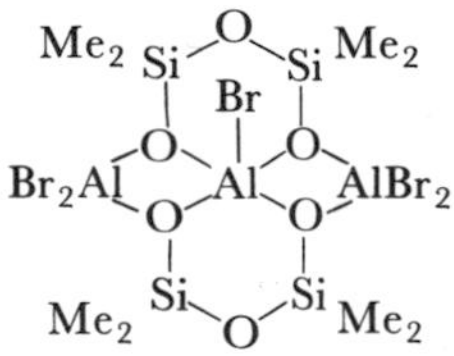

Fig. 3.66. The structure of $(Me_2Si)_4O_6Al_3Br_5$

The Si—O bonds incorporating the oxygen atoms involved in 4-membered rings are significantly larger (0.171 nm) (*cf.* 0.170 nm in $[Br_2AlOSiMe_3]_2$) than the others of the 6-membered rings only (0.162 nm). These compounds are the volatile precursors to the mineral aluminosilicates to be discussed in the next Chapter.

Miscellaneous oxyanion complexes

Many compounds are potential bidentate ligands and will react with the trialkyl derivatives of the Group III elements. These include sulphur dioxide, dimethylsulphoximine $Me_2S(O)NH$, diphenylphosphinic and dimethylarsonic acids. The products are formed either by addition of the metal–carbon bond across S=O or by alkane elimination, and are dimeric with puckered 8-membered rings.[217]

$$Et_3Ga \xrightarrow[-50°C]{SO_2} Et_2Ga(OS(Et)O)_2GaEt_2 \xrightarrow{HCl} [EtClGaO_2SEt]_2$$

(b.p. 111°C/3 mm)

$$Me_3Tl \xrightarrow{SO_2} [Me_2TlOS(Me)O]_2 \text{ (dimethylthallium ethylsulphinate)}$$

$$2Me_2S(O)NH + 2Me_3M \longrightarrow [Me_2MONSMe_2]_2$$

$$2Ph_2PO_2H + 2R_3Al \longrightarrow [R_2AlO_2PPh_2]_2$$

$$2Me_2AsO_2H + 2Me_3M \longrightarrow [Me_2MO_2AsMe_2]_2$$

By way of contrast, while monothiophosphinic acids give dimeric products, dithiophosphinic acids give monomers involving the 4-membered MS_2P ring, with the acid functioning as a bidentate chelate.

$$Me_2PS_2H + Me_3M \longrightarrow Me_2M(S)_2PMe_2 + CH_4$$

SULPHUR AND SELENIUM DERIVATIVES OF GROUP III

As with the oxygen derivatives of this group, only boron forms predominantly 3-co-ordinate compounds.[m, o, s] The other elements tend to form oligomeric 4-co-ordinate derivatives.

Boron sulphide B_2S_3 can be made by various methods, but is usually amorphous. These include roasting boric oxide and Al_2S_3, heating boron with sulphur or H_2S, and the thermal decomposition of $[HSBS]_2$ or $[HSBS]_3$. Crystalline samples do result, however, from the gentle pyrolysis of the silylthioborate, formed from BCl_3 and excess hexamethyldisilthiane.

$$BCl_3 + 6(Me_3Si)_2S \xrightarrow{-3Me_3SiCl} (Me_3SiS)_3B \longrightarrow \tfrac{1}{2}B_2S_3 + \tfrac{3}{2}(Me_3Si)_2S$$

Heating gallic oxide and an alkali metal carbonate with H_2S at 900°C forms the anion $Ga_2S_4^{2-}$. Indium and thallium form similar compounds, and the structures may well resemble that of $CaIn_2O_4$ already considered, in which 4-membered rings featured (p. 120).

Many cyclic derivatives involving boron–sulphur rings have now been synthesized. Thio analogues of the boroxines are normally used as precursors. Thiolysis of boron tribromide, or trisulphide yields $[HSBS]_2$.

Ring-expansion occurs at 90°C, and the HS substituents can be replaced using substituted boron compounds.

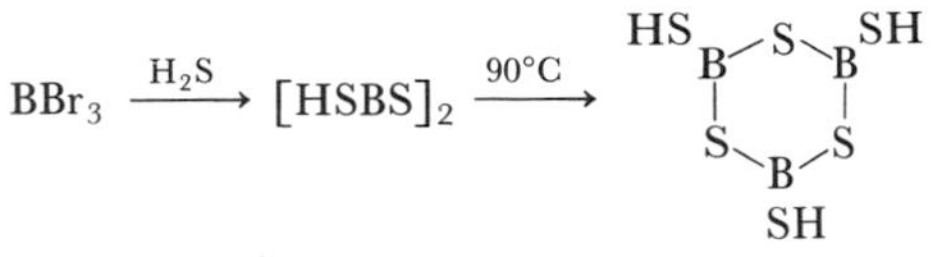

$$BBr_3 \xrightarrow{H_2S} [HSBS]_2 \xrightarrow{90°C} [HSBS]_3$$

(m.p. 142°C)

$$2[HSBS]_3 + 2BX_3 \longrightarrow 2[XBS]_3 + B_2S_3 + 3H_2S$$

(X = OMe, Me_2N, Cl, Br, Me)

$$2[HSBS]_3 + 2B(SEt)_3 \longrightarrow 3[EtSBS]_2 + B_2S_3 + 3H_2S$$

(m.p. 4°C)

The methyl compound can be oxidized to the methoxy ring, while the iodide and fluoride are formed using HI and AlF_3.

$$[MeBS]_4 \xleftarrow{56°C} [MeBS]_3 \xrightarrow{oxid.^n} [MeOBS]_3$$

$[MeBS]_3$ (m.p. 60°C); $[MeOBS]_3$ (m.p. 27.5°C)

$$[FBS]_3 \xleftarrow{AlF_3} [HSBS]_3 \xrightarrow{3HI} [IBS]_3$$

$[IBS]_3$ (m.p. 145°C)

The high stability of these compounds compared with similar oxygen compounds may well reflect much weaker π-bonding in the B—S bond. Indeed, the B—S bond length (0.185 nm) in $[BrBS]_3$ corresponds to the sum of the covalent radii.

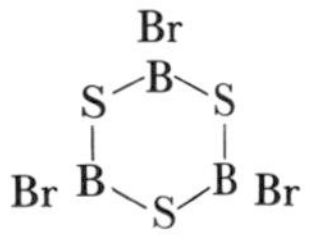

Fig. 3.67. Bromoborthiin

Direct aminolysis of the bromoborthiin is hindered by salt formation.

$$[BrBS]_3 + 3MeNH_2 \longrightarrow [MeNH_2BS]_3^{3+}3Br^-$$

The amino groups in the ring are readily quaternized. Hydrogen chloride readily cleaves the B—S bond before the B—N one, a sequence seen before with aminophosphinoboranes.

With sulphur, the bromoborthiin ring contacts to the 1,3,5-trithiadiborolane.[218]

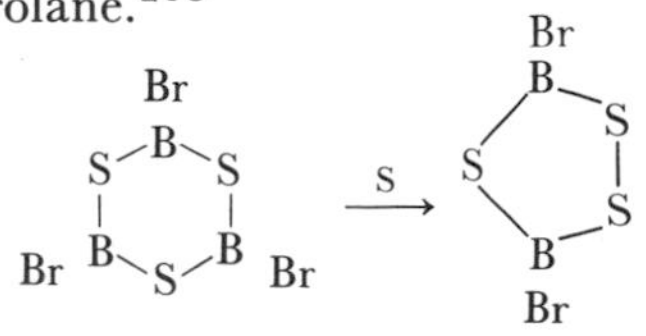

$$[BrBS]_3 \xrightarrow{S} Br_2B_2S_3$$

(b.p. 50°C/1.5 mm)

This 5-membered ring is thermally very stable but sensitive to oxygen and many protic compounds. Its facile synthesis by the routes indicated below has led to a variety of reactions including simple substitution and redox ones. The three halogeno derivatives result using hydrogen disulphide, and it is peculiar that the B—S—B link does not arise from H_2S as an intermediate, according to the p.m.r. spectrum of the reactants.[219]

$$BX_3 + H_2S_2 \xrightarrow{CS_2} \text{XB(S}_3\text{)BX (cyclic)} \xrightarrow{BI_3} \text{IB(S}_3\text{)BI (cyclic)}$$

(X = Cl, Br).

(X = Cl, b.p. 67°C/12 mm) (m.p. 27°C)

$$PhBCl_2 + H_2S_2 \xrightarrow{CS_2} \text{PhB(S}_3\text{)BPh (cyclic)}$$

While thiols are oxidized readily by iodine in aqueous solution, the process is reversible under anhydrous conditions. Consequently iodoboranes, polysulphides and sulphur can be used to prepare B—S compounds.[220]

$$2RS{-}SR + 2R'_2BI \rightleftharpoons 2R'_2BSR + I_2$$

$$2PhBI_2 + \tfrac{3}{8}S_8 \longrightarrow \text{PhB(S}_3\text{)BPh (cyclic)} + 2I_2$$

(m.p. 115–6°C)

$$BI_3 + \tfrac{3}{8}S_8 \xrightarrow[21°C]{CS_2} \text{IB(S}_3\text{)BI (cyclic)} \xrightarrow{BI_3} \text{(IBS)}_3 + I_2$$

$$\text{IB(S}_3\text{)BI (cyclic)} \xrightarrow{R_2S_2} \text{RSB(S}_3\text{)BSR (cyclic)}$$

The other boron halides are less electrophilic so H—S compounds have to be used. Phenylboron dibromide is a stronger Lewis acid than BBr_3, and with sulphur gives initially the *B*-phenyltrithiodiboralane but the bromine generated through the redox reaction cleaves the B—C bonds. Other products also form.

$$PhBBr_2 + \tfrac{3}{8}S_8 \longrightarrow \underset{\text{S—S}}{\text{Ph B}^{\diagup S \diagdown}\text{B Ph}} \xrightarrow{Br_2} \underset{\text{S—S}}{\text{BrB}^{\diagup S \diagdown}\text{BBr}}$$

Selenium will also oxidize the boron–iodine bond to iodine.[221] This 1,3,5-triselenadiboralane ring is oxidized by dimethyldisulphide not only at the B—I bonds but at the B—Se ones also, to give trimethylthioborate.

$$2PhBI_2 + 3Se \longrightarrow \underset{\text{Se—Se}}{\text{Ph B}^{\diagup Se \diagdown}\text{B Ph}} \quad \text{(m.p. 122–4°C)}$$

$$2BI_3 + 3Se \longrightarrow \underset{\text{Se—Se}}{\text{IB}^{\diagup Se \diagdown}\text{BI}} \quad \text{(m.p. 72°C; b.p. 125°C/0.15 mm)}$$

$$\downarrow Me_2S_2$$

$$(MeS)_3B + Se + I_2$$

Heating borane adducts of trimethylamine and triethylamine with H_2S at 200°C gives two thermally stable B—S ring compounds.[222]

$$R_3N{\cdot}BH_3 + H_2S \xrightarrow{200°C} \frac{1}{n}[R_2NBS]_n + RH + 2H_2$$

$$(R = Et, \text{ b.p. } 65°C/2 \text{ mm}) \quad (n = 2, R = Et; n = 3, R = Me)$$

The less-hindered compound is trimeric, and both are sublimable solids. $[Et_2NBS]_2$ is one of the rare examples of the dithiadiboralane ring, but unlike $[HSBS]_2$ it is highly stable thermally. Attempts to prepare this ring starting from a 4-membered cyclodisilthiane proved futile, the 6-membered borthiin ring being formed.[223]

$$6PhBCl_2 + 3[Me_2SiS]_2 \longrightarrow 2[PhBS]_3 \longleftarrow 2[Me_2SiS]_3 + 6PhBCl_2$$

These 3-co-ordinate boron–sulphur compounds are all sensitive to moisture and protonated compounds in general. This is similar to the reactivity of oligomeric $[RSBH_2]_n$ compounds, and contrasts with the phosphinoboranes which are stable to such reagents.

Diborane forms thermally unstable adducts with thiols. The polymers so formed yield trimers on heating, which are readily distillable liquids bridged through sulphur.[224, 225]

$$RSH{\cdot}BH_3 \xrightarrow{-H_2} [RSBH_2]_n \longrightarrow [RSBH_2]_3$$

$$(R = Me, Et, Pr^n, Bu^n)$$

An excess of the mercaptan gives dimeric $(RS)_2BH$ and eventually trithioborates, while excess diborane gives the mixed bridged methylthiodiborane, with a structure involving both methylthio and hydride bridges,[225] as observed with $Me_2NB_2H_5$.

$$[MeSBH_2]_x + B_2H_6 \xrightarrow{90°C} MeSB_2H_5$$

$$(\text{m.p. } -100°C, \text{ b.p. } 53°C \text{ (extrapolated)})$$

These compounds readily hydrolyze, and the trimeric thioboranes can be depolymerized with amines.

$$[MeSBH_2]_x + xMe_3N \longrightarrow xMeSBH_2{\cdot}NMe_3 \xrightarrow{B_2H_6} MeSB_2H_5 + Me_3N{\cdot}BH_3$$

Alkylthiodihalogenoboranes readily form from the trihalide and the thiol or disulphide (with BI_3).[226]

$$2RSH{\cdot}BX_3 \longrightarrow [RSBX_2]_2 \longleftarrow 2BI_3 + R_2S_2$$

The products are dimeric if substituents are small, and the oligomeric nature is normally reflected in their being solids, for example $[EtSBBr_2]_2$ (m.p. 108°C),[218] $[MeSBI_2]_2$ (m.p. 148–9°C), Bu^n_2BSMe (b.p. 27°C/1 mm), $PhSBI_2$, b.p. 90°/0.1 mm). In this respect they resemble the aminoboron dihalides.

Heating $MeSBI_2$ to 200°C causes dissociation into iodomethane and the iodoborthiin,

$$3[MeSBI_2]_2 \longrightarrow 2[IBS]_3 + 6MeI$$

while dimethyldisulphide is reduced to the thioborate.

The selenium compound studied to date, $MeSeBBr_2$, exhibits monomer–dimer equilibrium.

$$2MeSeBBr_2 \rightleftharpoons Br_2B(\mu\text{-}SeMe)_2BBr_2$$

Normally, 4-co-ordinate compounds result from trimethylaluminium dimer and a thiol. These are dimeric in solution, but the methyl compound is polymeric in the solid state.[p]

$$2Me_3M + 2MeSH \longrightarrow [Me_2MSMe]_2$$
$$(M = Al, Ga, In)$$

They are readily dissociated by stronger bases such as amines. Selenium compounds react similarly.

$$[Me_2AlSMe]_2 + 2Me_3N \underset{\text{vacuum}}{\rightleftharpoons} 2Me_2Al(SMe)NMe_3$$

Diethylaluminium and gallium azides are trimeric with azide bridges. This bridging occurs through the nitrogen atom already bonded as it is a better nucleophile on an electropositive metal. This structure (D_{3h}) is supported by infrared data, and $[Et_2MSCN]_3$ is analogous. This is formed from Et_3M (M = Al, Ga, In) and thiocyanogen.[227]

$$3[Et_3M]_2 + 6(SCN)_2 \longrightarrow 2[Et_2MSCN]_3 + 6EtSCN$$

The high C—N and low C—S vibrational frequencies support the resonance structure indicated, while the aluminium compound readily disproportionates above 180°C.

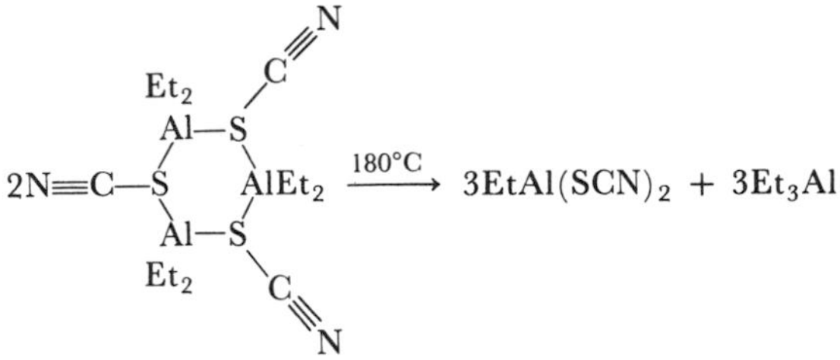

$$2\,N{\equiv}C{-}S\,(Et_2Al{-}S{-}C{\equiv}N)(Et_2Al{-}S{-}C{\equiv}N)AlEt_2 \xrightarrow{180^\circ C} 3EtAl(SCN)_2 + 3Et_3Al$$

CYCLIC GROUP III HALOGEN COMPOUNDS

The trihalides of all five Group III elements are known, apart from the tri-iodide of thallium(III): coincidentally, TlI_3 is the thallium(I) derivative of the tri-iodide ion I_3^-. Those of boron are monomers, while the chlorides, bromides and iodides of the heavier elements are dimeric, with bridging halogen atoms.[228] Reasons for the monomeric nature of the boron halides are still a little confusing. The small size of boron and the subsequent difficulty in bridging, has been proposed as a reason, but the ions BX_4^- ($X = F \rightarrow I$) are all known. Stabilization through π-bonding from the halogens has also been invoked, but there is evidence for such bonding even in the bridging halide atoms of the aluminium halides. Perhaps the most plausible explanation would be that as boron is more electronegative than the other Group III elements, the M^{III}—X bond polarity is smaller for boron. This will reduce the tendency to dimerize.

Most of the bridging halides of aluminium, gallium and indium are dimeric in the solid and vapour phase, though aluminium tri-iodide is almost completely dissociated in the vapour. The bridge is readily broken. An interesting comparison of relative bridge stability occurs with ethylethoxyaluminium chloride, which at room temperature comprises 90% ethoxy-bridged trimer. This ring contracts to the chloro-bridged dimer on heating.[229]

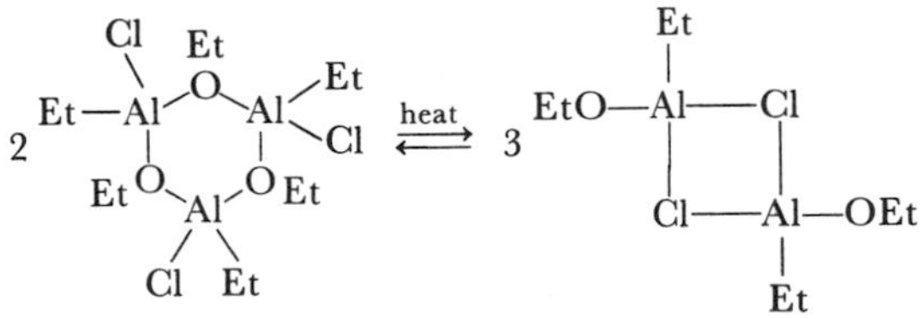

$$2\,[EtAl(Cl)(\mu\text{-}OEt)]_3 \underset{}{\overset{\text{heat}}{\rightleftharpoons}} 3\,[EtAl(OEt)(\mu\text{-}Cl)]_2$$

It is interesting to compare this situation with that of the alkoxygallium halides. These result by the routes indicated, and are dimeric and may involve alkoxy bridging.

$$[MeGaCl_2]_2 + 2MeOH \longrightarrow [MeOGaCl_2]_2 + 2CH_4$$

$$\underset{\text{(polymer)}}{2(MeO)_3Ga} + 4MeCOCl \longrightarrow 2[MeOGaCl_2]_2 + 4MeOCOMe$$

While $(Pr^iO)_2GaX$ (X = Cl, Br) are trimeric, alkylalkoxygallium halides are tetrameric in solution but dimeric in the vapour, in support of an alkoxy-bridged cube structure which cleaves on vaporization.[230]

$$2[Me_2GaCl]_2 + MeOH \longrightarrow [Me(MeO)GaCl]_4 \longrightarrow$$

Me Ga Cl
MeO OMe
Ga
Cl Me

Dimethylgallium hydroxide $[Me_2GaOH]_4$ is reported to have an 8-membered ring structure, while the dihalogenogallium hydroxides which result from the alcohol–trihalide complex are high melting tetramers which may well possess the cube structure, with hydroxide bridges.

$$4ROH{\cdot}GaX_3 \xrightarrow{>120°C} [X_2GaOH]_4 + 4RX \ (R = Me, Bu^n)$$

(X = Cl, m.p. 120°C;
X = Br, m.p. 175°C)

The halogen can also influence the co-ordination number of the Group III element. Only aluminium and gallium form hexahalide anions, namely the fluorides. Indium forms $InCl_6^{3-}$, and thallium $Tl_2Cl_9^{3-}$, comprising two face-bonded octahedra. Gallium forms the ion $Ga_3Cl_{10}^-$ in which the gallium atoms are thought to be present in two co-ordination states as shown.[231]

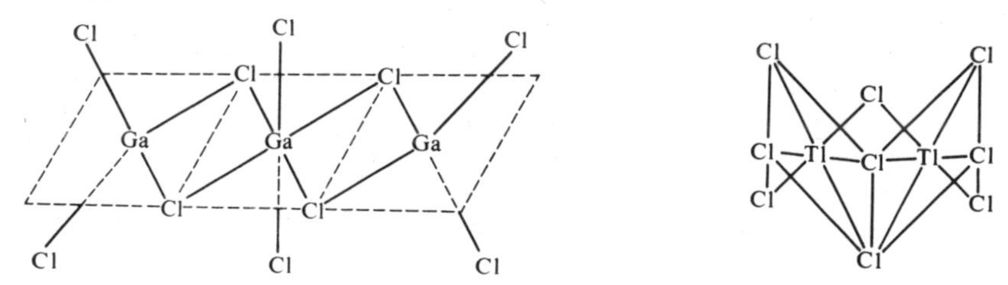

Fig. 3.68. The proposed structures of $Ga_3Cl_{10}^-$ and $Tl_2Cl_9^{3-}$

B-Halogenated borolanes and boraindanes[f]

The nature of organohalogen compounds of boron especially proves to be most exceptional, for while simple alkyl and aryl derivatives are monomeric, a peculiar monomer/oligomer/polymer equilibrium exists for the

B-halogenoborolanes and boraindanes. Their synthesis from *B*-alkyl derivatives is catalyzed by borohydrides, *e.g.*,

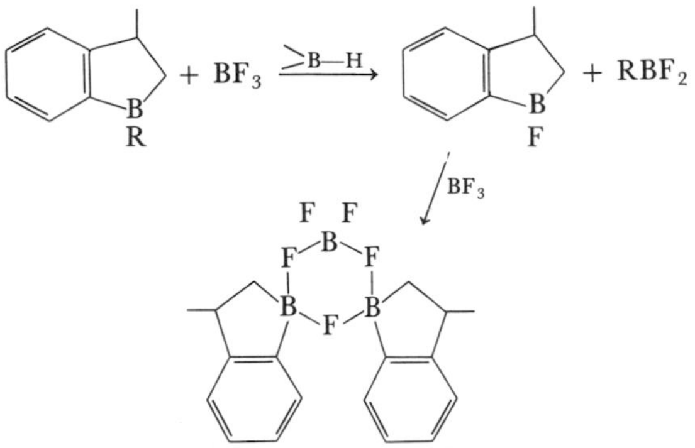

$$\text{B-R boraindane} + BF_3 \xrightarrow{>B-H} \text{B-F boraindane} + RBF_2$$

$$\downarrow BF_3$$

Alkylboron difluorides can be distilled out, leaving the associated fluoroborolanes or boraindanes as dimers or trimers, or complexed with BF_3.

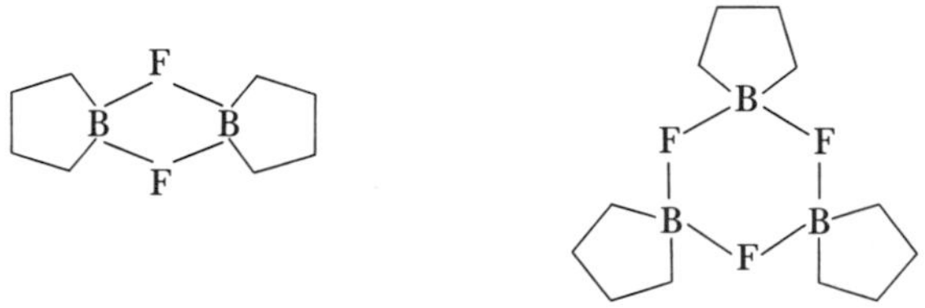

Fig. 3.69. Dimeric and trimeric fluoroborolane

B-Fluoroborolane forms an interesting equilibrium mixture with excess BF_3, illustrating symmetrical and unsymmetrical bridging.

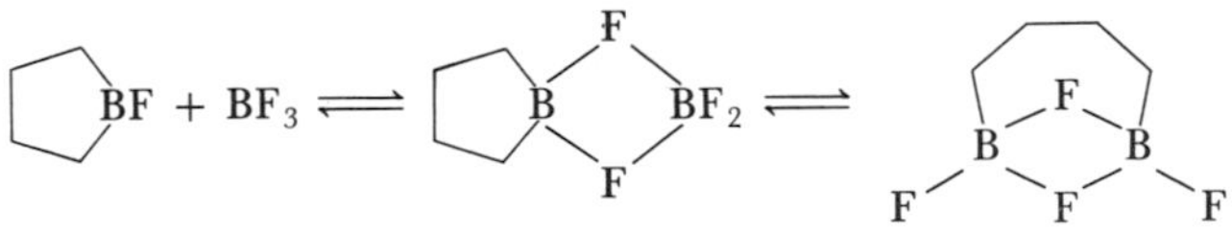

$$\text{BF} + BF_3 \rightleftharpoons \text{B(F)}_2\text{BF}_2 \rightleftharpoons$$

The *B*-chloroborolanes can be prepared similarly, this monomer/polymer exchange occurring at lower temperatures. The parent compound can be distilled at 96°C at atmospheric pressure, but crystallizes on standing, while the 3-methyl homologue distils at 110°C and becomes viscous on cooling, due to polymeric ring and chain compounds. Depolymerization occurs on heating, or in the presence of B—H compounds, while reduction with lithium aluminium hydride forms the hydride-bridged bis-borolane.

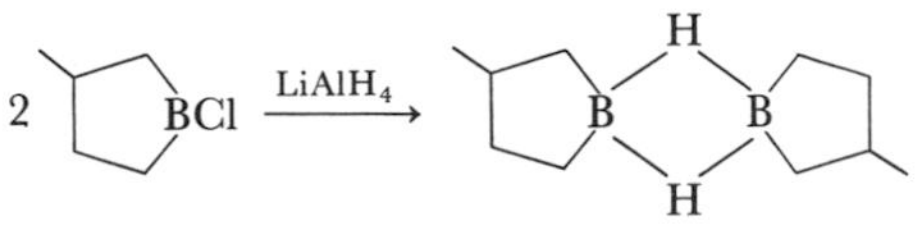

$$2\ \text{BCl} \xrightarrow{LiAlH_4} \text{B(H)}_2\text{B}$$

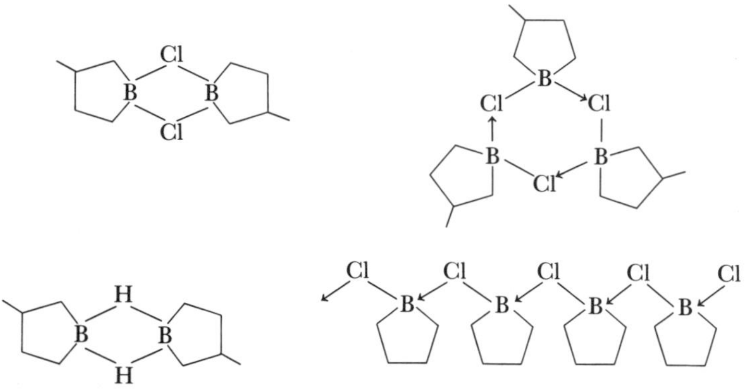

Fig. 3.70. Chloroborolanes

Both alkyl and aryl chlorides of aluminium, gallium and indium are dimeric, as are the bromides. Less obvious synthetic routes include heating triethyl indium with chloroform and bromoform,[232] and methylating gallium trichloride using methyl–silicon compounds.

$$\underset{\text{(m.p. 168°C)}}{[Et_2InBr]_2} \xleftarrow{CHBr_3} Et_3In \xrightarrow{CHCl_3} \underset{\text{(m.p. 202°C)}}{[Et_2InCl]_2}$$

This reaction has already been mentioned in connection with the synthesis of siloxygallanes.[233] Tetramethylsilane readily yielded methylgallium dichloride with gallium trichloride while trimethylsilane reduced the latter to dimeric dichlorogallane, with no evidence of $[MeGaCl_2]_2$ or dimethylchlorosilane.

$$[MeGaCl_2]_2 \xleftarrow{Me_4Si} [GaCl_3]_2 \xrightarrow{Me_3SiH}$$

$$\underset{\text{(m.p. 29°C)}}{[HGaCl_2]_2} \xrightarrow{30°C} H_2 + Ga^{I}Ga^{III}Cl_4$$

Dichlorogallane decomposes to hydrogen and "gallium dichloride", $Ga^{I}Ga^{III}Cl_4$, and probably contains a chloride bridge which, like the trihalides, is readily cleaved by amines. The Ga—H bond readily adds to olefins and the carbonyl group.

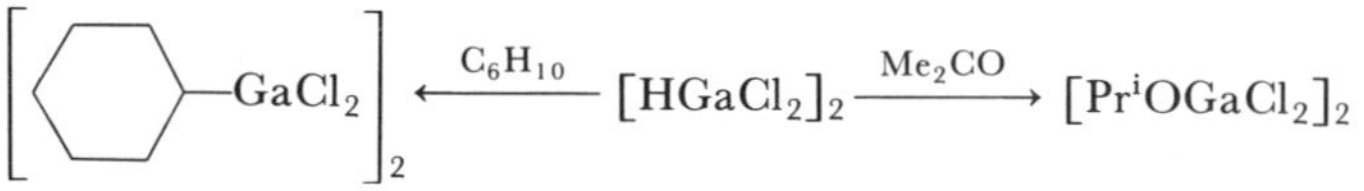

$$[C_6H_{11}\text{—}GaCl_2]_2 \xleftarrow{C_6H_{10}} [HGaCl_2]_2 \xrightarrow{Me_2CO} [Pr^{i}OGaCl_2]_2$$

Trimethylethylsilane and trimethylphenylsilane give a mixture of products with gallium trichloride, but yields show that the phenyl group is preferentially transferred (Me_4Si and Et_4Si give yields of $[RGaCl_2]_2$ in excess of 90%).

$$\underset{(72\%)}{[MeGaCl_2]_2} \xleftarrow{EtSiMe_3} [GaCl_3]_2 \xrightarrow{PhSiMe_3} \underset{(64\%)}{[PhGaCl_2]_2}$$

Gallium tribromide reacts similarly, while tetramethylgermane is more reactive than the silane (Ge—C bond is weaker).

So far no mention has been made of the organofluoro derivatives of these elements. Those of aluminium can be made by pyrolyzing the dimeric dimethyl chloride with sodium fluoride, or heating the adduct formed between trimethylsilicon fluoride and the trialkylalane.[234]

$$2[Me_2AlCl]_2 \xrightarrow{NaF} [Me_2AlF]_4 \quad \text{(b.p. 98°C/80 mm)}$$

$$Me_3SiF + AlEt_3 \longrightarrow Me_3SiFAlEt_3 \xrightarrow{35°C} [Et_2AlF]_4 \quad \text{(b.p. 55°C/0.016 mm)}$$

While the methyl and ethyl compounds are fluoride-bridged tetramers, the n-propyl and isobutyl compounds are trimeric.

While fluorosilanes will not fluorinate trialkylgallanes, polymeric trimethyltin fluoride will, as will boron trifluoride.[235] The products appear to be trimeric but on standing the methyl homologue ring expands to the tetramer.

$$Et_3Ga + \frac{1}{x}[Me_3SnF]_x \longrightarrow \tfrac{1}{3}[Et_2GaF]_3 + Me_3SnEt \quad \text{(b.p. 92°C/1 mm)}$$

$$Me_3Ga{\cdot}OEt_2 + Et_2O{\cdot}BF_3 \longrightarrow Me_3B + 2Et_2O + \tfrac{1}{3}[Me_2GaF]_3$$

$$[Me_2GaF]_3 \xrightarrow[10°C]{3\ weeks} [Me_2GaF]_4$$

(m.p. 20–2°C, 75°C/20 mm) (m.p. 27.5°C)

The bridging in these compounds is strong enough to withstand ether cleavage, unlike the other halides, so the reaction can be carried out in this solvent.

Although thallium trihalides readily lose one molecule of halogen, and dialkyl- and diaryl-thallium halides readily ionize instead of forming bridged dimers, strongly electronegative substituents will stabilize Tl^{III} in this manner. They result from thallic chloride and pentafluorophenylmagnesium bromide.[236]

$$2TlCl_3 + 4C_6F_5MgBr \longrightarrow [(C_6F_5)_2TlBr]_2 \xrightarrow[AgX]{MeOH} [(C_6F_5)_2TlX]_2 \quad (X = Cl, F)$$

The three halides are dimeric, with halogen bridges which are readily broken by acetone, phosphines and arsines. The relative instability of

Tl^{III} renders these compounds useful agents for oxidative trans-pentafluorophenylations. This applies to Hg^{I}, Sn^{II}, Ni^{0}, Pd^{0}, Pt^{0}, Ph^{I}, Ir^{I} and Pd^{II}, *e.g.*,

$$2SnCl_2 + [(C_6F_5)_2TlBr]_2 \longrightarrow 2(C_6F_5)_2SnCl_2 + 2TlBr$$

$$(Ph_3P)_2Ni(CO)_2 + [(C_6F_5)_2TlBr]_2 \longrightarrow (Ph_3P)_2Ni(C_6F_5)_2 + 2TlBr + 2CO$$

Dicobalt octacarbonyl and indous bromide form an insertion product using THF as the solvent. This is dimeric and probably contains a bromide bridge.[237]

$$2Co_2(CO)_8 + 2InBr \longrightarrow 2[(CO)_4Co]_2InBr{\cdot}THF \longrightarrow$$

Br

$[(CO)_4Co]_2In$ $In[Co(CO)_4]_2$

Br

In benzene, however, the compound $In_3Br_3Co_4(CO)_{15}$ forms, which possesses a 6-membered $[InBr]_3$ ring with $Co(CO)_4$ residues bonded to each indium atom and $Co(CO)_3$ straddling the ring.

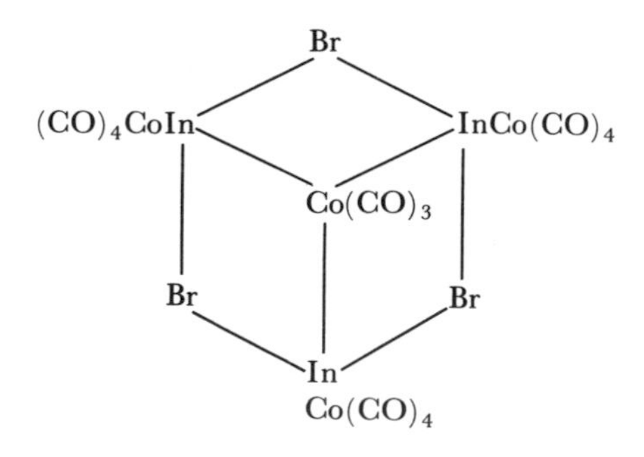

Fig. 3.71. The proposed structure of $In_3Br_3Co_4(CO)_{15}$

Bibliography

General

a The Nomenclature of Boron Compounds, *Inorg. Chem.*, 1968, **7,** 1945.

b 'Boron, Metallo–Boron Compounds and Boranes', ed. R. M. ADAMS, Interscience, New York, 1964.

c A. F. WELLS, 'Structural Inorganic Chemistry', Oxford University Press, 3rd Edn., 1962, p. 820; B. ARONSSON, T. LUNDSTRÖM and S. RUNDQUIST, 'Borides, Silicides and Phosphides', Methuen and Co. Ltd., London, 1965.

d A. STOCK, 'Hydrides of Boron and Silicon', Cornell University Press, Ithaca, New York, 1933, Vol. 12.

e H. C. BROWN, 'Hydroboration', W. A. Benjamin Inc., New York, 1962.

f R. KÖSTER, *Adv. Organometallic Chem.*, 1964, **2,** 257.
g T. ONAK, *ibid.*, 1965, **3,** 263 Academic Press, London and New York.
h 'Organometallic Chemistry', (A.C.S. Monograph No. 147), ed. H. ZEISS, Reinhold, New York, 1960.
i W. N. LIPSCOMB, 'Boron Hydrides', W. A. Benjamin Inc., New York, 1963; R. L. HUGHES, I. C. SMITH and E. W. LAWLESS, in 'Production of Boranes and Related Research', ed. R. T. HOLZMANN, Academic Press, New York, 1967; 'Chemistry of Boron and its Compounds', ed. E. L. MUETTERTIES, John Wiley and Sons Inc., New York, 1967; E. L. MUETTERTIES and W. H. KNOTH, 'Polyhedral Boranes', Marcel Dekker Inc., New York and Edward Arnold Ltd., London, 1968; G. R. EATON and W. N. LIPSCOMB, 'N.M.R. Studies of Boron Hydrides and related Compounds', W. A. Benjamin Inc., New York, 1969.
j M. L. H. GREEN, 'Organometallic Compounds', Methuen and Co. Ltd., London, 3rd Edn., Vol. II, 1967, p. 109.
k E. K. MELLON, JR. and J. J. LAGOWSKI, *Adv. Inorg. Chem. Radiochem.*, 1963, **5,** 259.
l H. R. ALLCOCK, 'Heteroatom Ring Systems and Polymers', Academic Press, New York, 1967.
m 'Chemistry of Boron and its Compounds', ed. E. L. MUETTERTIES, John Wiley and Sons, Inc., New York, 1967; H. STEINBERG and R. J. BROTHERTON, 'Organoboron Chemistry', Interscience, New York, 1966, Vol. 2.
n H. STEINBERG and R. J. BROTHERTON, 'Organoboron Chemistry', Interscience, New York, 1964, Vol. 2.
o H. STEINBERG, 'Organoboron Chemistry', Interscience, New York, 1964, Vol. 1.
p G. E. COATES, M. L. H. GREEN and K. WADE, 'Organometallic Compounds', Methuen and Co. Ltd., London, 3rd Edn., Vol. I, 1967.
q D. C. BRADLEY, *Progr. Inorg. Chem.*, 1960, **2,** 303.
r H. SCHMIDBAUR, *Angew. Chem. Internat. Edn.*, 1965, **4,** 201.
s R. H. CRAGG and M. F. LAPPERT, *Organometallic Chem. Rev.*, 1966, **1,** 43.

References

1 N. N. GREENWOOD, R. V. PARISH and P. THORNTON, *Quart. Rev.*, 1966, **20,** 441.
2 D. S. URCH, 'Orbitals and Symmetry', Penguin Press, Harmondsworth, 1970.
3 A. STOCK and C. MASSENEZ, *Ber.*, 1912, **45,** 3539; *Chem. Abs.*, 1913, **7,** 1334; A. STOCK and E. KUSS, *Ber.*, 1923, **56,** 789; *Chem. Abs.*, 1923, **17,** 2241.
4 H. J. EMELÉUS and J. S. ANDERSON, 'Modern Aspects of Inorganic Chemistry', Routledge and Kegan Paul, London, 1960; W. L. JOLLY, 'Synthetic Inorganic Chemistry', Prentice-Hall, Inc., New York, 1962, p. 158; G. F. FREEGUARD and L. H. LONG, *Chem. and Ind.*, 1965, 471.
5 B. J. DUKE, J. R. GILBERT and I. A. READ, *J. Chem. Soc.*, 1964, 540.
6 W. C. PRICE, *J. Chem. Phys.*, 1947, **15,** 614; 1948, **16,** 894; K. HEDBERG and V. SCHOMAKER, *J. Amer. Chem. Soc.*, 1951, **73,** 1482; L. S. BARTELL and B. L. CARROLL, *J. Chem. Phys.*, 1965, **42,** 1135.
7 J. KEELY JR., J. RAY and R. A. OGG JR., *Phys. Rev.*, 1954, **94,** 767; R. A. OGG JR., *J. Chem. Phys.*, 1954, **22,** 1933; J. N. SCHOOLERY, *Discuss. Faraday Soc.*, 1955, **19,** 215.
8 W. G. EVANS, C. E. HOLLOWAY, K. SUKUMARABANDHU and D. H. MCDANIEL, *Inorg. Chem.*, 1968, **7,** 1746.
9 H. I. SCHLESINGER and A. O. WALKER, *J. Amer. Chem. Soc.*, 1935, **57,** 621; H. I. SCHLESINGER, L. HORVITZ and A. B. BURG, *ibid.*, 1936, **58,** 407.
10 B. L. CARROLL and L. S. BARTELL, *Inorg. Chem.*, 1968, **7,** 219.
11 H. I. SCHLESINGER, N. W. FLODIN and A. B. BURG, *J. Amer. Chem. Soc.*, 1939, **61,** 1078; I. J. SOLOMON, M. J. KLEIN and K. HATTARI, *ibid.*, 1958, **80,** 4520.
12 B. M. MIKHAILOV, V. A. DOROKHOV, *Zhur. obshchei. Khim.*, 1961, **31,** 4020; E. WIBERG, J. E. F. EVANS and H. NÖTH, *Z. Naturforsch.*, 1958, **13b,** 263.

13 H. C. BROWN and co-workers, *J. Amer. Chem. Soc.*, 1959, **81,** 6423, 6428; 1960, **82,** 4708; 1961, **83,** 2544; 1962, **84,** 1478; 1963, **85,** 2066, 2072; 1964, **86,** 393; T. J. LOGAN and T. J. FLAUTT, *ibid.*, 1960, **82,** 3446; H. C. BROWN and G. J. KLENDER, *Inorg. Chem.*, 1962, **1,** 204.
14 R. KÖSTER, *Angew. Chem.*, 1960, **72,** 626; R. KÖSTER and K. IWASAKI, *Adv. Chem. Ser.*, 1964, **42,** 148.
15 L. H. LONG and M. G. H. WALLBRIDGE, *Chem. and Ind.*, 1959, 295; T. WARTIK and R. PEARSON, *J. Inorg. Nuclear Chem.*, 1958, **5,** 250; R. KÖSTER and G. BENEDIKT, *Angew. Chem. Internat. Edn.*, 1963, **2,** 219.
16 H. G. WEISS, W. J. LEHMAN and I. SHAPIRO, *J. Amer. Chem. Soc.*, 1962, **84,** 3840; H. H. LINDNER and T. ONAK, *ibid.*, 1966, **88,** 886; R. KÖSTER and G. BENEDIKT, *Angew. Chem. Internat. Edn.*, 1963, **2,** 323.
17 N. DAVIES, C. A. SMITH and M. G. H. WALLBRIDGE, *J. Chem. Soc. (A)*, 1970, 342; H. I. SCHLESINGER, H. C. BROWN and G. W. SCHAEFFER, *J. Amer. Chem. Soc.*, 1943, **65,** 1786.
18 T. WARTIK and H. I. SCHLESINGER, *ibid.*, 1953, **75,** 835.
19 K. ZIEGLER, W. R. KROLL, W. LARBIG and O. W. STEUDEL, *Annalen*, 1960, **629,** 53.
20 E. WIBERG and R. BAUER, *Z. Naturforsch.*, 1950, **5b,** 396, 397; L. H. ZAKHARKIN and I. M. KHORLINA, *Zhur. obshchei. Khim.*, 1963, **32,** 2783. *Chem. Abs.*, 1963, **58,** 9110g.
21 W. KOTLENSKY and R. J. SCHAEFFER, *J. Amer. Chem. Soc.*, 1958, **80,** 4517.
22 A. STOCK and E. POHLAND, *Ber.*, 1926, **59B,** 2223.
23 T. WARTIK, V. LINEVSKY and H. BOWKLEY, *Nucl. Sci. Abs.*, 1955, **9,** 862; J. L. BOONE and A. B. BURG, *J. Amer. Chem. Soc.*, 1958, **80,** 1519.
24 R. K. PEARSON and J. W. FRAZER, presented at the 132nd A.C.S. meeting, New York, 1958.
25 K. H. LUDLUM, *Diss. Abs.*, 1961, **22,** 97.
26 J. A. DUPONT and R. SCHAEFFER, *J. Inorg. Nuclear Chem.*, 1960, **15,** 310.
27 R. W. PARRY and L. J. EDWARDS, *J. Amer. Chem. Soc.*, 1959, **81,** 3554; W. R. DEEVER and D. M. RITTER, *Inorg. Chem.*, 1969, **8,** 2461; F. M. MILLER and D. M. RITTER, *ibid.*, 1970, **9,** 1284; B. M. GRAYBILL and J. K. RUFF, *J. Amer. Chem. Soc.*, 1962, **84,** 1062.
28 W. D. PHILLIPS, H. C. MILLER and E. L. MUETTERTIES, *ibid.*, 1959, **81,** 4496; M. A. RING, E. F. WITUCKI and R. C. GREENHOUGH, *Inorg. Chem.*, 1967, **6,** 395.
29 B. C. HARRISON, I. J. SOLOMON, R. D. HITES and M. J. KLEIN, *J. Inorg. Nuclear Chem.*, 1960, **14,** 195; R. E. WILLIAMS and F. J. GERHART, *J. Organometallic Chem.*, 1967, **10,** 168.
30 J. R. SPIELMAN and A. B. BURG, *Inorg. Chem.*, 1963, **2,** 1139; M. D. LAPRADE and C. E. NORDMAN, *ibid.*, 1969, **8,** 1669; L. F. CENTOFANTI, G. KODAMA and R. W. PARRY, *ibid.*, 1969, **8,** 2072.
31 A. D. NORMAN and R. SCHAEFFER, *J. Amer. Chem. Soc.*, 1966, **88,** 1143.
32 L. V. MCCARTY and P. A. DIGIORGIO, *J. Amer. Chem. Soc.*, 1951, **73,** 3138.
33 D. F. GAINES, *ibid.*, 1966, **88,** 4528; D. F. GAINES and J. A. MARTENS, *Inorg. Chem.*, 1968, **7,** 704.
34 T. ONAK and R. E. WILLIAMS, *ibid.*, 1962, **1,** 106.
35 T. ONAK, G. B. DUNKS, I. W. SEAREY and J. SPIELMAN, *ibid.*, 1967, **6,** 1465; P. M. TUCKER, T. ONAK and J. B. LEACH, *ibid.*, 1970, **9,** 1430.
36 D. F. GAINES and T. V. IORNS, *J. Amer. Chem. Soc.*, 1967, **89,** 3376.
37 *Idem*, *ibid.*, 1967, **89,** 4250; 1968, **90,** 6617.
38 D. F. GAINES, *ibid.*, 1969, **91,** 1230; D. F. GAINES and T. V. IORNS, *Inorg. Chem.*, 1968, **7,** 1041.
39 A. B. BURG and H. I. SCHESINGER, *J. Amer. Chem. Soc.*, 1933, **55,** 4009.
40 I. J. SOLOMON, M. J. KLEIN, R. G. MAGUIRE and K. HATTORI, *Inorg. Chem.*, 1963, **2,** 1136.

41 A. B. BURG and J. R. SPIELMAN, *J. Amer. Chem. Soc.*, 1959, **81,** 3479.
42 J. L. BOONE and A. B. BURG, *ibid.*, 1959, **81,** 1766.
43 R. E. WILLIAMS, S. G. GIBBINS and I. SHAPIRO, *J. Chem. Phys.*, 1959, **30,** 333.
44 H. D. JOHNSON II, S. G. SHORE, L. L. MOCK and J. C. CARTER, *J. Amer. Chem. Soc.*, 1969, **91,** 2131.
45 D. F. GAINES and R. SCHAEFFER, *Proc. Chem. Soc.*, 1953, 267; *Inorg. Chem.*, 1964, **3,** 438.
46 J. DOBSON and R. SCHAEFFER, *Inorg. Chem.*, 1968, **7,** 402; J. F. DITTER, J. R. SPIELMAN and R. E. WILLIAMS, *ibid.*, 1966, **5,** 118; J. DOBSON, D. F. GAINES and R. SCHAEFFER, *J. Amer. Chem. Soc.*, 1965, **87,** 4072.
47 A. B. BURG and R. KRATZER, *Inorg. Chem.*, 1962, **1,** 725; J. F. DITTER, J. R. SPIELMAN and R. E. WILLIAMS, *ibid.*, 1966, **5,** 118; J. DOBSON and R. SCHAEFFER, *ibid.*, 1968, **7,** 402.
48 L. E. BENJAMIN, S. F. STAFIEJ and E. A. TAKACS, *J. Amer. Chem. Soc.*, 1963, **85,** 2674; B. M. GRAYBILL, A. R. PITOCHELLI and M. F. HAWTHORNE, *Inorg. Chem.*, 1962, **1,** 626; J. DOBSON, P. C. KELLER and R. SCHAEFFER, *ibid.*, 1968, **7,** 399.
49 M. J. HILLMAN, D. J. MARIGOLD and J. H. NORMAN, *J. Inorg. Nuclear Chem.*, 1963, **24,** 1565.
50 M. F. HAWTHORNE and J. J. MILLER, *J. Amer. Chem. Soc.*, 1958, **80,** 754; R. W. ATTEBERRY, *J. Chem. Phys.*, 1958, **62,** 1457; I. SHAPIRO, M. LUSTIG and R. E. WILLIAMS, *J. Amer. Chem. Soc.*, 1959, **81,** 838; R. E. WILLIAMS, *J. Inorg. Nuclear Chem.*, 1961, **20,** 198.
51 G. A. GUTER and G. W. SCHAEFFER, *J. Amer. Chem. Soc.*, 1956, **78,** 3546; P. H. WILKS and J. C. CARTER, *ibid.*, 1966, **88,** 3441; E. L. MUETTERTIES, *Inorg. Chem.*, 1963, **2,** 647.
52 N. N. GREENWOOD and J. A. MCGINNETY, *Chem. Comm.*, 1965, 331; *J. Chem. Soc.* (*A*), 1966, 1090; N. N. GREENWOOD and N. F. TRAVERS, *Inorg. Nuclear Chem. Letters*, 1966, **2,** 169; *Chem. Comm.*, 1967, 217; *J. Chem. Soc.* (*A*), 1967, 881; 1968, 15; N. N. GREENWOOD and D. N. SHARROCKS, *ibid.*, 1969, 2334.
53 R. SCHAEFFER, *J. Amer. Chem. Soc.*, 1957, **79,** 1006; W. KNOTH and E. L. MUETTERTIES, *J. Inorg. Nuclear Chem.*, 1961, **20,** 66; W. R. HERTLER and E. L. MUETTERTIES, *Inorg. Chem.*, 1966, **5,** 160; M. F. HAWTHORNE and A. R. PITOCHELLI, *J. Amer. Chem. Soc.*, 1959, **81,** 5519; D. E. HYATT, D. A. OWEN and L. J. TODD, *Inorg. Chem.*, 1966, **5,** 1749; M. D. MARSHALL, R. M. HUNT, G. T. HEFFERAN, R. M. ADAMS and J. M. MAKHLOUF, *J. Amer. Chem. Soc.*, 1967, **89,** 3361.
54 E. B. MOORE JR., L. L. LOHR JR. and W. N. LIPSCOMB, *J. Chem. Phys.*, 1961, **35,** 1329; C. J. FRITCHIE JR., *Inorg. Chem.*, 1967, **6,** 1199.
55 H. C. LONGUETT-HIGGINS and M. DE V. ROBERTS, *Proc. Roy. Soc.*, 1955, **A230,** 110; A. R. PITOCHELLI and M. F. HAWTHORNE, *J. Amer. Chem. Soc.*, 1960, **82,** 3228.
56 V. D. AFTANDILIAN, H. C. MILLER, G. W. PARSHALL and E. L. MUETTERTIES, *Inorg. Chem.*, 1962, **1,** 734; R. M. ADAMS, A. R. SIEDLE and J. GRANT, *ibid.* , 1964, **3,** 465.
57 N. J. BLAY, F. DUNSTAN and R. L. WILLIAMS, *J. Chem. Soc.*, 1960, 430, 5006.
58 M. HILLMAN, *J. Amer. Chem. Soc.*, 1960, **82,** 1096; A. SEQUEIRA and W. C. HAMILTON, *Inorg. Chem.*, 1967, **6,** 1281; M. H. G. WALLBRIDGE and R. L. WILLIAMS, *J. Chem. Soc.*, (*A*), 1967, 132; A. E. MESSNER, *Analyt. Chem.*, 1958, **30,** 547.
59 W. N. LIPSCOMB, M. F. HAWTHORNE and A. R. PITOCHELLI, *J. Amer. Chem. Soc.*, 1959, **81,** 5833.
60 W. H. KNOTH, H. C. MILLER, D. C. ENGLAND, G. W. PARSHALL, J. SAUER and E. L. MUETTERTIES, *J. Amer. Chem. Soc.*, 1962, **84,** 1056; *Inorg. Chem.*, 1964, **3,** 159.
61 B. L. CHAMBERLAND and E. L. MUETTERTIES, *Inorg. Chem.*, 1964, **3,** 1450; B. G. DEBOER, A. ZALKIN and D. H. TEMPLETON, *ibid.*, 1968, **7,** 1085; M.

HAWTHORNE, R. L. PILLING and P. F. STOKELY, *J. Amer. Chem. Soc.*, 1965, **87,** 1893; *Idem* and P. M. GARRETT, *ibid.*, 1963, **85,** 3705; A. KACZMARCZYK, *Inorg. Chem.*, 1968, **7,** 164.
62 W. H. KNOTH, J. C. SAUER, H. C. MILLER and E. L. MUETTERTIES, *J. Amer. Chem. Soc.*, 1964, **86,** 115.
63 W. H. KNOTH, J. C. SAUER, D. C. ENGLAND, W. R. HERTLER and E. L. MUETTERTIES, *J. Amer. Chem. Soc.*, 1964, **86,** 3973; W. R. HERTLER and M. S. RAASCH, *ibid.*, 1964, **86,** 3661; W. H. KNOTH, J. C. SAUER, J. H. BALTHIS, H. C. MILLER and E. L. MUETTERTIES, *ibid.*, 1967, **89,** 4842.
64 R. J. WIERSEMA and R. L. MIDDAUGH, *J. Amer. Chem. Soc.*, 1967, **89,** 5078; *Idem, Inorg. Chem.*, 1969, **8,** 2075.
65 F. KLANBERG, D. R. EATON, L. J. GUGGENBERGER and E. L. MUETTERTIES, *Inorg. Chem.*, 1967, **6,** 1271; L. J. GUGGENBERGER, *ibid.*, 1968, **7,** 2260; 1969, **8,** 2771; F. KLANBERG and E. L. MUETTERTIES, *ibid.*, 1966, **5,** 1955.
66 H. I. SCHLESINGER, R. T. SANDERSON and A. B. BURG, *J. Amer. Chem. Soc.*, 1939, **61,** 536; A. E. FINHOLT, A. C. BOND JR. and H. I. SCHLESINGER, *ibid.*, 1947, **69,** 1199; H. I. SCHLESINGER, H. C. BROWN and E. R. HYDE, *ibid.*, 1953, **75,** 209.
67 P. H. BIRD and M. G. H. WALLBRIDGE, *J. Chem. Soc.*, 1965, 3923; N. A. BAILEY, P. H. BIRD and M. G. H. WALLBRIDGE, *Chem. Comm.*, 1967, 286; P. C. MAYBURY, J. C. DAVIS JR. and R. A. PATZ, *Inorg. Chem.*, 1969, **8,** 160.
68 S. BAUER, *J. Amer. Chem. Soc.*, 1950, **72,** 622; P. C. MAYBURY and J. E. AHNELL, *Inorg. Chem.*, 1967, **6,** 1286.
69 J. W. TURLEY and H. W. RINN, *Inorg. Chem.*, 1969, **8,** 18.
70 D. F. GAINES, R. SCHAEFFER and F. TEBBE, *Inorg. Chem.*, 1963, **2,** 526; R. SCHAEFFER, F. TEBBE and C. PHILLIPS, *ibid.*, 1964, **3,** 1475.
71 H. BEALL, C. HACKETT BUSHWELLER, W. J. DEWKETT and M. GRACE, *J. Amer. Chem. Soc.*, 1970, **92,** 3484.
72 F. KLANBERG and L. J. GUGGENBERGER, *Chem. Comm.*, 1967, 1293; *Idem* and E. L. MUETTERTIES, *Inorg. Chem.*, 1968, **7,** 2272; L. J. GUGGENBERGER, *ibid.*, 1970, **9,** 367; S. J. LIPPARD and K. M. MELMED, *ibid.*, 1969, **8,** 2755.
73 H. C. MILLER, N. E. MILLER and E. L. MUETTERTIES, *J. Amer. Chem. Soc.*, 1963, **85,** 3885; 1964, **86,** 1033; B. M. GRAYBILL, J. K. RUFF and M. F. HAWTHORNE, *ibid.*, 1961, **83,** 2669.
74 P. BINGER, *Tetrahedron Letters*, 1966, 2675; *Angew. Chem. Internat. Edn.*, 1968, **7,** 286; T. P. ONAK and G. T. F. WING, *J. Amer. Chem. Soc.*, 1970, **92,** 5226.
75 D. A. FRANZ and R. N. GRIMES, *J. Amer. Chem. Soc.*, 1970, **92,** 1438.
76 T. P. ONAK, R. P. DRAKE and G. B. DUNKS, *ibid.*, 1965, **87,** 2505; *Angew. Chem. Internat. Edn.*, 1966, **5,** 258.
77 T. P. ONAK, G. B. DUNKS, J. R. SPIELMAN, F. J. GERHART and R. E. WILLIAMS, *J. Amer. Chem. Soc.*, 1966, **88,** 2062.
78 M. A. GRASSBERGER, E. G. HOFFMANN, G. SCHOMBURG and R. KÖSTER, *ibid.*, 1968, **90,** 56; R. KÖSTER, G. BENEDIKT and R. A. GRASSBERGER, *Annalen*, 1968, **719,** 187.
79 R. KÖSTER and G. BENEDIKT, *Angew. Chem. Internat. Edn.*, 1964, **3,** 515.
80 T. P. ONAK, R. P. DRAKE and G. B. DUNKS, *Inorg. Chem.*, 1964, **3,** 1686; T. P. ONAK, D. MARYNICK, P. MATTSCHEI and G. B. DUNKS, *ibid.*, 1968, **7,** 1754.
81 T. P. ONAK and G. B. DUNKS, *ibid.*, 1966, **5,** 439; J. R. SPIELMAN, R. WARREN, G. B. DUNKS, J. E. SCOTT JR. and T. P. ONAK, *ibid.*, 1968, **7,** 216.
82 J. R. SPIELMAN, G. B. DUNKS and R. WARREN, *ibid.*, 1969, **8,** 2172.
83 R. N. GRIMES and W. J. RADEMAKER, *J. Amer. Chem. Soc.*, 1969, **91,** 6499.
84 C. L. BRAMLETT and R. N. GRIMES, *J. Amer. Chem. Soc.*, 1966, **88,** 4269; *Idem, ibid.*, 1967, **89,** 2557; *Idem* and R. L. VANCE, *Inorg. Chem.*, 1968, **7,** 1066; D. A. FRANZ, J. W. HOWARD and R. N. GRIMES, *J. Amer. Chem. Soc.*, 1969, **91,** 4010.
85 F. N. TEBBE, P. M. GARRETT and M. F. HAWTHORNE, *J. Amer. Chem. Soc.*, 1966,

88, 607; *Idem, ibid.*, 1964, **86,** 4222; R. A. WIESBOECK and M. F. HAWTHORNE, *ibid.*, 1642.
86 P. M. GARRETT, T. A. GEORGE and M. F. HAWTHORNE, *Inorg. Chem.*, 1969, **8,** 2008.
87 D. VOET and W. N. LIPSCOMB, *ibid.*, 1967, **6,** 113.
88 P. M. GARRETT, J. C. SMART and M. F. HAWTHORNE, *J. Amer. Chem. Soc.*, 1969, **91,** 4707; *Idem* and G. S. DITTA, *Inorg. Chem.*, 1969, **8,** 1907.
89 I. SHAPIRO, D. GOOD and R. E. WILLIAMS, *J. Amer. Chem. Soc.*, 1962, **84,** 3837; J. F. DITTER, *Inorg. Chem.*, 1968, **90,** 1748.
90 R. N. GRIMES, *J. Amer. Chem. Soc.*, 1966, **88,** 1070.
91 I. SHAPIRO, B. KEILIN, R. E. WILLIAMS and R. D. GOOD, *ibid.*, 1963, **85,** 3378.
92 T. P. ONAK, F. J. GERHART and R. E. WILLIAMS, *J. Amer. Chem. Soc.*, 1963, **85,** 3378; R. A. BEAUDET and R. L. PAYNTER, *ibid.*, 1964, **86,** 1258; *Idem*, T. P. ONAK and G. B. DUNKS, *ibid.*, 1966, **88,** 4622; R. KÖSTER, *Tetrahedron Letters*, 1965, 777; R. KÖSTER and M. A. GRASSBERGER, *Angew. Chem. Internat. Edn.*, 1966, **5,** 580.
93 G. B. DUNKS and M. F. HAWTHORNE, *Inorg. Chem.*, 1968, **7,** 1038; R. E. WILLIAMS and F. J. GERHART, *J. Amer. Chem. Soc.*, 1965, **87,** 3513.
94 H. HART and W. N. LIPSCOMB, *Inorg. Chem.*, 1968, **7,** 1070.
95 T. F. KOETZLE, F. A. SCARBROUGH and W. N. LIPSCOMB, *Inorg. Chem.*, 1968, **7,** 1076; G. B. DUNKS and M. F. HAWTHORNE, *ibid.*, 1970, **9,** 893.
96 F. N. TEBBE, P. M. GARRETT, D. C. YOUNG and M. F. HAWTHORNE, *J. Amer. Chem. Soc.*, 1966, **88,** 609; J. A. POTENZA and W. N. LIPSCOMB, *Inorg. Chem.*, 1966, **5,** 1301.
97 F. N. TEBBE, P. M. GARRETT and M. F. HAWTHORNE, *J. Amer. Chem. Soc.*, 1968, **90,** 869.
98 C. TSAI and W. E. STREIB, *ibid.*, 1966, **88,** 4513.
99 M. F. HAWTHORNE and D. A. OWEN, *ibid.*, 1968, **90,** 5912.
100 H. A. SCHROEDER, *Inorg. Macromol. Rev.*, 1970, **1,** 45.
101 D. GRAFSTEIN and J. DVORAK, *Inorg. Chem.*, 1963, **2,** 1128; S. PAPETTI and T. L. HEYLING, *J. Amer. Chem. Soc.*, 1964, **86,** 2295.
102 W. N. LIPSCOMB and co-workers, *Inorg. Chem.*, 1966, **5,** 1471; 1967, **6,** 874; *J. Amer. Chem. Soc.*, 1966, **88,** 628; H. A. SCHROEDER and co-workers, *Inorg. Chem.*, 1965, **4,** 107; 1967, **6,** 572; L. I. ZAKHARKIN and N. A. OGORODINKHOVA, *J. Organometallic Chem.*, 1968, **12,** 13.
103 P. M. GARRETT, F. N. TEBBE and M. F. HAWTHORNE, *J. Amer. Chem. Soc.*, 1964, **86,** 5016.
104 M. F. HAWTHORNE and P. A. WEGNER, *J. Amer. Chem. Soc.*, 1965, **87,** 4392; *Idem, ibid.*, 1968, **90,** 896; J. S. ROSCOE, S. KONGPRICHA and S. PAPETTI, *Inorg. Chem.*, 1970, **9,** 1561.
105 N. K. HOTA and D. S. MATTESON, *ibid.*, 1968, **90,** 3570; H. D. SMITH JR., C. O. OBENLAND and S. PAPETTI, *Inorg. Chem.*, 1966, **5,** 1013.
106 D. E. HYATT, D. A. OWEN and L. J. TODD, *ibid.*, 1966, **5,** 1749.
107 D. E. HYATT, F. R. SCHOLER, L. J. TODD and J. L. WARNER, *ibid.*, 1967, **6,** 2229.
108 W. H. KNOTH, *J. Amer. Chem. Soc.*, 1967, **89,** 1274.
109 G. POPP and M. F. HAWTHORNE, *ibid.*, 1968, **90,** 6553.
110 R. L. VOORHEES and R. W. RUDOLPH, *ibid.*, 1969, **91,** 2173.
111 L. J. TODD, A. R. BURKE, H. T. SILVERSTEIN, J. L. LITTLE and G. S. WIKHOLM, *ibid.*, 1969, **91,** 3376.
112 J. L. LITTLE, J. T. MORAN and L. J. TODD, *ibid.*, 1967, **89,** 5495; L. J. TODD, I. C. PAUL, J. L. LITTLE, P. S. WELCKER and C. R. PETERSON, *ibid.*, 1968, **90,** 4489; L. J. TODD, J. L. LITTLE and H. T. SILVERSTEIN, *Inorg. Chem.*, 1969, **8,** 1698.
113 W. J. MIDDLETON, *J. Amer. Chem. Soc.*, 1966, **88,** 3842; W. R. HERTLER, F. KLANBERG and E. L. MUETTERTIES, *Inorg. Chem.*, 1967, **6,** 1696.

114 L. J. TODD, *Adv. Organometallic Chem.*, 1970, **8,** 87. Academic Press, London and New York.
115 L. F. WARREN JR. and M. F. HAWTHORNE, *J. Amer. Chem. Soc.*, 1968, **90,** 4823.
116 R. M. WING, *ibid.*, 1968, **90,** 4828.
117 A. ZALKIN, D. H. TEMPLETON and T. E. HOPKINS, *ibid.*, 1965, **87,** 3988; *Idem, Inorg. Chem.*, 1966, **5,** 1189; *Idem, ibid.*, 1967, **6,** 1911; R. M. WING, *J. Amer. Chem. Soc.*, 1967, **89,** 5599.
118 M. F. HAWTHORNE and co-workers, *J. Amer. Chem. Soc.*, 1967, **89,** 7115; 1968, **90,** 1662; 1969, **91,** 5475; *Inorg. Chem.*, 1969, **8,** 1799.
119 M. F. HAWTHORNE and A. D. PITTS, *J. Amer. Chem. Soc.*, 1967, **89,** 7116.
120 J. W. HOWARD and R. N. GRIMES, *ibid.*, 1969, **91,** 6499.
121 J. N. FRANCIS and M. F. HAWTHORNE, *ibid.*, 1968, **90,** 1663; A. ZALKIN, D. J. ST. CLAIR and D. H. TEMPLETON, *Inorg.Chem.*, 1969, **8,** 2080; M. R. CHURCHILL, A. H. REIS JR., J. N. FRANCIS and M. F. HAWTHORNE, *J. Amer. Chem. Soc.*, 1970, **92,** 4993.
122 P. A. WEGNER, L. J. GUGGENBERGER and E. L. MUETTERTIES, *J. Amer. Chem. Soc.*, 1970, **92,** 3473.
123 F. KLANBERG, P. A. WEGNER, G. W. PARSHALL and E. L. MUETTERTIES, *Inorg. Chem.*, 1968, **7,** 2072; A. R. SIEDLE and T. A. HILL, *J. Inorg. Nuclear Chem.*, 1969, **31,** 3875.
124 G. URRY, T. WARTIK and H. T. SCHLESINGER, *J. Amer. Chem. Soc.*, 1952, **74,** 5809; A. G. MASSEY, D. S. URCH and A. K. HOLLIDAY, *J. Inorg. Nuclear Chem.*, 1966, **28,** 365; J. KANE and A. G. MASSEY, *Chem. Comm.*, 1970, 378.
125 G. URRY, A. G. GARRETT and H. I. SCHLESINGER, *Inorg. Chem.*, 1963, **2,** 396.
126 M. ATOJI and W. N. LIPSCOMB, *J. Chem. Phys.*, 1959, **31,** 601; R. A. JACOBSEN and W. N. LIPSCOMB, *ibid.*, 605; *Idem, J. Amer. Chem. Soc.*, 1958, **80,** 5571; G. S. PAWLEY, *Acta Cryst.*, 1966, **20,** 631.
127 G. F. LANTHIER and A. G. MASSEY, *J. Inorg. Nuclear Chem.*, 1970, **32,** 1807.
128 J. A. FORSTNER, T. E. HAAS and E. L. MUETTERTIES, *Inorg. Chem.*, 1964, **3,** 155.
129 M. F. LAPPERT and M. K. MAJUMDAR, *Proc. Chem. Soc.*, 1963, 88.
130 C. R. RUSS and A. G. MACDIARMID, *Angew. Chem. Internat. Edn.*, 1964, **3,** 509; P. GEYMAYER, E. G. ROCHOW and U. WANNAGAT, *ibid.*, 633.
131 H. HESS, *ibid.*, 1967, **6,** 975 and private communication.
132 J. S. THAYER, *Organometallic Chem. Rev.*, 1966, **1,** 157.
133 P. I. PAETZOLD and W. M. SIMSON, *Angew. Chem. Internat. Edn.*, 1966, **5,** 842.
134 P. I. PAETZOLD, P. P. HABEREDER and R. MÜLLBAUER, *J. Organometallic Chem.*, 1967, **7,** 51.
135 P. I. PAETZOLD, *Angew. Chem. Internat. Edn.*, 1967, **6,** 572.
136 A. STOCK and E. POHLAND, *Chem. Ber.*, 1926, **59B,** 2215; E. WIBERG and A. BOLZ, *ibid.*, 1940, **73,** 209.
137 H. NÖTH, *Z. Naturforsch.*, 1961, **16b,** 618.
138 B. R. CURRELL, W. GERRARD and M. KHODABOCUS, *J. Organometallic Chem.*, 1967, **8,** 411.
139 N. WIBERG, F. RASCHIG and K. H. SCHMIDT, *ibid.*, 1967, **10,** 29; *Idem, Angew. Chem. Internat. Edn.*, 1965, **4,** 715; P. I. PAETZOLD, *Z. anorg. Chem.*, 1963, **326,** 47; J. G. HAASNOOT and W. L. GROENEVELD, *Inorg. Nuclear Chem. Letters*, 1967, **1,** 597.
140 R. PRINZ and H. WERNER, *Angew. Chem. Internat. Edn.*, 1967, **6,** 91.
141 H. WERNER, R. PRINZ and E. DECKELMANN, *Chem. Ber.*, 1969, **102,** 95; E. O. FISCHER, E. LOUIS and C. G. KREITER, *Angew. Chem. Internat. Edn.*, 1969, **8,** 377; E. DECKELMANN and H. WERNER, *Helv. Chim. Acta*, 1970, **53,** 139.
142 N. E. S. CHAMPION, R. FOSTER and R. K. MACKIE, *J. Chem. Soc.*, 1961, 5060.
143 W. HARSBARGER, G. LEE, R. F. PORTER and S. H. BAUER, *Inorg. Chem.*, 1969, **8,** 1683.
144 A. W. LAUBENGAYER, P. W. MOEWS JR. and R. F. PORTER, *J. Amer. Chem. Soc.*, 1961, **83,** 1337.

145 R. I. WAGNER and J. L. BRADFORD, *Inorg. Chem.*, 1962, **1,** 99; V. GUTMANN, A. MELLER and R. SCHLEGEL, *Monatsh.*, 1964, **95,** 314.
146 H. S. TURNER and R. J. WARNE, *Proc. Chem. Soc.*, 1962, 69.
147 P. T. CLARKE and H. M. POWELL, *J. Chem. Soc.*, (*B*), 1966, 1172.
148 K. LONSDALE, *Nature*, 1959, **184,** 1060.
149 B. R. CURRELL, W. GERRARD and M. KHODABOCUS, *Chem. Comm.*, 1966, 77.
150 J. IDRIS JONES and W. S. MCDONALD, *Proc. Chem. Soc.*, 1962, 366; T. R. R. MCDONALD and W. S. MCDONALD, *ibid.*, 1963, 382.
151 J. K. GILBERT and J. D. SMITH, *J. Chem. Soc.* (*A*), 1968, 233.
152 E. WIBERG and K. SCHUSTER, *Z. anorg. Chem.*, 1933, **213,** 77, 89.
153 F. C. GUNDERLOY and C. E. ERICKSON, *Inorg. Chem.*, 1962, **1,** 349.
154 A. J. BANISTER, N. N. GREENWOOD, B. P. STRAUGHAN and J. WALKER, *J. Chem. Soc.*, 1964, 995; H. NÖTH and H. VAHRENKAMP, *Chem. Ber.*, 1967, **100,** 3353.
155 A. C. HAZELL, *J. Chem. Soc.* (*A*), 1966, 1392.
156 N. N. GREENWOOD and J. WALKER, *ibid.*, 1967, 959.
157 J. R. JENNINGS, I. PATTISON and K. WADE, *ibid.*, 1969, 565.
158 T. C. BISSOT and R. W. PARRY, *J. Amer. Chem. Soc.*, 1955, **77,** 3481; M. P. BROWN, R. W. HESELTINE and L. H. SUTCLIFFE, *J. Chem. Soc.* (*A*), 1968, 612; M. P. BROWN, R. W. HESELTINE and D. W. JOHNSON, *ibid.*, 1967, 597; O. T. BEACHLEY JR., *Inorg. Chem.*, 1967, **6,** 870; D. F. GAINES and R. SCHAEFFER, *J. Amer. Chem. Soc.*, 1963, **85,** 395.
159 G. E. COATES and J. G. LIVINGSTONE, *J. Chem. Soc.*, 1961, 1000; J. E. LLOYD and K. WADE, *ibid.*, 1964, 1649.
160 V. A. DOROKHOV and M. F. LAPPERT, *Chem. Comm.*, 1968, 250; *J. Chem. Soc.* (*A*), 1969, 433.
161 S. G. SHORE and co-workers, *Inorg. Chem.*, 1963, **2,** 639; *J. Amer. Chem. Soc.*, 1966, **88,** 4396; *J. Chem. Soc.* (*A*), 1969, 1580.
162 L. M. TREFONAS and W. N. LIPSCOMB, *J. Amer. Chem. Soc.*, 1959, **81,** 4435; *Idem* and F. S. MATTHEWS, *Acta Cryst.*, 1961, **14,** 273.
163 D. Y. YEE and R. EHRLICH, *J. Inorg. Nuclear Chem.*, 1965, **27,** 2681; G. E. COATES and R. A. WHITCOMBE, *J. Chem. Soc.*, 1956, 3351; O. T. BEACHLEY, G. E. COATES and G. KOHNSTAM, *ibid.*, 1965, 3248.
164 N. DAVIDSON and H. C. BROWN, *J. Amer. Chem. Soc.*, 1942, **64,** 316; J. K. GILBERT and J. D. SMITH, *J. Chem. Soc.* (*A*), 1968, 233; T. A. GEORGE and M. F. LAPPERT, *Chem. Comm.*, 1966, 463; *Idem, J. Chem. Soc.* (*A*), 1969, 992; K. WADE and co-workers, *ibid.*, 1965, 5083; 1967, 1339, 1608; 1969, 1121; 1970, 380.
165 H. SCHMIDBAUR and M. SCHMIDT, *Angew. Chem. Internat. Edn.*, 1962, **1,** 327; N. WIBERG, F. RASCHIG and K. H. SCHMIDT, *ibid.*, 1965, **4,** 715.
166 J. R. JENNINGS and K. WADE, *J. Chem. Soc.* (*A*), 1967, 1333; I. PATTISON and K. WADE, *ibid.*, 1968, 2618.
167 A. MAZZEI, S. CUCMELLA and W. MARCONI, *Inorg. Chim. Acta*, 1968, **2,** 305.
168 J. L. ATWOOD and G. D. STUCKY, *J. Amer. Chem. Soc.*, 1970, **92,** 285; R. J. GILLESPIE, *Angew. Chem. Internat. Edn.*, 1967, **6,** 819.
169 K. DEHNICKE and co-workers, *J. Organometallic Chem.*, 1966, **6,** 298; 1967, **7,** P1; 1968, **12,** 37; *Chem. Ber.*, 1965, **98,** 1173; *Z. anorg. Chem.*, 1966, **348,** 261.
170 D. F. CLEMENS, W. S. BREY JR and H. H. SISLER, *Inorg. Chem.*, 1963, **2,** 1251.
171 K. NIEDENZU, H. BEYER and J. W. DAWSON, *ibid.*, 1962, **1,** 738; H. NÖTH and W. REGNET, *Z. Naturforsch.*, 1963, **18b,** 1138; *Idem. Chem. Ber.*, 1969, **101,** 167; K. NIEDENZU, *Angew. Chem. Internat. Edn.*, 1964, **3,** 90.
172 H. NÖTH and W. REGNET, *Chem. Ber.*, 1969, **102,** 2241.
173 J. J. MILLER and F. A. JOHNSON, *J. Amer. Chem. Soc.*, 1968, **90,** 218; *Idem, Inorg. Chem.*, 1970, **9,** 69; P. C. THOMAS and I. C. PAUL, *Chem. Comm.*, 1968, 1130.
174 K. NIEDENZU, P. FRITZ and H. JENNE, *Angew. Chem. Internat. Edn.*, 1964, **3,** 506; H. NÖTH and G. ABELER, *ibid.*, 1965, **4,** 522; P. FRITZ, K. NIEDENZU and

J. W. DAWSON, *Inorg. Chem.*, 1965, **4,** 886; H. NÖTH and W. REGNET, *Z. anorg. Chem.*, 1967, **352,** 1.

175 I. GEISLER and H. NÖTH, *Chem. Comm.*, 1969, 775; H. NÖTH and G. ABELER, *Chem. Ber.*, 1968, **101,** 969.

176 P. I. PAETZOLD and co-workers, *ibid.*, 1968, **101,** 2870–2888.

177 J. H. MORRIS and P. G. PERKINS, *J. Chem. Soc.* (*A*), 1966, 576.

178 C. H. CHANG, R. F. PORTER and S. H. BAUER, *Inorg. Chem.*, 1969, **8,** 1677.

179 P. C. KELLER, *J. Amer. Chem. Soc.*, 1969, **91,** 1231; A. R. YOUNG and R. EHRLICH, *ibid.*, 1964, **86,** 5359.

180 H. SCHMIDBAUR and G. JONES, *Chem. Ber.*, 1968, **101,** 1271; *Idem. Angew. Chem. Internat. Edn.*, 1967, **6,** 449; H. SCHMIDBAUR, W. WOLFSBERGER and H. KRÖNER, *Chem. Ber.*, 1967, **100,** 1023.

181 H. SCHMIDBAUR and W. WOLFSBERGER, *Angew. Chem. Internat. Edn.*, 1967, **6,** 448.

182 H. SCHMIDBAUR, W. WOLFSBERGER and K. SCHWIRTEN, *Chem. Ber.*, 1969, **102,** 556.

183 W. GEE, R. A. SHAW, B. C. SMITH and G. D. BULLEN, *Proc. Chem. Soc.*, 1961, 432; A. D. TEVEBAUGH, *Inorg. Chem.*, 1964, **3,** 302; W. GEE, R. A. SHAW and B. C. SMITH, *J. Chem. Soc.*, 1964, 4180.

184 K. ISSLEIB and H-J. DEYLIG, *Z. Naturforsch.*, 1962, **17b,** 198; N. N. GREENWOOD, E. J. F. ROSS and A. STORR, *J. Chem. Soc.* (*A*), 1966, 706; O. T. BEACHLEY and G. E. COATES, *J. Chem. Soc.*, 1965, 3241.

185 A. B. BURG and G. BRENDEL, *J. Amer. Chem. Soc.*, 1958, **80,** 3198; W. GEE, J. B. HOLDEN, R. A. SHAW and B. C. SMITH, *J. Chem. Soc.*, 1965, 3171; R. I. WAGNER and C. O. WILSON JR., *Inorg. Chem.*, 1966, **5,** 1009.

186 H. NÖTH and W. SCHRÄGLE, *Z. Naturforsch.*, 1961, **16b,** 473; C. R. RUSS and A. G. MACDIARMID, *Angew. Chem. Internat. Edn.*, 1966, **5,** 418; E. W. ABEL, R. A. N. MCLEAN and I. H. SABHERWAL, *J. Chem. Soc.* (*A*), 1968, 2371.

187 H. NÖTH and W. SCHRÄGLE, *Angew. Chem. Internat. Edn.*, 1962, **1,** 457; G. FRITZ and G. TRENCZEK, *ibid.*, 1963, **2,** 482.

188 A. B. BURG and P. J. SLOTA JR., *J. Amer. Chem. Soc.*, 1960, **82,** 2145; R. H. BIDDULPH, M. P. BROWN, R. C. CASS, R. LONG and H. B. SILVER, *J. Chem. Soc.*, 1961, 1822.

189 A. B. BURG and co-workers, *J. Amer. Chem. Soc.*, 1953, **75,** 3872; 1958, **80,** 3198; 1967, **89,** 1040.

190 W. C. HAMILTON, *Acta Cryst.*, 1955, **8,** 199.

191 G. J. BULLEN and P. R. MALLINSON, *Chem. Comm.*, 1969, 132.

192 M. H. GOODROW, R. I. WAGNER and R. D. STEWART, *Inorg. Chem.*, 1964, **3,** 1212; W. GEE, J. B. HOLDEN, R. A. SHAW and B. C. SMITH, *J. Chem. Soc.*, 1965, 3171; *J. Chem. Soc.* (*A*), 1967, 1545.

193 L. BARTON, F. A. GRIMM, C. PERRIN and R. F. PORTER, *Inorg. Chem.*, 1966, **5,** 1477, 2077.

194 C. H. CHANG, R. F. PORTER and S. H. BAUER, *ibid.*, 1969, **8,** 1689.

195 S. K. WASON and R. F. PORTER, *ibid.*, 1966, **5,** 161; L. BARTON, *J. Inorg. Nuclear Chem.*, 1968, **30,** 1683; M. W. RATHKE and H. C. BROWN, *J. Amer. Chem. Soc.*, 1966, **88,** 2606.

196 J. ECONOMY and R. ANDERSON, *Inorg. Chem.*, 1966, **5,** 989.

197 W. GERRARD, E. F. MOONEY and W. G. PETERSON, *J. Inorg. Nuclear Chem.*, 1967, **29,** 943.

198 T. MOLE, *Austral. J. Chem.*, 1966, **19,** 373, 381; T. MOLE and E. A. JEFFREY, *ibid.*, 1968, **21,** 2683.

199 H. SCHMIDBAUR and co-workers, *Angew. Chem. Internat. Edn.*, 1965, **4,** 152, 201, 876; 1966, **5,** 312, 313; *Chem. Ber.*, 1966, **99,** 2178; 1967, **100,** 1521.

200 H. SCHMIDBAUR and W. FINDEISS, *ibid.*, 1966, **99,** 2187; H. SCHMIDBAUR and B. ARMER, *ibid.*, 1968, **101,** 2256.

201 M. BONAMICO and G. DESSY, *J. Chem. Soc.* (*A*), 1967, 1786.
202 A. F. REID, *Inorg. Chem.*, 1967, **6,** 631.
203 H. SCHMIDBAUR and M. SCHMIDT, *Angew. Chem. Internat. Edn.*, 1962, **1,** 328, 549.
204 V. J. SHINER JR., D. WHITTAKER and V. P. FERNANDEZ, *J. Amer. Chem. Soc.*, 1963, **85,** 2318; T. SAEGUSA and T. VESHIMA, *Inorg. Chem.*, 1967, **6,** 1679; S. R. BINDAL, V. K. MATHUR and R. C. MEHROTRA, *J. Chem. Soc.* (*A*), 1969, 863.
205 A. C. AYRES, M. BARNARD and M. R. CHAMBERS, *J. Chem. Soc.* (*B*), 1967, 1385.
206 T. SAEGUSA, S. TOMITA and T. VESHIMA, *J. Organometallic Chem.*, 1967, **10,** 362.
207 L. F. FIESER and M. FIESER, 'Organic Chemistry', Reinhold, New York, 1956.
208 D. C. AYRES and M. R. CHAMBERS, *J. Chem. Soc.*, (*B*), 1967, 1388.
209 T. N. HUCKERBY, J. G. OLIVER and I. J. WORRALL, *Chem. Comm.*, 1968, 918; *Inorg. Nuclear Chem. Letters*, 1969, **5,** 749.
210 R. C. MEHROTRA and M. M. AGRAWAL, *Chem. Comm.*, 1968, 469; J. G. OLIVER and I. J. WORRALL, *J. Chem. Soc.* (*A*), 1970, 845.
211 H. SCHMIDBAUR, H. HUSSEK and F. SCHINDLER, *Chem. Ber.*, 1964, **97,** 255.
212 M. BONAMICO, G. DESSY and C. ERCOLANI, *Chem. Comm.*, 1966, 24.
213 H. SCHMIDBAUR and W. FINDEISS, *Chem. Ber.*, 1966, **99,** 2187.
214 P. A. MCCUSKER and T. OSTDICK O.S.B., *J. Amer. Chem. Soc.*, 1959, **81,** 5550.
215 D. CORDISCHI, A. MELE and A. SOMOGYI, *J. Chem. Soc.*, 1964, 5281; C. ERCOLANI, A. CAMILLI and L. DE LUCA, *ibid.*, 1964, 5278.
216 M. BONAMICO, *Chem. Comm.*, 1966, 135; M. BONAMICO and G. DESSY, *J. Chem. Soc.*, (*A*), 1968, 291.
217 J. WEIDLEM, *Z. anorg. Chem.*, 1969, **366,** 22; A. G. LEE, *J. Chem. Soc.* (*A*), 1970, 467; H. SCHMIDBAUR and G. KAMMEL, *J. Organometallic Chem.*, 1968, **14,** P28; T. J. HURLEY, M. A. ROBINSON, J. A. SCRUGGS and S. I. TROTZ, *Inorg. Chem.*, 1967, **6,** 1310; G. E. COATES and R. N. MUKHERJEE, *J. Chem. Soc.*, 1964, 1295.
218 M. SCHMIDT and W. SIEBERT, *Chem. Ber.*, 1969, **102,** 2752.
219 *Idem*, *Angew. Chem. Internat. Edn.*, 1964, **3,** 637; *Z. anorg. Chem.*, 1966, **345,** 87.
220 *Idem*, *Angew. Chem. Internat. Edn.*, 1966, **5,** 597; *Idem* and F. R. RITTIG, *Chem. Ber.*, 1968, **101,** 281.
221 M. SCHMIDT, W. SIEBERT and E. GAST, *Z. Naturforsch.*, 1967, **22b,** 557.
222 J. A. FORSTNER and E. L. MUETTERTIES, *Inorg. Chem.*, 1966, **5,** 164.
223 E. W. ABEL, D. A. ARMITAGE and R. P. BUSH, *J. Chem. Soc.*, 1965, 3045.
224 A. B. BURG and R. I. WAGNER, *J. Amer. Chem. Soc.*, 1954, **76,** 3307.
225 E. L. MUETTERTIES, N. E. MILLER, K. J. PACKER and H. C. MILLER, *Inorg. Chem.*, 1964, **3,** 870; B. Z. EGAN, S. G. SHORE and J. E. BONNELL, *ibid.*, 1024.
226 W. SIEBERT, F. R. RITTIG and M. SCHMIDT, *J. Organometallic Chem.*, 1970, **22,** 511.
227 K. DEHNICKE, *Angew. Chem. Internat. Edn.*, 1967, **6,** 947.
228 P. A. RENES and C. H. MACGILLAVRY, *Rec. Trav. chim.*, 1945, **64,** 275; J. D. FORRESTER, A. ZALKIN and D. H. TEMPLETON, *Inorg. Chem.*, 1964, **3,** 63; S. C. WALLWORK and I. J. WORRALL, *J. Chem. Soc.*, 1965, 1818.
229 H. SCHERER and G. SEYDAL, *Angew. Chem.*, 1963, **75,** 846.
230 L. MÖGELE, *Angew. Chem. Internat. Edn.*, 1967, **6,** 986; *Z. Naturforsch.*, 1968, **23b,** 1013; R. C. MEHROTRA and co-workers, *Inorg. Chem.*, 1968, **7,** 384; *J. Chem. Soc.* (*A*), 1969, 863.
231 E. COLTON and M. M. JONES, *Z. Naturforsch.*, 1956, **11b,** 491; R. S. NYHOLM and K. ULM, *J. Chem. Soc.*, 1965, 4199.
232 K. YASUDA and R. OKAWARA, *Inorg. Nuclear Chem. Letters*, 1967, **3,** 135.
233 H. SCHMIDBAUR and co-workers, *Angew. Chem. Internat. Edn.*, 1964, **3,** 696 ; 1965, **4,** 152; 1966, **5,** 312; *Chem. Ber.*, 1966, **99,** 2187; 1967, **100,** 1129.
234 A. W. LAUBENGAYER and G. F. LENGNICK, *Inorg. Chem.*, 1966, **5,** 503; H. SCHMIDBAUR and H. F. KLEIN, *Angew. Chem. Internat. Edn.*, 1966, **5,** 726; J. WINDLEIN and H. KRIEG, *J. Organometallic Chem.*, 1968, **11,** 9; 1970, **21,** 281.
235 H. SCHMIDBAUR and co-workers, *Angew. Chem. Internat. Edn.*, 1967, **6,** 806; *Chem. Ber.*, 1968, **101,** 2268, 2278.

236 G. B. DEACON, J. H. S. GREEN and R. S. NYHOLM, *J. Chem. Soc.*, 1965, 3411; *Idem, ibid.*, 6107; R. S. NYHOLM, *Quart. Rev.*, 1970, **24,** 1.
237 D. J. PATMORE and W. A. G. GRAHAM, *Inorg. Chem.*, 1966, **5,** 1586; *Idem*, P. D. CRADWICK and D. HALL, *Chem. Comm.*, 1968, 872.

APPENDIX

Much work has been published in the past eighteen months, so while some is briefly summarized, it has been found necessary to limit certain topics to a mere list of references. Two books have recently appeared.[1]

Boron allotropes have been synthesized and the introduction of methane and bromoform gives $(B_{12})_4B_2C$, $(B_{12})_4B_2C_2$ and eventually $B_{13}C_2$ and B_4C[2]. The crystal structure of $MgAlB_{14}$ indicates the presence of isolated B and Al atoms in a lattice of B_{12} icosahedra.[3]

Bonding in boron hydrides

Two schools of thought are offered to explain the stability of boron hydrides. These are Lipscomb's 'three-centred bond' qualitative theory using topology, and the more quantitative molecular orbital treatment of Longuet-Higgins and Roberts utilizing symmetry.[4] Both methods have been incorporated to explain the electronic structures of cage boron hydride derivatives, and are found to give similar results. Basket molecules are also considered.[5] Lipscomb has simplified his topological approach by excluding the open BBB three-centered bond,[6] though he does point out that they should not necessarily be excluded from carboranes.[7]

The structure of $LiBMe_4$ contains the first reported case of carbon functioning as a bridging group (see p. 68) to boron in both a bent and linear arrangement.[8]

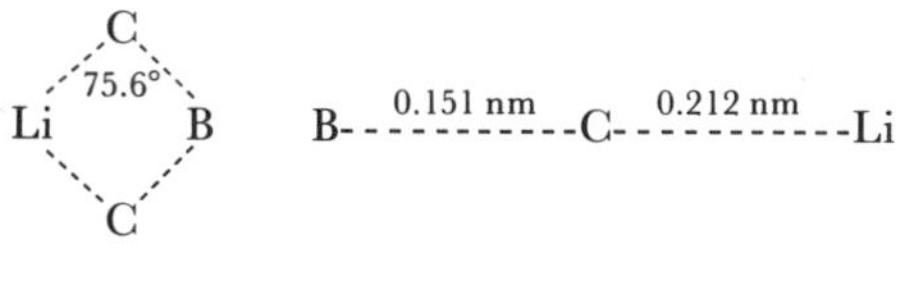

C
75.6°
Li B
C

B---- 0.151 nm ----C---- 0.212 nm ----Li

The prefixing of carboranes as *closo* $(C_2B_nH_{n+2})$, *nido* $(C_2B_nH_{n+4})$ and *arachno* $(C_2B_nH_{n+6})$ is extended to boranes with *nido*boranes $(B_nH_{n+4} - B_6H_{10}, B_{10}H_{14})$ and *arachno*boranes $(B_nH_{n+6} - B_4H_{10}, B_5H_{11})$. These boranes, along with B_6H_{12}, B_8H_{14} and B_9H_{15} are probably all icosahedral fragments. The stability of these and the isoelectronic carboranes is discussed in terms of tautomeric structures[9] and electron pairs,[10] and with respect to the number of hydrogen atoms[11] and charge distribution.[12]

Reactions of boron hydrides

The application of organosubstituted boron hydrides to hydroboration is further examined[13] while an electron diffraction study of $(Me_2AlH)_2$ shows a hydride bridge.[14]

Nido- and *arachno*boranes can be readily deprotonated at the bridging position with NH_3. The anions often further decompose unless stabilized by a large tetra-alkylammonium cation.[15]

$$\begin{array}{lcll} B_4H_{10} & & NH_4^+B_4H_9^- & & H_2B(NH_3)_2^+B_3H_8^- \\ B_5H_{11} & \xrightarrow{NH_3} & NH_4^+B_5H_{10}^- & \longrightarrow & H_2B(NH_3)_2^+B_4H_9^- \\ B_6H_{10} & & NH_4^+B_6H_9^- & & H_2B(NH_3)_2^+B_4H_9^- \end{array}$$

These anions readily deprotonate other boron hydrides and the acidity sequence $B_{10}H_{14} > B_4H_{10} > B_5H_9$ has been established.[16] In addition,

$$B_4H_9^- + B_5H_{11} \longrightarrow B_5H_{10}^- + B_4H_{10}$$

B_5H_{11} reacts similarly. With diborane(6) these anions provide a convenient synthetic route to B_6H_{12} and B_5H_{11}.[17]

Interest in the valence tautomerism of the $B_3H_8^-$ ion is still maintained.[18] Bisphosphine platinum dichloride and $Cs^+B_3H_8^-$ yield a complex $(R_3P)_2PtB_3H_7$ which has higher thermally and aquatic stability than $B_3H_8^-$ complexes. The n.m.r. spectra support π-borallyl structure.[19]

Substitution at the bridging position of B_5H_9 by silyl groups has been further examined,[20] while with dimethylchloroborane, $B_5H_8^-$ gives the bridged product[21] which isomerises in ether to 2,3-$Me_2B_6H_8$.

Coinage metal complexes of anionic boron hydrides $B_5H_8^-$, $B_6H_9^-$ etc,[22] while one carbonyl group in nickel carbonyl[23] is readily displaced by 1-$(CF_3)_2PB_5H_8$. n-$B_{18}H_{20}$ complexes with nickel and cobalt carbonyls[24] while zinc and thallium yield $Zn(B_{10}H_{12})_2^{2-}$ and $Me_2TlB_{10}H_{12}^-$ with decaborane(14).[25]

$B_{10}H_{12}PPh$ results from $B_{10}H_{14}$ with (1) NaH and (2) $PhPCl_2$[26]. Base yields $B_{10}H_{11}PPh^-$ which with (1) NaH and (2) $Mn(CO)_5Br$ gives the complex $(B_{10}H_{10}PPh)\ Mn(CO)_3^-$.

The rest of the boron hydride work is listed through the references at the end of the Appendix.

Carboranes

Most of the work on carboranes is summarized through the reference list, and includes σ- and π-complexes of transition metals.

New heterocarboranes

Aluminium can be readily incorporated as an icosahedral member using $B_9C_2H_{11}^{2-}$ and $EtAlCl_2$.[27] The product, 1,2-$B_9C_2H_{11}AlEt$ also results from Et_3Al and isomerizes at 410°C.[28] The low melting intermediate probably contains the Me_2Al group bonded to the open pentagonal face

$$B_9C_2H_{11}^{2-} \xrightarrow[2.\ Et_3Al]{1.\ 2H^+} 1,2\text{-}B_9C_2H_{12}AlEt_2 \quad \text{m. pt } 35°C$$

$$\downarrow 77°C$$

$$1,7\text{-}B_9C_2H_{11}AlEt \xleftarrow{410°C} 1,2\text{-}B_9C_2H_{11}AlEt \quad \text{m. pt } 97\text{–}9°$$

m. pt 100–102°

of the B_9C_2 cage through hydride bridges, since it resides outside the projection of the pentagonal face.[28]

Dialkyl beryllium reacts with the acidified carbollide cage, incorporating Be into the icosahedron.[29] Germanium, tin and lead are readily included using $B_9C_2H_{11}^{2-}$ and germanium di-iodide, stannous chloride or lead di-acetate.[30]

$C_2B_4H_6^{2-}$ is formally a 6π electron donor so it is not surprising that $C_2B_4H_6$ will function as a 4π electron donor to the $Fe(CO)_3$ residue.[31] The reaction generating this $C_2B_4H_6Fe(CO)_3$ complex also produces π-$C_2B_3H_7Fe(CO)_3$. The ligand contains 2 bridging hydrogen atoms and is probably a 4π electron donor (cf. butadiene).

The anions $B_9H_9CH^-$ and $B_{11}H_{11}CH^-$ result when $B_{10}H_{12}CH^-$ is heated.[32] It also yields $B_{10}H_{10}CH^{3-}$ with base, and will stabilize higher oxidation states than $B_9C_2H_{11}^{2-}$ anion eg. Mn^{IV}, Ni^{IV}.

Halogenated boron cages

The photoelectronic spectrum of B_4Cl_4 supports B—Cl π-bonding,[33] while heating B_2Cl_4 at 80°C for 3 days yields B_nCl_n ($n = 8 - 12$).[34] B_8Cl_8 melts at 185°C. B_2Br_4 decomposes to B_7Br_7 and B_9Br_9.[35]

Oligomeric amino compounds

Butylboron dichloride and cyanogen chloride give dimeric $Bu^nClBNCCl_2$.[36] A similar 4 membered ring results from heating the adduct $K^+(Me_3AlSCNAlMe_3)^-$, formed from potassium thiocyanate and trimethyl aluminium.[37] $(R_2N)_3Al$, Me_2NAlMe_2,[39] $Ph_2C{=}NAlCl_2$,[40] $Me_2NAlEtBr$[41] and $(R_2N)_2AlH$,[38] are also dimeric, along with $RNHGaH_2$[42] (R is branched alkyl). If R is normal, the products are trimeric, as is aziridylGaH_2.[43] Stereoisomers of $(Me_2AlNHMe)_3$ have been identified.[44]

Miscellaneous nitrogen rings

The $(BN)_3$ ring in trimeric $(BCl_2N_3)_3$ is thought to be a skew boat.[45] Complexes of the tetrazaborolane ring with Lewis acid chlorides have been characterized.[46] Synthetic routes to mixed borazines, sulphur substituted borazines[47] and boron heterocycles containing the diaza groups have also been devised.[48] $(Me_2Al)_2BH_2(NMe_2)_3$ is cyclic[49] and $(AlCl)_4(NMe)_4$-$(NMe)_2$ a cage compound.[50] $LiAl(NCBu^t_2)_4$ contains a 4-membered ring incorporating Li and Al along with 2 nitrogen atoms.[50a]

Borazines

These have been synthesized from borates, aluminium, hydrogen and primary amines[51] and via primary amine/BCl_3 adducts.[52] The parent borazine $B_3N_3H_6$ results when $NaBH_4$ and NH_4Cl are heated[53] while exchange readily occurs at organo-B-substituted borazines using the Grignard reagent.[54] Borazines will readily reduce halides of titanium, mercury and tin, and are readily monosubstituted by silver pseudo-halides.[55]

The vibrational spectra of various borazines have been interpreted by using derivatives isotopically substituted with H/D and $^{10}B/^{11}B$.[56] New evidence is presented to support the highest occupied orbital of borazine being π rather than σ.[57] While the ring in $(Me_2NBN)_3$ is planar,[58] that in $Et_6B_3N_3Cr(CO)_3$ is puckered with Cr—N 0.222 nm, Cr—B 0.231 nm.[59] The B—N distance is 0.144 nm but one bond is much shorter (0.136 nm). Similar deviations arise in $Me_6C_6Cr(CO)_3$.

Oxygen derivatives

Trimethylboron and oxygen produce $Me_2B_2O_3$ which is believed to be a trioxadiborolane.[60] Trimethylsiloxyalane is dimeric[61] as is triphenoxy-aluminium,[62] while n.m.r. spectra data shows $(RO)_3Al$ ($R = PhCH_2$) to be tetrameric.[63] Alcohols react with "divalent" gallium halides $Ga^IGa^{III}X_4$ to yield oligomeric alkoxygallium halides.[64] In_2Br_4 is oxidised by methyl bromide[65] to $MeIn_2Br_5$ while SO_2 inserts into Me_3In giving the dimeric sulphinate.[66]

Sulphur and selenium derivatives

Notable among the synthetic routes found for cyclic boron–sulphur[67] and boron–selenium[68] compounds is the cleavage of the C—S as well as the S—S bond of $Bu^t_2S_2$,[69] and the isolation of iodine pentasulphide[70] when preparing 1,3,4-trithia-2,5-diboralanes. Boron tribromide and

$$Bu^t_2S_2 + RBI_2 \xrightarrow{-Bu^tI} RB\langle S{-}S,\ S\rangle BR \longleftarrow S_8 + RBI_2$$

selenols[71] give dimeric selenoboron halides $(RSeBBr_2)_2$.

References

1 R. N. GRIMES, 'Carboranes', Academic Press, New York, 1970. K. WADE, Electron Deficient Compounds, T. Nelson and Sons Ltd., Sunbury-on-Thames, 1971.
2 E. AMBERGER and co-workers, *J. Less Common Metals*, 1971, **23,** 21, 33, 43.
3 V. I. MATKOVICH and J. ECONOMY, *Acta Cryst.*, 1970, **B26,** 616.
4 H. C. LONGUET-HIGGINS and M. DE V. ROBERTS, *Proc. Roy. Soc.*, 1954, A, **224,** 336, *idem.*, *ibid.*, 1955, A, **230,** 110; R. HOFFMANN and W. N. LIPSCOMB, *J. Chem. Phys.*, 1962, **36,** 2179.
5 S. F. A. KETTLE and MISS V. TOMLINSON, *J. Chem. Soc.*, (A), 1969, 2002 and 2007.
6 I. R. EPSTEIN and W. N. LIPSCOMB, *Inorg. Chem.*, 1971, **10,** 1921.
7 I. R. EPSTEIN, J. A. TOSSELL, E. SWITKES, R. M. STEVENS and W. N. LIPSCOMB, *ibid.*, 171.
8 D. GROVES, W. RHINE and G. D. STUCKY, *J. Amer. Chem. Soc.*, 1971, **93,** 1553.
9 R. E. WILLIAMS, *Inorg. Chem.*, 1971, **10,** 210.
10 K. WADE, *Chem. Comm.*, 1971, 792.
11 L. PAULING, *J. Inorg. Nuclear Chem.*, 1970, **32,** 3745.
12 C-C. S. CHEUNG, R. A. BEAUDET and G. A. SEGAL, *J. Amer. Chem. Soc.*, 1970, **92,** 4158.
13 H. C. BROWN and E. I. NEGISHI, *J. Organometallic Chem.*, 1971, **26,** C67 and 1971, **28,** C1.
14 G. A. ANDERSON, A. ALMENNINGEN, F. R. FORGAARD and A. HAALAND, *Chem. Comm.*, 1971, 480.
15 G. L. BRUBAKER, M. L. DENNISTON, S. G. SHORE, J. CARTER and F. SWICKER, *J. Amer. Chem. Soc.*, 1970, **92,** 7216, H. D. JOHNSON II and S. G. SHORE, *ibid.*, 7586.
16 A. C. BOND and M. L. PINSKY, *ibid.*, 1970, **92,** 7585. G. KODAMA, J. E. DUNNING and R. W. PARRY, *ibid.*, 1971, **93,** 3372.
17 H. D. JOHNSON II and S. G. SHORE, *ibid.*, 1971, **93,** 3798.
18 H. BEALL, C. HACKET BUSHWELLER and M. GRACE, *Inorg. Nuclear Chem. Letters.*, 1971, **7,** 641, *idem.*, W. J. DEWHETT and H. S. BILOFSKY, *J. Amer. Chem. Soc.*, 1971, **93,** 2145.
19 A. R. KANE and E. L. MUETTERTIES, *ibid.*, 1971, **93,** 1042.
20 T. C. GEISLER and A. D. NORMAN, *Inorg. Chem.*, 1970, **9,** 2167. D. F. GAINES and T. V. IORNS, *ibid.*, 1971, **10,** 1094.
21 *Idem, J. Amer. Chem. Soc.*, 1970, **92,** 4571.
22 E. L. MUETTERTIES, W. G. PEET, P. A. WEGNER and C. W. A. LEGRANTI, *Inorg. Chem.*, 1970, **9,** 2447, V. T. BRICE and S. G. SHORE, *Chem. Comm.*, 1970, 1312.
23 A. B. BURG and I. B. MISHRA, *J. Organometallic Chem.*, 1970, **24,** C33.
24 R. L. SNEATH, J. L. LITTLE, A. R. BURKE and L. J. TODD, *Chem. Comm.*, 1970, 693.
25 N. N. GREENWOOD, J. A. MCGINNETY and J. D. OWEN, *J.C.S.*(A), 1971, 809, N. N. GREENWOOD, N. F. TRAVERS and D. W. WAITE, *Chem. Comm.*, 1971, 1027.
26 J. L. LITTLE and A. C. WONG, *J. Amer. Chem. Soc.*, 1971, **93,** 522.
27 B. M. MIKHAILOV, T. V. POTAPOVA, *Izv. Acad. Nauk. SSSR. Ser khim*, 1968, 1153. *Chem. Abs.* 1969, **70,** 4185g.
28 D. A. T. YOUNG, G. R. WILLEY, M. F. HAWTHORNE, M. R. CHURCHILL and A. H. REES JR., *J. Amer. Chem. Soc.*, 1970, **92,** 6663; *idem, Chem. Comm.*, 1971, 298.
29 G. POPP and M. F. HAWTHORNE, *Inorg. Chem.*, 1971, **10,** 391.
30 R. W. RUDOLPH, R. L. VOORHEES and R. E. COCHEY, *J. Amer. Chem. Soc.*, 1970, **92,** 3351.
31 R. N. GRIMES, *J. Amer. Chem. Soc.*, 1971, **93,** 261.
32 W. H. KNOTH, *Inorg. Chem.*, 1971, **10,** 598.
33 D. R. LLOYD and N. LYNAUGH, *Chem. Comm.*, 1971, 627.
34 G. F. LANTHIER, J. KANE and A. G. MASSEY, *J. Inorg. Nuclear Chem.*, 1971, **33,** 1569.
35 J. KANE and A. G. MASSEY, *ibid.*, 1971, **33,** 1195.

36 A. MELLER and W. MARINGGELE, *Monatsch*, 1970, **101,** 387.
37 J. L. ATWOOD, P. A. MILTON and S. K. SEALE, *J. Organometallic Chem.*, 1971, **28,** C29.
38 R. A. KOVAR and E. C. ASHBY, *Inorg. Chem.*, 1971, **10,** 893.
39 H. HESS, A. HINDERER and S. STEINHAUSER, *Z. anorg. Chem.*, 1970, **377,** 1.
40 R. SNAITH, C. SUMMERFORD, K. WADE and B. K. WYATT, *J. Chem. Soc.* (A), 1970, 2635.
41 K. GOSLING, A. L. BHUIYAN and K. R. MOONEY, *Inorg. Nucl. Chem. Letters*, 1971, **7,** 913.
42 A. STORR and A. D. PENLAND, *J. Chem. Soc.*, 1971, 1237.
43 W. HARRISON, A. STORR and J. TROTTER, *Chem. Comm.*, 1971, 1101.
44 K. GOSLING, G. M. MCLAUGHLIN, G. A. SIM and J. D. SMITH, *ibid.*, 1970, 1617.
45 U. MILLER, *Z. anorg. Chem.*, 1971, **382,** 110.
46 B. HESSETT, J. H. MORRIS and P. G. PERKINS, *J. Chem. Soc.* (A) 1971, 2466.
47 H. NÖTH and M. J. SPRAGUE, *J. Organometallic Chem.*, 1970, **23,** 323.
48 J. J. MILLER, *ibid.*, 1970, **24,** 595, D. NÖLLE and H. NOTH, *Angew Chem. Internat. Edn.*, 1971, **10,** 126.
49 R. E. HALL and E. P. SCHRAM, *Inorg. Chem.*, 1971, **10,** 192.
50 U. THEWALT and I. KAWADA, *Chem. Ber.*, 1970, **103,** 2754.
50a H. M. M. SHEARER, R. SNAITH, J. D. SOWERBY and K. WADE, *Chem. Comm.*, 1971, 1275.
51 E. C. ASHBY and R. A. KOVAR, *Inorg. Chem.*, 1971, **10,** 1524.
52 J. BLACKBOROW, J. BLACKMORE and J. C. LOCKHART, *J. Chem. Soc.*, (A) 1971, 49.
53 V. V. VOLKOV, G. I. BAGRYANTSEV and K. G. MYAKISHEV, *Russian J. Inorg. Chem.*, 1970, **15,** 1570.
54 J. L. ADCOCK and J. J. LAGOWSKI, *Inorg. Nuclear Chem. Letters*, 1971, **7,** 473.
55 G. A. ANDERSON and J. J. LAGOWSKI, *Inorg. Chem.*, 1971, **10,** 1910, O. T. BEACHLEY JR., *J. Amer. Chem. Soc.*, 1971, **93,** 5066.
56 A. KALDOR and R. F. PORTOR, *Inorg. Chem.*, 1971, **10,** 775; K. E. BLICK, K. NIEDENZU, W. SAWODNY, M. TAKASUKA, T. TOTANI and H. WATANABE, *ibid.*, 1133; K. E. BLICK, I. A. BOENIG and K. NIEDEUZU, *ibid.*, 1917.
57 H. BOCK and W. FUSS, *Angew. Chem. Internat. Edn.*, 1971, **10,** 182.
58 H. HESS and B. REISER, *Z. anorg. Chem.*, 1971, **381,** 91.
59 G. HUTTNER and B. KRIEG, *Angew. Chem. Internat. Edn.*, 1971, **10,** 512.
60 L. BARTON and G. T. BOHN, *Chem. Comm.*, 1971, 77.
61 C. B. ROBERTS and D. D. TONER, *Inorg. Chem.*, 1970, **9,** 2361.
62 J. LUKASIAK, L. A. MAY, I. Y. STRAUSS and R. PIEKOS, *Rocz. Chem.*, 1970, **44,** 1675.
63 J. G. OLIVER and I. J. WORRALL, *J. Chem. Soc.* (A) 1970, 1389.
64 *Idem, ibid.*, 1971, 2315.
65 L. WATERWORTH and I. J. WORRALL, *Inorg. and Nuclear Chem. Letters*, 1971, **7,** 403.
66 A. T. T. HSIEH, *J. Organometallic Chem.*, 1971, **27,** 293.
67 M. SCHMIDT and F. R. RITTIG, *Z. Nat.* 1970, **25b,** 1062. W. SIEBERT, K-J. SCHAPER and M. SCHMIDT, *J. Organometallic Chem.*, 1970, **25,** 315.
68 M. SCHMIDT and E. KIEWERT, *Z. Nat.*, 1971, **26b,** 613.
69 M. SCHMIDT and F. R. RITTIG, *Angew. Chem. Internat. Edn.*, 1970, **9,** 738.
70 W. SIEBERT, E. GAST and M. SCHMIDT, *J. Organometallic Chem.*, 1970, **23,** 329.
71 M. SCHMIDT and H.-D. BLOCK, *Z. anorg. Chem.*, 1970, **377,** 305.

N.M.R. Spectra for Boron Hydrides

J. B. LEACH, T. ONAK, J. SPIELMAN, R. R. REITZ, R. SCHAEFFER and L. G. SNEDDON, *Inorg. Chem.*, 1970, **9,** 2170 (small boron hydrides).

J. D. ODOM, P. D. ELLIS and H. C. WALSH, *J. Amer. Chem. Soc.*, 1971, **93,** 3530 (^{11}B n.m.r. of B_5H_9).

C. B. MURPHY JR. and R. E. ENRIONE, *J. Inorg. Nuclear Chem.*, 1971, **33,** 584 (Solvolysis of 1-BrB_5H_8 followed by ^{11}B n.m.r.).

G. M. BODNER and L. G. SNEDDON, *Inorg. Chem.*, 1970, **9,** 1421, G. M. BODNER, F. R. SCHOLER, L. J. TODD, L. E. SENOR and J. C. CARTER, *ibid.*, 1971, **10,** 942 ($B_9H_{13}L$ and $B_{10}H_{14}$).

Mass Spectra and Ion Cyclotron Resonance

S. J. STECK, G. A. PRESSLEY JR., F. E. STAFFORD, J. E. DOBSON and R. SCHAEFFER, *Inorg. Chem.*, 1970, **9,** 2452.

R. C. DUNBAR, *J. Amer. Chem. Soc.*, 1971, **93,** 4167.

Carboranes

T. ONAK and J. B. LEACH, *Chem. Comm.*, 1971, 76 (CB_5H_7).

D. A. FRANZ and R. N. GRIMES, *J. Amer. Chem. Soc.*, 1971, **93,** 387, (2-$MeC_3B_3H_6$ and $Me_2C_3B_3H_5$ isomers).

R. R. OLSEN and R. N. GRIMES, *Inorg. Chem.*, 1971, **10,** 1103 and I. R. EPSTEIN, T. F. KOETZLE, R. M. STEVENS and W. N. LIPSCOMB, *J. Amer. Chem. Soc.*, 1970, **92,** 7019 ($B_4C_2H_6$ isomers).

C. G. SAVORY and M. G. H. WALLBRIDGE, *Chem. Comm.*, 1971, 622 (μ-trimethylsilyl derivative of 4,5-dicarbanidohexaborane(8)) *idem* and J. S. MCAVORY, *J. Chem. Soc.* (A) 1971, 3038 (Halogen substitution in $C_2B_4H_8$).

R. R. OLSEN and R. N. GRIMES, *J. Amer. Chem. Soc.*, 1970, **92,** 5072 and R. WARREN, D. PAQUIN, T. ONAK, G. DUNKS and J. R. SPEILMAN, *Inorg. Chem.*, 1970, **9,** 2285 (2,4-$C_2B_5H_7$-reactions).

P. M. GARRETT, G. S. DITTA and M. F. HAWTHORNE, *ibid.*, 1970, **9,** 1947; T. F. KOETZLE and W. N. LIPSCOMB, *ibid.*, 2279 ($Me_2C_2B_8H_8$).

R. SCHAEFFER and R. R. RIETZ, *J. Amer. Chem. Soc.*, 1971, **93,** 1263. P. M. GARRET, G. S. DITTA and M. F. HAWTHORNE, *ibid.*, 1265. (Two new nido-carboranes $B_7C_2H_{11}$ and $B_8C_2H_{12}$).

M. F. HAWTHORNE, D. A. OWEN and J. W. WIGGINS, *Inorg. Chem.*, 1971, **10,** 1304 (dicarba-closo-undecaboranes).

R. W. RUDOLPH, J. L. PFLUG, C. M. BOCK and M. HODGSON, *Inorg. Chem.*, 1970, **9,** 2274; R. K. BOHM and M. D. BOHM, *ibid.*, 1971, **10,** 350; D. A. T. YOUNG, T. E. PAXSON and M. F. HAWTHORNE, *ibid.*, 786 (o-, m-, and p-carboranes).

R. G. ADLER and M. F. HAWTHORNE, *J. Amer. Chem. Soc.*, 1970, **92,** 6174, (properties of carboranes and derivatives by ^{19}F n.m.r.).

Metal Derivatives of Carboranes

(a) *σ-Bonded*

D. A. OWEN and H. F. HAWTHORNE, *J. Amer. Chem. Soc.*, 1970, **92,** 3194; *idem, ibid.*, 1971, **93,** 873, *idem*, J. C. S. SMART and P. M. GARRETT, *ibid.*, 1362, (complexes of carboranes σ-bonded to metal and metal carbonyl residues through carbon).

MISS C. M. MITCHELL and F. G. A. STONE, *Chem. Comm.*, 1970,1263, (a gold-carborane complex).

(b) *π-Bonded*

L. J. TODD, A. R. BURKE, A. R. GARBER, H. T. SILVERSTEIN and B. N. STORHOFF, *Inorg. Chem.*, 1970, 2175 [C_5H_5-Co(1,2-B_9H_9CHAs)].

M. F. HAWTHORNE, L. F. WARREN JR., K. P. CALLAHAN and N. F. TRAVERS, *J. Amer. Chem. Soc.*, 1971, **93,** 2407, (protonation of Fe^{II} carbollide complexes).

W. E. GEIGER JR. and D. E. S. SMITH, *Chem. Comm.*, 1971, 8. (Evidence for unreported low oxidation states of transition metal dicarbollide complexes.)

M. F. HAWTHORNE and co-workers, *J. Amer. Chem. Soc.*, 1970, **92,** 7214, *ibid.*, 1971, **93,** 2541 and 3063 and *Inorg. Chem.*, 1971, **10,** 863, (complexes of closo-carborane anions—$B_6C_2H_8^{2-}$, $B_6C_2H_8^{4-}$, $B_8C_2H_{10}^{4-}$ (3,6)-1,2-dicarbacanastide from Spanish-Canasta, basket-bidentate like $B_6C_2H_8^{4-}$ and $B_{10}C_2H_{12}^{2-}$ giving a 13 membered polygon when complexed).

M. K. KALOUSTIAN, R. J. WIERSEMA and M. F. HAWTHORNE, *J. Amer. Chem. Soc.*, 1971, **93,** 4912 (Heating $CpCo^{III}B_9C_2H_9R_2$ induces migration of carbon around the icosahedron).

4

Silicon, Germanium, Tin and Lead

In considering any feature of the chemistry of the Group IV elements, the inclusion of carbon for comparative purposes is imperative. Carbon catenation, both linear and cyclic, provides the yardstick for comparison with the rest of the Group elements. Likewise, the absence of p_π–p_π bonding in compounds that can be isolated is also a feature alien to carbon chemistry.

So although the cyclic compounds mentioned in this book exclude carbon as an integral part of the ring, those of the heavier elements are better understood if comparisons are made. Thus, Wurtz reactions normally employed to prepare olefins yield cyclopoly-anes of silicon, germanium and tin, illustrating instability in p_π–p_π multiple bonds relative to chain formation.

$$Ph_2MCl_2 \xrightarrow[\text{solvent}]{\text{sodium}} Ph_2MMPh_2 \ (M = C) \text{ or } [Ph_2M]_n \ (M = Si)$$

The monomer formulation of the cyclopolysiloxanes and silicones corresponds to that of the ketones.

THE STRUCTURES OF THE ELEMENTS

The macromolecular two- and three-dimensional structures of graphite and diamond provide not only one of the best examples of allotropy but also the starting point for a comparison of the structures of the Group IV elements.

Graphite has already been compared with the laminated form of polymeric boron nitride, but this provides the only analogue—none of the other Group IV elements form graphite-like allotropes. This is not surprising as the bond shortening required for effective p_π–p_π bonding between silicon and heavier atoms would not compensate for the increase in repulsion between the non-bonding electrons.[1]

The diamond structure is adopted by silicon, germanium and the grey-, or α-form of tin.

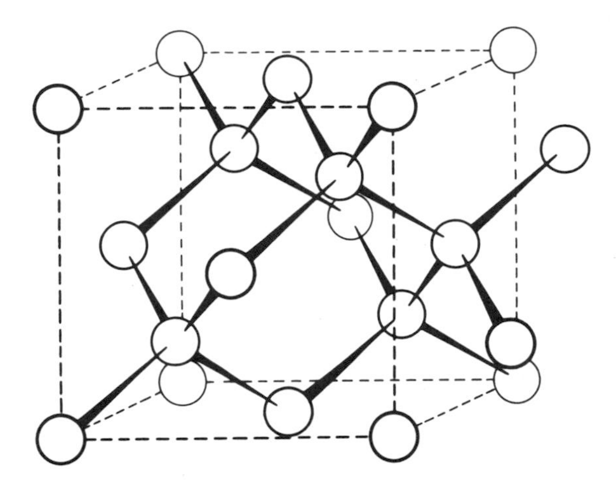

Fig. 4.1. The diamond structure adopted by silicon, germanium and α-tin[a]

The bond lengths correspond to those expected for single bonds,[a] (Si 0.234; Ge 0.244; α-Sn 0.280), the value for α-Sn being close to that found in $[Ph_2Sn]_6$,[2] to be considered later (0.277 nm) (p. 164).

Above 13.2°C, α-tin changes reversibly into β-tin, which is more metallic in nature. Each tin atom has four nearest neighbours at 0.302 nm in a flattened tetrahedron, and two at 0.318 nm. This is indicative of divalent character to tin (covalent radius of Sn^{II} 0.163 nm) and indeed, hydrochloric acid gives both tin(II) and tin(IV) hydrated chlorides.[3]

Lead exists only in a metallic state with a cubic close-packed array with 12 nearest neighbours.

ANIONIC DERIVATIVES OF THE ELEMENTS

The alkali and alkaline earth metals react with the Group IV elements at high temperature in an inert atmosphere to give crystalline derivatives of various compositions. The general formula of the products varies with the metal considered, and as silicon provides the widest range of silicides it will be considered first and used for comparison.

The alkali metal silicides M^ISi and barium disilicide $BaSi_2$ possess the anion Si_4^{4-}. This has been shown to be tetrahedral[4] like P_4 of white phosphorus, and the Si—Si bond lengths vary from 0.234 to 0.248 nm. While the shortest length corresponds to a single Si—Si bond,[a] the larger values may indicate a degree of electron migration due to cation polarization. With the isoelectronic P_4, the P—P lengths are shorter than the single bonds, which is consistent with the more electronegative, uncharged P_4 molecule.

Sodium germanicide and plumbicide possess similar anionic tetrahedra with Ge—Ge 0.253–0.258 nm[5] and Pb—Pb 0.315–0.316 nm. This Pb—Pb distance is much larger than in hexamethyldilead[7] (0.288 nm), and corresponds to a bond order of a half.[6]

The presence of these lead anions in sodium/lead alloys may account for reaction with alkyl chlorides. Alkyl radicals generated from the chloride and sodium may attack a lead radical-anion generated through Na^+ polarization.

$$Pb_4^{4-} + Na^+ \longrightarrow Na + Pb_4^{3-}$$

$$RCl + Na \longrightarrow R\cdot + NaCl$$

$$Pb_4^{3-} + R\cdot \longrightarrow RPb_4^{3-}$$

The substituted anions may then disproportionate into tetra-alkyl-leads.

Though silicides with a higher proportion of metal ions are not cyclic, brief mention can be made to complete the structural picture. The mono-silicides of calcium, strontium and barium contain an infinite zig-zag chain of silicon atoms,[8] while with Mg_2Si and Ca_2Si isolated silicide anions probably exist.[9]

The anion present in Li_2Si is diatomic Si_2^{4-}. The Si—Si bond length (0.237 nm) is close to that of a single bond,[10] and a comparison shows carbon to give acetylides directly with both lithium and sodium (M_2C_2). The C—C bond length in CaC_2 shows triple-bond character in contrast to the diatomic Si_2^{4-}. This would be expected to have double-bond character, but the apparent inability of post first period elements to give p_π–p_π bonds and the polarizing power of Li^+ probably results in the bond lengthening observed.

Polymeric anionic arrays occur with $CaSi_2$ and $SrSi_2$. The calcium compound involves a two-dimensional puckered honeycomb array of fused Si_6 rings.[4]

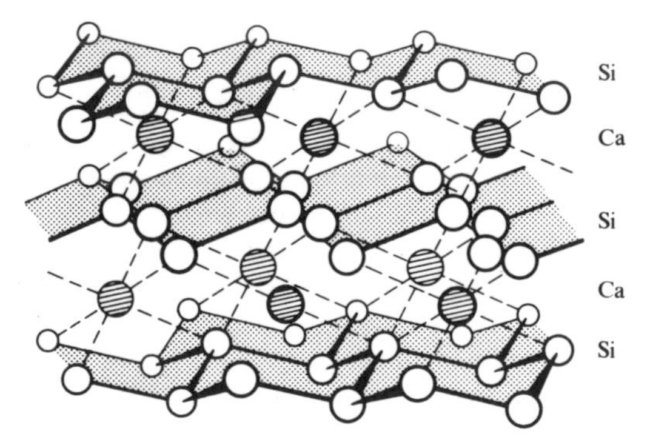

Fig. 4.2. The crystal structure of $CaSi_2$[a]

These are all present in the chair form, and so the compound can be crudely considered as a graphite-like inclusion compound, but with puckered instead of planar rings.[11]

This laminated form of calcium disilicide reacts with iodine monochloride in CCl_4. The product is a brown form of silicon known as lepidoidal silicon.

$$Si_{2n}^{2n-} + 2nICl \longrightarrow nI_2 + 2Cl^- + Si_{2n}$$

Thus, it closely resembles $CaSi_2$ with Ca^{2+} ions extracted,[12] and possesses electron-free sites on each silicon atom. So it enflames in air and reacts spontaneously with alcohols and water, unlike the "diamond" allotrope of silicon already considered.

An excess of ICl produces a layered monochloride of silicon. This is yellow, and like lepidoidal silicon is a scaly product with a chlorine atom at each silicon site. Hence each Si_6 ring has six axial chlorine atoms, three above the layer and three below.

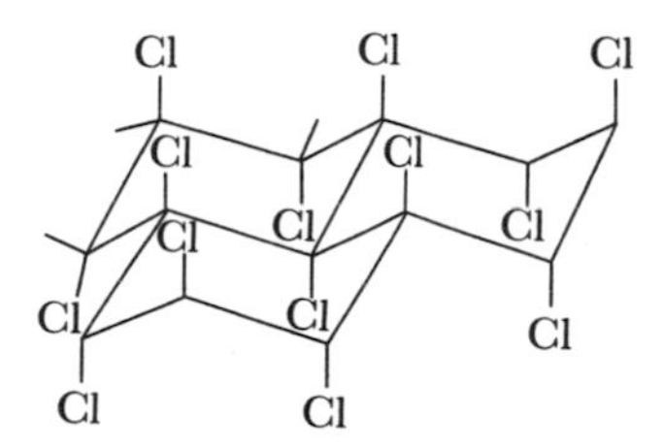

Fig. 4.3. Polymeric silicon monochloride

The steric hindrance is such that with methanol only 80% of the silicon sites exchange.

Strontium disilicide, like several of the rare earth silicides, possesses a three-dimensional open array of silicon atoms, occluding the cations.[13] The Si—Si bond lengths are close to a single bond value, and the structure resembles the borides mentioned in the last Chapter (p. 41).

These structures emphasize the similarity between boron and silicon, supporting the diagonal relationship proposed to generalize the properties of some of the first and second period elements found to the left of the Periodic Table.

A series of polyatomic anions result from potentiometric titrations of a liquid ammonia solution of sodium with various metals and metalloid elements. With tin and lead, the compounds Na_4Sn_9 and Na_4Pb_9 are formed. Though neither has been fully characterized,[14] both are thought to have structures analogous to Bi_9^{5+} to be mentioned in the next Chapter (p. 269).[15] This involves face-bonding bismuth atoms into a trigonal prism of bismuth atoms, and Pb_9^{4-} is isoelectronic with Bi_9^{5+}.

Dodecaphenylcyclohexatin is cleaved by the amide ion in liquid ammonia. The products include tin(IV) and tin(II) amides, along with Sn_9^{4-}.[16]

$$11(Ph_2Sn)_6 + 84NH_2^- + 132NH_3 \longrightarrow$$
$$33Sn(NH_2)_6^{2-} + 6Sn(NH_2)_3^- + 3Sn_9^{4-} + 132C_6H_6$$

CYCLOPOLY-SILANES, -GERMANES AND -STANNANES

Valencies two less than the Group valency are quite common for the heavier Main Group elements, especially in those bonded to electronegative substituents. This is the case with the thallium(I) alkoxides, and for a time compounds now formulated as $[R_2Sn]_n$ were considered to be monomeric ($n = 1$). They are now known to be oligomeric with $n = 4$, 5, 6 and 9, though monomeric diphenyltin Ph_2Sn can be stabilized when bonded to magnesium. So a brief mention will be made of these pseudocarbenes.

The triethylamine complex of ethylmagnesium bromide reacts with triphenylstannane to form the stannylmagnesium compound (equation (*4.1*)). This is an oil which yields hexaphenylditin with triphenyltin chloride.[17]

$$[EtMgBr{\cdot}Et_3N]_2 + 2Ph_3SnH \xrightarrow{Et_2O} 2Ph_3SnMgBr{\cdot}Et_3N + 2C_2H_6$$

$$Ph_3SnMgBr{\cdot}Et_3N + Ph_3SnCl \longrightarrow Ph_6Sn_2 + MgClBr + Et_3N \qquad (4.1)$$

Heating at 50°C for 5 h causes the stannylmagnesium compound to solidify into an unsolvated dimer analyzing for $[Ph_3SnMgBr]_2$ (equation (*4.2*)). The reaction with water and triphenyltin chloride establishes the presence of a Ph—Mg bond, while the presence of a Ph_2Sn residue was supported by the reactions of this dimer with methyl iodide and dicobalt octacarbonyl.

$$2Ph_3SnMgBr{\cdot}Et_3N \longrightarrow \begin{matrix} Ph_2Sn & & Br & & Ph \\ & \diagdown & \diagup\diagdown & \diagup & \\ & Mg & & Mg & \\ & \diagup & \diagdown\diagup & \diagdown & \\ Ph & & Br & & SnPh_2 \end{matrix}$$

$$\xrightarrow{H_2O} C_6H_6 \qquad \xrightarrow{Ph_3SnCl} Ph_4Sn \qquad \xrightarrow{MeI} Ph_2SnMe_2 \qquad \xrightarrow{Co_2(CO)_8} Ph_2Sn[Co(CO)_4]_2 \qquad (4.2)$$

This indicates a 1–2 migration of the phenyl group from tin to magnesium, and is therefore an inorganic example of the Wagner–Meerwein rearrangement.[18]

Evidence for the transitory existence of silylenes, the silicon analogues of carbenes, is now becoming quite copious. They are generated through a 1–2 migration of a methoxy group in 1,2-dimethoxytetramethyldisilane above 220°C.[19, 20]

$$MeOSiMe_2SiMe_2OMe \xrightarrow{220°C} (MeO)_2SiMe_2 + Me_2Si$$

The dimethylsilylene reacts readily with acetylenes and will insert into the Si—O bond producing polysilanes.[22,23]

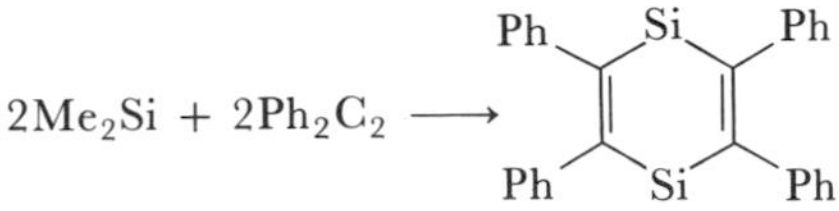

$$2Me_2Si + 2Ph_2C_2 \longrightarrow \text{(1,4-disilacyclohexa-2,5-diene ring: Si, Si with Ph at each carbon)}$$

$$MeOSiMe_2SiMe_2OMe + Me_2Si \longrightarrow MeO[SiMe_2]_nOMe \ (n = 3, 4 \text{ or } 5)$$

Phenylcyclopolysilanes

Cyclopolysilanes were first isolated by Kipping during the 1920's. The phenyl-substituted ones were isolated from the reaction of sodium with diphenylsilicon dichloride in toluene, and the exact nature of three of the products is now known.[b] The historical development of the proposals for the structures of these compounds, described by Kipping as A, B and C, is an interesting story in itself.[21]

$$Ph_2SiCl_2 + 2Na \xrightarrow{\text{toluene}} \frac{1}{n}[Ph_2Si]_n + 2NaCl$$

Compound A was thought to be linear octaphenyltetrasilane with tervalent terminal silicon atoms, due to its high reactivity in comparison to compound B. As this had the same molecular weight as A, it was thought to be a cyclic tetrasilane.

However, A has now been shown conclusively to be the cyclotetrasilane from various derivatives, and results in better yields from Ph_2SiCl_2 and lithium.[22]

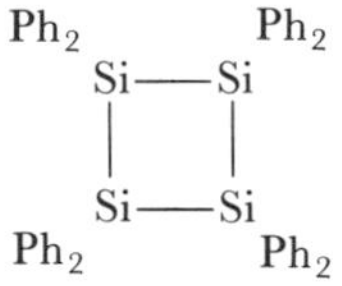

$[SiPh_2]_4$ (four-membered Si ring, each Si bearing Ph_2)

A

(m.p. 321–3°C)

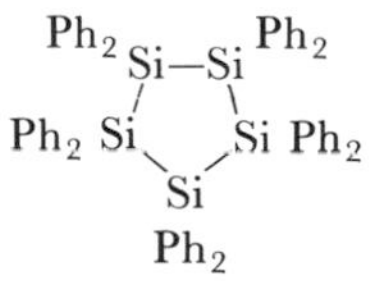

$[SiPh_2]_5$ (five-membered Si ring, each Si bearing Ph_2)

B

(m.p. 472–4°C)

$[SiPh_2]_6$ (six-membered Si ring, each Si bearing Ph_2)

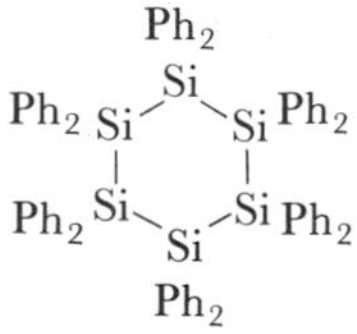

C

(m.p. 500–5°C)

$$\underset{\text{(m.p. 274–6°C)}}{I[Ph_2Si]_4I} \xleftarrow[C_6H_6]{I_2/60°C} [Ph_2Si]_4 \xrightarrow[\text{piperidine}]{H_2O} Ph_2Si(OH)_2$$

$I[Ph_2Si]_4I \xrightarrow{H_2O}$ (cyclo-Ph_2Si–Ph_2Si–Ph_2Si–O–$SiPh_2$) $\xrightarrow{O_2}$ (cyclo-Ph_2Si–O–$SiPh_2$–Ph_2Si–O–$SiPh_2$) (m.p. 221°C)

$I[Ph_2Si]_4I \xrightarrow{RNH_2}$ (cyclo-Ph_2Si–Ph_2Si–Ph_2Si–NR–$SiPh_2$) (R = Et, Ph)

$$[Ph_2Si]_4 \xrightarrow{(CHCl_2)_2} \underset{\text{(m.p. 179–81°C)}}{Cl[Ph_2Si]_4Cl} \xrightarrow{PhLi} \underset{\text{(m.p. 358–60°C)}}{Ph_{10}Si_4}$$

$$Cl[Ph_2Si]_4Cl \xrightarrow{MeLi} \underset{\text{(m.p. 223°C)}}{Me[Ph_2Si]_4Me}$$

$$[Ph_2Si]_4 \xrightarrow[THF]{Li} Li[Ph_2Si]_4Li \xrightarrow{(MeO)_3PO} Me[Ph_2Si]_4Me$$

$Li[Ph_2Si]_4Li \xrightarrow{MCl_2}$ (cyclo-Ph_2Si–Ph_2Si–Ph_2Si–M–$SiPh_2$)

(M = PhP, S, Ph_2Si, Ph_2Ge)

Many of these derivatives of A are new heterocyclic compounds involving four silicon atoms.

Compound B[23] also results from Ph_2SiCl_2 and lithium, and is less reactive than the cyclotetrasilane which has been used to synthesize compound B.

$$\underset{\text{(m.p. 189–90°C)}}{Me[Ph_2Si]_5Me} \xleftarrow{MeMgI} \underset{\text{(m.p. 207–9°C)}}{Br[Ph_2Si]_5Br} \xrightarrow{LiAlH_4} \underset{\text{(m.p. 151°C)}}{H[Ph_2Si]_5H}$$

$$Li[Ph_2Si]_5Li \xrightarrow{(MeO)_3PO} Me[Ph_2Si]_5Me$$

$$Li[Ph_2Si]_5Li \xleftarrow[THF]{Li} [Ph_2Si]_5 \xrightarrow[C_6H_6]{Br_2} Br[Ph_2Si]_5Br \xrightarrow{H_3O^+} \underset{\text{(m.p. 172–4°C)}}{HO[Ph_2Si]_5OH}$$

While compound A is readily oxidized by iodine and by tetrachloroethane, the cyclopentasilane resists attack under the same conditions. Likewise, A can be titrated with bromine, but not B.

1,5-Dilithiodecaphenylpentasilane reacts with Ph_2SiCl_2 to give a 3% yield of C, $[Ph_2Si]_6$, along with A and B. It is even more stable than either of these oligomers, resisting cleavage by oxidizing agents. Lithium will cleave the Si—Si bond, and subsequent methylation affords 1,6-dimethyldodecaphenylhexasilane, (m.p. 195–200°C).[24]

The molecular weights of compounds A, B and C have all been established by mass spectrometry, the parent ion being present in all the compounds.[25]

Rather surprisingly, chlorination with the gas is slow even for the cyclotetrasilane.

$$[Ph_2Si]_4 + Cl_2 \longrightarrow Cl[Ph_2Si]_4Cl + Cl[Ph_2Si]_2Cl$$

The α,ω-dichlorosilane formed will decompose further in CCl_4 to 1,2-dichlorotetraphenyldisilane. With the cyclopentasilane, α,ω-dichlorosilanes again form, and it appears that chlorine cleavage does not occur in the 1–2 position of the chain.[26]

$$[Ph_2Si]_5 + Cl_2 \longrightarrow Cl[SiPh_2]_5Cl + Cl[SiPh_2]_{2+3}Cl$$

Methylcyclopolysilanes

Despite the interest aroused by Kipping in the 1920's in arylcyclopolysilanes, the first alkyl one was not reported until 1949,[27] and only in the past six years has a prolonged study been made of the permethylcyclopolysilanes.

These are best synthesized from dimethyldichlorosilane using a Na/K alloy in THF.[28] The hexameric silane is the major product, but pentamer and heptamer can be isolated conveniently if work up is immediate.[29] Delays cause the yields of these oligomers to fall.

$$Me_2SiCl_2 + Na/K \xrightarrow{THF} \frac{1}{n}[Me_2Si]_n + NaCl + KCl$$

(n = 5, m.p. 188–90°C; n = 6, m.p. 252–5°C; n = 7, m.p. 228–32°C)

All are crystalline solids and the pentamer is oxygen-sensitive, giving the oxopentasilane with a characteristic Si—O stretching frequency in the infrared spectrum at 1050 cm^{-1}.

Permethylcyclohexasilane can be obtained in 60% yields from lithium using triphenylsilyl-lithium as a catalyst.[30] The synthesis is believed to involve a chain-lengthening process which terminates with the extrusion of $[Me_2Si]_6$ through the formation of the triphenylsilyl anion as the leaving group.

$$Ph_3SiLi + Me_2SiCl_2 \longrightarrow Ph_3SiMe_2SiCl$$

$$Ph_3SiMe_2SiCl + 2Li \longrightarrow Ph_3SiMe_2SiLi$$

$$\longrightarrow Ph_3Si[Me_2Si]_6Li$$

(Cyclic structure: Ph_3Si—$SiMe_2$ in a six-membered chain of Me_2Si units ending in $Si^{\ominus}$, with arrows indicating ring closure)

$$\longrightarrow [Me_2Si]_6 + Ph_3Si^-$$

Triphenylsilyl-lithium catalyzes the disproportionation of α,ω-bis-(triphenylsilyl)permethylpolysilane, $Ph_3Si[Me_2Si]_nSiPh_3$ (n = 1–6), yielding the cyclopenta- and cyclohexa-silanes along with hexaphenyldisilane. The shorter of these polysilanes (n = 1–4) give higher proportions of the cyclopentasilane relative to the cyclohexasilane, but the latter forms in larger quantities on standing. With n = 5 and 6, a long reaction time gives mainly cyclohexasilane.

These results support a base-catalyzed equilibrium between the two rings, with the larger being the more stable.

$$6[Me_2Si]_5 \xrightleftharpoons{Ph_3SiLi} 5[Me_2Si]_6$$

The mode of formation probably involves intramolecular nucleophilic attack by a terminal silyl anion.[31]

$$Ph_3Si[Me_2Si]_nSiPh_3 + Ph_3SiLi \longrightarrow Ph_6Si_2 + Ph_3Si[Me_2Si]_nLi$$

$$Ph_3Si[Me_2Si]_nLi \longrightarrow Ph_3SiLi + [Me_2Si]_n \text{ (when } n = 5 \text{ or } 6)$$

Polymeric permethylpolysilanes give the cyclohexasilane on heating with caustic potash in diglyme.[32]

Catalyzed ring-contraction of permethylcyclohexasilane occurs with aluminium chloride in benzene.[33]

$$[Me_2Si]_6 \xrightarrow[C_6H_6]{AlCl_3} Me_3Si{-}Si(Me)\text{(cyclo-}Si_5\text{, Me groups on Si)}$$

The product is trimethylsilyl-nonamethylcyclopentasilane, and the isomerization is an inorganic example of the long known cyclohexane conversion into methylcyclopentane.[34]

These cyclopolysilanes are cleaved quite readily by chlorinating agents,[35]

$$[Me_2Si]_6 \xrightarrow[Cl_2CHCHCl_2]{PCl_5} Cl[Me_2Si]_nCl \ (n = 2\text{–}6)$$

but further cleavage of the terminal linear dichloropolysilanes so formed is more difficult, and again, as with the perphenyl compounds already mentioned, no monosilanes were isolated, indicating that the terminal Si—Si bonds resist halogenation.

The three permethylcyclopolysilanes mentioned ($[Me_2Si]_n$, n = 5, 6 and 7) all form radical anions with sodium/potassium alloy in monoglyme/THF.[29,31,36] The e.s.r. spectrum shows that all protons interact equally with the odd electron, which must therefore be completely delocalized, probably through the empty d orbitals of the cyclosilane ring.[37]

Cyclopolygermanes

The first cyclopolygermane was reported in 1930, and was probably octaphenylcyclotetragermane.[38] It was formed from diphenylgermanium dichloride and sodium, but not studied further until 1963. Since then, the three germanium analogues of Kipping's phenylcyclosilanes A, B and C have been synthesized.

The cyclotetragermane is conveniently prepared by irradiating or heating the polymer produced from diphenylgermane and diethylmercury.[39]

$$Ph_2GeH_2 + Et_2Hg \longrightarrow [Ph_2GeHg]_n \longrightarrow [Ph_2Ge]_4$$

The higher oligomers are synthesized by the more conventional route involving lithium and diphenylgermanium dichloride in THF.[40] A separation of $[Ph_2Ge]_5$ and $[Ph_2Ge]_6$ can be conveniently carried out as the former is soluble in benzene from which it can be precipitated with ether, while the hexamer is only partially soluble in benzene.

The reactions of these compounds, in so far as they have been studied, show some similarities to the cyclopolysilanes, though the cyclotetragermane is less air-sensitive than the tetrasilane.[39]

$$Ph_{10}Ge_4 \xleftarrow[PhBr]{Li} \begin{array}{c} Ph_2Ge—GePh_2 \\ | \quad\quad | \\ Ph_2Ge—GePh_2 \end{array} \xrightarrow{Li/trace\ PhBr} [Ph_2Ge]_5$$

(m.p. 238°C)

$[Ph_2Ge]_4 \xrightarrow{PhLi} Ph_{10}Ge_4$

$$[Ph_2Ge]_4 \xrightarrow{I_2} I[Ph_2Ge]_4I \xrightarrow{H_2O} \text{cyclo-}(Ph_2Ge)_4O$$

(m.p. 206–7°C; ν_{GeOGe} = 855 cm^{-1})

Perhaps the most interesting reaction which occurs with the three cyclopolygermanes involves the dimetallation with sodium naphthalenide in monoglyme.[41] Methylation yields dimethyldiphenylgermane.

$$[Ph_2Ge]_n \xrightarrow[\text{[MeOCH}_2]_2]{Na^+ C_{10}H_8^-} Ph_2GeNa_2 \xrightarrow{Me_2SO_4} Ph_2GeMe_2$$

(n = 4, 5 or 6)

Permethylcyclotetra- and hexa-germane are by-products from the methylation of the ether complex of germanochloroform $HGeCl_3(Et_2O)_2$.

$$HGeCl_3\cdot 2Et_2O \xrightarrow{MeMgBr} Me_{2n+2}Ge_n + [Me_2Ge]_4 + [Me_2Ge]_6$$

$Me_{2n+2}Ge_n$ (~90%); $[Me_2Ge]_4$ (b.p. 60–70°C/0.3 mm, m.p. 89–90°C); $[Me_2Ge]_6$ (b.p. 130–155°C/1.5 mm)

Though telomers are the major products, the cyclogermanes can be isolated by fractional distillation.[42]

Cyclopolystannanes

The synthesis of diethyltin was reported in 1852, and the historical context in which this, one of the first organometallic compounds, was discovered, is perhaps best appreciated when it is realized that carbon was then still considered to have an atomic weight of 6. It was made from ethyl iodide and a 14% sodium–tin alloy.[43]

$$2Na/Sn + 2EtI \longrightarrow Et_2Sn + 2NaI$$

Subsequent routes involved either the coupling of diethyltin dibromide with sodium or reacting stannous chloride with ethyl-lithium.[44] The product analyzed approximately for Et_2Sn, but halogen degradation showed the presence of Et_3Sn and EtSn groups. Also, the colour of the oils varied from red to colourless. This branching severely limits the use of these synthetic methods, and, indeed, only small yields of dodecamethylcyclohexatin result from the dichloride and sodium in liquid ammonia.[45]

Similarly, both phenyl-lithium and phenyl Grignard reagents react with stannous halides to form yellow solids analyzing for Ph_2Sn. However, subsequent halogenation shows the presence of mono-, di- and tri-functional products, indicating chain-branching.[46] The monofunctional residue may result from Ph_2Sn and PhLi giving Ph_3SnLi, while PhSnCl may result from partial arylation of stannous chloride.

$$\left.\begin{array}{l} Ph_2Sn + PhLi \longrightarrow Ph_3SnLi \\ SnCl_2 + PhLi \longrightarrow PhSnCl \end{array}\right\} \rightarrow PhSnSnPh_3 \xrightarrow{Ph_3SnLi} (Ph_3Sn)_2Sn(Ph)Li$$

$$(Ph_3Sn)_2Sn(Ph)Li \xrightarrow{Ph_2Sn} LiSnPh_2Sn(Ph)(SnPh_3)_2$$

$$LiSnPh_2Sn(Ph)(SnPh_3)_2 \xrightarrow{X_2} Ph_3SnCl + Ph_2SnCl_2 + PhSnCl_3$$

The only cyclostannanes obtained pure from an electropositive metal compound, apart from $[Me_2Sn]_6$ just mentioned, are $[Bu^t_2Sn]_4$ and $[Bu^i_2Sn]_6$. t-Butylmagnesium chloride reduces di-t-butyltin dichloride, and the product cleaves readily with iodine.[47]

$$Bu^t_2SnCl_2 + 2Bu^tMgCl \longrightarrow \underset{\text{(yellow crystals)}}{[Bu^t_2Sn]_4} \xrightarrow{I_2} I[Bu^t_2Sn]_4I$$

Magnesium reduces di-i-butyltin dichloride to the cyclohexastannane in good yields, showing already the variation in ring sizes encountered among these cyclopolystannanes.[48]

$$6Bu^i_2SnCl_2 + 6Mg \longrightarrow [Bu^i_2Sn]_6 + 6MgCl_2$$

The difficulties found in synthesizing these compounds have now largely been overcome by using dihydrides as the precursors. Though silicon dihydrides have never been used directly to synthesize cyclopolysilanes, diphenylgermane gives octaphenylcyclotetragermane in the presence of diethylmercury.[39]

Diorganotin dihydrides decompose even more readily, especially in the presence of a basic catalyst, giving cyclopolytins with $n = 4, 5, 6, 7$ or 9.

$$nR_2SnH_2 \xrightarrow{\text{amine}} [R_2Sn]_n + nH_2$$

The mechanism by which the Sn—Sn bond is formed has not been studied for the cyclopolytins, but tin hydrides react with both tin amines and tin–oxygen compounds[49] to form the tin–tin bond, and the mechanism of this reaction is probably closely related.

$$Me_2Sn(NEt_2)_2 + 2Et_3SnH \longrightarrow Et_3SnMe_2SnSnEt_3 + 2Et_2NH$$

$$(Bu^n_3Sn)_2O + Bu^n_2SnH_2 \xrightarrow[\text{temp.}]{\text{room}} 2Bu^n_3SnH + [Bu^n_2SnO] \xrightarrow[-H_2O]{100°C} Bu^n_8Sn_3$$

The rate-determining step is believed to be base protonation, followed by nucleophilic attack of the tin anion thus generated.

$$R_3SnNR_2 + R_3SnH \xrightarrow{\text{slow}} R_3Sn\overset{+}{\underset{H}{N}}R_2 + R_3Sn^-$$

$$R_3Sn^- + R_3Sn\overset{+}{\underset{H}{N}}R_2 \xrightarrow{\text{fast}} R_3SnSnR_3 + R_2NH$$

Cyclopolytins produced this way result in high yield, and the products are crystalline solids melting at about 200°.[48,49]

$$R_2SnH_2 + R_2Sn(NEt_2)_2 \longrightarrow [R_2Sn]_n + 2Et_2NH$$

($R = Et$, $n = 6$; $R = Bu^n$, $n = 6$; $R = Bu^i$, $n = 9$; $R =$ cyclohexyl, $n = 5$).

Both aliphatic and aromatic tin dihydrides catalytically dehydrogenate in the presence of base or a base–tin halide complex.

$$nR_2SnH_2 \xrightarrow{\text{catalyst}} [R_2Sn]_n + nH_2$$

The polystannanes synthesized this way are tabulated below,[49,50] and are normally obtained in good yields.

Table 4.1. Catalysts for the Preparation of Cyclopolystannanes

R	n	Catalyst
Et	9	pyridine Et_2SnCl_2
Bu^n	6	$Bu^n_2SnCl_2$/NaOMe
Bu^i	9	pyridine $Bu^i_2SnCl_2$
$PhCH_2$	4	$(PhCH_2)_2SnCl_2$–DMF
β-$C_{10}H_7$	6	DMF
o-tolyl	6	DMF

Pentameric and hexameric diphenyltin result from the dihydride dehydrogenation, the base solvent determining the product obtained.

With methanol, polymerization but not cyclization results, though pyridine will cyclize the linear hexastannane.[51]

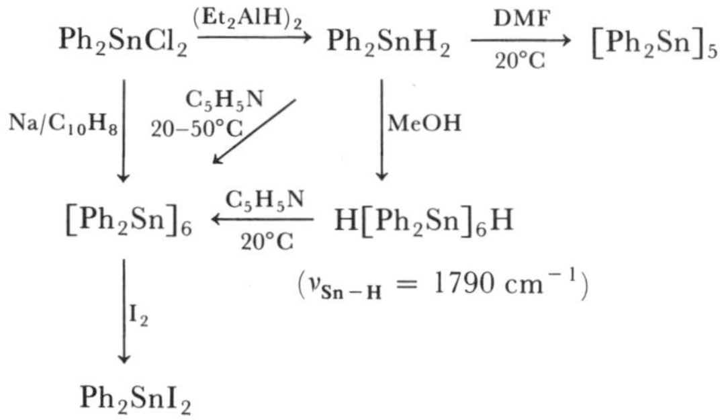

$$Ph_2SnCl_2 \xrightarrow{(Et_2AlH)_2} Ph_2SnH_2 \xrightarrow[20°C]{DMF} [Ph_2Sn]_5$$

$$Ph_2SnCl_2 \xrightarrow{Na/C_{10}H_8} [Ph_2Sn]_6$$

$$Ph_2SnH_2 \xrightarrow[20-50°C]{C_5H_5N} [Ph_2Sn]_6$$

$$Ph_2SnH_2 \xrightarrow{MeOH} H[Ph_2Sn]_6H \quad (\nu_{Sn-H} = 1790\ cm^{-1})$$

$$H[Ph_2Sn]_6H \xrightarrow[20°C]{C_5H_5N} [Ph_2Sn]_6$$

$$[Ph_2Sn]_6 \xrightarrow{I_2} Ph_2SnI_2$$

The crystal structure of dodecaphenylcyclohexatin shows a chair conformation for the Sn_6 ring,[52] with Sn—Sn bond lengths of 0.277 nm, close to the length found in α-tin.[a]

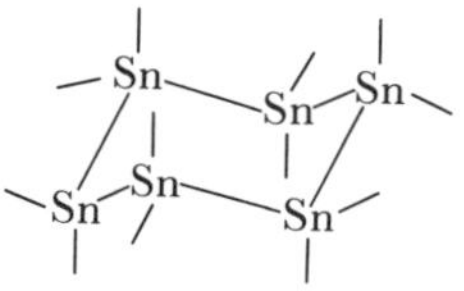

Sn—Sn 0.277 nm

Fig. 4.4. The structure of $[Ph_2Sn]_6$

Very few reactions of cyclopolytins have been studied. Halogenation has been used for analytical purposes, as has already been mentioned with polymeric diethyltin. Organotin compounds are now being widely used as p.v.c. stabilisers, insecticides, fungicides, *etc.*, and this has led to new synthetic methods. The major one involves tin and an organic halide, and resembles the Rochow process used for organosilicon halides. Organic halides also react with cyclopolystannanes, and an examination of the reaction may provide an insight into the tin/organic halide process.

The products formed appear to depend upon the halide chosen and the conditions employed. Alkyl iodides appear to react quite cleanly at 140°, giving the simple addition products in about 50% yield.

$$RI + R'_2Sn \xrightarrow[3\ h]{140°C} R'_2RSnI$$

With alkyl bromides, a mixture results, indicating some disproportionation.

$$R_2Sn + R'Br \xrightarrow[3\ h]{140°C} R_3SnBr + R_2R'SnBr$$

Yields of the symmetrical tin bromide were generally less than those of the unsymmetrical one, but heating for 15 hours improved yields of R_3SnBr.

Diethyl- and di-n-propyl-tins give low yields of the trialkyltin chlorides with n-butyl chloride in the absence of a catalyst. With triethylamine or tetraethylammonium chloride, however, yields of the symmetrical chlorides improve, together with R_2Bu^nSnCl.

$$R_2Sn + Bu^nCl \xrightarrow[15\ h]{Et_3N/160°C} R_3SnCl\ (R = Et, Pr^n)$$

The conditions employed are conducive to Sn—Sn and Sn—C cleavage, and may account for the presence of polybutyltin compounds, (*e.g.*, RBu^n_2SnCl, Bu^n_3SnCl and $Bu^n_2R_2Sn$). The migration of alkyl and halogen groups has been widely studied by Kozeschkow, and this is rapid between stannic chloride and tetra-alkyltins, even at room temperature.[53]

The route involved in the formation of the wide variety of products formed from $[R_2Sn]_n$ and Bu^nCl is not yet known, but it seems likely that R_2SnCl_2 is formed. This is supported by the isolation of all the alkyltin monochlorides from dibutyltin and diethyltin dichloride.

$$[R_2Sn]_n + Bu^nCl \longrightarrow R_2SnCl_2 \xrightarrow{[R_2Sn]_n} R_3SnX + Sn$$

$$[Bu^n_2Sn]_n + Et_2SnCl_2 \xrightarrow[15\ h]{160°C}$$

$$\underset{12\%}{Bu^n_2EtSnCl} + \underset{7\%}{Bu^nEt_2SnCl} + \underset{5\%}{Bu^n_3SnCl} + \underset{1\%}{Et_3SnCl} + \underset{48\%\ yield}{Sn}$$

Heating diphenyltin polymers with n-alkyl bromides or iodides at 140°C again produces triphenyl- and diphenyl-alkyltin halides, along with smaller quantities of phenyldialkyltin halides and dihalogenotin compounds.[52] The occurrence of these dihalides provides the only major difference between the alkyl and aryltin polymers.

CYCLIC NITROGEN DERIVATIVES OF SILICON, GERMANIUM AND TIN

The wide field of commercial application found for polysiloxanes—the silicones—has led to a comprehensive study of silicon–nitrogen compounds, notably cyclic and low molecular weight linear ones, as precursors to polysilazanes. While these have the advantage of an additional valency site on nitrogen, which will increase branching in the polymer, the facile hydrolysis of the Si—N bond detracts from their potential, since such polymers would perish in moist conditions.

Cyclosilazanes $[RNSiR'_2]_n$ are now known with $n = 2$, 3 or 4. They will be considered in order of increasing ring size, and the more limited cyclic chemistry of germanium– and tin–nitrogen compounds compared afterwards. Five- and 6-co-ordinated nitrogen complexes will then be described.

Perchlorocyclosilazanes

The variation in ring size of the cyclosilazanes is conveniently focused in the chlorinated cyclosilazanes. The 4-, 6- and 8-membered rings have all been isolated for this system, the first two members only recently.

The 8-membered ring was first synthesized in 1953 from ammonia and $SiCl_4$ at 825°C. The cyclotetrasilazane is *N*-substituted with trichlorosilyl groups, and also results from a glow discharge of an equimolar mixture of N_2 and $SiCl_4$.[54]

$$SiCl_4 + NH_3 \xrightarrow[N_2/35\ hr]{825°C} (Cl_3Si)_2NH\ (b.p.\ 34°C/3\ mm) + [Cl_3SiNSiCl_2]_4$$

$$SiCl_4 + N_2 \xrightarrow[glow\ discharge]{40\ hr} [Cl_3SiNSiCl_2]_4$$

Cl2
Cl3Si Si SiCl3
N N
Cl2Si SiCl2
Cl3Si N N SiCl3
Si
Cl2

Passing ammonia into $SiCl_4$ in ether at −78°C leads to a variety of products, notably hexachlorodisilazane and hexachlorocyclotrisilazane.[55,56]

$$SiCl_4 + NH_3 \longrightarrow (Cl_3Si)_2NH + [Cl_2SiNH]_3\ (m.p.\ 164°C)$$

$$[Cl_2SiNH]_3 \xrightarrow{MeSiCl_3} Cl_2SiMeNHSiCl_3$$

$$[Cl_2SiNH]_3 \xrightarrow{Me_2NH} [(Me_2N)_2SiNH]_3\ (m.p.\ 94°C)$$

$$\xrightarrow{HBr} (BrCl_2Si)_2NH\ (b.p.\ 57°C/2\ mm)$$

The reluctance of these compounds to undergo condensation reactions involving the removal of HCl is a little surprising, especially since $SiCl_4$ reacts so readily with an excess of ammonia to give silicon nitride.

$$3SiCl_4 + \underset{(excess)}{NH_3} \longrightarrow 3Si(NH_2)_4 \longrightarrow 3Si(NH)_2 \xrightarrow{heat} Si_3N_4 + 2NH_3$$

Further evidence for the kinetic stability of the Si—Cl bond in these chlorosilazanes is revealed in the reaction of n-butyl-lithium or diethylzinc with hexachlorodisilazane. *N*-Metallated products result, rather than Si–alkyl ones, and this may be due to the strong acidity of the proton and to the comparatively low Si—Cl dipole moment. Indeed, the two metallated amines are thermally stable to about 100°C, decomposing into *N,N′*-bis-(trichlorosilyl)tetrachlorocyclodisilazane.[55, 57]

$$(Cl_3Si)_2NH \xrightarrow{Bu^nLi} (Cl_3Si)_2NLi \xrightarrow{SiF_4} (Cl_3Si)_2NSiF_3 \text{ (b.p. 42°C/15 mm)}$$

$$(Cl_3Si)_2NH \xrightarrow{Et_2Zn} (Cl_3Si)_2NZnEt \xrightarrow{100°C} [Cl_3SiNSiCl_2]_2$$

$$(Cl_3Si)_2NLi \xrightarrow{85°C \text{ or } SiCl_4} Cl_3SiN\langle SiCl_2 \rangle_2 NSiCl_3$$

Organo-substituted Cyclodisilazanes

Small rings, involving three and four atoms have attracted the attention of the practical and theoretical chemist for many years. In carbon chemistry, cyclobutadiene and isomeric tetrahedrane (tetrahedral C_4H_4) are still elusive, though tetrahedrane derivatives have been detected mass spectroscopically. The second of the two C_4 skeletons in these isomers is known for heavier elements, however, with P_4 and Si_4^{4-} providing examples.

Synthesis

The past decade has provided a comparatively intense study of the cyclodisilazanes, which possess a 4-membered skeleton of alternate Si and N atoms. This was first synthesized in 1961 by heating silylated cyclotrisilazane at 400°C, with the 4-membered ring substituted at nitrogen by trimethylsilyl groups.[58]

$$[Me_2SiNH]_3 \xrightarrow{3Bu^nLi} [Me_2SiNLi]_3 \xrightarrow{3Me_3SiCl} [Me_2SiNSiMe_3]_3 \xrightarrow{400°C} [Me_3SiNSiMe_2]_2 \quad \textbf{(4-I)}$$

The reaction of *N*,*N'*-dilithiotrisilazanes with dichlorosilanes provides a better route to these silylated cyclodisilazanes,[59] while lithio-bis-aminosilanes yield the non-silylated derivatives.[60, 61] The spirocyclosilazane results using silicon tetrachloride.

$$R_2Si(NLiR')_2 + R''_2SiCl_2 \xrightarrow{-LiCl} R_2Si\langle NR' \rangle_2 SiR''_2 \quad \textbf{(4-II)}$$

$$Me_2Si(NLiMe)_2 + SiCl_4 \longrightarrow Me_2Si\langle NMe \rangle_2 Si \langle NMe \rangle_2 SiMe_2$$

(m.p. 34°C, b.p. 39°C/4 mm)

N,N'-Bis-(trimethylsilyl)tetramethylcyclodisilazane (**4-I**) has also been synthesized by other routes involving lithiated trisilazanes,[59,62] obtained from octamethyltrisilazane (**4-III**).

$$2(Me_3SiNLi)_2SiMe_2 \xleftarrow{4Bu^nLi} 2(Me_3SiNH)_2SiMe_2 \ (\mathbf{4\text{-}III}) \xrightarrow{2Bu^nLi} 2Me_3SiNLiSiMe_2NHSiMe_3$$

$$2(Me_3SiNLi)_2SiMe_2 \xrightarrow{2Me_3SiCl} (\mathbf{4\text{-}I}) + 2(Me_3Si)_2NLi$$

$$2(Me_3SiNLi)_2SiMe_2 \xrightarrow{(Me_3Si)_2NSiMe_2Cl} (\mathbf{4\text{-}I}) \text{ and } (\mathbf{4\text{-}III})$$

$$2Me_3SiNLiSiMe_2NHSiMe_3 \xrightarrow{Me_2SiCl_2} (\mathbf{4\text{-}I}) \text{ and } (\mathbf{4\text{-}III})$$

Table 4.2. Physical Properties of the Cyclodisilazanes (**4-II**)

R	R′	R″	m.p. (°C)	b.p. (°C)	n_D^{20}
Me	Me_3Si	Me	38–9	85/7 mm	
Me	Me	Me	16–17	52/51 mm	1.4210
Me	Me	Ph		143/4 mm	1.5460
Ph	Ph	Me	180–1		

Chlorosilylamines are used in a more general method for synthesizing cyclodisilazanes, and through this route the first cyclodisilazane with an NH group was made.[63, 64]

$$Me_3SiN(SiMe_2Cl)_2 + RNH_2 \longrightarrow Me_3Si{-}N\langle(SiMe_2)_2\rangle NR \ (R = H, Me, Et, Ph)$$

A complex reaction sequence employing dimethyldichlorosilane with cyclopolysilazanes gives two major products, depending on the ratio of starting materials.

$$4Me_2SiCl_2 + [Me_2SiNH]_4 \xrightarrow{24\text{ hr}} 4(ClMe_2Si)_2NH \xrightarrow{(Me_3Si)_2NNa} Me_3SiN\langle(SiMe_2)_2\rangle NSiMe_3 \ (\mathbf{4\text{-}I})$$

$$4(ClMe_2Si)_2NH \xrightarrow{(Me_3Si)_2NNa} ClMe_2SiN\langle(SiMe_2)_2\rangle NSiMe_2Cl$$

$$12Me_2SiCl_2 + \left\{\begin{matrix} 8[Me_2SiNH]_3 \\ 6[Me_2SiNH]_4 \end{matrix}\right\} \longrightarrow ClMe_2SiN\langle(SiMe_2)_2\rangle NSiMe_2Cl \xrightarrow{MeMgI} Me_3SiN\langle(SiMe_2)_2\rangle NSiMe_3 \ (\mathbf{4\text{-}I})$$

With an excess of chlorosilane, 1,3-dichlorotetramethyldisilazane is formed,[65,66] while using similar proportions gives chlorosilylcyclodisilazanes.[67] Surprisingly, attack on the NH group does not occur with an excess of chlorosilane. Both chlorosilazanes can readily be converted to permethyldisilylcyclodisilazane (**4-I**).[68]

Various classes of silicon–nitrogen compounds form cyclodisilazanes on pyrolysis. Bis-arylaminosilanes readily deaminate on heating, and a useful variation on this involves trans-amination using a high boiling amine.[61,69]

$$2R_2Si(NHAr)_2 \xrightarrow{300°C} [R_2SiNAr]_2 + 2ArNH_2$$

$$2R_2Si(NHEt)_2 + ArNH_2 \longrightarrow 2[R_2SiNAr]_2 + 4EtNH_2$$

Indeed, polymers containing the cyclodisilazane ring in the skeleton have been synthesized this way.

$$Me_2Si(NHEt)_2 + H_2N\text{—}C_6H_4\text{—}C_6H_4\text{—}NH_2 \longrightarrow \left[\text{—}C_6H_4\text{—}C_6H_4\text{—}N{<}(SiMe_2)_2{>}N\text{—}\right]_n$$

(m.p. 273–6°C)

Phenyl group migration from tin to magnesium has already been mentioned (p. 156). A similar migration also occurs with triphenylsilyl azide, the phenyl group migrating from Si to N on pyrolysis.[70, 71]

$$2RR'PhSiN_3 \xrightarrow{heat} RR'Si{<}(NPh)_2{>}SiRR' + 2N_2$$

Group migration also occurs on pyrolyzing *N*-phenylpentamethylchlorodisilazane and *N*,*N*′-diphenyloctamethyltrisilazane. *N*,*N*′-Diphenyltetramethylcyclodisilazane results in each case.[61]

$$PhN(SiMe_3)(SiMe_2Cl) \xrightarrow{-Me_3SiCl} PhN{<}(SiMe_2)_2{>}NPh \xleftarrow{-PhN(SiMe_3)_2} (Me_3SiNPh)_2SiMe_2$$

Primary amines and silylamines all react with silanes at about 100°C in the presence of sodium hydride, again providing a convenient synthetic route to cyclodisilazanes.[65, 72]

$$2R'NH_2 + 2R_2SiH_2 \xrightarrow[NaH]{80°C} [R_2SiNR']_2 + 4H_2 \ (R' = \text{alkyl or } Bu^n_3Si)$$

Chemical Properties

A crystal structure determination on *N,N'*-bis-trimethylsilyltetramethylcyclodisilazane indicates a planar 4-membered ring.[73]

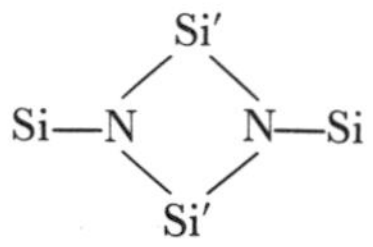

Si—N 0.1707 nm
Si'—N 0·1724 nm
NSiN 88 degrees
SiNSi 92 degrees

Fig. 4.5. The *N,N'*-disilylcyclodisilazane skeleton

The Si—N bond lengths are equal within experimental error, and are close to those observed for trisilylamine $(H_3Si)_3N$ and for $[(Me_3Si)_2NLi]_3$ (mentioned in Chapter 1, p. 5). This bond length indicates multiple bonding, while the down-field shift of the dimethylsilyl group in comparison to the trimethylsilyl group in the n.m.r. spectrum may show less π-bonding in the Si'—N bond.[74]

Cyclodisilazanes are thermally stable, but catalysts induce ring expansion. Thus, with ammonium bromide *N,N'*-dimethyltetramethylcyclodisilazane gives the trisilazane, while *N,N'*-dimethyltrimethylphenylcyclodisilazane gives mono- and di-phenylcyclotrisilazanes in equal proportions.[60]

$$3[Me_2SiNMe]_2 \xrightarrow[\text{heat}]{NH_4Br} 2[Me_2SiNMe]_3 \text{ (m.p. 36°, b.p. 91°C/12 mm)}$$

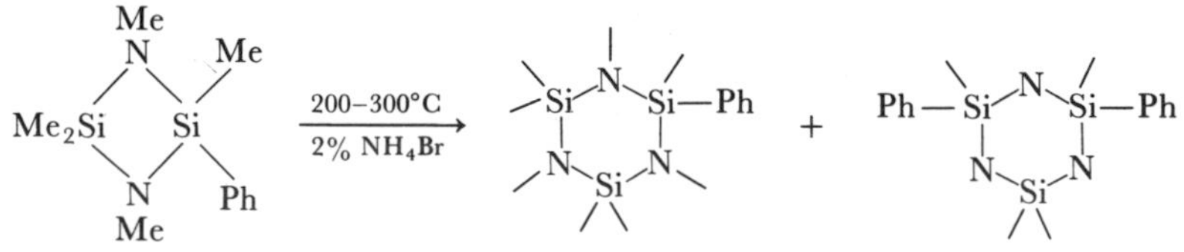

200–300°C, 2% NH_4Br

(b.p. 69°C/3 mm) (b.p. 116°C/1 mm) (b.p. 160°C/1 mm)

This ring expansion produced a significant rise of 70–80 cm^{-1} in the asymmetric Si—N—Si stretching frequency in the infrared spectrum of this compound. Aminosilylcyclodisilazanes are thermally stable above 200°C, but acid catalysts cause ring expansion.[67]

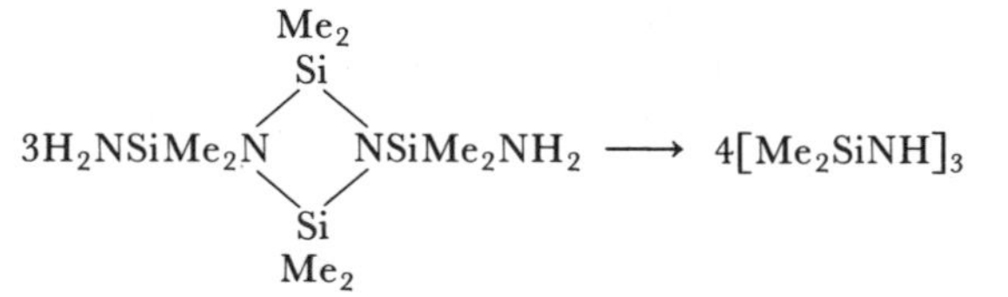

$$3H_2NSiMe_2N\langle(SiMe_2)_2\rangle NSiMe_2NH_2 \longrightarrow 4[Me_2SiNH]_3$$

$$MeNHSiMe_2N\langle SiMe_2 \rangle_2 NSiMe_2NHMe \longrightarrow [MeN-SiMe_2-NH-SiMe_2-NMe-SiMe_2-NH-SiMe_2]_{cyclic}$$

These reactions show the existence of *N*-protonated cyclotri- and tetrasilazanes. Only one cyclodisilazane with an NH group as a ring member has been synthesized to date. This is *N*-trimethylsilyltetramethylcyclodisilazane, and is conveniently prepared from bis-(chlorodimethylsilyl)-trimethylsilylamine and ammonia.[75]

$$Me_3SiN(SiMe_2Cl)_2 + 3NH_3 \longrightarrow Me_3Si{-}N\langle SiMe_2 \rangle_2 NH + 2NH_4Cl$$

The exceptional nature of the N—H group is confirmed by its high stretching frequency (3580 cm^{-1}) and low deformation mode (1080 cm^{-1}). Acids and bases catalyzed the dimerization to a disilazanylcyclodisilazane, with NH peaks at 3380 and 1178 cm^{-1}, while silylation occurs with bis-(trimethylsilyl)acetamide (BSA).

$$Me_3SiN\langle SiMe_2 \rangle_2 NH \longrightarrow Me_3SiNHSiMe_2N\langle SiMe_2 \rangle_2 NSiMe_2NHSiMe_3$$

Attempts to silylate via metallation resulted in ring opening, the first case of a Si—N cleavage by an alkyl-lithium compound.

$$Me_3SiN\langle SiMe_2 \rangle_2 NH \xrightarrow{BSA} [Me_3SiNSiMe_2]_2$$

$$[Me_2SiNEt]_2 + MeLi \longrightarrow Me_3SiNEtSiMe_2NLiEt \xrightarrow{Me_3SiCl} (Me_2SiNEt)_2SiMe_2$$

The stability of the exocyclic Si—N bond in these cyclodisilazanes to protonated compounds has led to the synthesis of various trisilylamines, while hydrolysis gives siloxazanes.[64, 75]

$$Me_3SiN(SiMe_2)_2NSiMe_3 \xrightarrow{ROH} Me_3SiN(SiMe_2OR)_2$$

$$Me_3SiN(SiMe_2)_2NSiMe_3 \xrightarrow{pentane/H_2O} [Me_3SiN(SiMe_2)_2O]_2 \text{ (8-membered ring: } Me_2Si\text{–O–}SiMe_2\text{–}NSiMe_3\text{–}SiMe_2\text{–O–}SiMe_2\text{–}NSiMe_3)$$

$$Me_3SiN(SiMe_2)_2NSiMe_3 \xrightarrow{HCl} Me_3SiN(SiMe_2Cl)_2 \xrightarrow{MeNH_2} Me_3SiN(SiMe_2)_2NMe$$

The dilithiated derivatives of perfluoroaromatic primary amines reacted with dichlorosilanes to give the first water-stable cyclodisilazanes. Diphenyldichlorogermane also gives cyclogermanazanes.[76]

$$C_6F_5NH_2 \xrightarrow[-70°C/THF]{2Bu^nLi} C_6F_5NLi_2 \xrightarrow{Me_2SiCl_2} [Me_2SiNC_6F_5]_2 \text{ (m.p. 305°C)}$$

$$C_6F_5NLi_2 \xrightarrow{Ph_2SiCl_2} [Ph_2SiNC_6F_5]_2 \text{ (m.p. 296°C)}$$

$$C_6F_5NLi_2 \xrightarrow{Ph_2GeCl_2} [Ph_2GeNC_6F_5]_2 \text{ (m.p. 331–2°C)}$$

Though no 4-membered tin–nitrogen rings are known, one involving Ge and Sn has been synthesized from a silylamine, and again illustrates the stability of Si—N and Ge—N bonds to alkyl-lithium reagents. The product contains nitrogen atoms bonded to Si, Ge and Sn atoms at the same time.[77]

$$Et_3SiNH_2 \xrightarrow{Bu^nLi} Et_3SiNHLi \xrightarrow{Me_2GeCl_2} (Et_3SiNH)_2GeMe_2 \quad \text{(b.p. 95°C/0.2 mm)}$$

$$(Et_3SiNH)_2GeMe_2 \longrightarrow (Et_3SiNLi)_2GeMe_2 \xrightarrow{Me_2SnCl_2} Et_3SiN{<}^{GeMe_2}_{SnMe_2}{>}NSiEt_3$$

(m.p. 1–3°C, b.p. 126°C/0.2 mm)

Cyclotri- and -tetra-silazanes

Synthesis

A wide variety of products result from the reaction of dichlorosilanes with primary amines and ammonia, the variety being normally determined by the groups attached to the silicon atom. With ammonia, dichlorosilane gives a polymer,[78] while methyldichlorosilane yields a mixture of cyclic and polymeric silazanes.[79] Ethyl and phenyldichlorosilanes, along with many diorganodichlorosilanes, give a mixture of cyclotri- and tetra-silazanes.[c]

$$RR'SiCl_2 + 3NH_3 \longrightarrow \frac{1}{n}[RR'SiNH]_n + 2NH_4Cl$$

(n = 3 and 4, R = Et, Ph, R′ = H; R = R′ = Me, Et, vinyl, allyl, phenyl)

With more hindered groups on silicon, monomeric products such as silane diamines and aminosilazanes result.[c] These have to be heated, often in the presence of an acid catalyst, to give the cyclosilazanes.

$$R_2Si(NH_2)_2 \longrightarrow (H_2NSiR_2)_2NH \longrightarrow [R_2SiNH]_{3\text{ and }4}$$

With primary amines, dichlorosilanes yield bis-aminosilanes, and only with methylamine and methyldichloro- or ethyldichloro-silane are cyclosilazanes produced.

$$MeHSiCl_2 + 3MeNH_2 \longrightarrow \frac{1}{n}[MeHSiNMe]_{3\text{ and }4} + 2MeNH_3Cl$$

(n = 3, b.p. 56°C/5 mm, n_D^{20} 1.4580; n = 4, b.p. 88°C/2 mm, n_D^{20} 1.4810)

These bis-aminosilanes are quite stable thermally, but a trace of $(NH_4)_2SO_4$ catalyzes deamination. With small groups on both nitrogen and silicon, good yields of cyclotrisilazanes result, but these fall as the group size increases.[80, 81]

$$Me_2Si(NHMe)_2 \longrightarrow [Me_2SiNMe]_3 \quad \text{(yield 90\%)}$$
(m.p. 33°C, b.p. 96°C/10 mm)

$$Me_2Si(NHCH_2Ph)_2 \longrightarrow [Me_2SiNCH_2Ph]_3 \quad (19\%)$$
(m.p. 57–60°C)

$$(p\text{-}MeOC_6H_4)_2Si(NHMe)_2 \longrightarrow [(p\text{-}MeOC_6H_4)_2SiNMe]_3 \quad (20\%)$$
(m.p. 280–5°C)

Thermolysis has been reported to give cyclic dimers, which themselves ring-expand in the presence of an acid catalyst. This occurs with all but the aniline derivative, which gives the cyclodisilazane even with the catalyst.[80, 82]

$$2Me_2Si(NHPh)_2 \xrightarrow[NH_4^+]{heat} [Me_2SiNPh]_2 + 2PhNH_2$$

Generally, however, cyclotri-[60] and tetra-silazanes[67] result.

Heating methyldisilylamine at 400°C produces *N,N′,N″,N‴*-tetramethylcyclotetrasilazane,[83] while condensing trisilylamine or reacting ammonia with silylchloride gives *N,N′,N″*-trisilylcyclotrisilazane.[84] In each case the silyl group has been deprotonated, as previously shown in the cyclodisilazane syntheses.

$$4MeN(SiH_3)_2 \longrightarrow [MeNSiH_2]_4 + SiH_4$$

$$3(H_3Si)_3N \xrightarrow{-3SiH_4} [H_2SiNSiH_3]_3 \longleftarrow 6NH_3 + 12H_3SiCl$$

Lithiation has been used to prepare cyclosilazanes from linear aminosilanes and from *N*-protonated cyclosilazanes. This work has led to an interesting series of ring size changes. Aminodisilazanes react readily with n-butyl-lithium, and subsequent reactions with dichlorosilanes give cyclotrisilazanes.[60, 85]

$$\text{cyclo-}[Me_2SiNMe\,SiMe_2NMe\,MCl_nNMe] \xleftarrow[AsCl_3]{SnCl_4} (Me_2SiNLiMe)_2NMe \xrightarrow{RR'SiCl_2} \text{cyclo-}[Me_2SiNMe\,SiMe_2NMe\,SiRR'NMe]$$

R = R′ = Cl, m.p. 55–8°C, b.p. 71°C/3 mm
R = R′ = Ph, m.p. 62–4°C, b.p. 151°C/3 mm
(M = Sn, *n* = 2, m.p. 38°C, b.p. 109°C/1 mm)
(M = As, *n* = 1, m.p. 2–3°C, b.p. 78°C/2 mm)

Other difunctional halides of the elements B, Ge, Sn, As and Sb react similarly to give ring-substituted cyclotrisilazanes.[75]

Lithium derivatives of *N*-protonated cyclotrisilazanes are white powders that are thermally stable in an inert atmosphere.[86] The partially-substituted ones are believed to be polymeric in non-polar solvents,[87] with mixed bridges similar to those discussed in Chapter 1 (p. 5).

Table 4.3. Physical Properties of Isomeric Cyclodi- and trisilazanes

Cyclotrisilazane ring: $(Me_2Si)_3(NX)_2(NY)$	**(4-IV)** X = Y = H	**(4-V)** X = H, Y = Me_3Si	**(4-VI)** Y = H, X = Me_3Si	**(4-VII)** X = Y = Me_3Si
b.p. °C/mm	111–2/85 188/756	112–3/10 255/760	107/1.1	126/0.4
m.p.°C	−10			54–7
n_D^{20}	1.4448	1.4596^{26} 1.4613	1.4725	
d_4^t	0.9196^{20}	0.931^{27}		

$Me_3SiN\langle Si_2\rangle NSiMe_2NXY$	**(4-VIII)** X = Y = H	**(4-IX)** X = H, Y = Me_3Si	**(4-X)** X = Y = Me_3Si
b.p. °C/mm		82/2 260/760 (est.[d])	
m.p. °C	36	−74	69–70
n_D^{20}		1.4422	

N-Silylated products result with chlorosilanes, but their structure depends on the reaction conditions.

Thus, mono-, bis- and tris-(trimethylsilyl)hexamethylcyclotrisilazanes have all been isolated. The monosilylated compound was first synthesized by heating $[Me_2SiNH]_3$ with sodium and styrene in dioxan.[88] It also results through lithiation and silylation in diglyme between −60°C and 60°C.[89]

$$2[Me_2SiNH]_3 \xrightarrow[\text{dioxan}]{\text{Na/styrene}} \underline{|(Me_2SiNH)_2Me_2SiN}Na + Ph(CH_2)_4Ph$$

$$2[Me_2SiNH]_3 \xrightarrow[\text{2. } Me_3SiCl]{\text{1. Bu}^nLi \text{ in diglyme}} \underline{|(Me_2SiNH)_2Me_2SiN}SiMe_3$$

The di- and tri-substituted compounds result in a similar way[89] with n-butyl-lithium at −60°C, and the physical properties are recorded in Table 4.3. The p.m.r. spectra are in accordance with the structures.

$$\text{(4-VI)} \xleftarrow[-60^\circ C]{Me_3SiCl} [(Me_2SiNLi)_2Me_2SiNH] \xrightarrow[60^\circ C]{\text{diglyme}} Me_3SiN(SiMe_2)_2NSiMe_2NHSiMe_3 \ \text{(4-IX)}$$

(4-VI) **(4-IX)**

Isomeric silazanylcyclodisilazanes (**4-IX**) and (**4-X**) result on silylation at 60°C in hydrocarbon solvents, but not when the monolithium amide is silylated.

$$Me_3Si{-}N(SiMe_2)_2NSiMe_2Cl \xrightarrow{NH_3} Me_3SiN(SiMe_2)_2NSiMe_2NH_2$$

Only the cyclotrisilazane results, since the cyclodisilazane, which can be synthesized directly, readily undergoes a base-catalyzed rearrangement to the cyclotrisilazane at 20°C in the presence of traces of n-butyl-lithium.[89]

(**4-VI**) and (**4-IX**) both equilibriate with 1 mole of Bu^nLi, giving mainly the lithiocyclotrisilazane at low temperatures, since silylation gives mainly (**4-VII**).

$$\textbf{(4-IX)} \xrightarrow[2.\ Me_3SiCl]{1.\ Bu^nLi/-60^\circ C} \underset{(78\%)}{\textbf{(4-VII)}} + \underset{(11\%)}{\textbf{(4-X)}}$$

The infrared spectra of these isomers show distinguishing features, with characteristic bands in that of the cyclodisilazane at about 890 cm^{-1} ($\nu_{(SiNSi)_{as.}}$) and 1030 cm^{-1}, while the cyclotrisilazanes show a strong absorption at between 900 and 950 cm^{-1} ($\nu_{(SiNSi)_{as.}}$).[87] Likewise, the p.m.r. spectra can be interpreted on the basis of a cyclodisilazane.

$$\text{(cyclo-Si}_3\text{N}_3\text{ anion)} \longrightarrow Si{-}N(Si)_2NSi\overset{\ominus}{N}H \xrightarrow{Si{-}Cl} Si{-}N(Si)_2NSiNHSi$$

$$2Me_2Si(NH{-}SiMe_2)_2NLi + 2Me_3SiCl \longrightarrow [Me_2SiNH]_3 + Me_3SiNHSi(Me)_2N(SiMe_2)_2NSiMe_3$$

Lithiation of (**4-IV**) in a non-polar solvent gives a polymeric amide. Disproportionation and isomerization occur on silylation, producing (**4-IV**) and (**4-IX**).[90] The mechanism is thought to involve a trans-annular attack by the amide group, giving the cyclodisilazane. Its boiling point and refractive index are lower than the isomeric cyclotrisilazane. The more-hindered monolithium amide of hexaphenylcyclotrisilazane readily forms a monosilyl derivative.

In order to establish this ring contraction, methyldiphenylchlorosilane was used with the dilithium amide. The cyclodisilazane was formed (m.p. 66–7°C), and the proposed structure confirmed by X-ray analysis.[91] The p.m.r. spectrum gives methyl group peak intensities in the ratio 3 : 6 : 12 : 3 as expected.

$$[Me_2SiNH]_3 \xrightarrow[2.\ 2MePh_2SiCl]{1.\ 2Bu^nLi} MePh_2SiNHSiMe_2N\langle(SiMe_2)_2\rangle NSiPh_2Me$$

Again trans-annular nucleophilic attack is thought to occur after the initial silylation at one of the nitrogen atoms.[87] This ring-contraction reaction also occurs with cyclosiloxazanes, and will be discussed later in the Chapter (p. 192).

The co-proportionation of dichlorosilanes and cyclosilazanes has already been briefly mentioned with reference to the synthesis of cyclodisilazanes.[65] Nonamethylcyclotrisilazane and dimethyldichlorosilane react similarly, providing a terminal dichlorosilazane as a synthetic intermediate for unsymmetrical cyclotrisilazanes.[92]

$$2[Me_2SiNMe]_3 + 3Me_2SiCl_2 \longrightarrow 3(ClMe_2SiNMe)_2SiMe_2$$

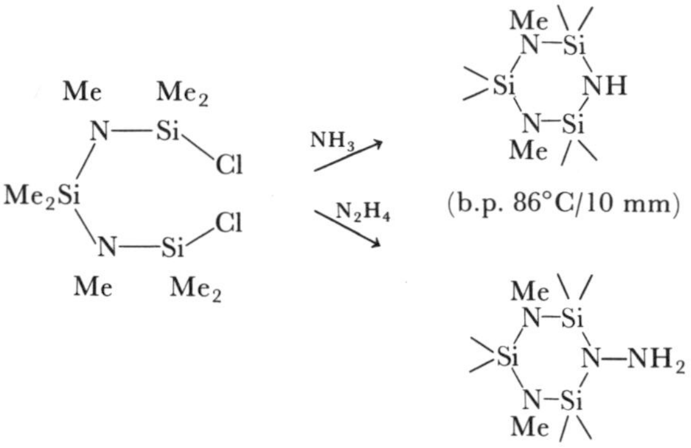

(b.p. 86°C/10 mm)

(m.p. 45°C, b.p. 57°C/0.05 mm)

Structures

An electron diffraction study of hexamethylcyclotrisilazane in the gas phase shows that the ring is puckered with Si—N bond lengths close to the single bond value of 0.180 nm.[93] Lack of appreciable delocalization is also supported by the constancy of the Si—Me p.m.r. absorption position even on replacement of an N—Me group by a methylene group in the cyclotrisilazane ring $[Me_2SiNMe]_3$.[85]

Both hexamethylbenzene and borazine form "piano stool" complexes with the $Cr(CO)_3$ residue.[94]

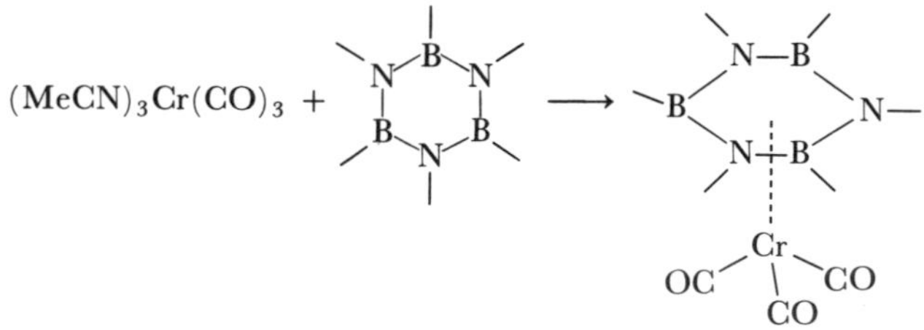

$(MeCN)_3Cr(CO)_3$ + [borazine] ⟶ [borazine–$Cr(CO)_3$ complex]

The bonding is considered to involve the ring π-orbitals and vacant d orbitals on the metal. Attempts to bond hexamethylcyclotrisilazane in a similar way proved unsuccessful, though it will function as a tridentate ligand with titanium and vanadium trichlorides, which are good "a"-type acids and readily bond to amines through σ bonds. The visible spectrum shows d–d transitions typical of a tri-amine complex of the halides.[95]

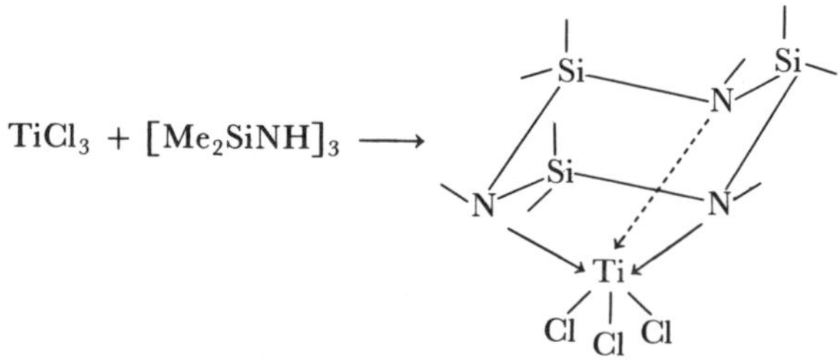

$TiCl_3 + [Me_2SiNH]_3 \longrightarrow$ [tridentate complex]

Hexamethyl 2,4,6-triphenyl- and tri-*p*-anisylcyclotrisilazanes result from dimethyldichlorosilane and the aromatic amine. Two separable geometric isomers result for each compound, which can be readily distinguished by p.m.r. spectroscopy.[80]

cis *trans*

R = aryl, CH_3 occupies unspecified valency sites.

cis	*trans*
R = phenyl, m.p. 153–4°C	m.p. 84–8°C
τ 9.52 Si–Me singlet	9.51, 9.48 (2 : 1)
τ 7.73 N–Me singlet	7.77
R = *p*-anisyl. m.p. 176–7°C	m.p. 114–6°C
τ 9.57	9.60, 9.57 (2 : 1)
τ 7.75	7.82

Fig. 4.6. The *cis* and *trans* isomers of hexamethyl-2,4,6-triarylcyclotrisilazane

A similar phenomenon is observed with the cyclosiloxanes[96] and cyclosilthianes.

Octamethylcyclotetrasilazene melts at 17.5°C, and a gas-phase electron-diffraction study shows the ring to be puckered.[93] An X-ray study has revealed a most singular crystal structure. Two fundamentally different geometrical isomers are present in the lattice in equal proportions and in an ordered array.[97] One is a chair form, the other a cradle.

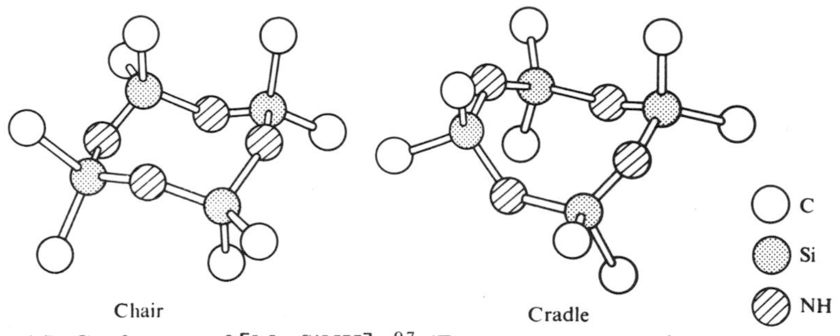

Fig. 4.7. Conformers of $[Me_2SiNH]_4$.[97] (From G. S. SMITH and L. E. ALEXANDER, *Acta. Cryst.*, 1963, **16,** 1015.

The Si—N bond lengths average to 0.1734 and 0.1722 nm, which are significantly, but not greatly, shorter than the single bond length. They are close to the values found already, and the side Si—N—Si bond angles in both isomers of 130 degrees also support some multiple-bonding between N and Si.

Chemical Properties

While the cyclosilazanes are quite stable thermally, rearrangements, polymerization and elimination reactions readily occur at high temperature and in the presence of catalysts, both acidic and basic. Thus, polymerization reactions often lead to cross-linked products through the occurrence of simultaneous rearrangements.

Small quantities of H_2SO_4 will catalyze ring expansion of the cyclotrisilazane at room temperature, and Lewis acids (*e.g.*, $AlCl_3$, $TiCl_4$ and $SnCl_4$) do so at 200°C. This reaction can be reversed at 350–450°C, and the contraction of permethylcyclotri- and tetra-silazanes to the cyclodisilazane has already been considered (p. 167).

$$4[Me_2SiNH]_3 \underset{350-450°C}{\overset{H_2SO_4}{\rightleftarrows}} 3[Me_2SiNH]_4$$

The mechanism of the ring expansion probably involves ring cleavage followed by recombination, through protonation of nitrogen.[98]

$$[Me_2SiNH]_3 \xrightarrow{H_2SO_4} HOSO_2O(SiMe_2NH)_2SiMe_2NH_2$$

$$\downarrow$$

$$HOSO_2OSiMe_2NH_2 + HOSO_2O(SiMe_2NH)_2H \quad (-H_2SO_4)$$

$$\longrightarrow 2H_2SO_4 + [Me_2SiNH]_4$$

Support for this mechanism involving silylsulphate intermediates was found in the reaction of bis-(trimethylsilyl)sulphate and triethylaminosilane. Sulphuric acid elimination gives the mixed disilylamine, and the acid catalyzes aminosilane self-condensation.

$$(Me_3SiO)_2SO_2 + 2H_2NSiEt_3 \longrightarrow 2Me_3SiNHSiEt_3 + H_2SO_4$$

$$2Et_3SiNH_2 + H^+ \longrightarrow (Et_3Si)_2NH + NH_4^+$$

$$H_2SO_4 + 2NH_4^+ \longrightarrow (NH_4)_2SO_4 + 2H^+$$

Similarly, with H_2SO_4, 2,2-diethyl-4,4,6,6-tetramethylcyclotrisilazane produces a random mixture of all the cyclotri- and tetra-silazanes, supporting random ring cleavage by the acid.

While hexaphenylcyclotrisilazane eliminates benzene at 450°C to give a highly cross-linked polymer, methylcyclotri- and tetra-silazanes eliminate methane on heating with 1% KOH. Products include cyclic oligomers and brittle polymers.[c]

$[Me_2SiNH]_3$ —(200°C, $-CH_4$)→

H
Si N Si
Si N Me N Si
Si
H N Si N Si N H

(m.p. 165–7°C)

$[Me_2SiNH]_3$ —(330°C, $-CH_4$)→

H Me
Si N Si N
H N Si N Si

At 160°C with NH_4^+, these cyclosilazanes yield simpler cross-linked polymers in 7 hr. They are waxes or elastomers, depending on the degree of polymerization.[99]

$[Me_2SiNH]_{3\ \text{or}\ 4}$ —(160°C, NH_4X)→

Si N Si N Si N Si N
Si N Si Si N Si
H H

The cyclotri- and tetra-silazanes $[R_2SiNH]_{3\ \text{and}\ 4}$ readily undergo trans-aminations with primary amines. Aniline gives the bis-aniline and various bis-alkylamino derivatives $R_2Si(NHR)_2$,[100] while the cyclodisilizane $[Me_2SiNPh]_2$ (m.p. 252.5°C) results from methylcyclosilazanes.[c] n-Pentylamine mono-substitutes the ring, while the phenyl compound can be prepared directly.[82]

$[Me_2SiNH]_3$ —($n\text{-}C_5H_{11}NH_2$)→

$n\text{-}C_5H_{11}$
Si N Si
HN Si NH

(b.p. 85–7°C/1 mm)

$3Me_2SiCl_2 + 8NH_3 + PhNH_2$ ⟶

Ph
Si N Si
HN Si NH

(m.p. 37°C)

Aromatic diamines readily give low molecular weight polymers,[101] while aliphatic diamines yield cross-linked polymers believed to involve polymethylenediamine chains cross-linked through cyclotrisilazane residues.[102]

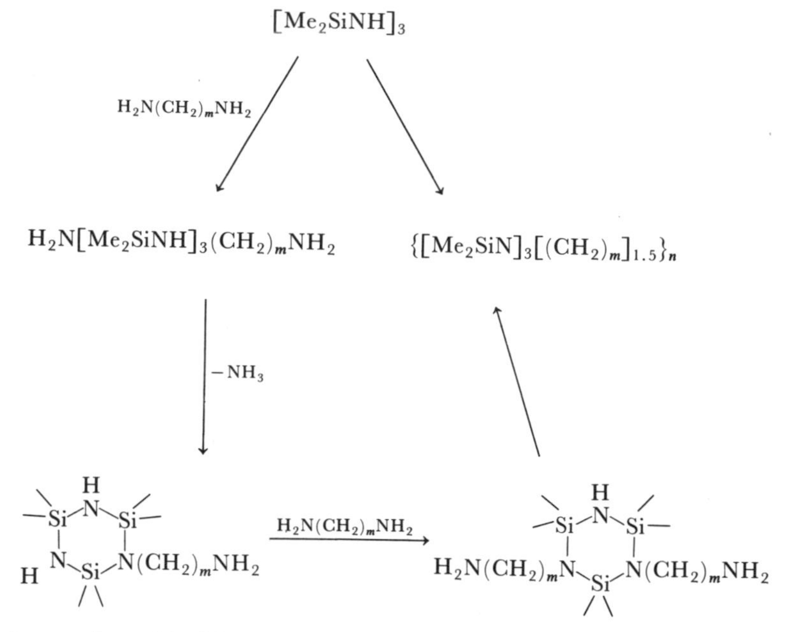

Ammonium chloride catalyzes the equilibration of hexamethyldisilazane and cyclotrisilazane to octamethyltrisilazane,[79] while $[Me_2SiNH]_3$ with ammonia under pressure yields polymeric linear silazanes.[103]

$$3(Me_3Si)_2NH + [Me_2SiNH]_3 \xrightarrow{NH_4Cl} 3Me_3SiNHSiMe_2NHSiMe_3$$

$$5[Me_2SiNH]_3 \xrightarrow{NH_3/140°C} H_2N[SiMe_2NH]_{15}H$$

The cyclopolysilazanes readily equilibrate with dichlorosilanes, and regulating the ratio of reactants gives either cyclodisilazanes or sym-dichlorotetramethyldisilazane. This has already been considered before in the discussion of cyclodisilazanes[65, 66, 67] (p. 168).

Controlled cleavage of $[Me_2SiNH]_{3\text{ and }4}$ with HX (X = F, Cl or Br) gives the dihalogenodisilazanes.[65]

$$[Me_2SiNH]_4 + 6HX \longrightarrow 2(XSiMe_2)_2NH + 2NH_4X$$

(X = F,Cl or Br; b.p. 120°C, 64°C/11 mm and 49°C/0.1 mm respectively)

Fluorinated linear polysilazanes can also be formed from cyclosilazanes[104] using the oldest of all Si—N compounds, the adduct $SiF_4{\cdot}2NH_3$, which was first synthesized by John Davy, brother of Sir Humphrey, and by Gay-Lussac.[105] The products $F[SiMe_2NH]_nSiMe_2F$ have all been isolated for $n = 0 \rightarrow 3$.

Examples have been given of the way acids and Lewis acids (*e.g.*, $AlCl_3$,

$SnCl_4$) cause rearrangements on cyclosilazanes, and how co-proportionation occurs with chlorosilanes (pp. 168, 170). While stannic chloride catalyzes the ring-expansion of hexamethylcyclotrisilazane at 195°C, it will also complex with the halide forming a compound similar to those of VCl_3 and $TiCl_3$ already mentioned.

$$[Me_2SiNH]_{3\text{ and }4} + SnCl_4 \longrightarrow [Me_2SiNH]_{3\text{ and }4}\cdot SnCl_4 \quad \begin{pmatrix}\text{sublimable}\\ \text{white solid}\end{pmatrix}$$

Presumably the cyclosilazane functions as a bidentate ligand.[106]

With the boron halides, ring cleavage occurs, the products depending on the proportions of reactants.[107]

$$Pr^n_2BCl + [Me_2SiNH]_3 \longrightarrow Pr^n_2B(NHSiMe_2)_3Cl \text{ (b.p. 125–30°C/0.1 mm)}$$

$$\downarrow 2Pr^n_2BCl$$

$$Me_2SiCl_2 \xleftarrow{Pr^n_2BCl} Pr^n_2BNHSiMe_2Cl \text{ (b.p. 77°C/0.1 mm)}$$

Cyclosilazanes are relatively stable to water and alkali, but are readily cleaved by acids and alcohols.[c] The alcoholysis can be controlled, and the various step-wise breakdown products isolated.[108] Terminal alkoxytri- and -disilazanes are formed first, bis-(alkoxy)silanes being the final product.[109] Organic acids give bis-(acyl)silanes.

Stability to hydrolysis, which yields the amine and the silanol, increases with the size of the substituents on both silicon[c] and nitrogen,[80] and is greater for the cyclotetrasilazane than the cyclotrisilazane.

$$[Me_2SiNH]_3 \xrightarrow[NH_4^+]{EtOH} (EtOSiMe_2NH)_2SiMe_2 \text{ (b.p. 92°C/11 mm)} + (EtOSiMe_2)_2NH \text{ (b.p. 70°C/11 mm)} \longrightarrow Me_2Si(OEt)_2$$

$$[Me_2SiNH]_3 \xrightarrow{PhOH} (PhOSiMe_2)_2NH \text{ (b.p. 119°C/1 mm)} + (PhO)_2SiMe_2$$

$$[Me_2SiNH]_3 \xrightarrow{Pr^iOH} (Pr^iOSiMe_2)_2NH \text{ (b.p. 80°C/11 mm)} + (Pr^iO)_2SiMe_2 \text{ (b.p. 29°C/11 mm)}$$

The physical properties of the commercially-useful silicones vary according to the ratio of di- and tri-halogenosilanes hydrolyzed. Likewise, the properties of polysilazanes produced by ammonolysis vary with the chlorosilane ratio and with the dilution. In benzene, high yields of soluble polymeric products result, along with about 5% of bicyclic compounds.[110]

$$2R'R''SiCl_2 + 2R'''SiCl_3 + 16NH_3 \longrightarrow$$

R″ H R‴
N
R′—Si Si—NH₂
HN NH
R‴—Si Si—R′
N N
H₂ H R″

$+$ polymer $+ 10NH_4Cl$

R″ H
N R‴
R —Si Si
HN HN NH
Si Si—R″
R‴ N
H R′

($R' = R'' = Me$; $R''' = Bu^n$, b.p. 194–7°C/4 mm)

($R' = Me$; $R'' = Et$; $R''' = Bu^n$, b.p. 156–61°C/1 mm)

Using alkyltrichlorosilanes gives cage compounds analyzing for $[RSi(NH)_{1.5}]_{6\text{ or }8}$. With small groups such as methyl or ethyl, only small yields result (10–15%), but with n-hexyl → n-nonyl, yields increase, presumably because of increased hindrance to cross-linking. Hexamers normally form, but with n-octyltrichlorosilane, the octamer results. The structures[111] of $[MeSi(NH)_{1.5}]_6$ (m.p. 260°C) and $[n\text{-}C_8H_{17}Si(NH)_{1.5}]_8$ (b.p. 345–50°C/1 mm) probably resemble the siloxane cages[112] $[RSiO]_{6\text{ and }8}$ which comprise fused 6- and 8-membered rings.

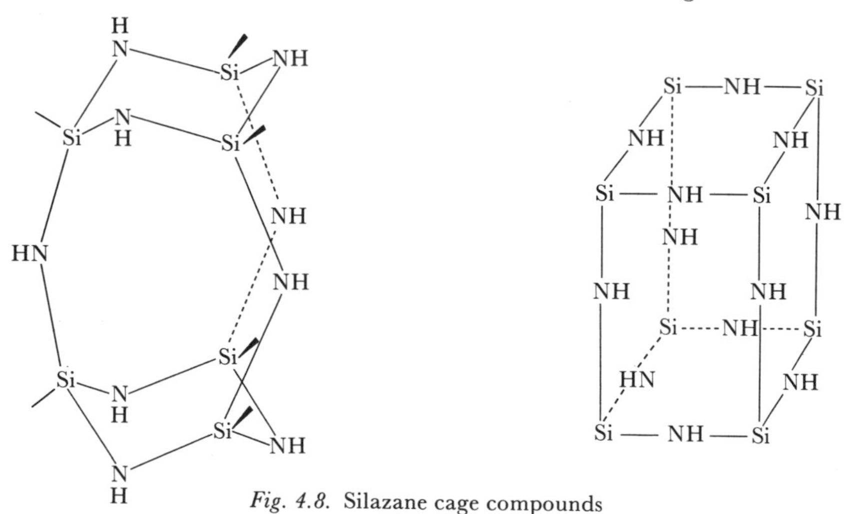

Fig. 4.8. Silazane cage compounds

The Si—N bond will add to both CO_2 and CS_2, and cyclotri- and -tetrasilazanes will also add to phenylisocyanates across the C=N bond.[113] Two moles of the isocyanate add to each Me_2SiNH residue, the structures being established by X-ray analysis[114] and chemical degradation.

$$1/n[Me_2SiNH]_n + 2PhNCO \xrightarrow[\text{toluene}]{90°C} \text{cyclo-}[Me_2Si(NPh)C(=O)NHC(=O)(NPh)] \text{ (m.p. 188–190°C)}$$

$$\text{cyclo-}[Me_2Si(NPh)C(=O)NHC(=O)(NPh)] \xrightarrow{2HCl} (PhNHCO)_2NH + Me_2SiCl_2$$

$(PhNHCO)_2NH$ (m.p. 212°C)

$$1/n[Me_2SiNH]_n \xrightarrow{p\text{-}BrC_6H_4NCO} \text{cyclo-}[Me_2Si(NC_6H_4Br)C(=O)NHC(=O)(NC_6H_4Br)] \text{ (m.p. 215°C dec}^n\text{.)}$$

Cyclic germanium–nitrogen and tin–nitrogen compounds

The chemistry of cyclic germanium–nitrogen compounds has received little attention, partly because of the difficulty and expense involved in synthesizing the required halides. The first compound, $[Et_2GeNH]_3$, was made from diethylgermanium dibromide and liquid ammonia in 1932.[115]

$$3Et_2GeBr_2 + 9NH_3 \longrightarrow [Et_2GeNH]_3 + 6NH_4Br$$

With the less-hindered dimethyldichlorogermane, ammonia reacts to give the trigermanylamine,[116] and not the cyclotrigermanazane as is obtained in the silicon system.

$$3Me_2GeCl_2 + 4NH_3 \longrightarrow \underset{\text{(m.p. 62°C)}}{(Me_2ClGe)_3N} + 3NH_4Cl$$

With methylamine, however, the trigermanazane is formed, and, unlike the analogous cyclotrisilazane, is readily cleaved by methyl-lithium at the Ge—N bond. The amide so formed has been used as a synthetic intermediate in the synthesis of unsymmetrical tertiary amines.[117, 118]

$$Me_2GeCl_2 + MeNH_2 \longrightarrow \underset{\text{(b.p. 80°C/2 mm)}}{[MeNGeMe_2]_3} \xrightarrow{3MeLi} Me_3GeN(Li)Me$$

$$Me_3GeN(Li)Me \xrightarrow{Me_3PbCl} \underset{\text{(b.p. 49°C/2 mm)}}{Me_3Pb\text{—}N(Me)\text{—}GeMe_3}$$

Germanium tetrachloride and methylamine form a hygroscopic cyclotrigermanazane,[119] with chlorine atoms as the germanium substituents.

$$9MeNH_2 + 3GeCl_4 \longrightarrow \underset{\text{(m.p. 133–135°C)}}{[MeNGeCl_2]_3} + 6MeNH_3Cl$$

Lithium salts of amines and hydrazines are used as intermediates in the synthesis of cyclic Ge—N and Sn—N compounds. The lithium adduct of azobenzene forms a hydrazine derivative with diphenyldichlorogermane.[120]

$$2\ (Ph)(Li)N\text{—}N(Ph)(Li) + 2Ph_2GeCl_2 \longrightarrow Ph_2Ge\langle N(Ph)\text{—}N(Ph)\rangle_2GePh_2$$

(m.p. 306–7°C)

Bis-(dimethylamino)dimethyltin results in a similar way, and readily undergoes trans-aminations in the presence of an excess of methyl- or ethylamine.[121] The cyclotristannazanes resulting from this are highly-reactive, air-sensitive liquids which are readily attacked by both moisture and CO_2.

$$Me_2SnCl_2 + 2LiNMe_2 \longrightarrow \underset{\text{(b.p. 138°C)}}{Me_2Sn(NMe_2)_2}$$

$$Me_2Sn(NMe_2)_2 \xrightarrow{MeNH_2} \underset{\text{(b.p. 114°C/0.2 mm)}}{[MeNSnMe_2]_3}$$

$$Me_2Sn(NMe_2)_2 \xrightarrow{EtNH_2} \underset{\text{(b.p. 104°C/0.05 mm)}}{[EtNSnMe_2]_3}$$

While few reactions have been studied with these Ge—N and Sn—N rings, the reactions of linear amino-germanes and -stannanes with the constituents of air are worth noting so that due care can be taken in handling these compounds. The Si—N, Ge—N and Sn—N bonds are all readily hydrolyzed, and CO_2 will add to all three of the bonds. The silyl[122] and germyl[123] compounds are distillable liquids while the tin carbamate is a solid.[124]

$$Me_3SiNEt_2 \xrightarrow{CO_2/\text{trace } Et_2NH} \underset{\text{(b.p. 74°C/15 mm)}}{Me_3SiOCONEt_2}$$

$$\underset{\text{(b.p. 176°C)}}{Et_3GeNMe_2} \xrightarrow{CO_2} \underset{\text{(b.p. 115°C/5 mm)}}{Et_3GeOCONMe_2}$$

$$Me_3SnNMe_2 \xrightarrow{CO_2} \underset{\text{(m.p. 165°C)}}{Me_3SnOCONMe_2}$$

Miscellaneous silicon–nitrogen

Cyclic compounds in which the silicon and nitrogen atoms do not alternate around the ring are common, and can be conveniently synthesized from hydrazines and disilanes. They will be considered in order of increasing ring size.

Three-membered Ring

Heating the dilithium salt of octamethyltrisilazane with $(Me_3Si)_2$-$NSiMe_2Cl$ gives not only $[Me_2SiNSiMe_2]_2$ but also the cyclodisilanimine (**4-XI**).[62] This is very stable thermally, but can be conveniently cleaved at the Si—Si bond by bromine, to give the unsymmetrical trisilylamine.

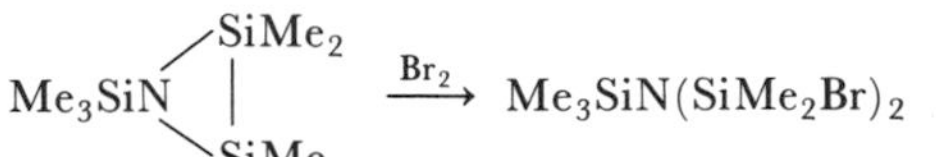

$$Me_3SiN\langle(SiMe_2)_2 \xrightarrow{Br_2} Me_3SiN(SiMe_2Br)_2$$

(**4-XI**) (b.p. 173°C) (b.p. 122–5°C/12 mm)

Five-membered Rings

These are known with both disila- and diaza- linkages.[125] Permethyl-2,5-diaza-1,3,4-trisilacyclopentane results from 1,2-dichlorotetramethyldisilane and the lithium salt of bis-(methylamino)dimethylsilane.

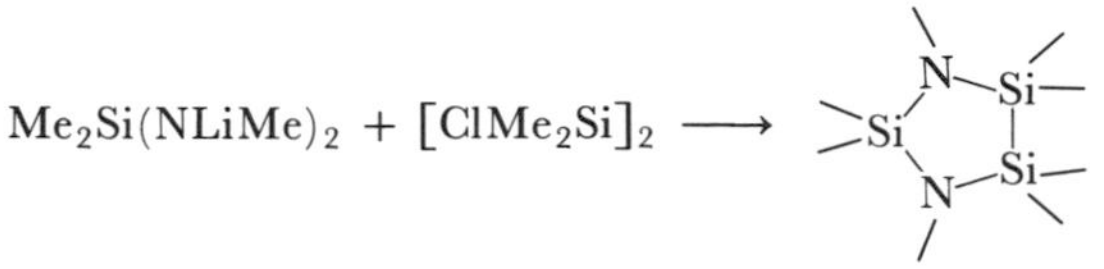

$$Me_2Si(NLiMe)_2 + [ClMe_2Si]_2 \longrightarrow Me_2Si(NMe)_2(SiMe_2)_2$$

(m.p. −12°C, b.p. 73°C/12 mm)

$$5N_2H_4 + 2MeN(SiMe_2Cl)_2 \longrightarrow MeN(SiMe_2)_2N{-}N(SiMe_2)_2NMe + HN_2H_4{\cdot}HCl$$

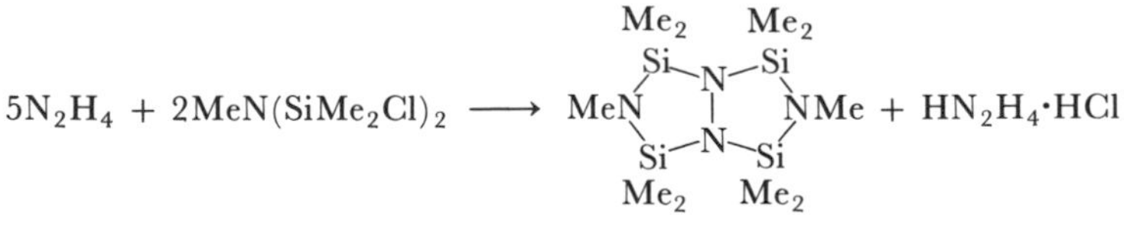

(m.p. 8–9°C, b.p. 62°C/1 mm)

The triazadisilacyclopentane ring appears more difficult to synthesize, but is reported to result from hydrazine and 1,3-dichloropentamethyldisilazane. The product is believed to have a fused bicyclopentane structure.

The azatetrasilacyclopentane ring has already been described in the Section devoted to $[Ph_2Si]_4$[126] (p. 158).

Six-membered Rings

Hydrazines[127] and their dilithium salts[120] react with dichlorosilanes to give 1,2,4,5-tetra-aza-3,6-disilacyclohexanes. Better yields are obtained using the dilithium derivatives, and all products show a characteristic peak in the infrared at 890 cm^{-1}.

$$RHNNR'H + R^2R^3SiCl_2 \longrightarrow RHNNR'SiR^2R^3NRNHR' \xrightarrow{-RHNNR'H} \text{cyclo-}[R^2R^3Si(NR\text{—}NR^1)]_2$$

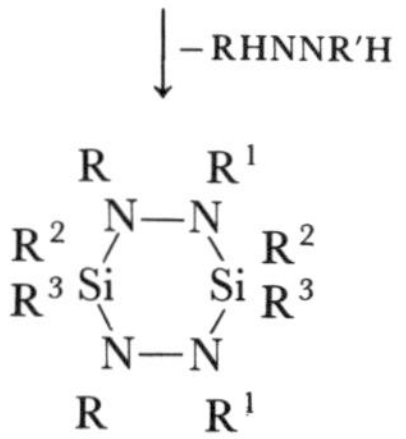

Both ammonia and methylamine react vigorously with 1,2-dichlorotetramethyldisilane, yielding tetrasilacyclohexanes.[125,128]

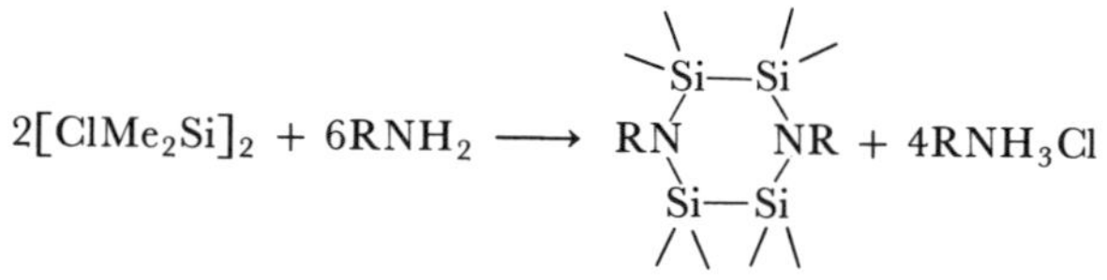

$$2[ClMe_2Si]_2 + 6RNH_2 \longrightarrow \text{cyclo-}[Me_2Si\text{—}SiMe_2\text{—}NR]_2 + 4RNH_3Cl$$

(R = H, m.p. 1°C, b.p. 61°C/2 mm; R = Me, m.p. 23°C, b.p. 62°C/3 mm)

Seven-membered Rings

1,5-Dichlorooctamethyltrisilazane and the lithium salt of sym-dimethylhydrazine give the expected tetra-azatrisilacycloheptane.[92]

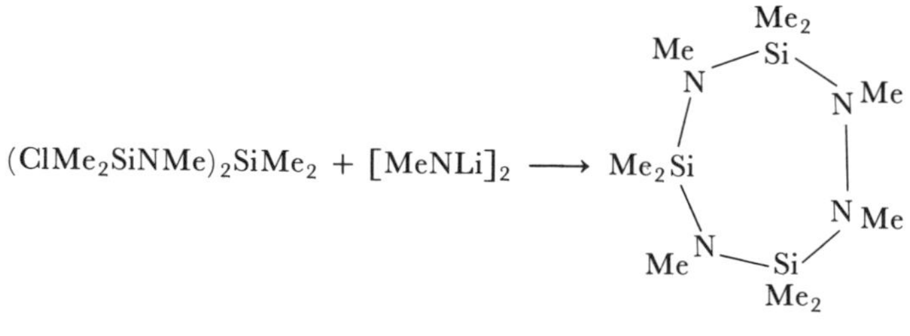

$$(ClMe_2SiNMe)_2SiMe_2 + [MeNLi]_2 \longrightarrow \text{cyclo-}[Me_2Si\text{—}NMe\text{—}SiMe_2\text{—}NMe\text{—}NMe\text{—}SiMe_2\text{—}NMe]$$

(m.p. 42–3°C, b.p. 67°C/0.1 mm)

Cyclosiloxazanes

These compounds contain the Si—N—Si and Si—O—Si groups as an integral part of the ring. They are normally synthesized by the aminolysis of a chlorinated siloxane.

Synthesis

Terminal dichlorosiloxanes and ammonia, or primary amines, yield cyclic monoazasiloxazanes, but in certain cases the dimer is produced.[129] The products are liquids or low melting solids.

$$Cl[SiMe_2O]_nSiMe_2Cl + 3RNH_2 \longrightarrow [RN(SiMe_2O)_nSiMe_2] + [RN(SiMe_2O)_nSiMe_2]_2$$

Table 4.4. Physical Properties of Cyclosiloxazanes

	$[HN(SiMe_2O)_nSiMe_2]$		
monomer	m.p. (°C)	b.p. (°C)	n_D^{20}
$n = 2$	9.5–10.5	151	1.4068
$n = 3$	20	190–1	1.4151
$n = 4$		115–9/13 mm	1.4125
$n = 5$		58–60/1 mm	1.4122
dimer			
$n = 1$	40	206–8	
$n = 2$	−12	148/2 mm	1.4269
	$[RN(SiMe_2O)_nSiMe_2]$		
monomer	m.p. (°C)	b.p. (°C)	n_D^{20}
R = Me, $n = 2$		60/17 mm	1.4100
$n = 3$		84/8 mm	1.4202
R = Et, $n = 2$		41/1.5 mm	1.4153
R = Bu^n, $n = 2$		78/5 mm	
dimer			
R = Me, $n = 1$	27–8	87/5 mm	1.4320^{30}
R = Et, $n = 1$		95/1 mm	1.4410
R = Bu^n, $n = 2$		95/0.25 mm	

Aminosiloxanes readily deaminate at 150–180°C in the presence of $(NH_4)_2SO_4$, a monoazasiloxazane resulting.[129, d]

$$\underset{\text{(b.p. 97°C/20 mm)}}{(MeNHSiMe_2O)_2SiMe_2} \xrightarrow[(NH_4)_2SO_4]{150\text{–}180°C} \text{cyclo-}[Me_2Si{-}O{-}SiMe_2{-}O{-}SiMe_2{-}N(Me)] + MeNH_2$$

The dilithium salt of pentafluoroaniline reacts with 1,3-dichlorotetramethyldisiloxane and 1,5-dichlorohexamethyltrisiloxane giving the expected 6-, 8- and 12-membered cyclic products.[130]

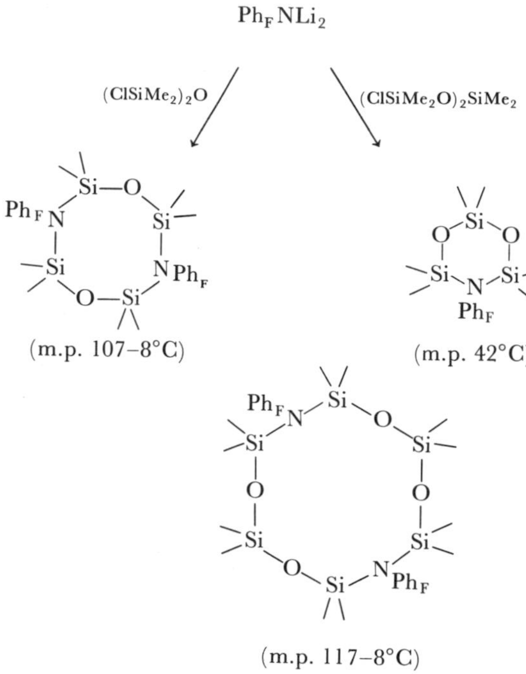

Ph_FNLi_2

$(ClSiMe_2)_2O$ ↙ ↘ $(ClSiMe_2O)_2SiMe_2$

Si—O
Ph_FN Si
Si NPh_F
O—Si

(m.p. 107–8°C)

Si
O O
Si N Si
Ph_F

(m.p. 42°C)

Ph_FN Si O
Si Si
O O
Si Si
O Si NPh_F

(m.p. 117–8°C)

A mixture of Me_2SiCl_2 and the dichlorosiloxane gives cyclosiloxazanes with ammonia, but yields are small due to polysilazane formation.[131] The products do contain at least two nitrogen atoms in the ring, and the novelty of these compounds will be seen later in their reactions (see pp. 192–3).

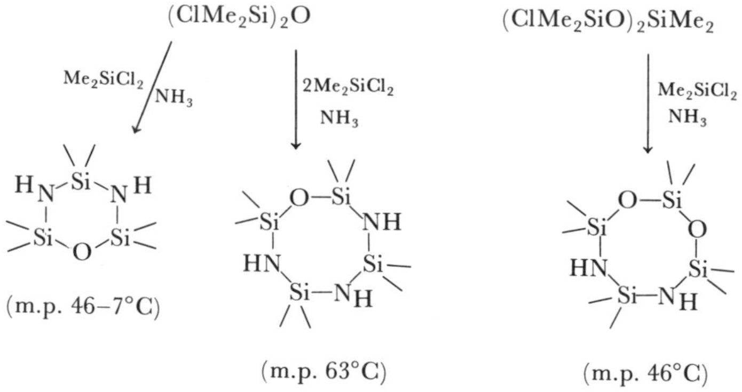

$(ClMe_2Si)_2O$ $(ClMe_2SiO)_2SiMe_2$

Me_2SiCl_2 NH_3 ↙ ↓ $2Me_2SiCl_2$ NH_3 ↓ Me_2SiCl_2 NH_3

HN Si NH
Si O Si

(m.p. 46–7°C)

O—Si
Si NH
HN Si
Si—NH

(m.p. 63°C)

O—Si
Si O
HN Si
Si—NH

(m.p. 46°C)

The hydrolysis of chlorodisilazanes as a synthetic route has been avoided because of complications through the hydrolysis of Si—N as well as Si—Cl.

However, in refluxing toluene, dry zinc oxide only attacks the Si—Cl bond.[132]

$$2(ClMe_2Si)_2NH + 2ZnO \longrightarrow [Me_2SiOMe_2SiNH]_2$$

(m.p. 37°C)

Reactions

Siloxazanes can be readily polymerized in the presence of acid or base catalysts, cross-linking occurring through the trivalent nitrogen.[c] A comparison of the hydrolytic stability of the 6- and 8-membered siloxazanes with isostructural cyclosilazanes shows that $[Me_2SiO(Me_2SiNH)_2]$ is less stable than $[Me_2SiNH]_3$, while $[Me_2SiOMe_2SiNH]_2$ is more stable hydrolytically than $[Me_2SiNH]_4$. With the former pair (6-membered rings), this difference in stability is thought to be due to the Si—N—Si angle of the siloxazane being close to the value in $[Me_2SiO]_3$ (133 degrees), while in $[Me_2SiNH]_3$ it is 117 degrees. A wide angle, however, will decrease the nucleophilic power of the nitrogen and the electrophilic power of silicon, and so should slow the hydrolysis, contrary to observations.

While organolithium compounds will cleave only the Si—N bond of cyclodisilazanes, phenyl-lithium cleaves the Si—O bond in both $[Me_2SiNHMe_2SiO]_2$[129, b] and $(Ph_3Si)_2O$.[133]

$$[Me_2SiNHMe_2SiO]_2 + 2PhLi \longrightarrow PhMe_2SiNHSiMe_2OLi \xrightarrow{Me_3SiCl} PhMe_2SiNHSiMe_2OSiMe_3$$

$$(Ph_3Si)_2O + PhLi \longrightarrow Ph_4Si + Ph_3SiOLi$$

With two moles of n-butyl-lithium at −60°C in dimethoxyethane both NH bonds of $[Me_2SiNHMe_2SiO]_2$ are metallated, and silylation gives the expected silyl siloxazane. Warming the di-anion to −30°C, followed by silylation, gives a siloxysilylcyclosiloxazane with ring contraction, while the same reaction at 0°C yields the bis-(siloxysilyl)cyclodisilazane. Such ring contractions have already been encountered with cyclosilazane derivatives[134] (pp. 176–7).

O—Si, N Li, Li N, Si—O (cyclic lithium amide)

−60°C   −30°C   0°C

O—Si, N—SiMe₃, Me₃Si N, Si—O   Me₃Si N, Si, N—SiMe₂OSiMe₃, Si, O, Si   SiMe₂OSiMe₃, Si—N, Me₃SiOSiMe₂N—Si

Analogously, the monolithium salt gives the contracted siloxysilylcyclosiloxazane at 0°C, while at −60°C this polymeric amide gives only the parent and bis-(trimethylsilyl)cyclosiloxazanes on silylation, as observed with the monolithium derivative of $[Me_2SiNH]_3$.

This ring contraction appears favourable only if the ring geometry brings the amide group close enough to a silicon atom sufficiently electrophilic to accept. Strong Si—O π-bonding in $[(Me_2SiO)_2(Me_2SiNH)]$ and $[(Me_2SiO)_2(Me_2SiNH)]$ reduces the electrophilic behaviour of likely Si atoms while $[Me_2SiO(Me_2SiNH)_2]$ may have the wrong geometry. In any case, all three amides give no ring contraction on silylation at 20°C.

Amides of silylated cyclosiloxazanes involving two nitrogen atoms and one oxygen atom will ring contract at room temperature, unlike the non-silylated analogue. This may be due to the extra Si atom bonded to nitrogen making the silicon ring atoms sufficiently electrophilic for attack. (See scheme on next page.)

The cyclosiloxazane ring $Si_4N_2O_2$ and the cyclotrisilazane rings both occur in the mixed oxynitride of silicon Si_2N_2O.[135]

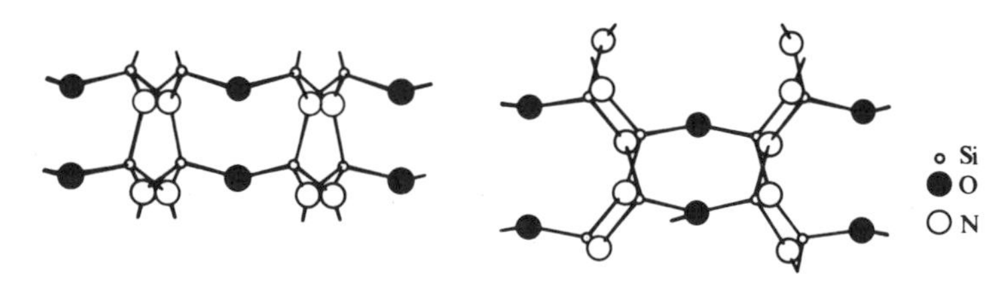

Fig. 4.9. The structure of Si_2N_2O showing the cyclotrisilazane and cyclotetrasiloxazane ring structure[135]

O—Si
Si NLi
HN Si
Si—O
Me_3SiCl
−60°C
O—Si SiMe₃
Si N
N Si
Me_3Si Si—O
+
O—Si
Si NH
HN Si
Si—O

0°C Me_3SiCl

Si $SiMe_2OSiMe_3$
HN N
Si O Si
Bu^nLi
Si $SiMe_2OSiMe_3$
LiN N
Si O Si

Me_3SiCl
20°C
Me_3SiCl −30°C

Si
Me_3SiOMe_2SiN $NSiMe_2OSiMe_3$
Si

Me_3Si Si $SiMe_2OSiMe_3$
N N
Si O Si

Si
Me_3SiN NLi
Si O Si
Me_3SiCl
20°C
Si
Me_3SiN $NSiMe_2OSiMe_3$
Si

It is formed by heating a silicon/silica mixture at 1450°C in an argon atmosphere containing 5% N_2 and is harder than quartz, SiO_2. The structure comprises a three-dimensional framework built up of SiN_3O tetrahedra, and the projection in the *xz* plane shows how the silazane and siloxazane rings are interlinked. The Si—O bond length of 0.1623 nm and the Si—O—Si angle of 148 degrees are close to those observed in quartz. The Si—N bond length of 0.172 nm is close to that found in Si_3N_4.

Co-ordination compounds

The chemistry of cyclic Group IV oxygen compounds illustrates a general point that the tendency of the element to become 5- and 6-co-ordinate increases with the atomic weight of the metal. This trend is by no means as prominent with nitrogen derivatives but examples are known.

Dimethylaminosilane Me_2NSiH_3 can be conveniently synthesized from silyl bromide and dimethylamine.[136] Unlike trimethylamine and the other silylamines $MeN(SiH_3)_2$ and $N(SiH_3)_3$, which all melt below $-100°C$, this has an exceptionally high melting point (3°C) and heat of sublimation, typical of an associated compound.[137] Molecular weight determinations show it to be pentameric, and an X-ray structure determination shows the presence of a 10-membered ring containing alternate Si and N atoms. These are arranged in a pentagon with the N—Si—N linkage occupying the edge of this regular pentagon. The Si—N bond lengths vary quite considerably, but average to a value of about 0.195 nm. This is larger than the sum of the covalent radii of the two elements, and tends to indicate the presence of multi-centre bonding in the two apex positions of the trigonal-bipyramid of ligands around silicon.

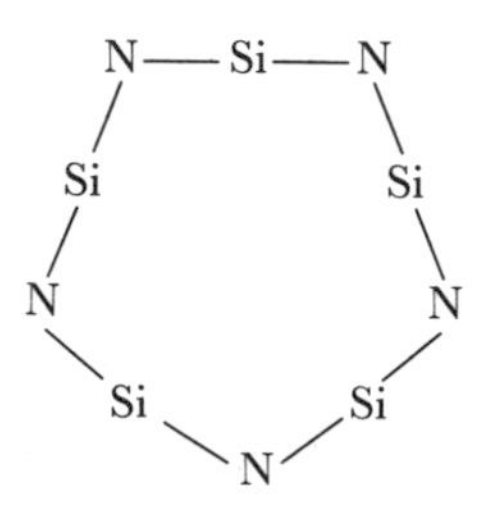

Fig. 4.10. The skeleton of pentameric Me_2NSiH_3

Azides of the Group IV elements have been widely studied, and one of their reactions involves stannic chloride.[138] Trimethylsilyl azide forms an azidostannyl compound which complexes with the azidosilane. This is dimeric and thought to have azide bridges, and therefore contains Si and Sn atoms in different co-ordination states. It decomposes with chlorosilane elimination to give a polymeric chloroazide of tin, probably with azide bridges. These have already been observed among the azides of the Group III elements.

$$2Me_3SiN_3 + 2SnCl_4 \longrightarrow [Cl_3SnN_3 \cdot Me_3SiN_3]_2 + 2Me_3SiCl$$

$$\downarrow 90°C$$

$$2/n[Cl_2Sn(N_3)_2]_n + 2Me_3SiCl$$

Me$_3$Si—(Cl, N$_3$ bridges)—Sn(Cl)(Cl)—(N$_3$ bridges)—Sn(Cl)(Cl)—(Cl, N$_3$ bridges)—SiMe$_3$

—Sn(Cl)(Cl)—(N$_3$ bridges)—Sn(Cl)(Cl)—(N$_3$ bridges)—Sn(Cl)(Cl)—

PHOSPHORUS DERIVATIVES OF SILICON, GERMANIUM AND TIN

Synthesis

While only a limited number of cyclic Group IV phosphorus compounds are known, along with a cyclic silicon–arsenic one, nevertheless they portray a cyclic architecture as varied as that of the Group IV nitrogen compounds.

Lithium methylarsenide and dimethyldichlorosilane react to give an 8-membered Si—As ring, permethylcyclotetrasilarsenane.[139]

$$Me_2SiCl_2 + 2LiAsHMe \longrightarrow \underset{(b.p.\ 60°C/2\ mm)}{Me_2Si(AsHMe)_2} \longrightarrow \underset{(b.p.\ 90°C/0.05\ mm)}{[Me_2SiAsMe]_4}$$

Cyclic silicon–phosphorus compounds result from similar starting materials. Passing phosphine into an excess of n-butyl-lithium gives a mixture of di- and tri-lithium phosphide. With diethyldichlorosilane, small ring compounds result.

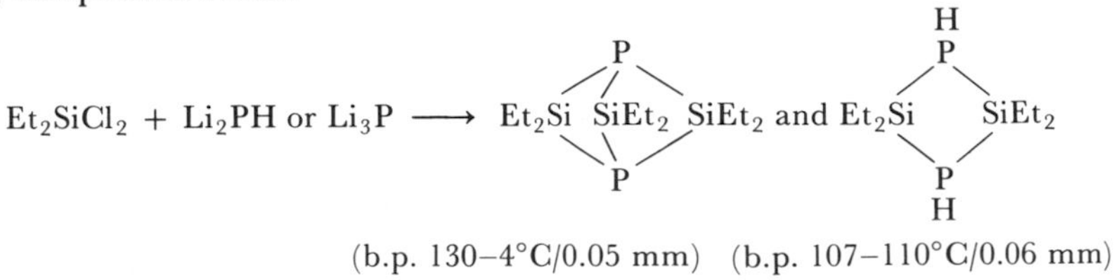

$$Et_2SiCl_2 + Li_2PH \text{ or } Li_3P \longrightarrow Et_2Si\ SiEt_2\ SiEt_2 \text{ (bridged by two P) and } Et_2Si(PH)_2SiEt_2$$

(b.p. 130–4°C/0.05 mm) (b.p. 107–110°C/0.06 mm)

Phenylphosphides give a similar cyclodisilaphosphane, while the trimethylsilyl-substituted product results through lithiation of trimethylsilyl phosphine.[140]

$$Me_3SiCl + LiPH_2 \longrightarrow \underset{(b.p.\ 69–73°C)}{Me_3SiPH_2} \longrightarrow Me_3SiPLi_2 \longrightarrow \underset{(b.p.\ 96–8°C/0.2\ mm)}{[Me_3SiPSiEt_2]_2}$$

Both phenyl-silicon and -germanium trichlorides react with dipotassium phenylphosphide to give oxygen-sensitive cage compounds in good yields.[141]

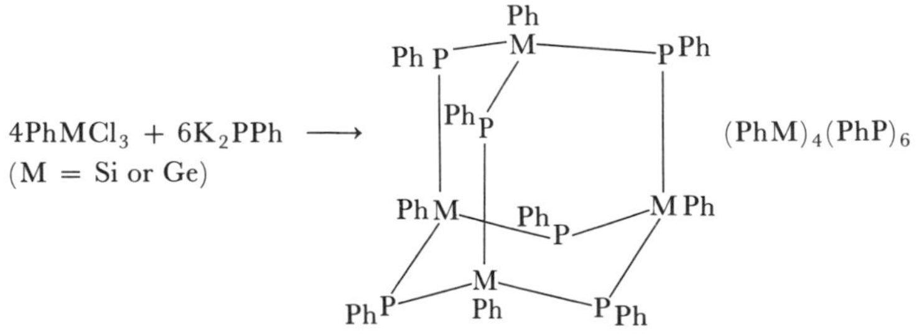

$$4PhMCl_3 + 6K_2PPh \longrightarrow (PhM)_4(PhP)_6$$

$(M = Si \text{ or } Ge)$

(Cage structure: four PhM units bridged by six PPh units.)

The structure resembles adamantane $(CH)_4(CH_2)_6$. By way of contrast, the reaction of n-butyl- or phenyl-trichlorostannanes with phenylphosphine gives lower yields of the "half-cage" structure $(RSn)_2(PhP)_3$ involving two tin and three phosphorus atoms.

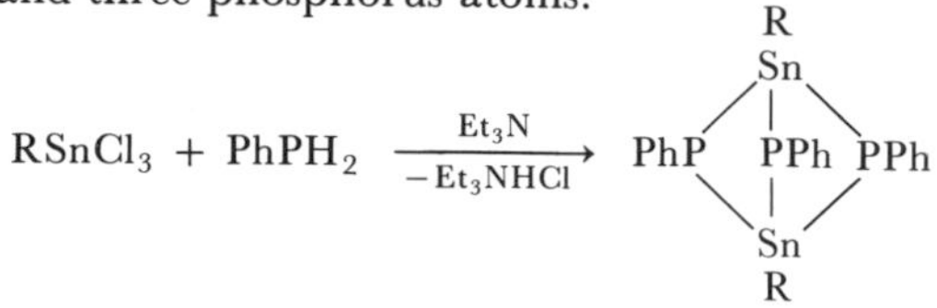

$$RSnCl_3 + PhPH_2 \xrightarrow[-Et_3NHCl]{Et_3N} \text{PhP} \; \text{PPh} \; \text{PPh} \text{ (bridged by two RSn groups)}$$

(R = n-Bu, m.p. 89–91°C; R = Ph, m.p. 133–5°C)

This is similar to the trigonal-bipyramidal structure of the Si—P cage, only with the Group IV/Group V elements inverted.[d]

This route has also been used to synthesize the usual cyclic products, and also "cubane"-type compounds, by varying the functionality of both phosphorus and the Group IV element. While phenylphosphine and diorganotin dichlorides give a cyclic trimer in the presence of triethylamine,[142] more forcing conditions have to be used to synthesize similar germanium compounds.

$$R_2SnCl_2 + PhPH_2 \xrightarrow[-Et_3NHCl]{Et_3N} [R_2SnPPh]_3 \; (R = Me, Bu^n, Ph)$$

The chlorogermanaphosphane is formed initially, and reacts with a phenylphosphide to give the cyclodi- and -trigermanaphosphanes.

$$Ph_2GeCl_2 + PhPH_2 \xrightarrow[-Et_3NHCl]{Et_3N} (Ph_2GeCl)_2PPh \; (\text{m.p. } 118\text{–}120°C)$$

$$(Ph_2GeCl)_2PPh \xrightarrow{PhPK_2}$$

$$Ph_2GeCl_2 + PhPK_2 \longrightarrow [Ph_2GePPh]_2 \; (\text{m.p. } 40\text{–}2°C) \text{ and } [Ph_2GePPh]_3 \; (\text{m.p. } 112\text{–}4°C)$$

With phosphine, tin halides give both cube and adamantane-like structures. Dichlorides yield the adamantane cage,[d] with phosphorus the trifunctional group in this case,

$$R_2SnCl_2 + PH_3 \xrightarrow[-Et_3NHCl]{Et_3N} (R_2Sn)_6P_4$$

while trichlorides give a 10% yield of tetrameric phenyltin phosphide.[143] This is thought to have a cube structure.

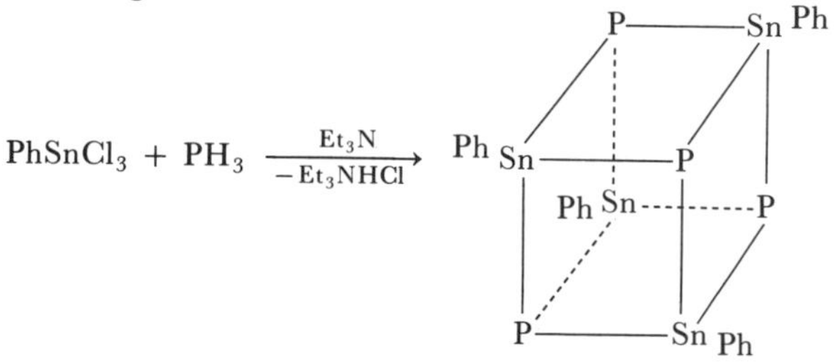

$$PhSnCl_3 + PH_3 \xrightarrow[-Et_3NHCl]{Et_3N} (PhSnP)_4 \text{ (cube of alternating PhSn and P)}$$

With sodium phosphide, $PhMCl_3$ (M = Si, Ge) gives heptameric products.[144]

$$PhMCl_3 + Na_3P \longrightarrow [PhMP]_7 \text{ (M = Si, Ge)}$$

These two routes are the most convenient for synthesizing Group IV phosphorus compounds. They involve only the cleavage of metalloid–halogen bonds, and have only been used comparatively recently. The first synthetic methods were more complicated, and involved (a) the cleavage of both P—Cl and P—Ph bonds by Ph_3SnLi,[145] and (b) pyrolyzing tetraphenyltin with red phosphorus.[146]

$$Ph_3SnLi + Ph_2PCl \longrightarrow \underset{\text{(m.p. 64°C)}}{[Ph_2SnPPh]_3} + Ph_3P + Ph_6Sn_2$$

$$Ph_3SnLi + PhPCl_2 \longrightarrow \underset{\text{(m.p. 99°C)}}{[Ph_3SnPSnPh_2]_3}$$

$$Ph_4Sn + P \xrightarrow{250°C} \underset{\text{(low yield)}}{[Ph_2SnPPh]_3}$$

Properties

While much attention has been devoted to studying the properties of the linear phosphorus compounds of this Group, little attention has been paid to the cyclic ones. It is important, however, to appreciate that many of the reactions undergone by the linear compounds apply equally to the cyclic ones.

All are sensitive to oxygen, and the alkyl-substituted products are also sensitive to water. The aryl ones appear to be water-stable in the absence of oxygen, especially those of tin.[d] While carbon dioxide appears not to react with Ge—P and Sn—P bonds, it will insert into the Si—P bond.[147] Due attention should be paid to these reactions when handling the compounds.

Unlike hexamethylcyclotrisilazane, nonaphenylcyclotrisilaphosphane will complex with metal carbonyl residues. Cycloheptatrienetricarbonyl molybdenum and the silicon–phosphorus ring give an air-sensitive yellow crystalline product.[148]

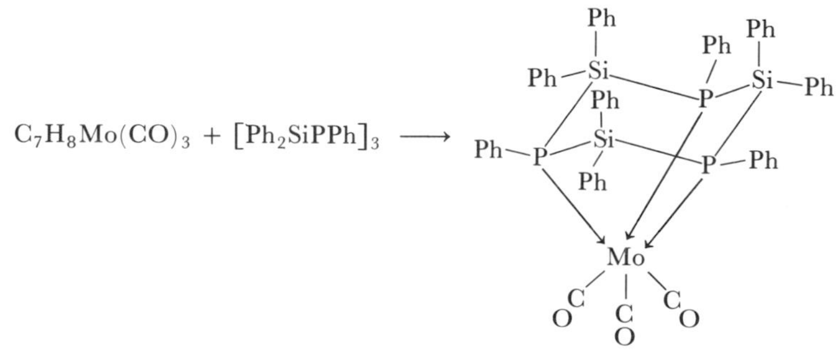

$$C_7H_8Mo(CO)_3 + [Ph_2SiPPh]_3 \longrightarrow$$

Ph Ph Ph
Ph—Si P Si—Ph
Ph
Ph—P Si P—Ph
Ph
Mo
C C C
O O O

The infrared spectrum shows two peaks typical of carbonyl stretches at 1931 and 1838 cm^{-1} and three Si—P stretches at 393, 333 and 305 cm^{-1}. These are typical of a molecule with C_{3v} point group symmetry.

C_{3v}	E	$2C_3$	$3\sigma_v$
Γ_{CO}	3	0	1
Γ_{Si-P}	6	0	0

$\Gamma_{CO} = A_1 + E$; $\Gamma_{Si-P} = A_1 + A_2$ (infrared inactive) $+ 2E$

The comparison with the cyclotrisilazane illustrates well the theory of hard and soft acids and bases, also known as "a"- and "b"-type acceptors and donors. The cyclotrisilazane is a hard, "a"-type donor, and so bonds as a tridentate ligand to Ti^{III} or V^{III} (hard acid), while the trisilaphosphane bonds as a tridentate, soft ligand to the soft "b"-type low-valency metal. It does not bond as a π-bonding ring like borazole.

The tin–phosphorus compounds do not possess high thermal stability and readily disproportionate. Thus, $[Me_2SnPPh]_3$ gives a cyclopolyphosphine, metallic tin and trimethyltin products, indicating methyl migration.[142]

$$5[Me_2SnPPh]_3 \longrightarrow 2[PhP]_5 + 5(Me_3Sn)_2PPh + 5Sn$$

CYCLIC OXYGEN DERIVATIVES OF SILICON, GERMANIUM AND TIN

Introduction

The reluctance of these Group IV elements to form 5- or 6-co-ordinate nitrogen derivatives has already been outlined. With oxygen, however, a wide variety of compounds with co-ordination number greater than four are known. This applies particularly to tin, with 1,3-dichlorotetra-alkyldistannoxanes $(ClR_2Sn)_2O$ being dimeric in the solid state and in solution, with a 4-membered $[Sn—O]_2$ ring.[149] The silicon analogue is a monomer, while the cyclosiloxanes, $[R_2SiO]_{3 \text{ and } 4}$, show no strong intermolecular bonding. The cyclostannoxanes however melt over a wide temperature range, and are believed to be cross-linked polymers involving 5-co-ordinate tin.[150] (See Figure 4.11 and appendix.)

The striking difference in the properties of the Si—O and Sn—O compounds is probably connected with the difference in the polarity of the two bonds, the Sn—O one being the more polar. This is reflected in the larger perturbation of the C—D stretching frequency in $CDCl_3$ caused by Sn—O than by Si—O compounds. This is believed to be due to the lessened effect of $p\pi$–$d\pi$ bonding in tin compounds. Thus, $(Me_3Si)_2O$ causes ν_{C-D} to fall by only 13 cm^{-1}, while $(Me_3Sn)_2O$ shifts the peak

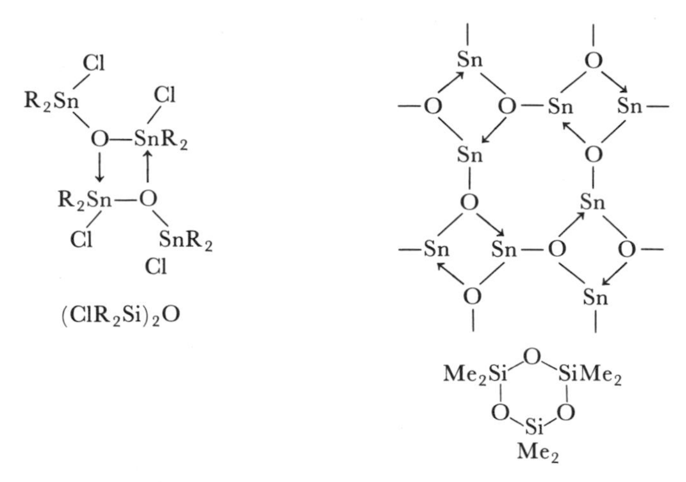

Fig. 4.11. Siloxanes and stannoxanes

84 cm^{-1}, showing the tin compound to be the stronger base.[151] Indeed, it will complex with tin halides, unlike siloxanes.

$$(Me_3Sn)_2O + Me_3SnBr \longrightarrow (Me_3Sn)_3O^+Br^-$$

As the Group IV elements occur naturally as oxygen derivatives, the mineral and structural chemistry of these compounds provides a large part of the cyclic oxygen section of this Chapter. This will be discussed first.

The silicates

Boron and silicon, diagonally related in the Periodic Table, resemble each other as closely as any other pair of Main Group elements. This is widely reflected in the physical properties and chemical nature of the elements and many of their compounds, not least the oxides. Both form strong bonds to oxygen which involve considerable multiple bonding. Consequently, borates and silicates occupy unique positions in Geology and Mineralogy, illustrating better than any other elements the intricate architecture of nature as displayed through structural Inorganic Chemistry.

Though two molecular units BO_3 and BO_4 are used to construct the borates, only one SiO_4 is used for the natural silicates, but leads to oligomeric and polymeric anions in one-, two- or three-dimensions. This SiO_4 is conveniently represented in projection as shown, and interbonding them leads to many of the structures encountered in natural minerals.

SiO_4^{4-} ≡ (Si tetrahedron with four O) ≡ (projection)

Since all but a few involve a cyclic structure, the whole field will be considered rather than sticking strictly to the terms of reference.[a, e, f]

Orthosilicates

The orthosilicates contain the anion SiO_4^{4-} found in the mineral olivine. This is olive-green, and is used for the gemstones chrysolite and peridot.

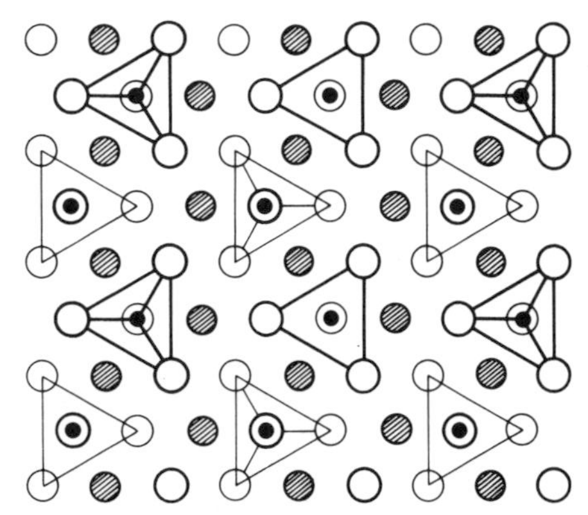

Fig. 4.12. Plan of the structure of Mg_2SiO_4. Small black circles represent Si, shaded circles Mg, and open ones, oxygen.[a] The light and heavy lines are used to distinguish between SiO_4 tetrahedra at different heights

The metals normally found include Mg^{2+}, Fe^{2+} and Mn^{2+}, and occupy octahedral holes in the close-packed array of oxygen atoms. Silicon fits into tetrahedral holes. Phenacite, Be_2SiO_4 has a structure different from olivine. The small Be atoms occupy tetrahedral holes, so the structure may be regarded as an infinite array of BeO_4 and SiO_4 tetrahedra sharing corners.

The garnets contain metal ions in two oxidation states $M^{II}_3M^{III}_2(SiO_4)_3$. They are quite hard and crystalline, so are used as abrasives and cut for gemstones, *e.g.*, cape ruby and carbuncle. Orthosilicates are also the major components of Portland cement.

Pyrosilicates

These silicates, which contain the ion $Si_2O_7^{6-}$, are rare, the anion being conveniently represented as two corner-linked SiO_4 tetrahedra. The best example is hemimorphite, a basic, hydrated silicate of zinc, $Zn_4(OH)_2$-$Si_2O_7 \cdot H_2O$.

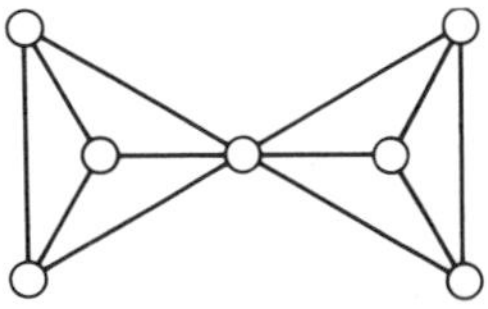

Fig. 4.13. Pyrosilicate ion $Si_2O_7^{6-}$

Cyclic Silicates

These comprise the SiO_4 residues bridging through two oxygen atoms. So many ring sizes could be produced, but only two are found naturally. These contain three and six SiO_4 units and are formulated as $Si_3O_9^{6-}$ and $Si_6O_{18}^{12-}$.

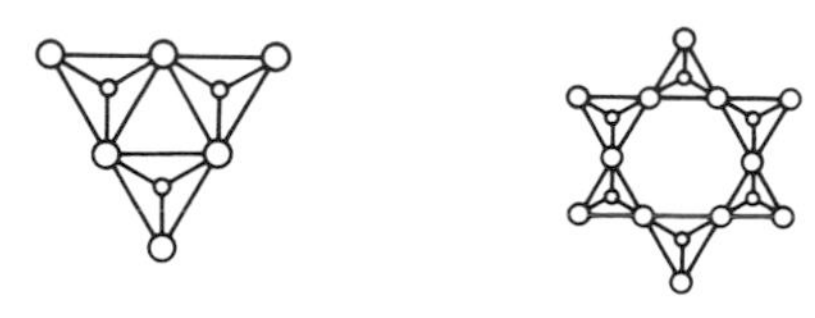

Fig. 4.14. Cyclic silicates

Though no cyclotrimetasilicates (*e.g.*, benitoite, $BaTiSi_3O_9$) have been used as semi-precious stones, the cyclohexametasilicate ion $Si_6O_{18}^{12-}$ occurs as beryl, $Be_3Al_2Si_6O_{18}$, in emeralds. Tourmaline is an example of a mixed borosilicate containing this and BO_3^{3-} anions. Again, these ions occur in basic salts, *e.g.*, $[Al_6(OH)_4(BO_3)_3Si_6O_{18}]^{7-}$. This ion occurs in sapphire (blue) and topaz (pale-yellow). The crystal structures show that the rings of $Si_6O_{18}^{12-}$ occupy sheets, these being bound to others by metal ions in between them.

Pyroxenes and Amphiboles

These represent two of the commonest classes of silicates, and are one-dimensional polymers. The pyroxenes are built up of chains of silicate groups linked through two oxygen atoms and are formulated as $[SiO_3]_n^{2n-}$.

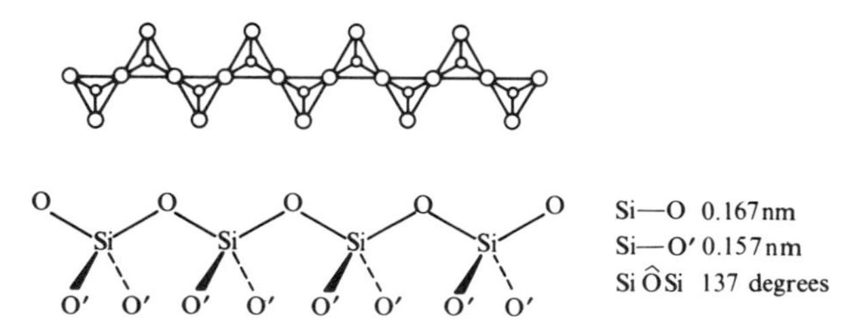

Fig. 4.15. Pyroxenes

Synthetic sodium metasilicate possess this chain structure, with the Si—O bonds much shorter than expected for a single bond (0.174 nm). This is thought to result through $p_\pi-d_\pi$ bonding between O and Si atoms, and occurs with both kinds of Si—O bond in pyroxenes, though more so in the terminal Si—O bonds. The wide Si—O—Si angle also reflects some π-bonding. Minerals include diopside, $CaMg(SiO_3)_2$, and spodumene, $LiAl(SiO_3)_2$, with the chains held together by metal ions.

Amphiboles comprise the double chain indicated and therefore involve inorganic rings. This one-dimensional polymer comprises the Si_6O_{18} residue and is formulated as $[Si_4O_{11}]_n^{6n-}$ as in the mineral tremolite $Ca_2Mg_5(OH)_2(Si_4O_{11})_2$.

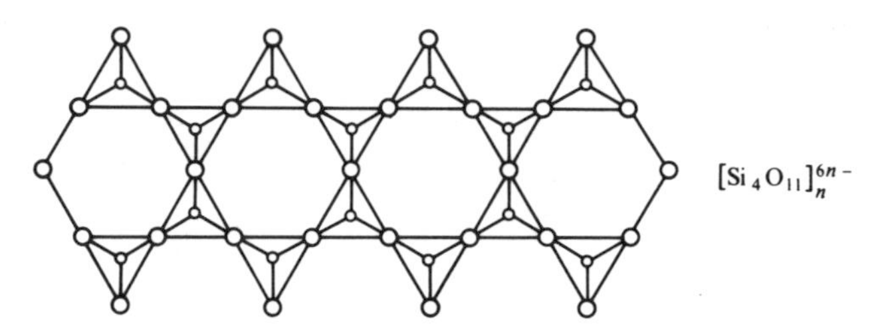

Fig. 4.16. Amphiboles

Asbestos was the term originally used to describe fibrous amphiboles, but it now incorporates many two-dimensional polymers encountered among the aluminosilicate polymers of clay. Thus chrysotile, once regarded as an amphibole $(OH)_6Mg_6Si_4O_{11}{\cdot}H_2O$, is actually $(OH)_4Mg_3Si_2O_5$, the Si_2O_5 anion having a layer structure. Replacing Mg by Al gives the aluminosilicate $(OH)_4Al_2Si_2O_5$ encountered in kaolin. These will be discussed in greater detail in the Section on layer silicates (p. 203).

Polysiloxanes with trifunctional Si atoms form ladder polymers and cyclic oligomers (which involve joining the ladder ends). The silicate $Si_6O_{15}^{6-}$ is an example of this, two 6-membered siloxane rings being bridged by three oxygen atoms, these providing the rungs of the ladder.[152]

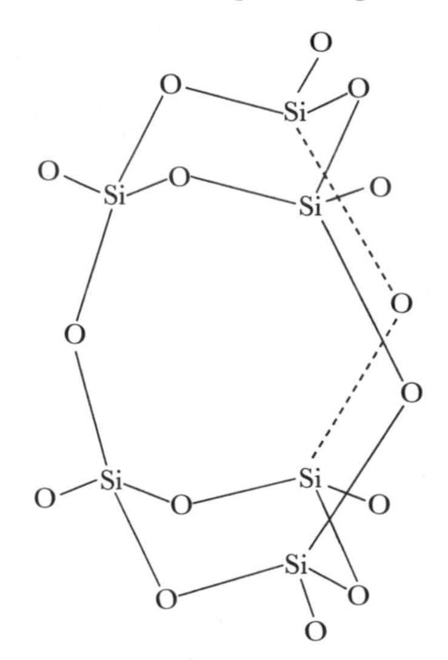

Fig. 4.17. $Si_6O_{15}^{6-}$

The Si—O bond lengths lie in the range 0.160–0.167 nm, the terminal ones being the shorter. These compare quite closely with those already discussed in the pyroxenes.

Layer-structured Silicates

These contain two-dimensional sheet polymers sharing three apices of the SiO_4 tetrahedra. Two arrangements have been found naturally and both are formulated as $(SiO_{1+(3/2)})^-$, *i.e.*, $[Si_2O_5]_n^{2n-}$.

The Si_4–Si_8 *ring layer structure.* This involves alternate Si_4O_4 and Si_8O_8 rings, but is rare. Apophyllite, $KF{\cdot}Ca_4Si_8O_{20}{\cdot}8H_2O$ has such a structure, with the terminal oxygen atoms of one Si_4O_4 ring directed to the opposite side of the layer to its Si_4O_4 neighbour. Cations hold the layers together.

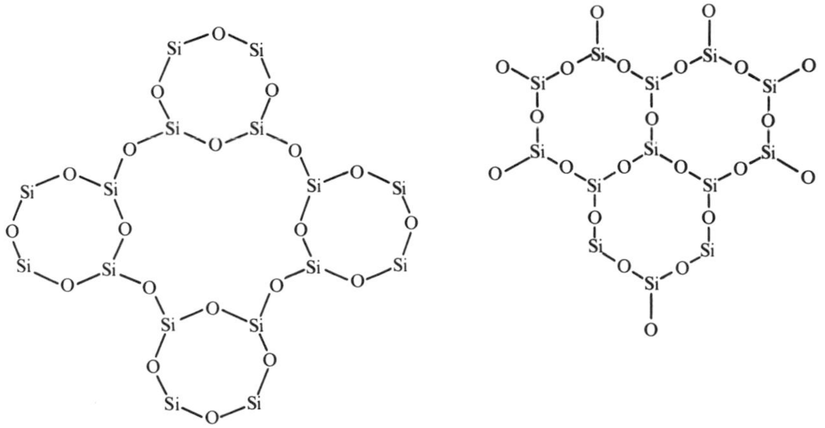

Fig. 4.18. Layer silicates $[Si_2O_5]_n^{2n-}$; (a) Si_4–Si_8, (b) Si_6

Si_6 *ring layer silicate.* These rings comprise six silicon and six oxygen atoms, and terminal oxygen atoms are all on one side of the ring. Consequently, it is theoretically possible to form a double sheet by bridging at these terminal positions, thereby obtaining a laminated form of silica.

Replacing a half of the Si atoms in such a structure by the isoelectronic Al^- ion leads to compounds of empirical formula $MAl_2Si_2O_8$ (M is Ca, Ba). The structure comprises these double layers bound together by 6-co-ordinate cations. (See Figure 4.19.)

Hydroxysilicates. The most important layer silicates are often interleaved by Mg or Al cations held through hydroxide ions.

These can ideally be formulated as $Mg_3(OH)_4Si_2O_5$ (chrysotile) and $Mg_3(OH)_2Si_4O_{10}$ (talc), together with the aluminium compounds. Partial replacement of Si atoms by Al^- ions gives charged layers which are neutralized by layers of alkali or alkaline earth metal ions, as in the micas.

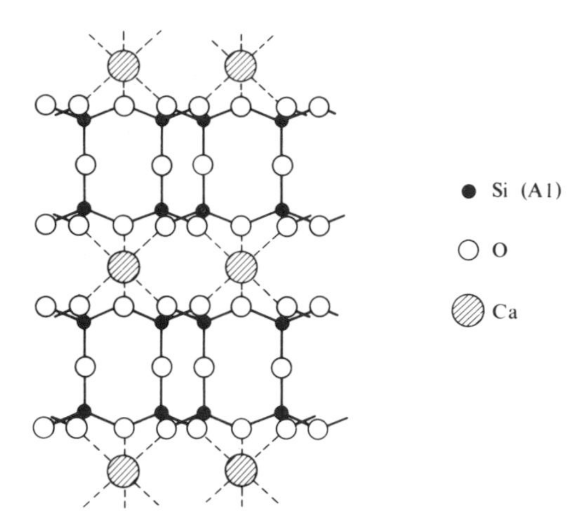

Fig. 4.19. Elevation of the structure of hexagonal $CaAl_2Si_2O_8$ showing the double layers interleaved with Ca^{2+} ions[a]

This neutralizing layer can also be hydrated ions or positively-charged Mg or Al hydroxides.

The variables in this structural array lead to property differences in these laminated compounds.

(i) The single silicate layers $Si_2O_5^{2-}$ interleaved with Mg or Al hydroxide residues is the structural unit present in china clay and kaolin minerals. This is normally formulated as $Al_2(OH)_4Si_2O_5$ (kaolinite).

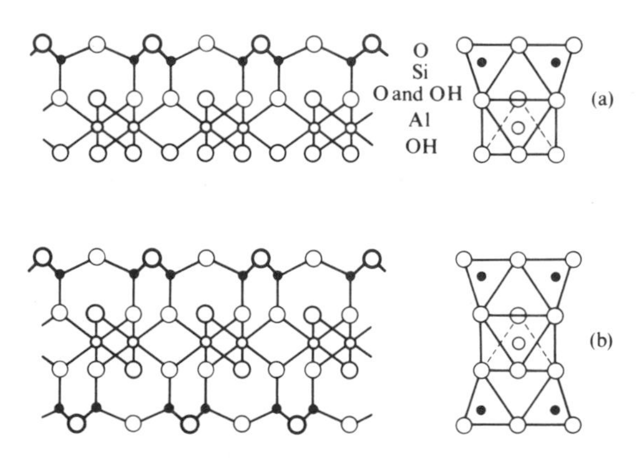

Fig. 4.20. (a) Chrysotile and (b) talc[a]

(ii) Those involving two silicate leaves held together by magnesium or aluminium hydroxides are, like kaolin, electrically neutral. They therefore readily cleave. Talc, $Mg_3(OH)_2Si_4O_{10}$, is widely used, therefore, as a lubricant (French chalk) and as a filler. Meershaum is a hydrated magnesium silicate resembling clay. After first soaking in tallow wax it can be made into pipes, and takes on an appealing red polish.

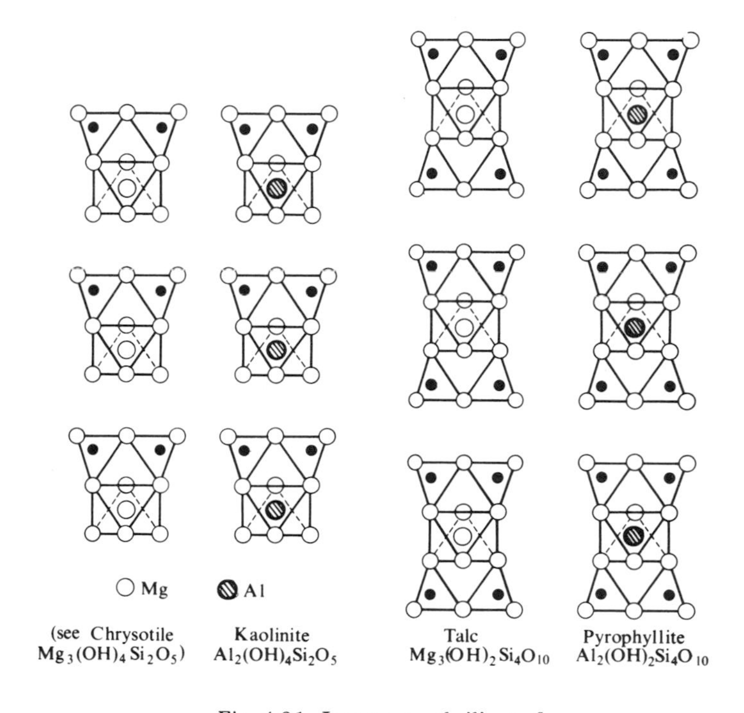

Fig. 4.21. Isostructural silicates[a]

(iii) The aluminium analogue, pyrophyllite, is like talc, and both can have up to a quarter of the Si atoms replaced by Al^- ions. An interleaving layer of cations neutralize the charge, with K^+ ions in the micas phlogopite, $KMg_3(OH)_2Si_3AlO_{10}$, and muscovite, $KAl_2(OH)_2$-Si_3AlO_{10}. Since K^+ ions occupy large holes with 12-fold co-ordination, the K^+O^- electrostatic bond is necessarily weak. So micas cleave readily along these layers. Further substitution with Al^- ions produces brittle micas, since a more highly charged cation is necessary to neutralize the extra charge. So the electrostatic bonding is stronger and the mica harder. These are widely used in the electrical industry.

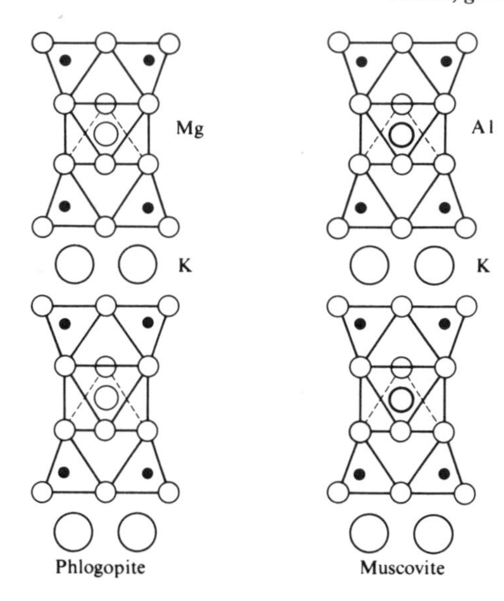

Fig. 4.22. The structures of the micas phlogopite and muscovite (diagrammatic)[a]

Table 4.5. Relative Hardness of Typical Hydroxysilicates

Hydroxysilicate	Formula	Hardness (on Mohs' Scale)
Talc	$Mg_3(OH)_2Si_4O_{10}$	1–2
Mica	$KMg_3(OH)_2Si_3AlO_{10}$	2–3
Brittle mica	$CaMg_3(OH)_2Si_2Al_2O_{10}$	$3\frac{1}{2}$–5

(iv) Interleaving with layers of hydrated cations gives hydrated micas which, with a much smaller cation charge density, cleave very easily. Further replacement of both the cation and Si atoms in talc, and interleaving with hydrated Mg^{2+} ions, produces vermiculite

$$[(Mg_{2.36}Fe^{III}_{0.48}Al_{0.16})(Si_{2.72}Al_{1.28})O_{10}(OH)_2]^{-0.64}[Mg_{0.32}(H_2O)_{4.32}]^{+0.64}.$$

This dehydrates readily to a talc-like structure, and is used widely as a soil conditioner and porous filler. (See Figure 4.23.)

Montmorillonite, $[Mg_{1/3}Al_{1\frac{2}{3}}Si_4O_{10}(OH)_2]^{-1/3}Na^{+1/3}_{1/3}$, results from pyrophyllite by replacing one sixth of the Al^{3+} ions by Mg^{2+} ions, ion for ion. Like many clay minerals, this can be readily hydrated and exhibits cation-exchange properties. It is an important constituent of fuller's earth, found widely in southern England. Fuller's earth was originally used for "fulling" or cleaning woollen fabrics and cloth, through its property to absorb grease and oil. It now finds wide use in refining oils and fats, the efficiency depending upon the aluminium

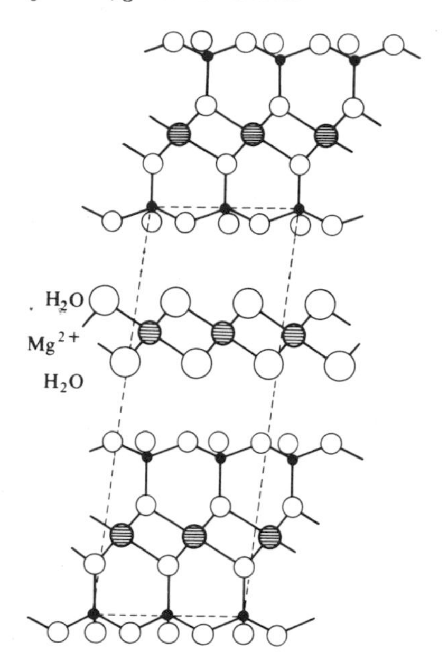

Fig. 4.23. Elevation of the structure of vermiculite[a]

silicate constituents. Montmorillonite has the property of forming gelatinous suspensions at low concentrations, which has given it wide industrial application, for example as a binder in foundry sand.

(v) With kaolin, talc and pyrophyllite the layers are uncharged, and so only weakly bound together, while the micas are held by cation layers, or hydrated cations (*e.g.*, vermiculite). Interleaving with charged hydroxide layers gives the chlorite minerals. Thus, the mica layers (composition $Mg_3(AlSi_3O_{10})(OH)_2^-$ to $Mg_2Al(Al_2Si_2O_{10})(OH)_2^-$) are held by $Mg_2Al(OH)_6^+$ ions. Thus, in phlogopite, $KMg_3(OH)_2Si_3AlO_{10}$, replacing the K^+ ion by the $Mg_2Al(OH)_6^+$ ion gives a chlorite mineral.

The structural changes discussed can be better represented pictorially, through cross-section projections of the alumino-silicates, these diagrams showing the relationship between the various minerals. (See Figure 4.24.)

Framework Silicates

Cross-linking the SiO_4 residues through all oxygen atoms would lead to silica, SiO_2, and so replacement of Si atoms by Al^- ions is required if anionic three-dimensional aluminosilicates are to result. Up to one half

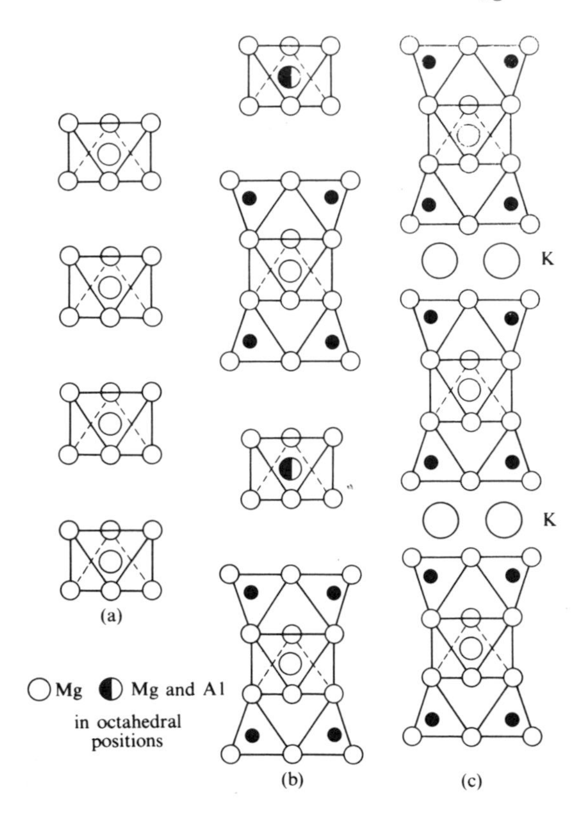

Fig. 4.24. The sequence of layers in (a) brucite, $Mg(OH)_2$, (b) chlorite and (c) phlogopite[a]

of the Si atoms can be replaced by Al^- ions, this charge being neutralized by the cations. Three such framework structures are known, the compact felspars and the open zeolites and ultramarines, which can accommodate water and various anions in the spaces in the structure.

The felspars are used in the manufacture of porcelain, and some occur in semi-precious minerals such as moonstone. The zeolites have a more open structure. They are hydrated and characterized by the ease with which they reversibly absorb water. Gases can be absorbed, and cation-exchange also readily occurs. Thus, Ca^{2+} ions readily replace Na^+ ions in sodium-containing zeolites (or "permutites"), and so the latter can be used for water-softening. The ultramarines are coloured silicates used for pigments. They are formulated as $Na_8Al_6Si_6O_{24}{\cdot}X$ where X is S_2^{2-}, $2Cl^-$ or SO_4^{2-}, and varying the anion and the cation produces a wide range of colours.

Silica, SiO_2

This possesses a structure comprising silicon atoms tetrahedrally surrounded by four oxygen atoms. Polymerization occurs in a variety of ways, and transitions between the various crystalline forms occur but with difficulty. The high-temperature forms have the more open structures, while the high-pressure ones are more compact, as would be expected from Le Chatelier's principle.

coesite	$\rho = 3.01$	Si 6-co-ordinate
↑ high pressure		
quartz	$\rho = 2.655$	α- and β-forms
↓ 870°C		
tridymite	$\rho = 2.30$	α- and β-forms
↓ 1470°C		
cristobalite	$\rho = 2.27$	α- and β-forms

While significant structural differences occur between these four forms of silica, those between α- and β-forms are small and generally only involve the rotation of a few Si—O bonds. This is supported by the same optical activity being present in both α- and β-quartz.

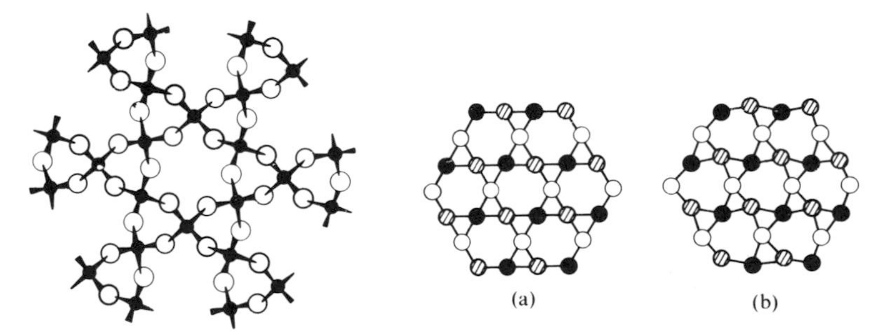

Fig. 4.25. Plan of the structure of β-quartz, with the 6- and 12-membered rings. The arrangement of the Si atoms in (a) β-quartz and (b) α-quartz[a]

From Figure 4.25, it will be seen that β-quartz has a more regular structure from the two projection diagrams, but both involve fused 6-membered and 12-membered rings. The structure therefore resembles the silicate sheet polymers. Impure forms of quartz are used in semi-precious jewellery, including amethyst, which probably owes its purple colour to manganese. Quartz sands are widely used in the building trade and as an abrasive, while quartz is employed in pottery and as a high-temperature lining to furnaces *etc*. It has also been widely used in short-wave radio apparatus, due to thin quartz plates possessing piezo-electric properties.

The more open structures of β-tridymite and β-cristobalite are analogous to the wurtzite and zinc blende forms of zinc sulphide. They are both metastable at room temperature, but their reluctance to undergo structural changes is shown by the natural occurrence of both tridymite (rhyolite in Antrim) and cristobalite (andesite in Mexico).

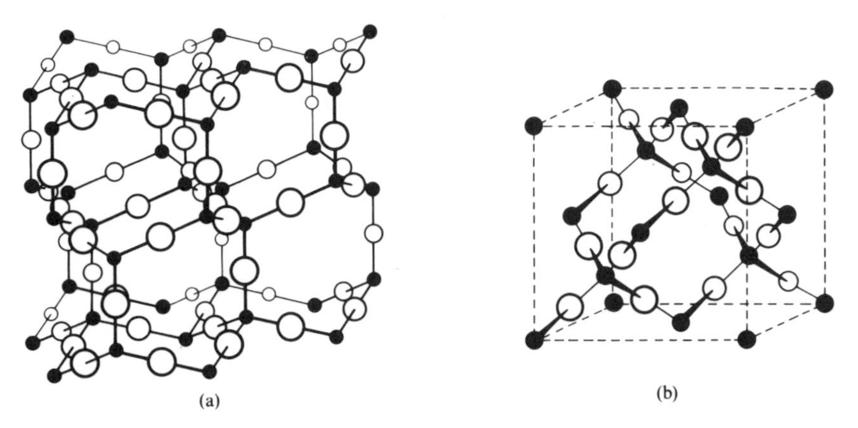

Fig. 4.26. The idealized structures of (a) β-tridymite and (b) β-cristobalite.[a] Small black circles represent Si atoms

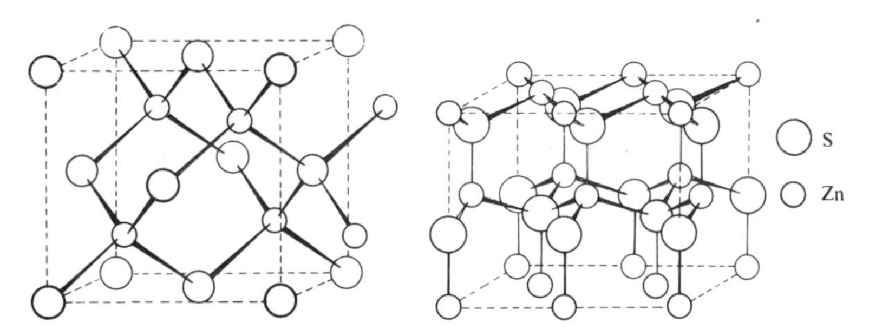

Fig. 4.27. The crystal structures of (a) zinc blende and (b) wurtzite[a]

Coesite and an even denser form stishovite[153] involve 6-co-ordinate Si atoms and have a rutile structure. Natural forms occur in meteors, and therefore establish the way high pressures cause contraction through an increase in co-ordination number. These forms of silica, unlike the others, are chemically highly-resistant, even to concentrated hydrogen fluoride. Rutile forms of silica are thought to occur in Moon rock,[154] and this again may be meteoric in origin.

Many of the cloudy minerals cut as semi-precious stones are forms of hydrated silica ($<10\%$ water). These are brilliantly coloured and include opal. A mixture of quartz and opal gives rise to a variety of materials known collectively as chalcodonic silica. These include onyx, jasper and agate, the latter being used for mortar and pestle combinations and for ornaments.

The Si—O bond distances in silica of about 0.160 nm and the Si—O—Si bond angles of 140–150 degrees support the presence of $p_\pi - d_\pi$ bonding in all Si—O bonds.

Orthorhombic Silica

While both silicon disulphide and diselenide have long been known to possess a structure involving SiX_4 tetrahedra edge-bridging in long chains (like $BeCl_2$), the stability induced in the Si—O bond by $p_\pi - d_\pi$ bonding, and the resultant widening of the Si—O—Si angle rather precluded the idea of a similar form of silica. Indeed the expected Si—O—Si bond angle (70 degrees 32 minutes), assuming no tetrahedral distortion and directed overlap, would be about half of that normally encountered in natural forms of silica.

This form has now been synthesized by heating silicon monoxide in oxygen or allowing it to disproportionate at 1200–1400°C.

$$SiO_2 + Si \xleftarrow{\text{heat}} 2SiO \xrightarrow{O_2} 2SiO_2$$

This orthorhombic silica forms needles, and is less dense (1.96) than the other forms. Heating to 1390°C causes a slow change to a fibrous cristobalite polymorph, while tridymite forms between 200 and 800°C.[155]

A crystal structure determination shows the edge-linked SiO_4 tetrahedra with an Si—O—Si bond angle of 88 degrees, while the Si—O bond length of 0.187 ± 0.009 nm is much longer than that encountered in other forms of silica. It indicates weak bonding which may account for the high reactivity of this form. Water will attack it even at room temperature, while other forms of silica are resistant.

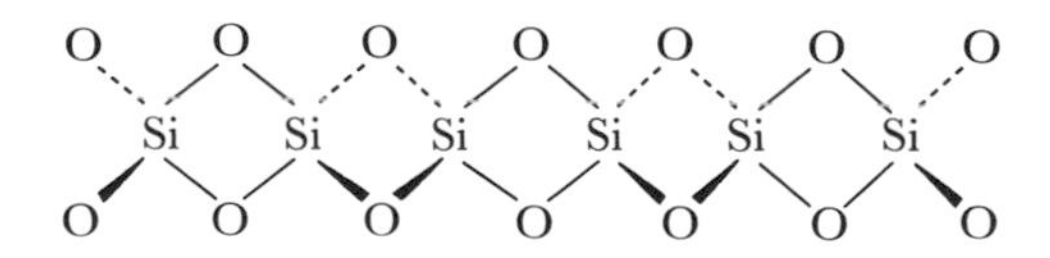

Fig. 4.28. The structure of orthorhombic silica

Hydrogen fluoride is used for etching quartz glass, and heating silica with lead difluoride will give SiF_4, thus providing a route to volatile silicon compounds. XeF_6 reacts similarly.

$$SiO_2 + 6HF \longrightarrow 2H_2O + H_2SiF_6$$

$$2PbF_2 + SiO_2 \longrightarrow 2PbO + SiF_4$$

Oxides of germanium, tin and lead

While these elements have the added complication of two stable valency states, they show none of the complex structural behaviour of silica and the silicates.

Germanium dioxide crystallizes with a rutile structure below 1033°C, but a cristobalite (4-co-ordinate) one above, as the radius ratio is close to that required for change from octahedral to tetrahedral co-ordination. With tin dioxide, the natural form, cassiterite, crystallizes with a rutile structure, as does PbO_2.

The few germanates encountered naturally all have silicate analogues. The two orthogermanates, Be_2GeO_4 and Zn_2GeO_4, are isomorphous with phenacite and willemite (Zn_2SiO_4), as is $Sc_2Ge_2O_7$ with thortveitite ($Sc_2Si_2O_7$). Likewise, benitoite has a cyclotrimetagermanate analogue in $BaTiGe_3O_9$.

Cyclosiloxanes

Replacement of the terminal Si—O$^-$ bonds of silicates by monovalent residues such as organic groups leads to the class of compounds known collectively as the silicones. These often involve the cyclosiloxane ring, *e.g.*, $[R_2SiO]_3$. Before discussing these, however, the perchlorocyclosiloxanes will be mentioned.

Perchlorosiloxanes

Like the perchlorocyclosilazanes, these can be synthesized by oxidizing $SiCl_4$ at 1000°C.

$$SiCl_4 \xrightarrow[O_2]{1000°C} [SiOCl_2]_n \ (n = 3, 4, 5)$$

The pentamer is formed in lower yields, and all three oligomers are crystalline solids with basic skeletal rings analogous to those of the silicones.[156] There is no dimeric $[SiOCl_2]_2$, unlike the case with sulphur $[SiSCl_2]_2$, emphasizing again the relative instability of the Si_2O_2 ring.

Silicones

These are perhaps the best known and most widely used class of inorganic polymers. They possess an Si—O skeleton similar to the silicates and so are very stable. The extra rigidity from reduced rotation provided by the partial multiple-bonding in Si—O bonds often maintains the physical properties of silicones consistent over a wide range of temperature. They are more air-stable than the hydrocarbon polymers, are water-resistant and have high thermal stability. The precursors to these polymers are often cyclic and cage compounds, but before discussing these, mention should be made of the synthesis of silicone precursors, the organochlorosilanes. Without an economic method to make these, the silicone industry could not have flourished as it has done.

Chlorosilanes. E. G. Rochow first reported and patented his method of synthesizing alkylchlorosilanes in 1945.[157] The process involves reacting an alkyl halide with silicon in the presence of a copper catalyst at between 300°C and 450°C.[g] Under these conditions, results are reproducible, but not in the absence of the catalyst.

The predominant reaction involves the formation of the dihalide,

$$Si + 2RX \xrightarrow{Si/Cu} R_2SiX_2$$

though many side-products result from the reactions listed below.

$$2Si + 4RX \longrightarrow R_3SiX + RSiX_3$$

$$Si + 3RX \longrightarrow RSiX_3 + 2R\cdot$$

$$Si + HX + RX \longrightarrow RHSiX_2$$

$$Si + 2X_2 \longrightarrow SiX_4$$

Thus, using methyl chloride results in the formation of Me_nSiCl_{4-n} (n = 0–4), $MeHSiCl_2$ and $HSiCl_3$ together with higher-boiling polysilanes, silylalkanes and siloxanes. The proportions vary with the conditions, and the catalyst is most effective if alloyed with silicon. Copper appears to be the best, and the alloy is made from cuprous chloride. It catalyzes through the formation of a reactive copper–methyl intermediate.

$$2Cu + MeCl \longrightarrow CuCl + CuMe$$

This methylates silicon at 250°C, while chlorination occurs with CuCl, thereby regenerating the copper.

$$Me + Si \longrightarrow MeSi \qquad CuCl + Si \longrightarrow Cu + SiCl$$

High yields of a particular product can be ensured if the conditions are carefully chosen, and Table 4.6 illustrates this point. The main product

Table 4.6. Reaction Conditions for the Formation of Alkylchlorosilanes

Main Product	Catalyst	Temp. (°C)	% Yield
Me_3SiCl	Ca	300	62
Me_2SiCl_2	Cu	280–350	90
$MeSiCl_3$	Fe	350–400	85
$MeHSiCl_2$	Pb	410	72

Ratio MeCl/HCl / Product	6:1	3:2	2:3	1:6
Me_2SiCl_2	78.3	41.7	26.5	7.4
$MeSiCl_3$	14.3	31.8	32.9	35.0
Me_3SiCl	1.2	0.8	0.3	0.7
$MeHSiCl_2$	4.5	15.1	17.7	11.0
$SiCl_4$	0.5	1.2	4.0	5.5
$HSiCl_3$	1.2	9.1	18.5	40.5

varies with both catalyst and temperature, while increasing the percentage of copper in the alloy produces more highly chlorinated silanes. The chlorine content also increases with the presence of HCl in the reactants and with temperature. This may be partly due to more extensive pyrolysis of methyl chloride which generates HCl and deposits carbon.

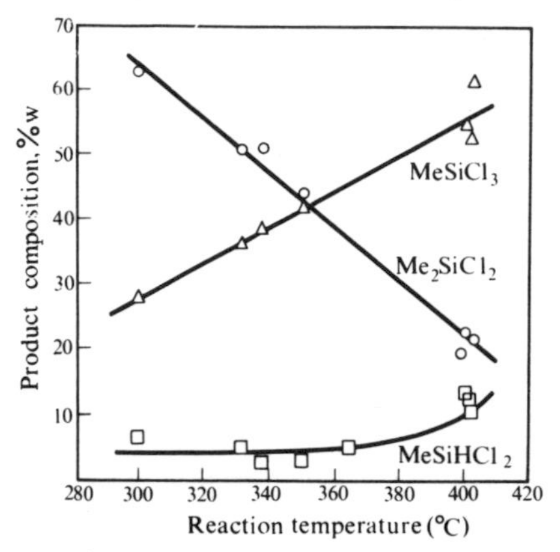

Fig. 4.29. Temperature-dependence of the composition of the product mixture in the reaction of methyl chloride with Si–Cu in a fixed bed[g]. (From R. J. H. VOORHOEVE, 'Organohalosilanes,' Elsevier, Amsterdam, 1967. Original source. *Angew. Chemie. Internat. Edn.*)

This carbon deposit slows down the reaction and can also lead to localized heating and loss of efficiency. The spent catalyst is pyrophoric but it need not be destroyed. With HCl, $SiCl_4$ and $HSiCl_3$ are generated. Adding fresh copper re-activates this catalyst, which with chlorobenzene gives excellent yields of the phenylchlorosilanes.

Synthesis of siloxanes. The inherent strength of the Si—O bond, and the electrophilic behaviour of silicon when bonded to chlorine, renders chlorosilanes excellent precursors to siloxanes. The hydrolysis of dichlorosilanes yields cyclosiloxanes which are widely used to synthesize the polymeric silicones.

$$R_2SiCl_2 + H_2O \longrightarrow \frac{1}{n}[R_2SiO]_n + 2HCl$$

Under neutral conditions, hydrolysis often gives the silanol.[e, h] Thus, with dimethyldichlorosilane, both oligomeric cyclosiloxanes $[Me_2SiO]_x$ and open-chain terminal dihydroxysiloxanes result. The latter condense further, and under dilute conditions give cyclosiloxanes.

$$HOSiMe_2[OSiMe_2]_xOSiMe_2OH \xrightarrow{-H_2O} [Me_2SiO]_{x+2}$$

The degree of polymerization varies with the hydrolytic conditions with

$n = 3$–9. These can be readily separated by fractional distillation, though the presence of acidic or basic catalysts can cause interconversions to occur.

$$EtHSiCl_2 \xrightarrow[EtOH]{\text{cold } H_2O} [EtHSiO]_n \ (n = 3, 48\%; n = 4, 33\%; n = 5, 11\%)$$

$$MeHSiCl_2 \xrightarrow[Et_2O]{H_2O} [MeHSiO]_n \ (n = 4\text{–}8)$$

$$\underset{70\%}{[Et_2SiO]_4} \xleftarrow[MeOH]{H_2O} Et_2SiCl_2 \xrightarrow[Et_2O]{\text{cold } H_2O} \underset{(n = 3, 41\%; n = 4, 31\%)}{[Et_2SiO]_n}$$

Cyclosiloxanes readily polymerize when heated to 300°C in a closed system, especially in the presence of a catalyst.[e, h, i] Heating to higher temperatures reverses this process so long as no cross-linking or alkyl decomposition has occurred.

$$\underset{(n = 3, 4, 5 \ldots)}{\frac{x}{n}[Me_2SiO]_n} \underset{>300°C}{\overset{250\text{–}300°C}{\rightleftharpoons}} \underset{\text{(high polymer)}}{\frac{n}{x}[Me_2SiO]_x}$$

Heating with a trace of a polychlorosilane results in telomerization, with the insertion of the cyclosiloxane into the Si—Cl bond.[158]

$$Me_2SiCl_2 + x[Me_2SiO]_n \longrightarrow ClSiMe_2[OSiMe_2]_{nx}Cl \ (n = 3 \text{ or } 4)$$

Polymerization to silicones. Both acid- and base-catalyzed polymerization of permethylcyclotri- and -tetrasiloxanes show the former to be the more reactive. With the acid-catalyzed polymerization, protonation is followed by nucleophilic attack at Si by X^-. This mechanism is one of several proposed.

$$(Si\text{—}OH\text{—}Si)^+ + X^- \longrightarrow Si\text{—}OH + Si\text{—}X$$

$$Si\text{—}OH + Si\text{—}X \longrightarrow Si\text{—}O\text{—}Si + HX$$

$$Si\text{—}OH + HO\text{—}Si \longrightarrow Si\text{—}O\text{—}Si + H_2O$$

$$Si\text{—}X + H_2O \longrightarrow Si\text{—}OH + HX$$

Lewis acids also cleave Si—O bonds. Thus, $SnCl_4$ yields an unstable stannosiloxane which polymerizes with climination of the catalyst. Titanium catalysts react similarly.

$$2[Me_2SiO]_4 + 2SnCl_4 \longrightarrow 2Cl[Me_2SiO]_4SnCl_3 \xrightarrow{-SnCl_4} Cl[Me_2SiO]_8SnCl_3$$

Base catalysis involves nucleophilic attack at silicon, with cleavage of the Si—O bonds and subsequent propagation.

$$[R_2SiO]_4 + OH^- \rightleftharpoons HO[SiR_2O]_3SiR_2O^-$$

$$\Big\Updownarrow n[R_2SiO]_4$$

$$HO[SiR_2O]_{3+4n}SiR_2O^-$$

The Grignard reagent will cleave polysiloxanes to give the magnesium siloxide. This hydrolyzes to the silanol, and so the reaction parallels that of a Grignard reagent with a ketone, from which the name silicone is reputedly derived.

$$\frac{1}{n}[Me_2SiO]_n + MeMgI \longrightarrow Me_3SiOMgI \xrightarrow{H_2O} Me_3SiOH$$

Before considering the hydrolysis products of trichlorosilanes, which generally yield cage compounds and high polymers, mention will be made of the combination of dichlorosilanes with tri- and tetra-halogenosilanes. Polycyclic and spirocyclic products result and D, T and Q will be used to represent the di-, tri- and quadri-functional silicon atoms.

Polycyclic Siloxanes[e]

As with organic polycyclic compounds, the bridgehead of two rings always involves at least a trifunctional atom. In the case of cyclosiloxanes involving two fused rings, the general formula of the compound will be $Si_nR_{2n-2}O_{(n-2)+3}(D_{n-2}T_2)$. D_3T_2, $(Si_5Me_8O_6)$, has been fully characterized, and comprises fused 6- and 8-membered siloxane rings. Several of these polycyclic compounds have been partially characterized. They are conveniently formed by the thermal rearrangement of the polymers obtained from the hydrolysis of Me_2SiCl_2 and $MeSiCl_3$.

Table 4.7. Cyclosiloxanes

Representation	Formula	Structure	m.p. (°C)	b.p. (°C)
D_3T_2	$Me_8Si_5O_6$	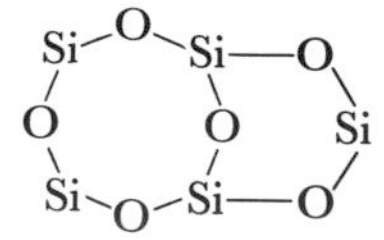	118	203
D_4Q	$Me_8Si_5O_6$	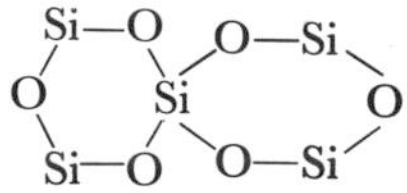	121	204
D_5Q	$Me_{10}Si_6O_7$	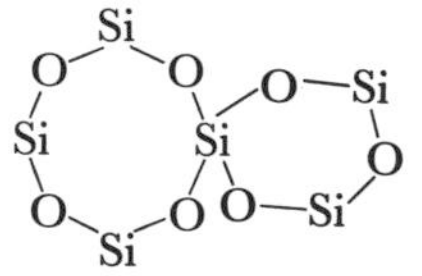	57	230
D_4T_2	$Me_{10}Si_6O_7$	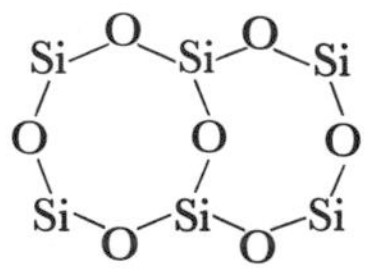	51	232

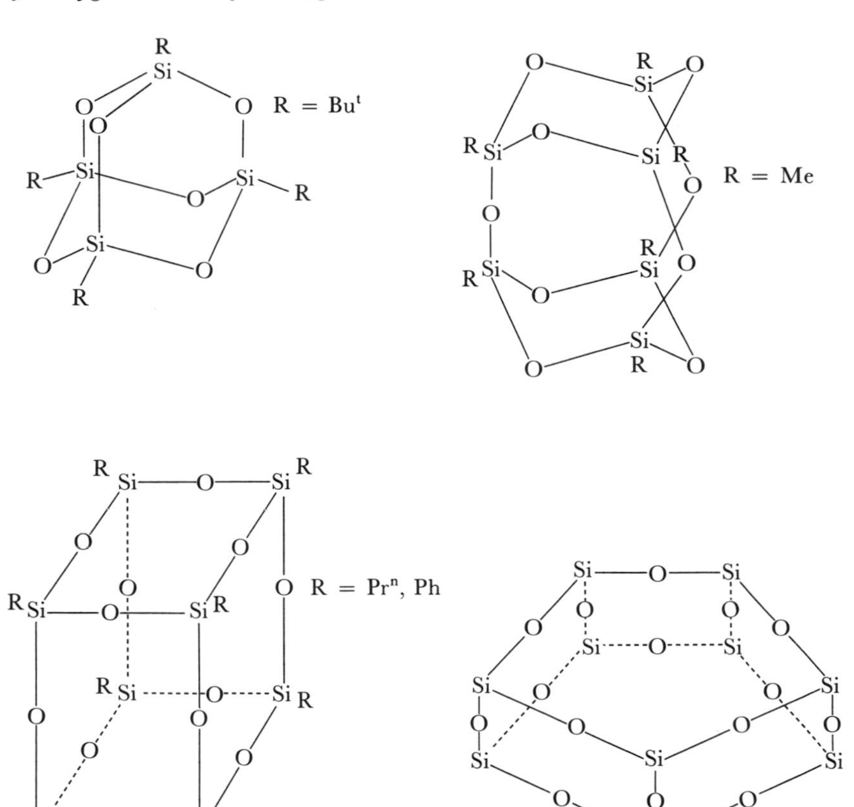

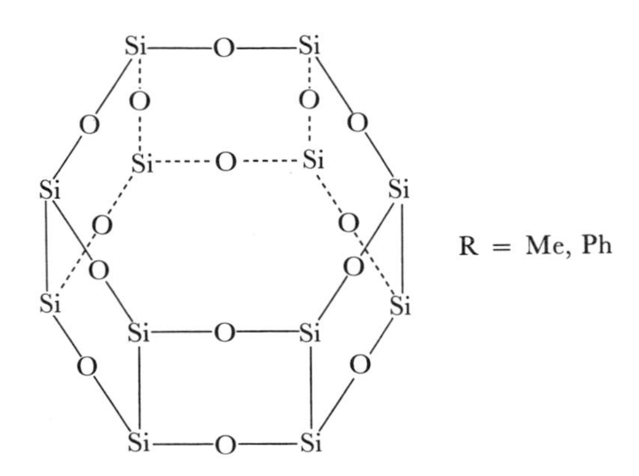

Fig. 4.30. Spherosiloxanes

Spirocyclic Siloxanes[e]

Polymers formed from the hydrolysis of Me_2SiCl_2 and $SiCl_4$ (traces) pyrolyze to spiro compounds formulated as $Si_nMe_{2n-2}O_{(n-1)+2}(D_{n-1}Q)$, with one Q site. D_4Q and D_5Q have been characterized. Isomeric D_4Q and D_3T_2 have similar physical properties, as with D_5Q and D_4T_2.

These spiro-compounds can be synthesized step-wise from 1,3-dihydroxysiloxanes and $SiCl_4$ in the presence of pyridine. D_4Q results this way, as do many arylspirosiloxanes.[159]

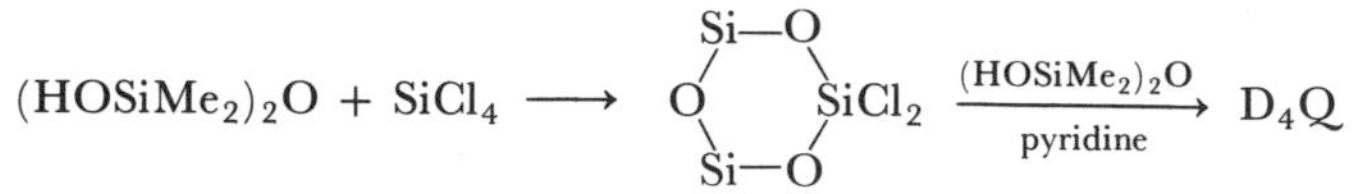

$$(HOSiMe_2)_2O + SiCl_4 \longrightarrow \text{(Si—O / O SiCl}_2\text{ / Si—O ring)} \xrightarrow[\text{pyridine}]{(HOSiMe_2)_2O} D_4Q$$

Spherosiloxanes

This is the term often used for the low molecular weight cage polymers resulting from the hydrolysis of trichlorosilanes under dilute conditions.

$$2RSiX_3 + 3H_2O \xrightarrow[\text{solvent}]{\text{inert}} \frac{1}{n}[(RSi)_2O_3]_n + 6HX$$

The products are normally sublimable solids, and the degree of polymerization appears to depend on the organic substituent.[e, 160] The compounds can be represented by T_{2n} (n = 2–6) and examples are known for each. Their high melting points reflect the symmetrical structures of these compounds (Figure 4.30 and Table 4.8).

Table 4.8. Spherosiloxanes

T_{2n} n =	2	3	4		5	6	
R	Bu^t	Me	Pr^n	Ph	Me	Me	Ph
m.p. (°C)	225	210–11	219–20	500	333–4	261	385

The phenyl compounds result when phenyltrihydroxysilane is heated in toluene under reflux, the structures resembling cyclic ladder polymers. A trace of hydroxide ion causes the cycloladderane to open and polymerize to the polyladderane T_{2n}.[161] (See appendix).

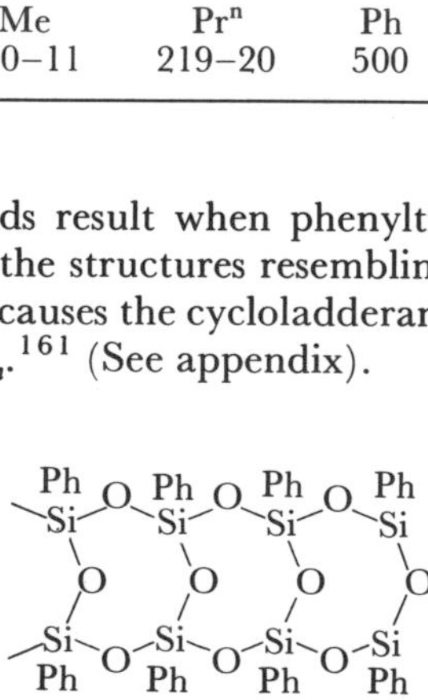

Ph O Ph O Ph O Ph
Si Si Si Si
O O O O
Si O Si O Si O Si
Ph Ph Ph Ph

Fig. 4.31. $[(PhSi)_2O_3]_n$

The Structure of Cyclosiloxanes

Hexamethylcyclotrisiloxane has a planar Si_3O_3 ring, while the spirosiloxane D_4Q octamethylspiro[5,5]pentasiloxane, possesses two such rings mutually perpendicular. The Si—O bond lengths are similar to those observed in silica and the silicates, where p_π–d_π multiple bonding is considered to account for the short Si—O bond. The bond angles at oxygen also support sp^2-hybridization with a p orbital available for p_π–d_π bonding.

The 8-membered ring in octamethylcyclotetrasiloxane is puckered, but the Si—O bond length of 0.165 nm, and the wide Si—O—Si angle of 143 degrees, supports multiple bonding. Such a wide angle is impossible in $[Me_2SiO]_3$, so there is possibly less strain in the tetramer which may account for the higher reactivity of the former to acid and base cleavage. One peculiar feature of the cyclotetrasiloxane $[Me_2SiO]_4$ is that it exists as two polymorphs, the transition occurring at $-16.3°C$.[162]

As with the cyclosilazanes, isomerism is encountered with the methylphenylcyclosiloxanes. Hydrolysis of the dichloride gives two cyclotrisiloxanes and four cyclotetrasiloxanes. The two trisiloxane isomers show different p.m.r. spectra, the *cis* one, with the higher melting point, having a single methyl peak, while the *trans* one shows two methyl resonances (ratio 2:1).

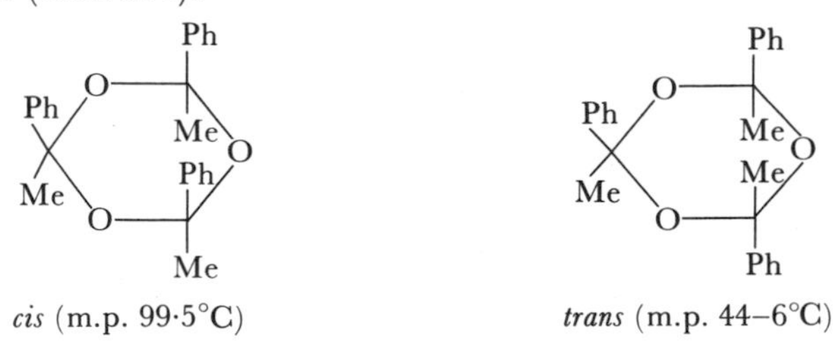

cis (m.p. 99·5°C) *trans* (m.p. 44–6°C)

Fig. 4.32. Isomers of $[PhMeSiO]_3$

This interpretation is confirmed by the synthesis of an equimolar isomer mixture from the *meso* form of $(HOSiMePh)_2O$ and $MePhSiCl_2$, while the racemic form gives only the low-melting *trans* isomer.[163]

HO–SiMePh–O–SiMePh–OH + $MePhSiCl_2$ ⟶ *cis* + *trans*

meso isomer (m.p. 110°C)

HO–SiMePh–O–SiPhMe–OH —($MePhSiCl_2$)⟶ *cis* + *trans*

racemic isomer (m.p. 100°C)

Four isomeric cyclotetrasiloxanes have been characterized, and again, the structures have been established by independent synthesis from the two diols and a *meso*/racemic mixture of $(ClSiPhMe)_2O$ in the same way as the cyclotrisiloxanes.

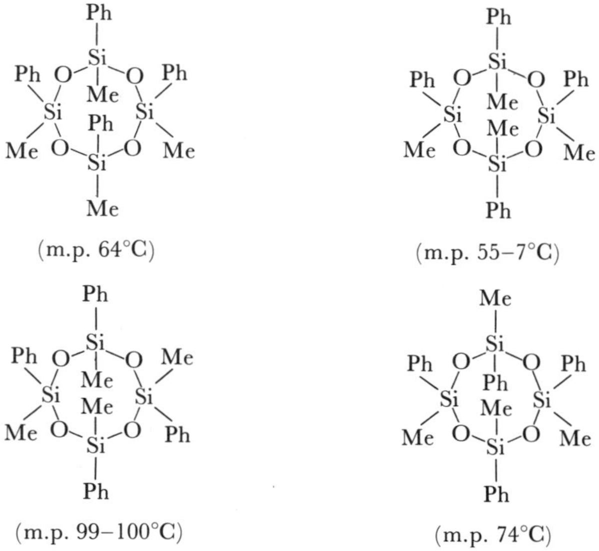

(m.p. 64°C) (m.p. 55–7°C)

(m.p. 99–100°C) (m.p. 74°C)

Fig. 4.33. Isomers of $[PhMeSiO]_4$

Cross-linked Siloxanes

These are the silicones which contain sufficient T and Q units with the D ones to give a polymeric two- or three-dimensional array. This can obviously occur in a wide variety of ways, and it is of interest to compare the physical properties of the polymer containing these cross-linked and fused siloxane rings.

Table 4.9. Polymethylsiloxanes with D *and* T *Units*

Ratio D/T	high	2	1	0.1	0.01
Ratio Me/Si	2	1.67	1.5	1.1	1
State	gums	liquids	resins	glassy and brittle	gel

Polyethylsilicones are softer than the methyl compounds of the same constitution, and so resinous products result with higher proportions of T and Q units. The larger alkyl groups render the polymers less thermally stable, while phenyl groups improve the mechanical, electrical and thermal properties of the silicone.

These polymers can be synthesized for specific requirements. Vulcanizing silicone gums by cross-linking can readily be achieved by high-temperature peroxidation. The resulting rubbers are of great commercial value, while the resins are used as adhesive and hydrophobic surface coatings. Solubility also enables them to be used as resin lacquers.[e]

Heterosiloxanes. Siloxane rings containing other atoms have been synthesized, primarily to examine their potential as polymer precursors. Aluminium and titanium have been tried, but boron and tin provide polymers with the more peculiar properties.

Phenylboronic acid readily condenses with diphenylsilanediol and other difunctional silanols to give cyclic products. Trichlorosilanes give ladder polymers.[e]

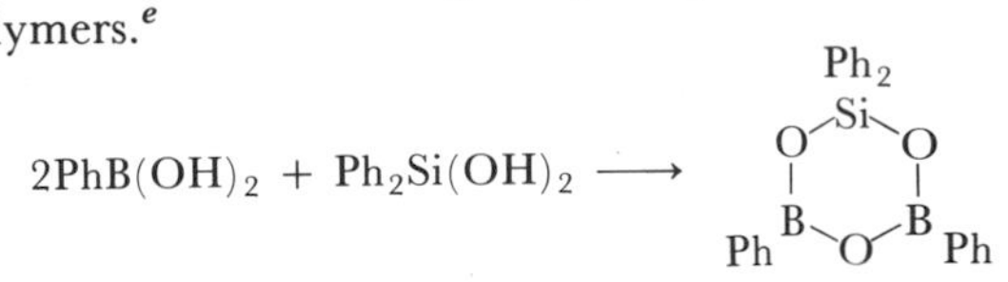

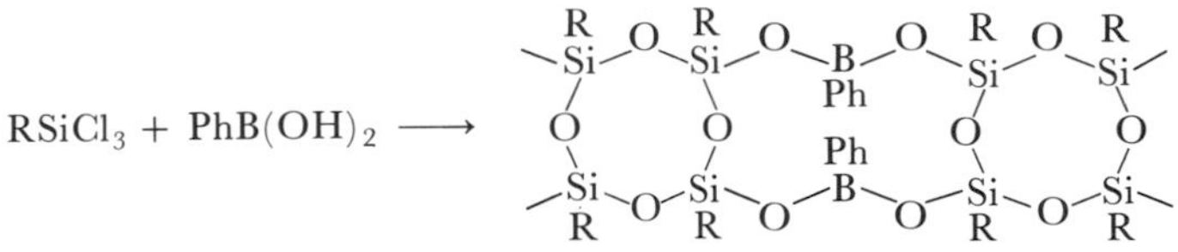

$$2PhB(OH)_2 + Ph_2Si(OH)_2 \longrightarrow \text{cyclo-}[Ph_2Si(O)\text{–}O\text{–}BPh\text{–}O\text{–}BPh\text{–}O]$$

$$RSiCl_3 + PhB(OH)_2 \longrightarrow \text{ladder polymer of } RSi\text{–}O\text{–}RSi\text{–}O\text{–}BPh\text{–}O\text{–}RSi\text{–}O\text{–}RSi \text{ chains linked by O bridges}$$

The incorporation of electrophilic boron into the siloxane provides a site for co-ordination, and so weak cross-linking occurs.

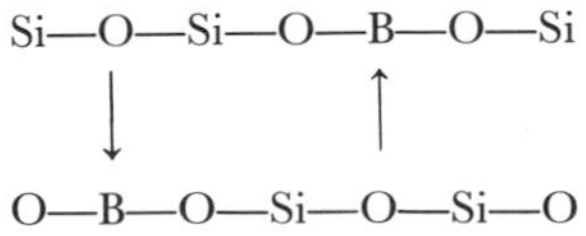

```
Si—O—Si—O—B—O—Si
   ↓        ↑
O—B—O—Si—O—Si—O
```

Such polymers resemble putty unless they are put under stress. Under these conditions the cross-linking gives the polymer pseudo-crystalline properties and resilience. This borosiloxane will readily bounce and shatter if struck with a hammer, to the obvious delight of children. However adults, especially scientists, find it much more entertaining. This has been rationalized by assuming that only a mature intellect can appreciate the apparent paradox of bouncing putty.[h]

Diphenylsilanediol and di-n-butyltin oxide condense to a cyclostannosiloxane,[164] and silicone polymers incorporating tin have a much higher viscosity, due to co-ordination. A small percentage of tin gives the polymer the bouncing putty properties of the borosiloxanes.[e]

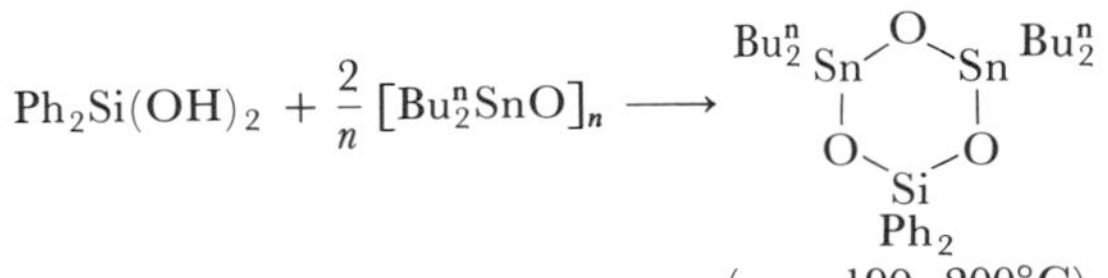

$$Ph_2Si(OH)_2 + \frac{2}{n}[Bu^n_2SnO]_n \longrightarrow \text{cyclo-}[Bu^n_2Sn\text{–}O\text{–}SnBu^n_2\text{–}O\text{–}SiPh_2\text{–}O]$$

(m.p. 190–200°C)

Cyclosiloxanes containing the Si—Si *bond.* A variety of cage and polycyclic compounds result from the hydrolysis of substituted disilanes.[165] The products are normally crystalline solids, and polycyclic compounds result when sym-di- and tetra-functional disilanes are co-hydrolyzed.

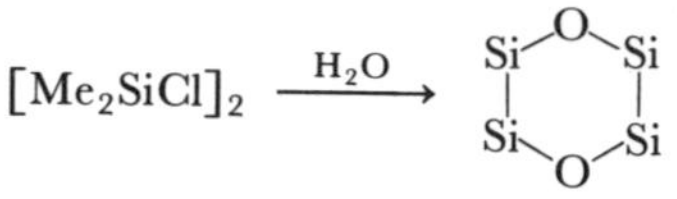

$$[Me_2SiCl]_2 \xrightarrow{H_2O} \text{Si}_4\text{O}_2 \text{ ring}$$

(m.p. 45°C)

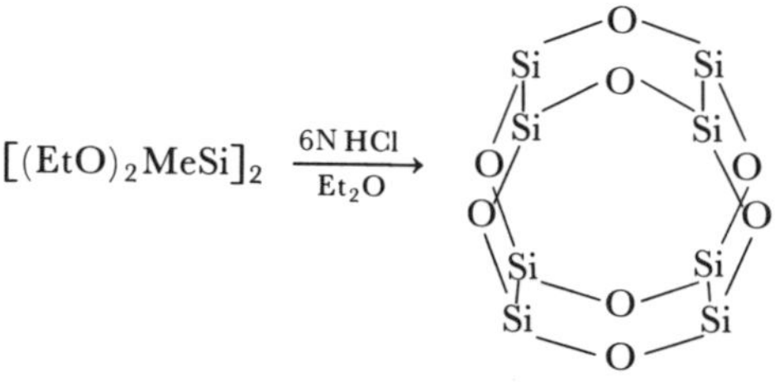

$$[(EtO)_2MeSi]_2 \xrightarrow[Et_2O]{6N\,HCl} \text{cage}$$

(m.p. 315–8°C)

$$2[Me_2(EtO)Si]_2 + [(EtO)_2MeSi]_2 \xrightarrow{6N\,HCl} \text{Si}_4\text{O}_2 \text{ ring} + \text{bicyclic}$$

(m.p. 56°C)

$$\downarrow Me_2Si(OEt)_2$$

Si₄O₃ seven-membered ring + tricyclic

(m.p. 126–7°C)

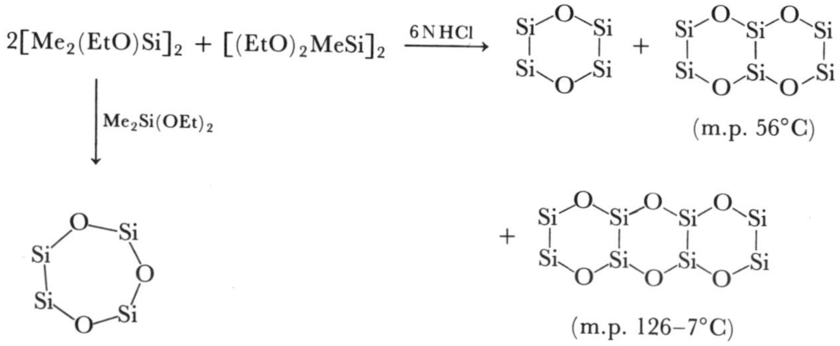

As with the silicones and siloxanes, acid-catalyzed redistributions readily occur. Thus, open-chain telomers are formed from hexamethyldisiloxane and octamethyl-1,4-dioxa-2,3,5,6-tetrasilacyclohexane in the presence of a trace of H_2SO_4.

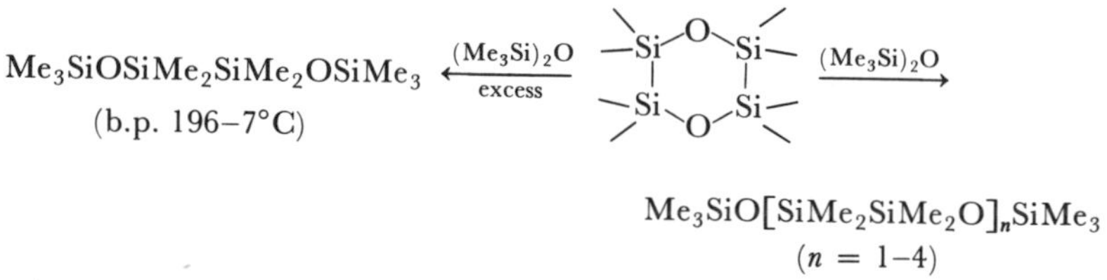

$$Me_3SiOSiMe_2SiMe_2OSiMe_3 \xleftarrow[\text{excess}]{(Me_3Si)_2O} \text{ring} \xrightarrow{(Me_3Si)_2O} Me_3SiO[SiMe_2SiMe_2O]_nSiMe_3$$

(b.p. 196–7°C) $(n = 1–4)$

Chlorination of cyclopolysilanes, followed by hydrolysis produces oxacyclopolysilanes. Some have already been discussed earlier in the Chapter (p. 159), but not the products of chlorination and further hydrolysis.[166] Isomeric chloro- and cyclo-siloxanes result at each stage.

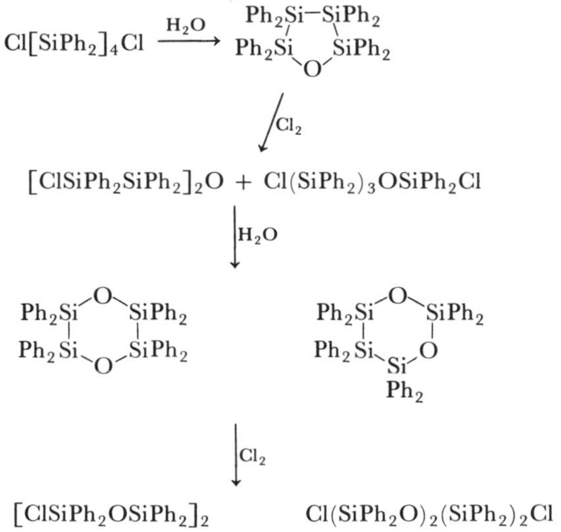

$Cl[SiPh_2]_4Cl \xrightarrow{H_2O}$ $Ph_2Si{-}SiPh_2$
$Ph_2Si \quad SiPh_2$
O

$\downarrow Cl_2$

$[ClSiPh_2SiPh_2]_2O + Cl(SiPh_2)_3OSiPh_2Cl$

$\downarrow H_2O$

$Ph_2Si{-}O{-}SiPh_2$ | $Ph_2Si{-}O{-}SiPh_2$
$Ph_2Si{-}O{-}SiPh_2$ | $Ph_2Si{-}Si(Ph_2){-}O$

$\downarrow Cl_2$

$[ClSiPh_2OSiPh_2]_2 \qquad Cl(SiPh_2O)_2(SiPh_2)_2Cl$

Siloxenes. These provide the purely inorganic examples of the ring systems involving oxygen and catenated silicon. The siloxene rings comprise nine atoms as three Si_2O fragments.

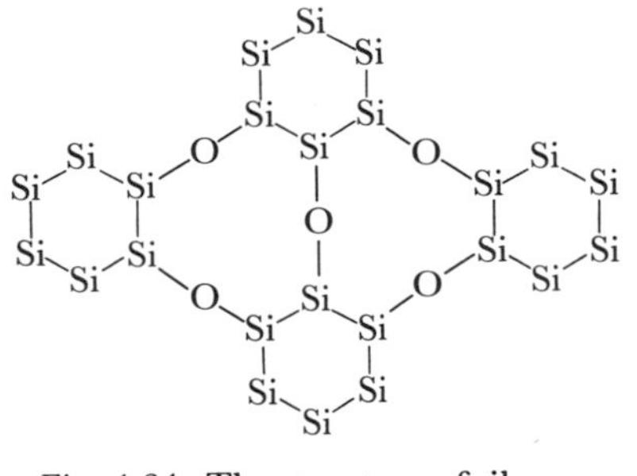

Si
Si Si
Si Si
Si Si Si O Si O Si Si
Si Si O Si Si
Si Si Si Si
Si O Si O Si Si
Si Si
Si Si
Si

Fig. 4.34. The structure of siloxene

Hydrolysis of $CaSi_2$ gives "siloxene", $Si_6O_3H_6$, as a pale green powder involving cyclohexasilane rings cross-linked by oxygen atoms.[167] The structure resembles that of $CaSi_2$, where again, layers of Si_6 rings prevail, and as with the hydrolysis of cyclotetrasilanes, hydrogen is evolved.

$$3Si_{2n}^{2n-} + 9nH_2O \longrightarrow [Si_6O_3H_6]_n + 6nOH^- + 3nH_2$$

The Si—H bonds are readily halogenated. Hydrogen bromide monosubstitutes, while bromine produces a variety of brominated products depending on the conditions employed. Di-substitution is readily achieved in CS_2, but further replacement is difficult as two substituents would be apical and on the same side of the Si_6 ring (chair conformer), with resulting steric clash.

With terminal diols, the dibromosiloxene $Si_6O_3H_4Br_2$ substitutes at only one OH group, whereas both do with a tribromosiloxene, showing that substitution in this case occurs with two bromine atoms on the same side of the ring.

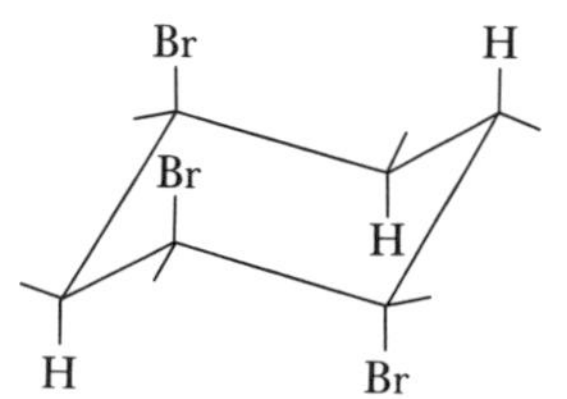

Fig. 4.35. Tribromosiloxene

Chlorine generally cleaves both Si—H and Si—Si bonds, giving hexachlorodisiloxane $(Cl_3Si)_2O$, while iodine only mono-substitutes. Steric clash will be greater with iodosiloxenes, and even di-iodosiloxene cannot be synthesized directly.

$$Si_6O_3H_4Br_2 \xrightarrow{Et_2NH} Si_6O_3H_4(NEt_2)_2 \xrightarrow{4HI} Si_6O_3H_4I_2$$

Sulphates. Both silicon and germanium form covalent cyclic sulphates incorporating S^{VI} in the ring. These are generally crystalline solids, and result from the halide or oxide with H_2SO_4 or SO_3.[168] Esters also give good yields of the sulphates.[169]

O_2
S—O
O Si—
$Me_2SiCl_2 + H_2SO_4 \longrightarrow$ | | $\longleftarrow [Me_2SiO]_n + SO_3$
—Si O
O—S
O_2

(m.p. 103–120°C, b.p. 148°C/1 mm)

R_2
Ge—O
O SO_2
$[R_2GeO]_n + H_2SO_4 \longrightarrow$ | | (R = Et, m.p. 116°C;
O_2S O R = Pr^n, m.p. 129°C)
O—Ge
R_2

$$Pr^n_2Ge(OCOMe)_2 \xrightarrow{H_2SO_4} [Pr^n_2GeOSO_2O]_2$$

Terminal dichloropolysilanes also form cyclic sulphates. They undergo many expected reactions, while an excess of triphenylsilyl lithium gives permethylcyclohexasilane.[170] The mechanism is thought to involve the linear octasilane intermediate which contracts and cyclizes in the presence of excess Ph_3SiLi.

$$Cl[SiMe_2]_nCl + H_2SO_4 \longrightarrow [Me_2Si]_n(O)_2SO_2$$

($n = 3$, m.p. 82–3°C; $n = 4$, b.p. 125–30°C/0.8 mm)

$$\text{cyclo-}[Si_3O_2SO_2] \xrightarrow{H_2O} \text{cyclo-}[Si_6O_2]$$

$$\text{cyclo-}[Si_3O_2SO_2] \xrightarrow{HCl/NH_4Cl} Cl[SiMe_2]_3Cl$$

$$\text{cyclo-}[Si_3O_2SO_2] \xrightarrow{EtMgBr} Et[SiMe_2]_3Et$$

$$\text{cyclo-}[Si_3O_2SO_2] \xrightarrow[-Li_2SO_4]{2Ph_3SiLi}$$

$$\underset{\textbf{(4-XII)}}{Ph_3Si(Me_2Si)_3SiPh_3} \xrightarrow[-Ph_6Si_2]{Ph_3SiLi} Ph_3Si[Me_2Si]_3Li$$

$$\xrightarrow[-Ph_3SiLi]{\textbf{(4-XII)}} Ph_3Si[Me_2Si]_6SiPh_3$$

$$\xrightarrow[-Ph_6Si_2]{Ph_3SiLi} Ph_3SiMe_2Si[Me_2Si]_4Me_2Si^-Li^+$$

$$Ph_3SiMe_2Si[Me_2Si]_4Me_2Si^-Li^+ \longrightarrow [Me_2Si]_6 + Ph_3SiLi$$

Cyclic germanium–oxygen compounds

The difficulties encountered in synthesizing organogermanium halides, and the inherent expense in extracting and purifying the element, has limited the amount of research devoted to germanium–oxygen compounds.

Dimethylgermanium oxide is formed by the alkaline hydrolysis of dimethylgermanium dichloride and is tetrameric in benzene. It is very soluble in water, probably due to diol formation,[171, j] which contrasts it strongly with hydrophobic silicones.

$$4Me_2GeCl_2 + 8OH^- \longrightarrow \underset{\text{(m.p. 91–2°C)}}{[Me_2GeO]_4} + 8Cl^- + 4H_2O$$

Evaporation of this solution gives polymeric $[Me_2GeO]_n$ which, like the tetramer, forms the cyclic trimer on heating to 210°C and cooling rapidly. This readily reverts to the tetramer at room temperature.[172] Ring contraction readily occurs on heating, and at high dilution,[173] a phenomenon observed with cyclic silicon–sulphur compounds.

$$4[Me_2GeO]_3 \rightleftharpoons 3[Me_2GeO]_4$$

Both rings are cleaved by methyl-lithium. The lithium salt of trimethylgermanol is formed, which is thermally unstable.[174]

$$[Me_2GeO]_4 + 4MeLi \longrightarrow 4Me_3GeOLi \xrightarrow{100°C} (Me_3Ge)_2O + Li_2O$$

Ethyl and n-propylcyclogermoxanes are polymeric, but can be isolated as an unstable cyclic tetramer and trimer respectively. A similar equilibrium exists with the phenyl compounds. Hydrolyzing diphenyldibromogermane with $AgNO_3$ gives a polymer.[175] Boiling in acetic acid yields the tetramer which distils as the trimer. The process can be reversed.

$$\underset{(m.p.\ 298°C)}{[Ph_2GeO]_n} \underset{EtOH/H_2O}{\overset{AcOH}{\rightleftharpoons}} \underset{(m.p.\ 218°C)}{[Ph_2GeO]_4} \underset{AcOH}{\overset{distn.}{\rightleftharpoons}} \underset{(m.p.\ 149°C)}{[Ph_2GeO]_3}$$

Hydrolysis of sym-dichlorodigermanes gives the expected cyclodioxatetragermanes.[176]

$$GeCl_4 + Et_6Ge_2 \longrightarrow [ClEt_2Ge]_2 \xrightarrow{H_2O} \begin{matrix} Et_2Ge\!-\!O\!-\!GeEt_2 \\ | \qquad\qquad | \\ Et_2Ge\!-\!O\!-\!GeEt_2 \end{matrix}$$

$$\underset{(X = H\ or\ Br)}{[XPh_2Ge]_2} \xrightarrow{OH^-} \begin{matrix} Ph_2Ge\!-\!O\!-\!GePh_2 \\ | \qquad\qquad | \\ Ph_2Ge\!-\!O\!-\!GePh_2 \end{matrix}$$

Cyclic tin–oxygen compounds

From the cyclic compounds discussed for silicon and germanium the absence of compounds with a co-ordination number greater than four is quite striking. With tin, however, structures readily form in which the oxygen atom bonded to tin functions as a bridging atom. Consequently a new range of cyclic compounds result which will be considered first through the alkoxy derivatives of tin.

Tin Alkoxides

Alkoxysilanes and germanes are usually distillable monomeric liquids. Those of tin, however, are generally associated. Trimethyltin methoxide is a crystalline solid and with cyclohexanone oxime, Me_3SnNEt_2 forms a dimeric tin oxime which involves tin–oxygen bridges.[177]

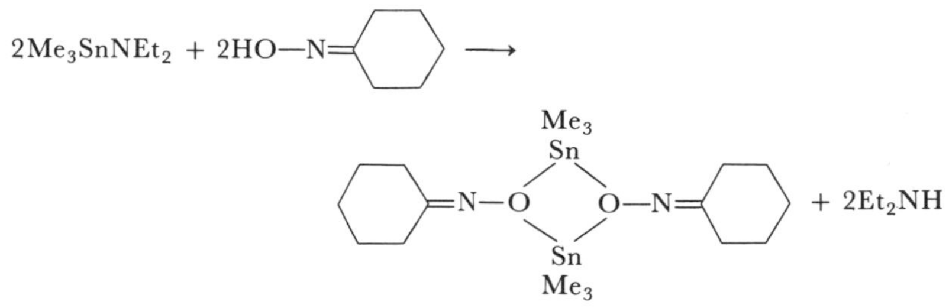

$$2Me_3SnNEt_2 + 2HO{-}N{=}C_6H_{10} \longrightarrow [Me_3Sn(ON{=}C_6H_{10})]_2 + 2Et_2NH$$

Consequently, tin is 5-co-ordinate, but this co-ordination can be increased to six if sufficient electronegative groups are bonded to tin.

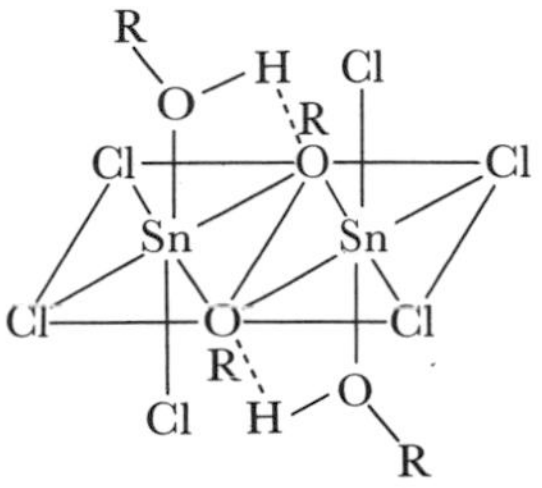

Fig. 4.36. An alcohol complex of dimeric alkoxytrichlorotins

Alkoxytrichlorotins are dimeric and readily add two moles of an alcohol while remaining dimeric.[178] Similarly, the partial alcoholysis of alkylhalobisacetylacetonato-tin gives a 6-co-ordinate alkoxytin complex.[179] The dihalide complexes, formed from $(acac)_2SnX_2$ and methanol, are also dimeric.

$$2(acac)_2SnX_2 \xrightarrow{2MeOH} [(acac)(MeO)SnX_2]_2 + 2CH_3COCH_2COCH_3$$
$$(X = Cl, Br \text{ or } I)$$

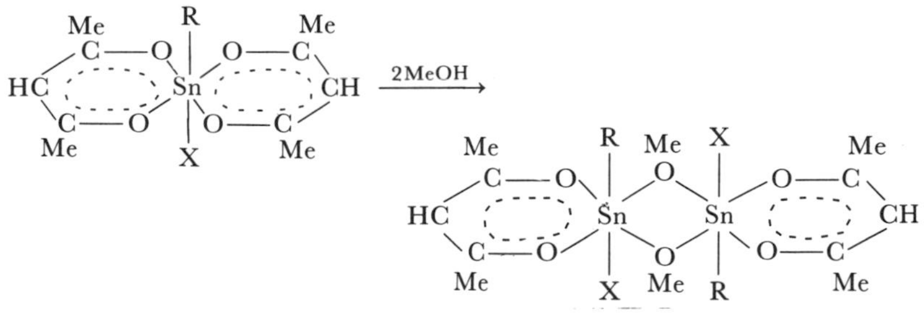

$$(acac)_2SnRX \xrightarrow{2MeOH} [(acac)SnRX(\mu\text{-}OMe)]_2$$

Di-n-butyltin dimethoxide is associated at high concentrations in CCl_4. It also shows a concentration-dependent infrared spectrum, one of the peaks associated with $\nu_{(SnOC)as.}$ disappearing at high dilution. This supports monomer–(alkoxy-bridged) dimer equilibrium, but with the dibutyltin derivatives of ethane-1,2-diol the equivalence of the ring protons as shown by p.m.r. spectroscopy supports the second equilibrium,[180] with a large ring and 4-co-ordinate tin.

$$\text{(CH}_2\text{O)}_2\text{SnBu}^n_2 \rightleftharpoons \text{dimer: } Bu^n_2Sn(OCH_2CH_2O)SnBu^n_2 \text{ with } O\rightarrow Sn \text{ bridges}$$

$$\text{(CH}_2\text{O)}_2\text{SnBu}^n_2 \rightleftharpoons Bu^n_2Sn\begin{matrix} O-(CH_2)_2-O \\ O-(CH_2)_2-O \end{matrix}SnBu^n_2$$

The alkoxyhalotindialkyls $R_2Sn(OR)X$ have attracted much attention. Again, these compounds are also dimeric at high solution concentrations but dissociate on dilution, though not in the case of the fluoride. They can be synthesized from the dialkoxide and an allyl halide, or by warming an equimolar mixture of the dihalide and dialkoxide.[180, 181]

$$2Bu^n_2Sn(OMe)_2 + 2BrCH_2CH{=}CH_2 \longrightarrow [Bu^n_2Sn(OMe)Br]_2 + 2\text{allylOMe}$$

$$R_2Sn(OMe)_2 + R_2SnX_2 \longrightarrow (XR_2SnOMe)_2$$

Compounds with the same functional groups have identical properties, irrespective of the synthetic pathway.

The exact nature of these compounds is still the subject of much research. They may involve the 4- and 6-co-ordination of a dialkoxide chelate complex of the dihalide, or alternatively be a halogen or alkoxy symmetrical-bridged compound with 5-co-ordinate tin. However, the addition of bipyridyl generates the dialkoxide and $R_2SnX_2{\cdot}$bipy, supporting a comparatively weak bridged structure for this dimer.

$$[R_2Sn(OMe)X]_2 + \text{bipy} \longrightarrow R_2Sn(OMe)_2 + R_2SnX_2{\cdot}\text{bipy}$$

This closely parallels the Schlenk equilibrium already discussed for the Grignard reagent (p. 31). The equilibria are represented below and the evidence to date strongly supports the right hand equilibrium, except when X is fluoride.

$$R_2Sn(\mu\text{-OMe})_2SnR_2X_2 \rightleftharpoons R_2Sn(OMe)_2 + R_2SnX_2 \rightleftharpoons$$

$$2R_2Sn(OMe)X \rightleftharpoons XR_2Sn(\mu\text{-OMe})_2SnR_2X$$

Thus, dialkyltin dichlorides and dibromides are monomeric, even in concentrated solution, so this, along with the associative tendency of the dialkoxides of tin(IV), supports an alkoxy-bridged dimer for $R_2Sn(OMe)X$ (X = Cl and Br).[182]

Me_2SnF_2 associates more readily than Me_2SnCl_2, as will be seen later (pp. 253–4), while Me_3SnF is a fluoride-bridged polymer. So the dimeric nature of $R_2Sn(OMe)F$ even at low concentrations supports a fluoride-bridged structure.

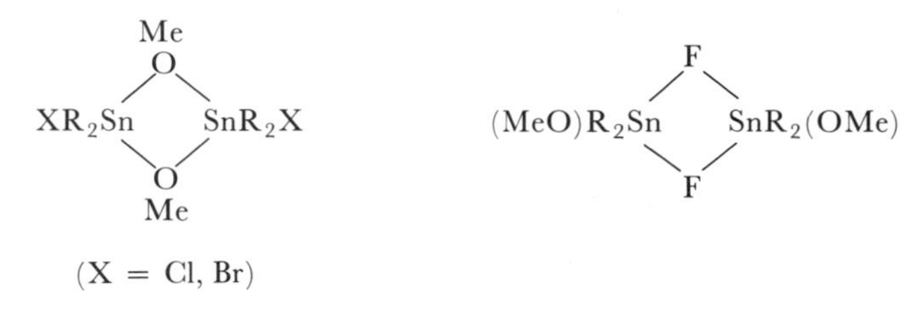

Fig. 4.37. Bridging in dimeric $R_2SnX(OR')$

This equilibrium is further supported by the isolation of two bipyridyl complexes on mixing dibutyltin dimethoxide with diethyltin dichloride.

$$Bu^n_2Sn(OMe)_2 + Et_2SnCl_2 \rightleftharpoons Bu^n_2Sn(OMe)Cl + Et_2Sn(OMe)Cl$$

$$Et_2SnCl_2 \xrightarrow{\text{bipy}} Et_2SnCl_2\cdot bipy$$

$$Et_2Sn(OMe)Cl \rightleftharpoons Bu^n_2SnCl_2 + Et_2Sn(OMe)_2$$

$$Bu^n_2SnCl_2 \xrightarrow{\text{bipy.}} Bu^n_2SnCl_2\cdot bipy$$

Organo-substituted Tin Hydroxides

These compounds can be readily prepared by hydrolyzing the appropriate tin halide with alkali. However, they become less stable, relative to the oxide, as the length of the n-alkyl chain increases. Chain branching increases stability to dehydration, while triaryltin hydroxides and oxides are readily interconverted.

The tendency to polymerize with tin having a co-ordination number greater than four, is clearly established only with trimethyltin hydroxide. It is dimeric in benzene and chloroform with both $\nu_{(Me_3Sn)_s}$ and $\nu_{(Me_3Sn)_{as.}}$ active in the infrared spectra. The non-planar Me_3Sn structure thus indicated may well be planar in the polymeric solid as $\nu_{(Me_3Sn)_s}$ is absent from the infrared spectrum.[183, 184] (See Figure 4.38.)

In aqueous solution, hydration of Me_3SnOH occurs,[185] and though it can be readily sublimed at 80°C, and melts at 118°C, disproportionation occurs above 100°C to Me_4Sn, dimethyltin oxide and water.[186] In this it resembles methoxytrimethyltin.[187]

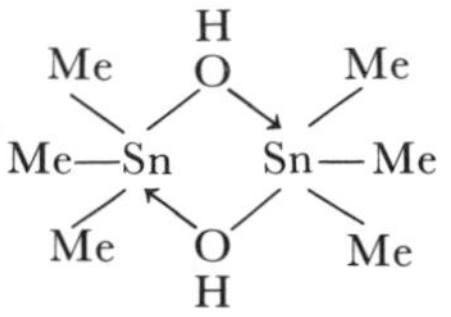

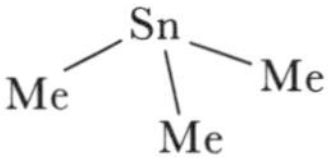

(local symmetry C_{3v})

	E	C_3	σ_v
Γ_{Sn-c}	3	0	1

i.e. $A_1 + E$

(Both infrared active)

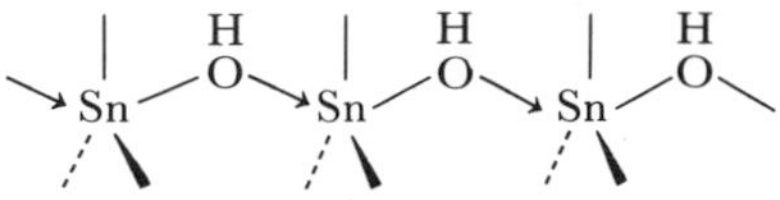

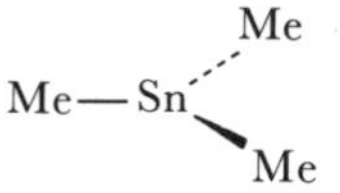

(local symmetry D_{3h})

	E	C_3	C_5	S_3	σ_h	σ_v
Γ_{Sn-c}	3	0	1	0	3	1

i.e $A_1' + E$

(Only E infrared active)

Fig. 4.38. Structure of trimethyltin hydroxide

$$2Me_3SnOH \xrightarrow{100°C} Me_4Sn + \frac{1}{n}[Me_2SnO]_n + H_2O$$

$$2Me_3SnOMe \xrightarrow{45°C} Me_4Sn + Me_2Sn(OMe)_2$$

Trimethyltin hydroxide forms adducts with Me_3SnX. These are solids, with ionic structures involving penta-co-ordinate tin.[188]

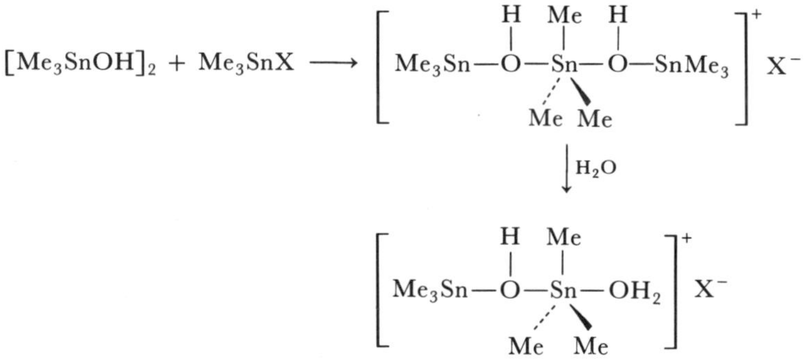

$$[Me_3SnOH]_2 + Me_3SnX \longrightarrow [Me_3Sn{-}O(H){-}SnMe_3(\text{penta}){-}O(H){-}SnMe_3]^+ X^-$$

$$\downarrow H_2O$$

$$[Me_3Sn{-}O(H){-}SnMe_3{-}OH_2]^+ X^-$$

Trimethylsilanol Me_3SiOH is also dimeric but probably through hydrogen bonding.

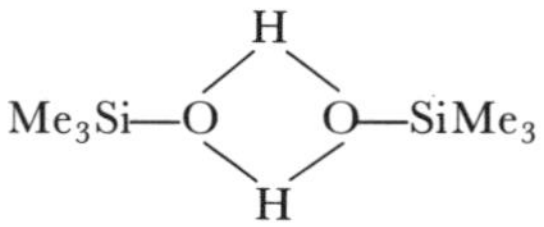

Fig. 4.39. Hydrogen-bridged structure of trimethylsilanol

This is particularly strong among the arylsilanols,[h] and while $p_\pi - d_\pi$ bonding between oxygen and silicon atoms will decrease the donating power of the remaining oxygen lone-pair, the proton becomes more acidic. The tin compound is hydroxide-bridged as oxygen is a better donor with $p_\pi - d_\pi$ bonding from oxygen to tin minimal.

With germanium–oxygen compounds, however, π-bonding is significant but less than with silicon, so the combined effect of an acidic proton and basic oxygen in Me_3GeOH may conveniently account for its non-existence to date.[189] Only the germoxane forms, and a germanol dimer involving both hydrogen and oxygen bridges would provide the low activation energy complex necessary for spontaneous dehydration.

$$\begin{array}{ccc} & H & \\ & | & \\ & O & \\ Me_3Ge & & H \\ & O & \\ & | & \\ & Ge & \\ & Me_3 & \end{array} \longrightarrow (Me_3Ge)_2O + H_2O$$

Peroxides

Just as with oxides and alkoxides, silyl and germyl peroxy-compounds appear to be monomeric, but the tin ones are associated. Alkylperoxytrialkyltins result from the peroxide attack on a tin halide, or from a hydroperoxide and base. However, alcohol or water displacement from a tin–oxygen compound by a hydroperoxide provides the most convenient route.[190, 191]

$$Et_3SnCl + NaOOBu^t \longrightarrow Et_3SnOOBu^t \quad \text{(b.p. 55°C/0.2 mm)}$$

$$Bu^n_3SnCl + HOOCMe_2Ph \xrightarrow{NH_3} Bu^n_3SnOOCMe_2Ph$$

$$Pr^n_3SnOMe + HOO\text{-}C_6H_9 \longrightarrow Pr^n_3SnOO\text{-}C_6H_9 + MeOH \quad \text{(b.p. 71°C/0.04 mm)}$$

$$(Ph_3Sn)_2O + 2Bu^tOOH \longrightarrow 2Ph_3SnOOBu^t + H_2O \quad \text{(m.p. 63–5°C)}$$

$$Me_3SnOH + Bu^tOOH \longrightarrow Me_3SnOOBu^t \text{ (b.p. 56°C/12 mm)} \xrightarrow{Bu^tOOH} Me_3SnOOBu^t{\cdot}HOOBu^t \text{ (m.p. 28°C)}$$

In the last case, an excess of the hydroperoxide leads to the formation of a crystalline complex which probably has a structure resembling the parent dimeric tin peroxide and the hydroperoxide,[192] though the tin peroxide would be expected to have a 4-membered ring. The mode of dimerization is similar to that proposed for Me_3MOH (M is Si, Ge, Sn).

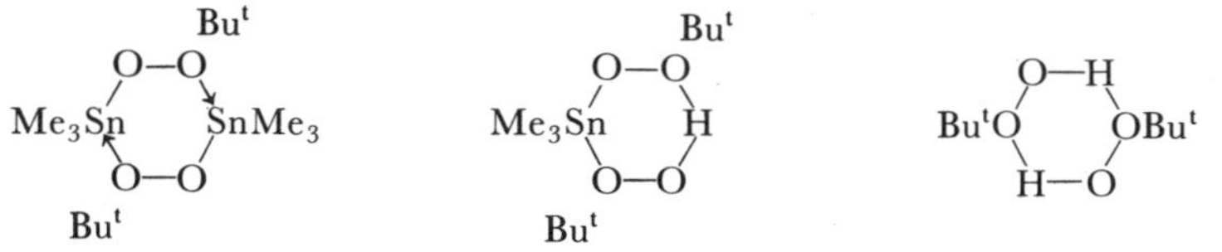

$$Me_3Sn(\mu\text{-}OOBu^t)_2SnMe_3 \qquad Me_3Sn(OOBu^t)\cdots HOOBu^t \qquad (Bu^tOOH)_2$$

Fig. 4.40. Dimeric peroxides

While the monoperoxy compounds appear to be quite stable thermally, attempts to substitute the tin further reduces the thermal stability of the products. With dialkyltin oxides and a hydroperoxide, the only stable product isolated was a 1,3-diperoxydistannoxane,[191, 193] a member of a class of compounds to be considered later (p. 233).

$$\frac{4}{n}[R_2SnO]_n + 4R'OOH \longrightarrow [R'OOSnR_2OSnR_2OOR']_2 + 2H_2O$$

Organotin hydroperoxides R_3SnOOH and ditin peroxides are crystalline solids, and probably oligomeric. The structures are not yet known but it seems likely that they will resemble the tin hydroxides. Trimethyltin hydroperoxide melts at 97–8°C and, like Me_3SnOH, is not thermally stable above this temperature.[194]

$$2Me_3SnOOH \longrightarrow Me_3SnOH + O_2 + \frac{1}{n}[Me_2SnO]_n + MeOH$$

Organo-substituted Distannoxanes

Dialkyltin oxides are polymeric, but can be depolymerized with organo-substituted tin compounds possessing monofunctional electronegative groups. With tri-n-butyltin compounds and the oxide, penta-alkyl-distannoxanes are formed, which decompose above 150°C.[195]

$$Bu^n_3SnX + \frac{1}{n}[R_2SnO]_n \underset{150°C}{\rightleftharpoons} Bu^n_3SnOSnR_2X \ (X = Cl, Br, CH_3CO_2)$$

The electronegative group X renders tin more electrophilic, so the compounds are slightly associated unlike $(R_3Sn)_2O$.

Association becomes more pronounced with X on each tin atom of the distannoxane, and these compounds provide an interesting structural and historical story in themselves.

The 1,1,3,3-tetra-alkyldistannoxanes. The part played by Löwig in isolating some of the first organotin compounds was mentioned in the Section on cyclopolystannanes (p. 162).[43] The same work with ethyl iodide and the sodium/tin alloy produced what was thought to be a tin free-radical Et_2SnI. Ten years later, in 1862, it was shown to be a di-iododistannoxane[196] more conveniently prepared by adding diethyltin oxide to the di-iodide.[197]

$$Et_2SnO + Et_2SnI_2 \longrightarrow ISnEt_2OSnEt_2I$$

This route is now used widely to prepare 1,3-substituted distannoxanes.

$$R_2SnO + R_2SnX_2 \longrightarrow XSnR_2OSnR_2X$$

The reaction can be conveniently carried out in toluene, heated under reflux if necessary.[198] While synthesis by the partial hydrolysis of R_2SnX_2 is unsatisfactory, it does show a reactivity sequence with groups attached to tin. The ability order for hydrolysis is CH_3CO_2, I > Br > Cl > F, so the hydrolysis of a mixed di-substituted stannane gives the stannoxane with the lighter halogen atom bonded to tin.

$$2Bu^n_2SnClI \xrightarrow{H_2O} ClSnBu^n_2OSnBu^n_2Cl + 2HI$$

These substituted distannoxanes are also formed from the polymeric dialkyltin oxide and an acid, or by metathetical exchange with a parent distannoxane.

Structure of distannoxanes. Both CO_2 and hydrogen sulphide react with hexa-alkyldistannoxanes, forming the carbonate and sulphide. 1,3-Dichlorodistannoxanes are inert to these reagents, possibly because they are oligomeric and involve tin in a co-ordination state greater than four. A series of papers between 1939 and 1948 argued the case for a trimeric structure with 4- and 5-co-ordinate tin atoms.[199] Each oxygen atom of the cyclotristannoxane ring is co-ordinated to an R_2SnX_2 molecule (Figure 4.41(a)).

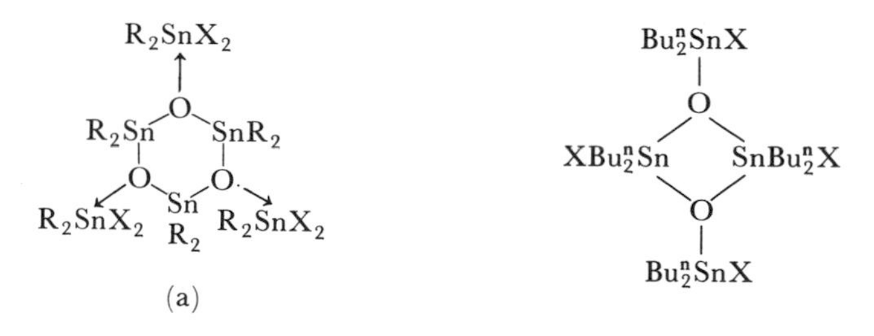

Fig. 4.41. The structure of halogenodistannoxanes

However, evidence of late shows the compounds to be dimeric with an oxygen bridge giving a 4-membered ring (Figure 4.41(b)). So tin is present in two co-ordination states, and the ^{119}Sn n.m.r. spectrum of dimeric $[XSnBu^n_2]_2O$ (X = Cl and Br) shows two resonance positions, one down-field (4-co-ordinate) and one up-field (5-co-ordinate) of tetra-n-butyltin.[k]

The presence of 4-membered rings is also supported by the X-ray structure of 1,3-bis-(trimethylsiloxy)tetramethyldistannoxane, with long bridge-bonds as indicated.

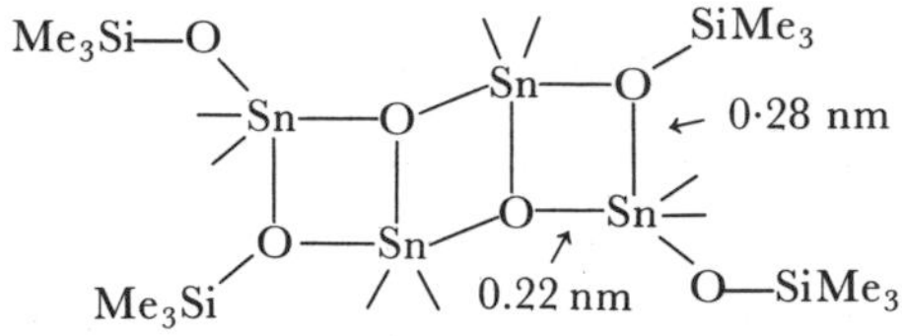

Fig. 4.42. The "ladder" structure of siloxydistannoxanes

Using an oxide and dihalide with different organic groups leads to the isolation of two isomers if the groups are interchanged on the starting materials.[200]

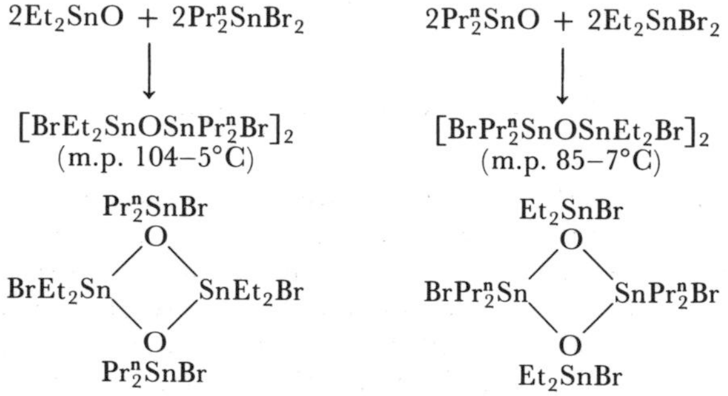

$$2Et_2SnO + 2Pr^n_2SnBr_2 \longrightarrow [BrEt_2SnOSnPr^n_2Br]_2 \quad \text{(m.p. 104–5°C)}$$

$$2Pr^n_2SnO + 2Et_2SnBr_2 \longrightarrow [BrPr^n_2SnOSnEt_2Br]_2 \quad \text{(m.p. 85–7°C)}$$

This supports the proposal that the polymeric dialkyltin oxide comprises 4-membered Sn_2O_2 rings, which are maintained during the above reaction. So monomeric stannoxanes are not formed at any stage.

Halohydroxydistannoxanes are the first hydrolysis product of the dihalo compounds.[201] Alkoxides are formed when recrystallization is effected from alcohol but the halogen atom is unaffected. These are dimeric and are thought to possess a "ladder" structure.[202]

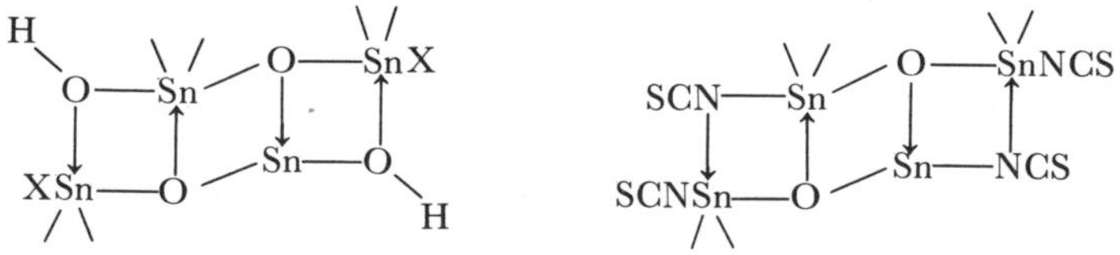

Fig. 4.43. The structure of stannoxanes $[XR_2SnOSnR_2Y]_2$

An analogous structure is proposed for bis-(isothiocyanato)- and hydroxyisothiocyanato-distannoxanes. The infrared spectra of both these compounds exhibit two N—C stretching bands near 2000 cm^{-1}, the lower one at 1960 cm^{-1} being among the lowest reported for isothiocyanates. This could indicate a weakening of the C—N bond through co-ordination.[203]

$$Sn{-}N{=}C{=}S \longleftrightarrow Sn{-}\bar{N}{-}C{\equiv}\overset{+}{S}$$

Polymeric diorganotin oxides. Comparisons of the Si—O and Sn—O bonds at the beginning of this Section showed that oxygen appears to be the better nucleophile if bonded to tin. Such considerations will explain the difference in the properties of cyclosiloxanes (volatile liquids) and polymeric dialkyltin oxides (semi-crystalline solids melting over a wide temperature range). The silicon compounds involve 4-co-ordinate silicon atoms but the electrophilic character of tin renders the cyclic tin–oxygen compound polymeric, with 5-co-ordinate tin atoms. The wide melting range is indicative of a macromolecular structure with strong intermolecular bonding, and these compounds, which result from the alkaline hydrolysis of the dihalides, possess Mössbauer spectra indicative of 5-co-ordinate tin.[201, 204]

$$R_2SnCl_2 + 2OH^- \longrightarrow \frac{1}{x}[R_2SnO]_x + 2Cl^-$$

The structure is thought to involve 4-membered rings from discussions in the previous Section, and these are thought to be linked together in 8-membered rings, providing a laminated lattice of the condensed rings.

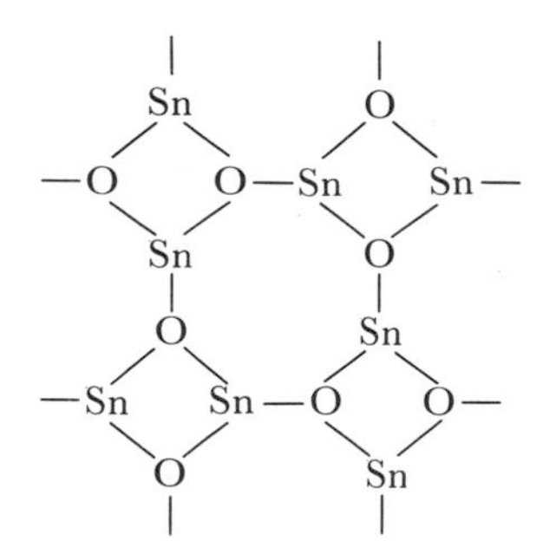

Fig. 4.44. The structure of polymeric dialkyltin oxide

The insertion of the dialkyltin oxide residue into the Sn—X bond has already been discussed (p. 232) for the two cases of R_3SnX and R_2SnX_2, both 1:1. It can be used in a wider way, to synthesize a wide variety of distannoxanes and polymeric tin–oxygen compounds. n-Butyltin trichloride gives the trichlorostannoxane with di-n-butyltin oxide, while polymers result by feeding more than one R_2SnO residue into the Sn—X bond.[205]

$$Bu^n_2SnO + Bu^nSnCl_3 \longrightarrow Cl_2Bu^nSnOSnBu^n_2Cl \quad \text{(m.p. 34–5°C)}$$

$$nBu^n_2SnO + Bu^n_2SnCl_2 \longrightarrow ClBu^n_2Sn[OSnBu^n_2]_nCl \quad (n = 1\text{–}6, 9, 12)$$

Di-n-butyltin dichloride forms linear oligomers, all but the last two having sharp melting points, and all possessing the characteristic peak in the infrared spectrum at 670–690 cm^{-1} for $\nu_{(SnOSn)_{as.}}$. Branched oligomers result with n-butyltin trichloride, and the products are low-melting solids or waxes ($x = y = 1$, m.p. 109–10°C; $x = y = 3$, m.p. 100–102°C).

$$Bu^nSnCl_3 + Bu^n_2SnO \longrightarrow Bu^nSn\begin{cases}[OSnBu^n_2]_xCl \\ [OSnBu^n_2]_yCl \\ [OSnBu^n_2]_yCl\end{cases}$$

All show the infrared band at 670–690 cm^{-1} and rearrange in the presence of bipyridyl as the di-n-butyltin dichloride complex is formed, *i.e.*, $Bu^n_2SnCl_2 \cdot bipy$.

The tetrahalodistannoxane formed from SnX_4 (X = Cl or Br) readily decomposes into dialkyltin dihalide and the oxyhalide of tin.

$$SnX_4 + R_2SnO \longrightarrow [XR_2SnOSnX_3] \longrightarrow R_2SnX_2 + OSnX_2$$

Tin esters will also add across the Sn—O bond of $[Et_2SnO]_x$, and the two products, with $n = 1$ or 2, have been isolated as crystalline solids.[206, 207]

$$2Et_2SnO + Et_2Sn(OCOPh)_2 \longrightarrow (PhCO_2)Et_2Sn[OSnEt_2]_nOCOPh$$

This addition is not limited to functional groups on tin alone. Many metallostannoxanes form using halides and esters of elements of Groups II, III and IV. Alkylsilyl, germyl and lead chlorides, arylmercury and thallium halides all undergo this reaction.[195]

$$Bu^n_2SnO + Me_2Si(OAc)_2 \longrightarrow AcOSiMe_2OSnBu^n_2OAc \quad \text{(oil)}$$

$$Me_2SnO + MeSiCl_3 \longrightarrow Cl_2SiMeOSnMe_2Cl \quad \text{(m.p. 80°C (dec.}^n\text{))}$$

$$Bu^n_2SnO + Bu^n_2PbCl_2 \longrightarrow ClPbBu^n_2OSnBu^n_2Cl \quad \text{(m.p. 120–30°C (dec.}^n\text{))}$$

$$Bu^n_2SnO + PhHgCl \longrightarrow PhHgOSnBu^n_2Cl \quad \text{(m.p. 206–8°C)}$$

Hydrogen sulphide and mercaptans react with these polymeric oxides to give cyclic or linear tin–sulphur compounds.[208]

$$R_2Sn(SR')_2 \xleftarrow{2R'SH} [R_2SnO]_x \xrightarrow{H_2S} [R_2SnS]_3 + H_2O$$

This shows one of the main differences between organosilicon and organotin chemistry. The "a"-type, hard bases of the first short period tend to bond more strongly to silicon than to tin, while the "b"-type bases, such as sulphur and phosphorus, bond more strongly to tin.

With nitric acid, these oxides form a complicated mixture of products.

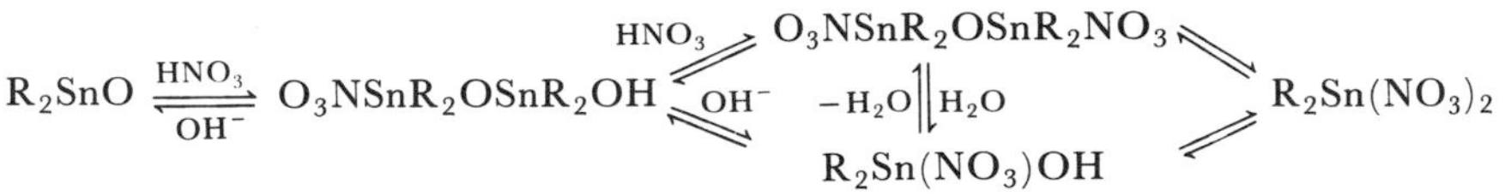

$$R_2SnO \underset{OH^-}{\overset{HNO_3}{\rightleftharpoons}} O_3NSnR_2OSnR_2OH \underset{OH^-}{\overset{HNO_3}{\rightleftharpoons}} O_3NSnR_2OSnR_2NO_3 \rightleftharpoons R_2Sn(NO_3)_2$$

$$O_3NSnR_2OSnR_2OH \rightleftharpoons R_2Sn(NO_3)OH \rightleftharpoons R_2Sn(NO_3)_2; \quad O_3NSnR_2OSnR_2NO_3 \underset{H_2O}{\overset{-H_2O}{\rightleftharpoons}} R_2Sn(NO_3)OH$$

All have been isolated, and the dinitratodistannoxane is dimeric with chelating nitrate groups.[209] The structure proposed involves equivalent tin atoms all 5-co-ordinate.

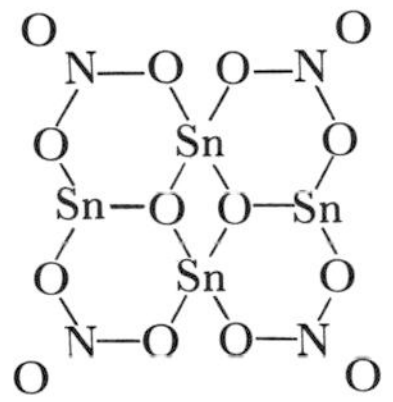

```
 O             O
  N—O   O—N
 /     \ /     \
O       Sn      O
 \     / \     /
  Sn—O   O—Sn
 /     \ /     \
O       Sn      O
 \     / \     /
  N—O   O—N
 O             O
```

Fig. 4.45. Nitratodistannoxane with chelating nitrate groups

This bridging appears stronger than that in 1,3-bis-(trimethylsiloxy)-distannoxane, which possesses the ladder structure, since the nitrate is dimeric in hot benzene but the siloxy compound monomeric in hot chloroform.[k]

Perchlorotin–oxygen compounds. Hexachlorocyclotrisiloxane results when $SiCl_4$ is oxidized at 1000°C by oxygen. Stannic chloride and chlorine monoxide form a similar crystalline solid $SnOCl_2$.[210] This is trimeric in $POCl_3$, but appears to complex rather readily with this solvent, since Cl_2O oxidizes a stannic chloride–phosphoryl chloride mixture to $[SnOCl_2{\cdot}2POCl_3]_2$. The 8-membered ring comprises 6-co-ordinate tin involving co-ordinated $POCl_3$.[211]

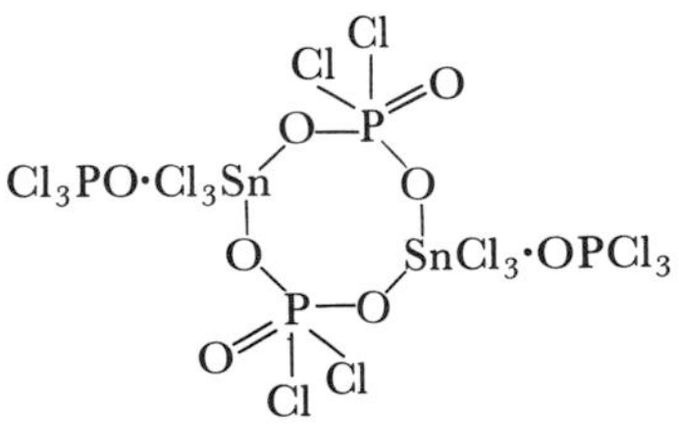

```
                  Cl
               Cl  |
                 \ | =O
                O—P
               /    \
Cl3PO·Cl3Sn          O
           |         |
           O          SnCl3·OPCl3
            \        /
             P—O
          O=/ | \
              Cl  Cl
```

Fig. 4.46. The phosphoryl chloride complex of dichlorotin oxide

Organo-tin and -lead salts in solution

Aqueous solutions of tin and lead halides contain cationic species involving the metal in bridged and caged structures. In strong acid, dimethyltin dihalides hydrolyze to the bridged cation $[Me_2Sn(OH)]_2^{2+}$. In weakly acid solution, $(Me_2Sn)_3(OH)_4^{2+}$ and $(Me_2Sn)_2(OH)_3^{+}$ result, while at pH 7 dimethyltindihydroxide is precipitated.[212] Though the structures are not known, they are all believed to involve the hydroxide bridges typified in trimethyltin hydroxide.

In basic solution, Pb^{II} yields two oxycations. The simpler, $Pb_4(OH)_4^{4+}$, is thought to involve the cube structure typified by elements of the first half of a period (*cf.* $[KOBu^t]_4$, $[EtMgOBu^t]_4$, $[TlOEt]_4$). Alternatively, this can be considered as a tetrahedron of lead atoms face-bridged by the hydroxide anions.[213]

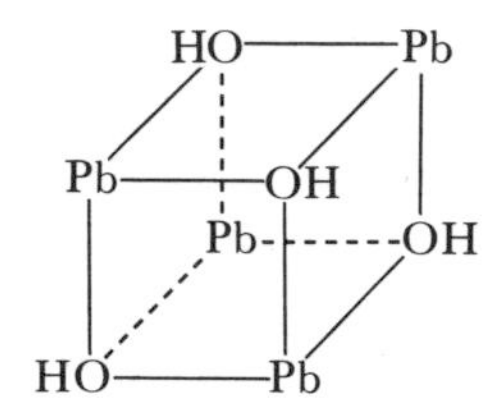

Fig. 4.47. The structure of the $Pb_4(OH)_4^{4+}$ ion

A second cationic lead(II) hydroxide, with six lead atoms, was thought to be $Pb_6(OH)_8^{4+}$, and to possess an octahedron of lead atoms each face possessing a central OH^- bridging three lead atoms. An X-ray structure showed the presence of two edge-bonded tetrahedra with the two faces forming this edge held by a 4-co-ordinate oxide ion, as found among basic beryllium salts. So this cation should be formulated as $Pb_6O{\cdot}(OH)_6^{4+}$, and the data obtained from the vibrational spectra fit this structure rather than the octahedral one.[214]

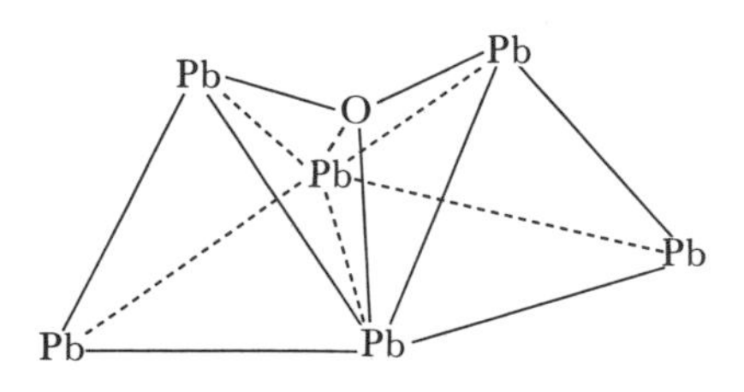

Fig. 4.48. The structure of $Pb_6O{\cdot}(OH)_6^{4+}$ (OH face-bonded)

COMPLEXES OF THE GROUP IV ELEMENTS

Chelated 6-co-ordinate complexes form for the Group IV elements (except carbon), the ligand normally comprising Group V and VI elements.

Diphosphinic imides $(R_2PO)_2NH$ readily complex with the tetrahalides of silicon, germanium and tin. The product is determined by the reactant ratios. Stannic chloride and bromide form 1 : 1 complexes which readily disproportionate and lose hydrogen halide, yielding the *trans* dihalide complex (**4-XIII**). This also results when the tin(II) complex of the imide is oxidized by a halogen, and can be protonated by perchloric acid.

Analogous complexes result with silicon and germanium halides and two moles of the imide, while an excess of the ligand forms a trichelate.[215]

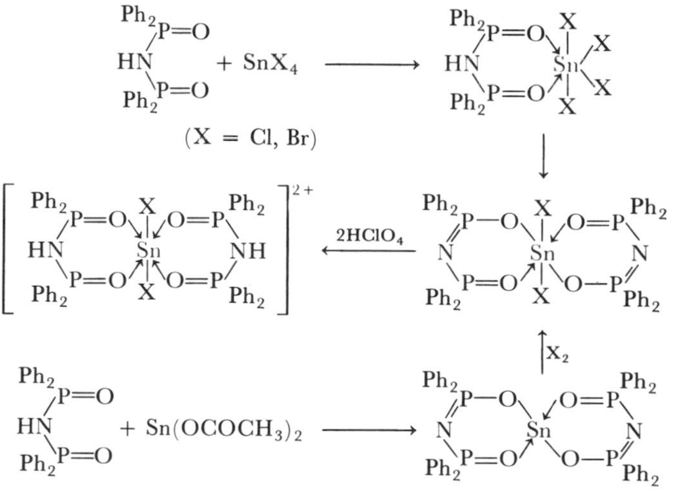

$HN(P(O)Ph_2)_2 + SnX_4 \longrightarrow [HN(Ph_2P{=}O)_2]SnX_4$

(X = Cl, Br)

$\downarrow$

$[(HN(Ph_2P{=}O)_2)_2SnX_2]^{2+} \xleftarrow{2HClO_4} Sn[Ph_2P(O)NP(O)Ph_2]_2X_2$ (**4-XIII**)

$\uparrow X_2$

$HN(P(O)Ph_2)_2 + Sn(OCOCH_3)_2 \longrightarrow Sn[Ph_2P(O)NP(O)Ph_2]_2$

$M[Ph_2P(O)NP(O)Ph_2]_2X_2 \xrightarrow{(Ph_2PO)_2NH} \{M[Ph_2P(O)NP(O)Ph_2]_3\}^{2+}$

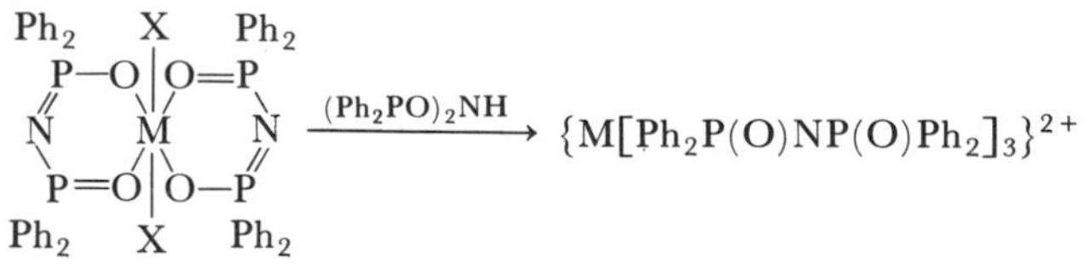

(M = Si, Ge; X = Cl, Br, I)

Thiophosphinic acid $R_2P(S)SH$ reacts with tin halides, forming esters. With dialkyltin dihalides, the bis-(dithiophosphinate) formed is monomeric, the phosphinate group forming a 4-membered chelate ring.[216]

$$2R_2PS_2^- + R'_2SnX_2 \longrightarrow R'_2Sn(S_2PR_2)_2 + 2X^-$$

Phosphine sulphides form 2:1 adducts while diphosphine disulphides form 1:1 chelate complexes.[217]

$$2R_3PS + SnX_4 \longrightarrow (R_3PS)_2SnX_4$$

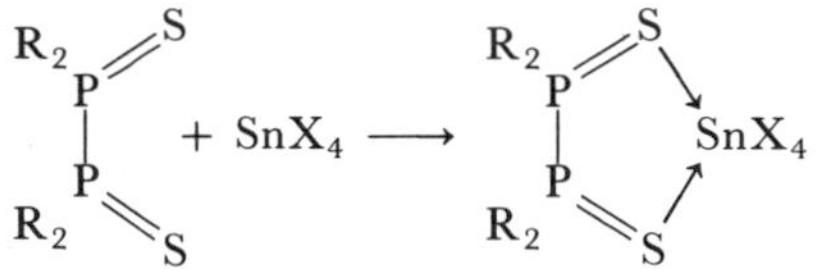

$$R_2P(=S)\text{—}P(=S)R_2 + SnX_4 \longrightarrow R_2P(=S\rightarrow)\text{—}P(=S\rightarrow)R_2\ SnX_4$$

Both tetra-alkyl- and -aryl-leads react with sulphur dioxide. Two of the C—Pb bonds are cleaved, and the disulphinates are polymeric involving bridging sulphinate groups, with lead 6-co-ordinate.[218]

$$R_4Pb + 2SO_2 \longrightarrow R_2Pb(SO_2R)_2$$

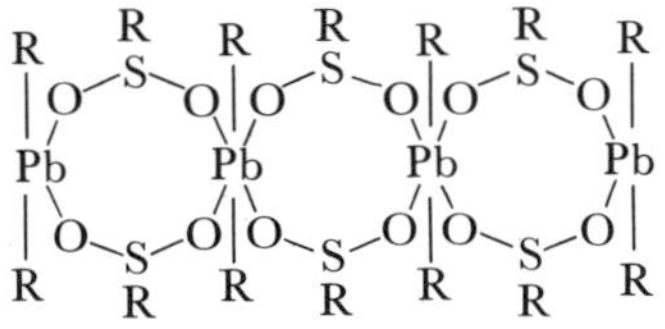

Fig. 4.49. Dialkyl-lead sulphinates

This kind of cyclic polymer also results from the reaction of R_2SnCl_2 with sulphonic acids $RSO_2{\cdot}OH$, the product involving 6-co-ordinate tin.[219]

$$Me_2SnCl_2 + HOSO_2R \longrightarrow$$

$(R = F, Cl, Me, Et, CF_3)$

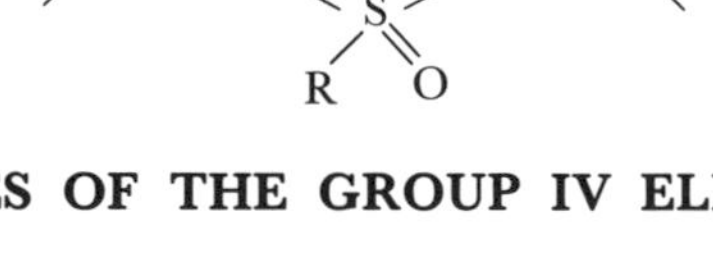

SULPHUR DERIVATIVES OF THE GROUP IV ELEMENTS

While most of the research work has concerned sulphur compounds, what little has been done on selenium and tellurium will also be included, and comparisons will be made with oxygen derivatives where appropriate.

MX_2 compounds (M = Group IV element, X = Group VI element)

Silicon disulphide can be prepared directly from the elements at high temperature, or by heating tetraethylorthothiosilicate with sulphur at

200°C.[220]

$$(EtS)_4Si + 2S \longrightarrow 2Et_2S_2 + SiS_2$$

It is readily hydrolyzed and will reduce CO_2 to CO and oxidize CO to COS.[l]

$$SiS_2 + 2H_2O \longrightarrow SiO_2 + 2H_2S$$

$$SiS_2 + 2CO_2 \longrightarrow SiO_2 + 2CO + 2S$$

$$SiS_2 + CO \longrightarrow SiS + COS$$

The structure, like $BeCl_2$, involves chains of silicon atoms bridged by pairs of sulphur ones, as with orthorhombic silica. Adjacent 4-membered Si_2S_2 rings lie in mutually perpendicular planes with the Si—S bond distance corresponding to that expected for a single bond (0.214 nm).

Germanium disulphide possesses a three-dimensional covalent lattice comprising 4-co-ordinate germanium. The GeS_4 tetrahedra are arranged in a manner resembling the quartz structure,[221] with the Ge—S bond distance that expected for a single bond (0.219 nm). GeO_2 has the ionic rutile structure, while stannic sulphide possesses the semi-ionic layer structure of CdI_2 with 6-co-ordinate tin.[a]

This trend from a covalent to anionic lattice is summarized in Table 4.10,

Table 4.10. Structure adopted by MX_2

MX_2	M = Si	Ge	Sn	Pb
X = O	C	R	R	R
S	C	C	L	
Se	C			
Te	L			

C = covalent
R = rutile
L = layer CdI_2

and it will be seen that ionic character increases with the atomic weight of M. The compounds absent reflect the inability of the Group VI element to stabilize the higher oxidation state.

An increase in covalency also occurs in going from oxides to sulphides, and by increasing the organic substituents on the Group IV element. Stannic oxide and sulphide illustrate this, as do the organotin oxides and sulphides R_2SnO and R_2SnS. The layer structure of the former, with 5-co-ordinate tin has already been discussed (p. 235), while the sulphides are distillable molecular trimers.[222]

Halogenocyclodisilthianes

While silica and SiF_4 give linear fluorosiloxanes (mainly $F_3SiOSiF_3$),[223] SiS_2 yields a cyclodisilthiane at 1000°C, which decomposes at 20°C.[224]

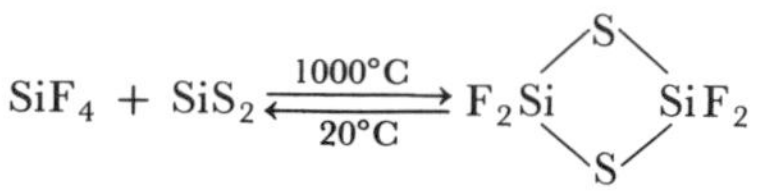

$$SiF_4 + SiS_2 \underset{20^\circ C}{\overset{1000^\circ C}{\rightleftharpoons}} F_2Si\langle S \rangle_2 SiF_2$$

The chloride is more stable, and has been prepared in various ways.

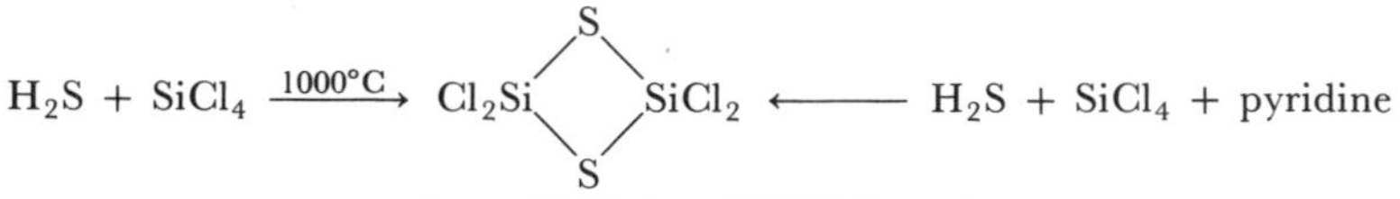

$$H_2S + SiCl_4 \xrightarrow{1000^\circ C} Cl_2Si\langle S \rangle_2 SiCl_2 \longleftarrow H_2S + SiCl_4 + \text{pyridine}$$

(m.p. 75°C, b.p. 92°C/22 mm)

It is soluble in common organic solvents, and is among the compounds formed by heating $SiCl_4$ and SiS_2 at 900°C.[l, 227]

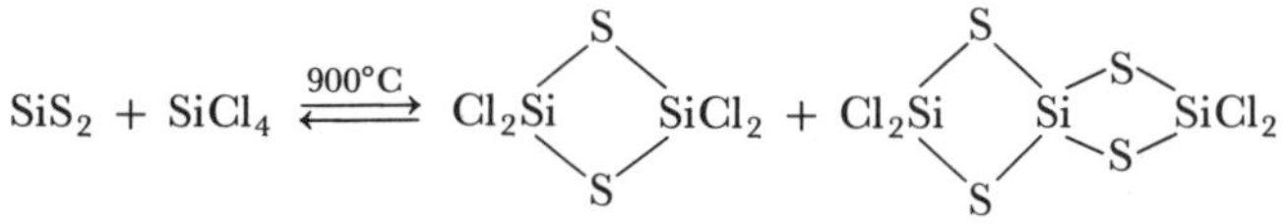

$$SiS_2 + SiCl_4 \overset{900^\circ C}{\rightleftharpoons} Cl_2Si\langle S \rangle_2 SiCl_2 + Cl_2Si\langle S \rangle_2 Si\langle S \rangle_2 SiCl_2$$

Sulphur and polymeric $[SiCl_2]_x$ give similar products at 180°C, along with silyl-substituted cyclodisilthianes.

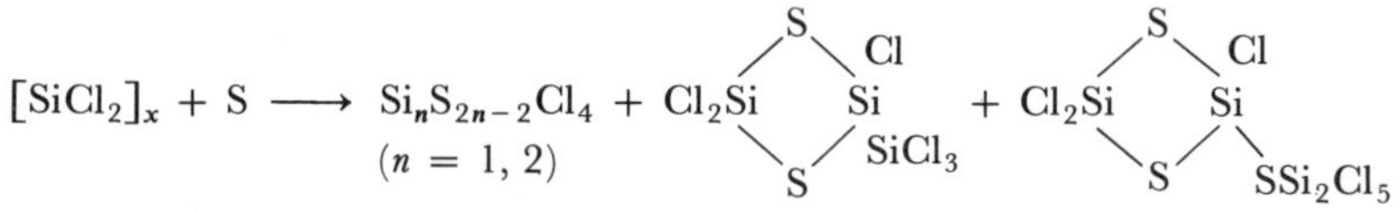

$$[SiCl_2]_x + S \longrightarrow \underset{(n = 1, 2)}{Si_nS_{2n-2}Cl_4} + Cl_2Si\langle S \rangle_2 Si(Cl)(SiCl_3) + Cl_2Si\langle S \rangle_2 Si(Cl)(SSi_2Cl_5)$$

(b.p. 92°C/2 mm) (b.p. 136–8°C/0.01 mm)

Bromine cleaves the Si—S bond in $Si_2S_2Cl_4$ giving dibromodichlorosilane and sulphur monobromide,

$$Si_2S_2Cl_4 + 3Br_2 \longrightarrow 2SiCl_2Br_2 + S_2Br_2$$

while heating H_2S and $SiBr_4$ with a trace of $AlBr_3$ produces the bromocyclodisilthiane.[226] Higher homologues result with the addition of SiS_2, all of which can be recrystallized from CS_2.

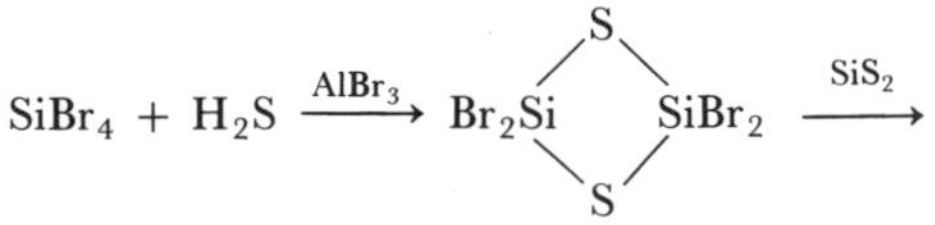

$$SiBr_4 + H_2S \xrightarrow{AlBr_3} Br_2Si\langle S \rangle_2 SiBr_2 \xrightarrow{SiS_2}$$

(m.p. 93°C, b.p. 150°C/18 mm)

$$Br_2Si\langle S \rangle_2 Si\langle S \rangle_2 SiBr_2 + Br_2Si\langle S \rangle_2 Si\langle S \rangle_2 Si\langle S \rangle_2 SiBr_2$$

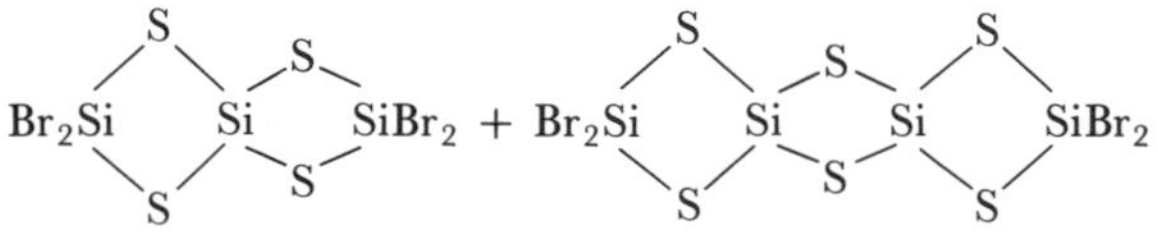

(m.p. 108°C, b.p. 164°C/12 mm) (m.p. 148°C, b.p. 171°C/12 mm)

Organo-group IV sulphides

The wide range of cyclic oxygen derivatives of the Group IV elements have been surveyed, and, from the reactions considered, the reactivity of the M—O bond to sulphur compounds increases with the atomic weight of the Group IV element. In general, a reverse trend is seen with the stability of the M—S bond to water and alcohols. Consequently, cyclosilthianes are normally prepared under strictly anhydrous conditions while germanium–, tin– and lead–sulphur compounds can be made in aqueous solution.

Synthesis of $[R_2MS]_{2 \text{ and } 3}$

Cyclotrisilthianes are synthesized from a dichlorosilane and hydrogen sulphide in the presence of a tertiary amine as base.[m, 228]

$$3R_2SiCl_2 + 3H_2S + 6C_5H_5N \xrightarrow{C_6H_6} [R_2SiS]_3 \xrightarrow{\text{heat}} [R_2SiS]_2$$
$$(R = Me, Et, Pr^n, Ph)$$

These are thermally unstable and the ring contracts to the cyclodisilthiane. The analogous Ge, Sn and Pb compounds can be made in alkaline aqueous or alcoholic solution using similar precursors.[l, m, 226]

Polymeric organotin oxides[229] react with S^{2-} in acid solution to give the tin sulphide, and trimeric diphenyl-lead sulphide[221] was first prepared this way in 1887,[230] using the diacetate.

$$[R_2SnO]_x + 3S^{2-} \xrightarrow[\text{solution}]{\text{acid}} [R_2SnS]_3$$

$$Ph_2Pb(OCOMe)_2 \xrightarrow[\text{acid}]{S^{2-}} [Ph_2PbS]_3$$

Using metal sulphides and selenides under anhydrous conditions also gives cyclic products.[m]

$$R_2SnX_2 + Na_2S \longrightarrow R_2SnS + 2NaX$$

$$R_2MX_2 + Na_2Se \longrightarrow R_2MSe + 2NaX$$
$$(M = Si, Ge, Sn)$$

Cyclic tin–sulphur compounds can be prepared from organotin oxides and CS_2. The reaction is thought to involve addition of Sn—O across the C=S bond, followed by Sn—O cleavage and the elimination of COS.[231]

$$[R_2SnO]_x + CS_2 \longrightarrow [R_2SnS]_3 + COS$$

$$\text{Sn—O—Sn} + CS_2 \longrightarrow \begin{matrix}\text{Sn—O} & \\ & \text{C=S} \\ \text{Sn—S} & \end{matrix} \longrightarrow \text{Sn—S—Sn} + OCS$$

The pyrolysis of linear compounds of these elements as a synthetic route to cyclic derivatives is limited by the small yields obtained. The products do show the relative stability of these compounds, and so will be considered in more detail than the synthetic applications justify.

Organosilicon sulphur compounds generally pyrolyze to the disulphide. This usually proceeds in a step-wise manner, the Si—S and C—S bonds cleaving before C—Si. The various ethylthiosilanes illustrate this well, decomposition normally occurring between 250 and 400°C.[220]

$$\left.\begin{array}{l} Et_3SiSEt \xrightarrow{300°C} Et_2S + (Et_3Si)_2S \\ Et_2Si(SEt)_2 \xrightarrow{350°C} 2Et_2S + [Et_2SiS]_2 \end{array}\right\} \longrightarrow Et_4Si + SiS_2$$

$$2Cl_2Si(SEt)_2 \xrightarrow{250°C} Et_2S + [Cl_2SiS]_2 \longrightarrow SiS_2 + SiCl_4$$

$$2Si(SEt)_4 \xrightarrow{250-300°C} 2Et_2S + [(EtS)_2SiS]_2 \longrightarrow 2Et_2S + 2SiS_2$$

Similar tin and lead compounds decompose at lower temperatures. Bis-(phenylthio)diphenyltin gives the cyclic sulphide quantitatively at 110°C,[232]

$$3Ph_2Sn(SPh)_2 \xrightarrow{110°C} \underset{\text{(m.p. 183–4°C)}}{[Ph_2SnS]_3} + 3Ph_2S$$

while the analogous lead compound appears to undergo cleavage at the Pb—C bond,[233] but not the C—S bond, since diphenyldisulphide results through disproportionation.

$$3Ph_2Pb(SPh)_2 \longrightarrow 2Ph_3PbSPh + (PhS)_2Pb + Ph_2S_2$$

So the increased lability of the M—C bond as M increases in atomic weight plays a significant part in determining pyrolysis products only with lead.

Sulphur readily inserts into the Sn—Sn bonds of cyclopolytins on heating, and will cleave tin–carbon bonds at 200°C.[m]

$$\frac{1}{x}[R_2Sn]_x + S \longrightarrow \tfrac{1}{3}[R_2SnS]_3$$

While cyclic trimers result in small yields using R_4Sn, thio-substituted tins give higher yields.

$$R_4Sn + S \xrightarrow{190-210°C} [R_2SnS]_3 \; (R = Bu^n, Ph)$$

$$Ph_3SnSPh + S \xrightarrow{190-210°C} [Ph_2SnS]_3 \xleftarrow{190-210°C} S + (Ph_3Sn)_2S$$

$$(Bu^n_3Sn)_2S + S \xrightarrow{150-190°C} [Bu^n_2SnS]_3$$

The reactions of sulphur with organic derivatives of the other Group IV elements have not been as widely studied. With Bu^n_4Ge, the cyclic trimer $[Bu^n_2GeS]_3$ forms, while the silane gives linear thiosilanes and silthianes. With lead, the lead(II) monosulphide is formed.

Cage Compounds

The synthetic routes employed for the cyclic sulphur compounds apply equally to cage compounds involving a trifunctional Group IV atom.

Both alkyl- and aryl-silicon trichlorides react with H_2S in the presence of an excess of tertiary amine to give compounds formulated as $(RSi)_4S_6$.[225]

$$4RSiCl_3 + 6H_2S + 12R_3N \longrightarrow (RSi)_4S_6 + 12R_3NHCl$$

Methylsilane and H_2X (X = S, Se) give similar products on heating,[234] while the Ge—S cage results from $MeGeBr_3$ and H_2S.[235]

$$4RSiH_3 + 6H_2X \xrightarrow{Al} (RSi)_4X_6 + 12H_2$$

The organotin sesquisulphides, as these compounds are often called, form by adding an organotin trihalide to an alkaline sulphide solution.[m, 236]

The compounds $(MeM)_4S_6$ (M = Si, Ge, Sn) are all isomorphous[235, 237] and possess a cage structure based on the adamantane–urotropine framework of four fused chair-conformer 6-membered rings, and not one involving two bridged 4-membered rings, as was first proposed.[225]

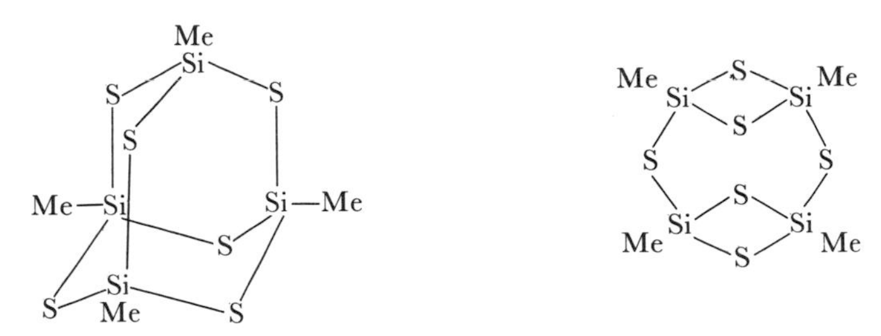

Fig. 4.50. The structural isomers of $(RSi)_4S_6$

This was justified by the synthesis of $(Pr^nSi)_4S_6$ from $[Pr^nClSiS]_2$.

$$[Pr^nClSiS]_2 \xrightarrow[Et_3N]{H_2S} (Pr^nSi)_4S_6$$

The Si—S bond length is not appreciably shorter than that expected for a single bond, but the Sn—S distance of 0.235 nm is less than expected for a single bond, though Pauling himself has thoroughly discussed the limitations of his bond length calculations.

Other Rings

The limited number of Si—S rings not involving alternate Si and S atoms result from chlorodisilanes or chlorosiloxanes, hydrogen sulphide and pyridine.[238, 239]

$$ClSiMe_2SiMe_2Cl + 2H_2S + 4C_5H_5N \longrightarrow$$

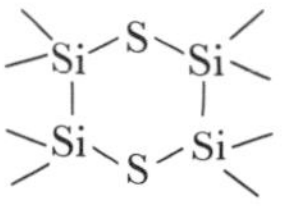

(m.p. 111–2°C)

$$(ClSiMe_2O)_2SiMe_2 + H_2S + 2C_5H_5N \longrightarrow$$

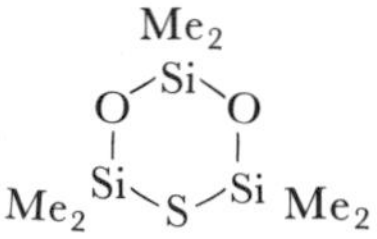

(b.p. 172–5°C)

$$(ClSiMe_2)_2O + 2H_2S + 4C_5H_5N \longrightarrow$$

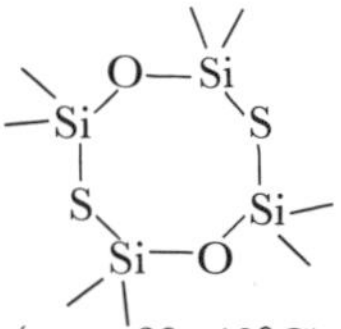

(m.p. 38–42°C)

Reactions of Organo–Group IV Sulphur and Selenium Rings

From the conditions used to synthesize these compounds it appears that silicon and germanium compounds are hydrolytically unstable.[m] Normally, the siloxane forms with cyclosilthianes, but alkaline hydrolysis of the methylcyclosilthianes yields sym-dihydroxytetramethyldisiloxane (m.p. 64–5°C).

$$[Me_2SiS]_2 + 3H_2O \longrightarrow (HOSiMe_2)_2O + 2H_2S$$

Hexamethylcyclotrigermthiane gives the tetrameric oxide with hydrogen peroxide and dil. H_2SO_4. Hot water will also hydrolyze this slowly, and $[Me_2GeSe]_3$ rapidly.

$$4[Me_2GeS]_3 + 12H_2O \xrightarrow{H^+} 3[Me_2GeO]_4 + 12H_2S$$

Strong acid or alkali is required to cleave Sn—S and Pb—S bonds.

Methyl-lithium cleaves both germanium–sulphur and –selenium rings, to give lithium salts of the thiol and selenol. The selenide decomposes at 65°C. With the tin–sulphur ring, complete methylation of tin occurs.[240]

$$[Me_2SnS]_3 + 6MeLi \longrightarrow 3Me_4Sn + 3Li_2S$$

The reactive C—O bonds of fluoroacetones, cyclic ethers and lactones all cleave the Si—S bond,[241, 242] to give cyclic and polymeric materials containing the —SiO(C)$_n$S— framework. Perhalo-acetones give 6-membered heterocycles after heating at 70°C for several days.

$$\tfrac{1}{3}[Me_2SiS]_3 + 2(CF_2X)(CF_2Y)CO \longrightarrow \text{(XF}_2\text{C)(YF}_2\text{C)C}\!-\!O\!-\!SiMe_2\!-\!O\!-\!C(CF_2X)(CF_2Y)\text{, ring closed by S}$$

(X = Y = Cl, b.p. 65°C/0.005 mm; X = Y = F, b.p. 38°C/10 mm; X = F, Y = Cl, b.p. 91°C/10 mm)

Ethylene and propylene oxides also open the silthiane ring, but $ZnCl_2$ is required as a catalyst. The products include both the heterocycle and polymeric products, resulting from the cleavage of one or both sulphur atoms bonded to silicon in the cyclosilthiane ring.

$$[Me_2SiS]_3 + \overline{(CH_2)_nO} \xrightarrow[70°C,\ 13\ days]{n = 2} \text{cyclic } Me_2Si\!-\!O\!-\!(CH_2)_2\!-\!S \text{ (b.p. 61°C/35 mm)} + [Me_2SiO(CH_2)_2S]_n$$

$$[Me_2SiS]_3 + \overline{(CH_2)_nO} \xrightarrow[70°C,\ 20\ days]{n = 3} \text{cyclic } Me_2Si\!-\!O\!-\!(CH_2)_3\!-\!S \text{ (b.p. 72°C/25 mm)} + [Me_2SiO(CH_2)_3S]_n + [Me_2SiO(CH_2)_3\text{-}S(CH_2)_3O]_n$$

An exothermic reaction occurs between β-propionolactone and $[Me_2SiS]_3$, giving a polymeric adduct quantitatively through the cleavage of all Si—S bonds.

$$\frac{n}{3}[Me_2SiS]_3 + n\ \overline{CH_2CH_2C(=O)O} \longrightarrow [Me_2SiOCOCH_2CH_2SCH_2CH_2CO{\cdot}O]_n$$

Many covalent halides of the Main Group elements[m] react with thio derivatives of the Group IV elements. The reactions are metathetical and the cyclosilthianes have been the widely used, especially in the synthesis of heterocycles.

Both tetramethylcyclodi- and hexamethylcyclotri-silthianes react with phenylboron dichloride to give the same B—S ring, *B*-triphenylborsulphole. No 4-membered boron–sulphur ring was isolated in this case, though they are known (see Chapter 3).

$$\tfrac{3}{2}\,[Me_2SiS]_2 + 3PhBCl_2 \longrightarrow [PhBS]_3 \longleftarrow [Me_2SiS]_3 + 3PhBCl_2$$

Terminal dibromoalkanes react with hexamethylcyclotrisilthiane slowly when heated under reflux. Dimethylsilicon dibromide and the thiocycloalkane are produced, while benzoyl chloride gives the thioanhydride.

$$3Br(CH_2)_nBr + [Me_2SiS]_3 \longrightarrow 3Me_2SiBr_2 + (CH_2)_nS \quad (n = 2, 3, 4)$$

$$6PhCOCl + [Me_2SiS]_3 \longrightarrow 3Me_2SiCl_2 + 3(PhCO)_2S$$

Phenyl- and diphenyl-chlorophosphines react exothermically with cyclosilthianes. Triphenylcyclotrithiophosphonous and tetraphenylthiophosphinous anhydrides result, while phenyldichloroarsine gives tetraphenylcyclotetra-arsthiane.

$$6PhPCl_2 + 3[Me_2SiS]_2 \longrightarrow 2[PhPS]_3 + 6Me_2SiCl_2$$

$$6Ph_2PCl + [Me_2SiS]_3 \longrightarrow 3(Ph_2P)_2S + 3Me_2SiCl_2$$

$$12PhAsCl_2 + 4[Me_2SiS]_3 \longrightarrow 3[PhAsS]_4 + 12Me_2SiCl_2$$

When both the halogen and sulphur are attached to atoms of the same element, then co-proportionation occurs. This has already been encountered with the tin–oxide/tin–halide system (p. 233); the tin–sulphur system behaves similarly.[243, 244] The halo-substituted stanthianes are oils or low-melting solids.

$$[R_2SnS]_3 + 3R_2SnX_2 \longrightarrow 3(XSnR_2)_2S$$
(*e.g.*, X = Cl, R = Bu^n, m.p. 35–7°C; X = Cl, R = Me, m.p. 59–61°C)

Unlike the stannoxanes, they are monomeric and yield their progenitors both on heating and, like the stannoxanes, on the addition of bipyridyl.

$$(XR_2Sn)_2S + \text{bipyridyl} \longrightarrow R_2SnX_2\cdot\text{bipy.} + \tfrac{1}{3}[R_2SnS]_3$$

As would be expected, trihalotins give the trihalodistanthiane.

$$[Bu^n_2SnS]_3 + PhSnCl_3 \longrightarrow ClBu^n_2SnSSnPhCl_2$$
(m.p. 38°C)

Complexes

Though there appears to be little tendency among these stanthianes for tin to become 5- or 6-co-ordinate, *o*-benzodithiols give high melting solids with stannic halides. These appear to be polymeric, and are typified by the toluene-3,4-dithiol complex.[245]

$$MeC_6H_3(SH)_2 + SnX_4 \longrightarrow [Sn(S_2C_6H_3Me)_2]_n$$
(m.p. 300–305°C)

The compound is insoluble in non-polar organic solvents,[246] and the polymerization is thought to occur through monomers cross-linking through S → Sn co-ordination. Tin is considered 6-co-ordinate with the two extra co-ordination sites occupied by sulphur from two adjacent planar units. So a system of 4-membered rings is constructed rather similar to SiS_2.

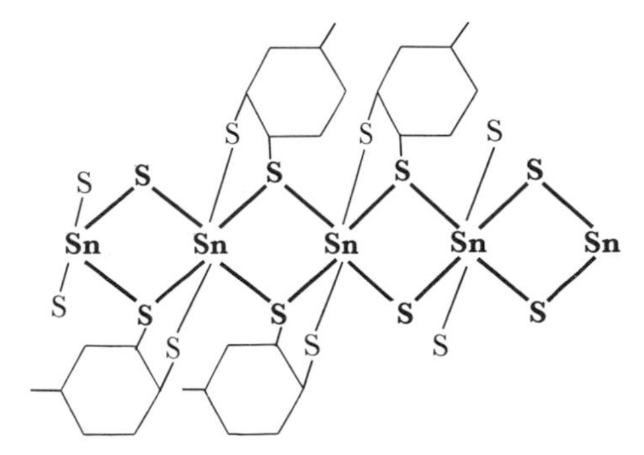

Fig. 4.51. The structure of the toluene-3,4-dithiol complex of tin

This polymer can be broken down by polar compounds, which crystallize as 2:1 complexes, *e.g.*, pyridine, supporting the presence of 6-co-ordinate tin.

The 6-co-ordinate phosphine sulphide complexes of tin have already been discussed (p. 240). Trimethylantimony sulphide forms similar complexes,[247] as shown by infrared and molecular weight data in toluene.

$$2Me_3SbS + R_2SnX_2 \longrightarrow (Me_3SbS)_2R_2SnX_2 \; (R = Me, Et; X = Cl, Br)$$

In polar solvents such as chloroform or DMF, dissociation occurs, giving a molecular weight of about half the toluene value. The infrared and p.m.r. spectra indicate a mixture of the antimony sulphide, dihalide and the cyclotritin sulphide, supporting an equilibrium mixture or complete dissociation. The former is suggested since $(Me_3SbS)_2R_2SnX_2$ can be recrystallized from a mixture of Me_3SbS, Me_3SbX_2 and $[R_2SnS]_3$ in a polar solvent.

$$(Me_3SbS)_2R_2SnX_2 \rightleftharpoons Me_3SbS + Me_3SbX_2 + \tfrac{1}{3}[R_2SnS]_3$$

With tri-cyclohexylantimony sulphide and R_2SnX_2, only exchange occurs, and stannic sulphide results quantitatively from $SnCl_4$ and $SnBr_4$.

$$2(C_6H_{11})_3SbS + R_2SnX_2 \longrightarrow \tfrac{1}{3}[R_2SnS]_3 + (C_6H_{11})_3SbS + (C_6H_{11})_3SbX_2$$

$$2R_3SbS + SnX_4 \longrightarrow SnS_2 + R_3SbS + R_3SbX_2$$

The Structures of Group IV–Sulphur Compounds

The 4-membered ring in tetramethylcyclodisilthiane is established as planar by electron diffraction, with bond lengths and angles close to those recorded for silicon disulphide.

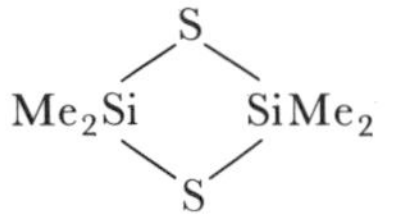

S
Me_2Si $SiMe_2$
S

Si–S 0.218 ± 0.003 nm

SiSSi 75 degrees

Fig. 4.52. The structure of tetramethylcyclodisilthiane

With the trimer, though, where an Si—S bond length of 0.215 ± 0.003 nm and bond angles of Si—S—Si 110 degrees and S—Si—S 115 degrees have been recorded, the ring structure has not been established as a boat or chair conformer.[m] Though diphenyltin sulphide trimer $[Ph_2SnS]_3$ has the boat configuration,[248] with Sn—S 0.242 ± 0.002 nm, the trimeric $[PhMeSiS]_3$ is assumed to have a chair conformation to explain the p.m.r. spectrum. Four conformations are possible (3, 2, 1 or 0 axial methyl groups).

Only one conformer was isolated, melting point 114°C, and the p.m.r. spectrum showed two methyl resonances at $\tau = 9.17$ and 9.44 (intensity 2:1) indicating two axial and one equatorial methyl group.[228]

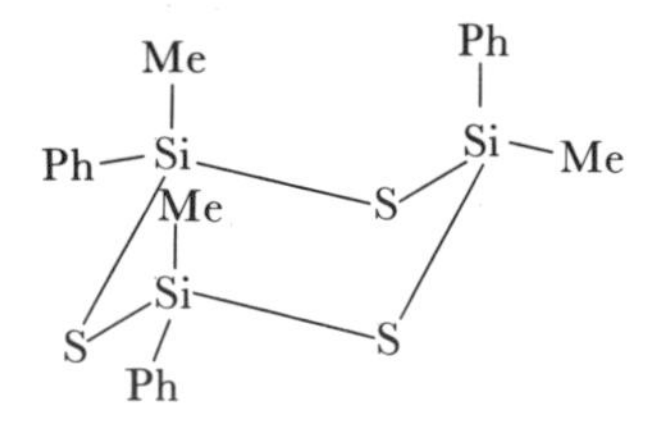

Fig. 4.53. The structure of $[PhMeSiS]_3$

The absence of the isomer with three axial methyl groups is surprising, while the hindrance of two axial phenyl groups probably explains why $[Ph_2SnS]_3$ has the less-hindred boat conformer.

EXCHANGE REACTIONS

The ease with which metathetical exchange occurs with Group IV sulphur compounds, *e.g.*,

P—Cl + Si—S ⟶ Si—Cl + P—S

is readily apparent from the reactions considered. This kind of reaction has been looked at from a more general standpoint by considering simpler systems (*e.g.*, Me_2SiX_2 and Me_2SiY_2). Exchange between X and Y can be random or non-random, and readily followed by p.m.r. spectroscopy. So before considering exchange in cyclic systems, that in the simpler, non-cyclic ones will be briefly considered first."

Random exchange. With two difunctional silanes, an equilibrium constant of 0.25 would be expected for random distribution.

$$2Me_2SiX_2 + 2Me_2SiY_2 \rightleftharpoons Me_2SiX_2 + 2Me_2SiXY + Me_2SiY_2$$

$$K = \frac{[Me_2SiX_2][Me_2SiY_2]}{[Me_2SiXY]^2} = \tfrac{1}{4}$$

Many pairs of functional groups exchange in this way, often requiring an acid catalyst, *e.g.*, Cl/Br, Cl/I, Br/NCO.

Non-random exchange. Bis-aminosilanes and dihalosilanes exchange very rapidly at room temperature to give the unsymmetrical, thermally-stable silane in high yields.

$$Me_2SiCl_2 + Me_2Si(NMe_2)_2 \longrightarrow 2Me_2Si(NMe_2)Cl$$

Naturally, K is very small for this reaction, and also for the halogen/alkoxy system. Non-random exchange occurs for germanium and tin, with the quantitative formation of $R_2Sn(OMe)X$ having already been discussed (p. 228).[181, 182]

Equilibria involving cyclic compounds. These can involve linear/cyclic or cyclic/cyclic combinations. The former involves the general equilibrium below,

$$R_2MX_2 + D_{1/2}MR_2D_{1/2} \rightleftharpoons 2XMR_2D_{1/2}(XMR_2DMR_2X)$$

where M = Si, Ge, or Sn, X = halogen and D is a difunctional group such as O, S, NH or NR. The proportions of starting materials determine the number of "monomer units" R_2MD which will insert in the M—X bond, and again, this can occur in a random or non-random fashion.

Random distribution normally occurs between compounds containing the same functional element. Thus, with methoxysilanes/cyclosiloxanes and alkylthiosilanes/cyclosilthians K is about 0.25.

Similar functional groups produce non-random exchange with linear/cyclic compounds as encountered with the linear/linear system. This is particularly so with chlorosilanes and cyclosilazanes, and its synthetic use has already been mentioned (pp. 168, 177), *e.g.*,

$$2[Me_2SiNMe]_3 + 3Me_2SiCl_2 \longrightarrow 3(ClMe_2SiNMe)_2SiMe_2$$

$$4Me_2SiCl_2 + [Me_2SiNH]_4 \longrightarrow 4(ClMe_2Si)_2NH$$

With Me_2SiCl_2 and cyclosiloxanes, the distribution is almost random and slow ($K = 0.11$) (several hours at 200°C in the presence of $AlCl_3$).[249] With germanium, exchange is faster and less random ($K = 0.021$), while the dimeric chlorostannoxanes are formed quantitatively and instantaneously.[250]

$$2R_2SnCl_2 + \frac{2}{n}[R_2SnO]_n \longrightarrow [ClR_2SnOSnR_2Cl]_2$$

The change in rate is probably related to the increase in donating power of oxygen as the metalloid increases in size, while the tendency of tin to become 5-co-ordinate when bonded to electronegative substituents probably encourages non-random exchange.

A similar trend occurs among thio derivatives of silicon and germanium. They equilibrate slowly with dichlorosilanes and dichlorogermanes (Si, $K = 0.13$ at 200°C; Ge, $K = 0.09$ at 120°C), while a rapid, non-random exchange occurs with tin, giving halodistanthianes quantitatively.[251]

Cyclic–cyclic mixtures. Here, the distribution between rings with different functional groups appears random, but where two oligomers are isolable the proportions are dependent not on random distribution but on their relative stability, the equilibrium constant and hence the temperature.

Distribution between Si—S and Si—N rings, Si—S and Ge—S, and between Ge—S and Ge—O rings appears random, all mixed and unmixed rings being detectable by p.m.r. spectroscopy.[252]

The equilibrium constant for the equilibrium

$$2\,[Me_2SiS]_3 \rightleftharpoons 3\,[Me_2SiS]_2$$

i.e., $K = \{[Me_2SiS]_3\}^2/\{[Me_2SiS]_2\}^3$, decreases as the temperature rises, showing that the 4-membered ring is preferred at higher temperatures (at 25°C, $K = 3.6 \times 10^3\ dm^3\ mol^{-1}$; at 120°C, $1.3 \times 10^2\ dm^3\ mol^{-1}$).

In solution, trimeric and tetrameric Me_2GeO co-exist.

$$3[Me_2GeO]_4 \rightleftharpoons 4[Me_2GeO]_3;\ K = \frac{\{[Me_2GeO]_4\}^3}{\{[Me_2GeO]_3\}^4}$$

At 35°C, $K = 6.0\ dm^3\ mol^{-1}$, while at 103°C it is 0.52, showing the trimer to be the more stable at high temperatures.[253]

HALIDES OF THE GROUP IV ELEMENTS

A comparison of the melting and boiling points of the simple organohalogeno derivatives of the Group IV elements shows breaks in continuity consistent with association. This is especially so with fluorides and chlorides, as Table 4.11 indicates.

The properties vary from those typical of monomeric covalent halides, through associated covalent ones, to ionic compounds (notably the lead fluorides). Among the covalent polymers, dimethyltin difluoride and stannic fluoride both have a layer lattice structure involving fluoride bridges and 6-co-ordinate tin. In the former, the Sn—F bond is 0.212 nm in length,[254] 0.01 nm longer than the analogous bond in SnF_4,[255] while the terminal Sn—F bonds, which occur in place of the Sn—Me bond of Me_2SnF_2 are 0.188 nm (in length).

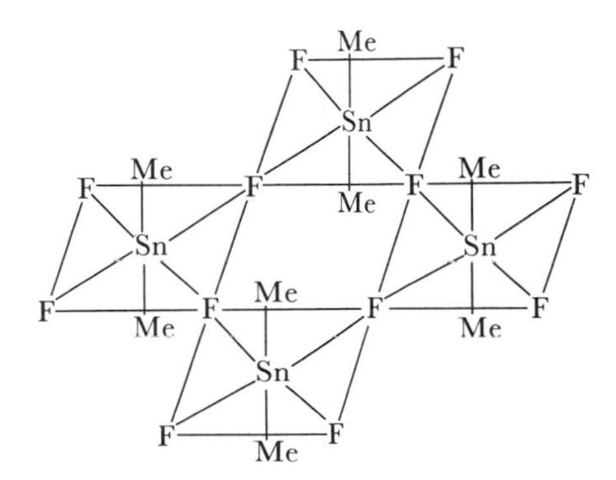

Fig. 4.54. The structure of dimethyltin difluoride

Only this last value is consistent with a single bond, the bridging Sn—F distances supporting a multi-centre bond involving the overlap of p orbitals on tin with a set on the four fluorine atoms around it. This gives linear three-centre bonds, and contrasts vividly with SiF_4, where the Si—F bond length of 0.156 nm is about 0.01 nm shorter than expected for a single bond.[a]

The lead fluorides appear to be ionic, while diphenyl-lead dichloride is a 6-co-ordinate linear polymer involving bent chloride bridges,[256] in contrast with the linear fluoride ones.

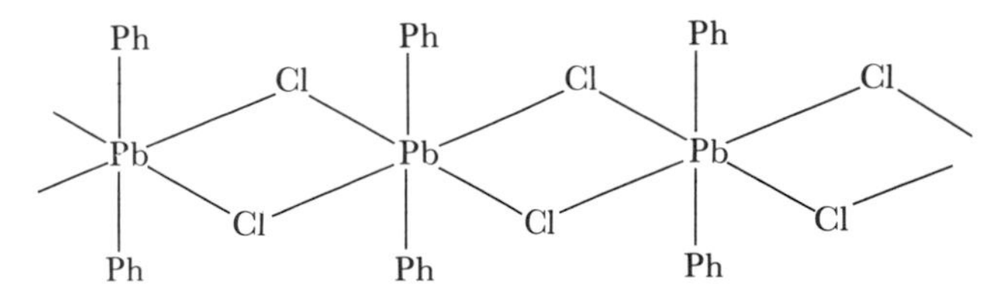

Fig. 4.55. The structure of diphenyl-lead dichloride

Table 4.11. Physical Properties of Methyl–Halogen Compounds of the Group IV Elements Illustrating Polymerization Trends

Compound \ M	Silicon		Germanium		Tin		Lead	
	m.p. (°C)	b.p. (°C)	m.p. (°C)	b.p. (°C)	m.p. (°C)	b.p. (°C)	m.p. (°C)	b.p. (°C)
Me_3MF	−74	16.4	1.9	76	375 (dec.″)		~305 (dec.″)	
Cl		57.3	−14	97	37	152–4	190 (dec.″)	
Br		79		113	27	165	133 (dec.″)	
I		107		136	3.4	170		
Me_2MF_2	−87.5	2.7		112	~360			
Cl_2		70	−22	124	106	190	155 (dec.″)	
Br_2		112		153	74	209	(dec.″)	
I_2		170		204	44	228		
$MeMF_3$	−73	−30	38	96.5				
Cl_3	−78	65.7		111	45–6			
Br_3	−28	131		168	53	211		
I_3		228–230	48–50	237	85			
MF_4		−65	sub. at m.p.			705 (sub.″)		saline
MCl_4	−68	57.6	−49.5	83.1	−30	114	−15	explodes above 105
MBr_4	5	153	26.1	186.5	31	202		
MI_4	120.5	290	144	dec.″	143.5	341 (sub.″ 180)		

With methyltin chlorides, however, no evidence for strong bridging in the solid phase exists.[257] Indeed, these compounds are monomeric over a wide concentration range in solution.[258] Bridging is observed, however, in the anion of $K_2Sn^{II}Cl_4 \cdot H_2O$, which involves $SnCl_6$ octahedra edge-bridged in an infinite chain,[a] like diphenyl-lead dichloride.

Bibliography

General

a A. F. WELLS, 'Structural Inorganic Chemistry', Oxford University Press, 3rd Edn., 1962.
b H. GILMAN and G. L. SCHWEBKE, *Adv. Organometallic Chem.*, 1964, **1,** 89.
c W. FINK, *Angew. Chem. Internat. Edn.*, 1966, **5,** 760.
d H. SCHUMANN, *ibid.*, 1969, **8,** 937.
e W. NOLL, 'Chemistry and Technology of Silicones', Academic Press, New York, 1968, p. 287.
f H. H. READ, 'Rutley's Mineralogy', Woodbridge Press, Guildford, 1947.
g R. J. H. VOORHOEVE, 'Organohalosilanes', Elsevier, Amsterdam, 1967.
h C. EABORN, 'Organosilicon Chemistry', Butterworths, London, 1960.
i H. R. ALLCOCK, 'Heteroatom Ring Systems and Polymers', Academic Press, New York, 1967.
j F. GLOCKING, 'The Chemistry of Germanium', Academic Press, London and New York, 1969.
k R. OKAWARA and M. WADA, *Adv. Organometallic Chem.*, 1967, **5,** 164.
l A. HAAS, *Angew. Chem. Internat. Edn.*, 1965, **4,** 1014.
m E. W. ABEL and D. A. ARMITAGE, *Adv. Organometallic Chem.*, 1967, **5,** 1.
n K. MOEDRITZER, *ibid.*, 1968, **6,** 171.

References

1 K. S. PITZER, *J. Amer. Chem. Soc.*, 1948, **70,** 2140.
2 D. H. OLSON and R. E. RUNDLE, *Inorg. Chem.*, 1963, **2,** 1310.
3 T. GELA, *J. Chem. Phys.*, 1956, **24,** 1009.
4 H. SCHÄFER, K. H. JANZON and A. WEISS, *Angew. Chem. Internat. Edn.*, 1963, **2,** 393.
5 J. WITTE and H. G. SCHNERING, *Z. anorg. Chem.*, 1964, **327,** 260; E. BUSMAN, *Naturwiss.*, 1960, **47,** 82; *Z. anorg. Chem.*, 1961, **313,** 90.
6 R. E. MARSH and D. P. SHOEMAKER, *Acta Cryst.*, 1953, **6,** 197.
7 H. A. SKINNER and L. E. SUTTON, *Trans. Faraday Soc.*, 1940, **36,** 1209.
8 G. ROCKTACHEL and A. WEISS, *Z. anorg. Chem.*, 1962, **316,** 231.
9 P. ECKERHIR and E. WOLFEL, *ibid.*, 1955, **280,** 321.
10 H. AXEL, H. SCHAFER and A. WEISS, *Angew. Chem. Internat. Edn.*, 1965, **4,** 358.
11 I. BOHM and O. HASSEL, *Z. anorg. Chem.*, 1927, **160,** 151.
12 E. HENNGE and G. SCHEFFLER, *Monatsh.*, 1964, **95,** 1450; *Naturwiss.*, 1963, **50,** 474.
13 K. JANSON, H. SCHAFER and A. WEISS, *Angew. Chem. Internat. Edn.*, 1965, **4,** 245.
14 E. ZINTL, J. GOUBEAU and W. DULLENKOPF, *Z. phys. Chem.*, 1934, **154,** 1.
15 D. BRITTON, *Inorg. Chem.*, 1964, **3,** 305.
16 O. SCHMITZ-DUMONT and H.-J. GÖTZE, *Z. anorg. Chem.*, 1969, **371,** 38.
17 H. M. J. C. CREEMERS, J. G. NOLTES and G. J. M. VAN DER KERK, *J. Organometallic Chem.*, 1968, **14,** 217.
18 D. J. CRAM and N. S. NEWMANN, 'Steric Effects in Organic Chemistry', John Wiley and Sons, Inc., New York, 1956, p. 249.

19 W. H. ATWELL and D. R. WEYENBERG, *J. Amer. Chem. Soc.*, 1968, **90,** 3438.
20 W. H. ATWELL, L. G. MALONE, S. F. HAYES and J. G. UHLMAN, *J. Organometallic Chem.*, 1969, **18,** 69.
21 F. S. KIPPING and co-workers, *J. Chem. Soc.*, 1921, 830; 1924, 229; 1927, 2719.
22 H. GILMAN, D. J. PETERSON, A. W. JARVIE and H. J. S. WINKLER, *J. Amer. Chem. Soc.*, 1960, **82,** 2076; 1961, **83,** 1921; E. HENGGE, R. PETZOLD and co-workers, *Z. Naturforsch.*, 1963, **18b,** 425; 1965, **20b,** 397.
23 H. GILMAN and G. SCHWEBKE, *J. Amer. Chem. Soc.*, 1964, **86,** 2693; 1963, **85,** 1016.
24 *Idem*, *J. Organometallic Chem.*, 1965, **3,** 382.
25 K. KUHLEIN and W. P. NEUMANN, *ibid.*, 1968, **14,** 317.
26 H. GILMAN, D. R. CHAPMAN and G. L. SCHWEBKE, *ibid.*, 1968, **14,** 267.
27 C. A. BURKHARD, *J. Amer. Chem. Soc.*, 1949, **71,** 963.
28 N. GRAF ZU STOLBERG, *Angew. Chem. Internat. Edn.*, 1963, **2,** 150.
29 E. CARBERRY and R. WEST, *J. Organometallic Chem.*, 1966, **6,** 583.
30 H. GILMAN and R. A. TOMASI, *J. Org. Chem.*, 1963, **28,** 1651.
31 M. KUMADA, M. ISHIKAWA, S. SAKAMOTO and S. MAEDA, *J. Organometallic Chem.*, 1969, **17,** 223.
32 M. KUMADA and K. TAMAO, *Adv. Organometallic Chem.*, 1968, **6,** 65.
33 M. ISHIKAWA and M. KUMADA, *Chem. Comm.*, 1969, 567.
34 G. A. RUSSELL, *J. Amer. Chem. Soc.*, 1959, **81,** 4834.
35 H. GILMAN and S. INOUE, *J. Org. Chem.*, 1964, **29,** 3418.
36 G. R. HUSK and R. WEST, *J. Amer. Chem. Soc.*, 1965, **87,** 3993; R. S. GOHLKE, *ibid.*, 1968, **90,** 2713.
37 E. CARBURY, R. WEST and G. E. GLASS, *ibid.*, 1969, **91,** 5440.
38 C. A. KRAUS and C. L. BROWN, *ibid.*, 1930, **52,** 4031.
39 W. P. NEUMANN and K. KUHLEIN, *Tetrahedron Letters*, 1963, **33,** 1541.
40 W. P. NEUMANN, *Angew. Chem. Internat. Edn.*, 1963, **2,** 555.
41 W. P. NEUMANN and K. KUHLEIN, *Annalen*, 1967, **702,** 13.
42 O. M. NEFEDOV and S. P. KOLESNIKOV, *Izvest. Akad. Nauk S.S.S.R., Ser. khim.*, 1966, **2,** 201; *Chem. Abs.*, 1966, **65,** 744a.
43 C. LÖWIG, *Annalen*, 1852, **84,** 308.
44 N. N. ZEMLYAUSKII, E. M. PANOV and K. A. KOCHESHKOV, *Doklady Akad. Nauk S.S.S.R.*, 1962, **146,** 1335; *Chem. Abs.*, 1963, **58,** 9110h.
45 T. L. BROWN and G. L. MORGAN, *Inorg. Chem.*, 1963, **2,** 736.
46 W. P. NEUMANN, *Angew. Chem. Internat. Edn.*, 1963, **2,** 165; H. GILMAN and F. K. CARTLEDGE, *J. Organometallic Chem.*, 1966, **5,** 48.
47 W. V. FARRAR and H. A. SKINNER, *ibid.*, 1964, **1,** 434.
48 W. P. NEUMANN, J. PEDAIN and R. SOMMER, *Annalen*, 1966, **694,** 9.
49 R. SOMMER, W. P. NEUMANN and B. SCHNEIDER, *Tetrahedron Letters*, 1964, 3875; A. K. SAWYER, *J. Amer. Chem. Soc.*, 1965, **87,** 537; H. M. J. C. CREEMERS, with Ph.D. Thesis, 1967.
50 W. P. NEUMANN and co-workers, *Angew. Chem. Internat. Edn.*, 1963, **2,** 555; 1964, **3,** 751; *Annalen*, 1964, **677,** 12.
51 W. P. NEUMANN and K. KONIG, *Annalen*, 1964, **677,** 1; *Angew. Chem. Internat. Edn.*, 1962, **1,** 212.
52 P. PFEIFFER, *Ber.*, 1911, **44,** 1269; K. SISIDO, S. KOZIMA and T. ISIBASI, *J. Organometallic Chem.*, 1967, **10,** 439; 1968, **11,** 281
53 K. A. KOZESCHKOW, *Chem. Ber.*, 1929, **62,** 996; 1933, **66,** 1661.
54 W. C. SCHUMB and L. H. TOWLE, *J. Amer. Chem. Soc.*, 1953, **75,** 6085; A. PFLUGMACHER and H. DAHMEN, *Z. anorg. Chem.*, 1957, **290,** 184.
55 U. WANNAGAT, P. SCHMIDT and M. SCHULZE, *Angew. Chem. Internat. Edn.*, 1967, **6,** 446, 447.
56 H. BURGER, M. SCHULZE and U. WANNAGAT, *Inorg. Nuclear Chem. Letters*, 1967, **3,** 43; 1969, **5,** 789.
57 U. WANNAGAT and P. SCHMIDT, *ibid.*, 1968, **4,** 331.

58 W. FINK, *Angew. Chem.*, 1961, **73,** 736.
59 W. FINK, *Chem. Ber.*, 1963, **96,** 1071.
60 K. LEINHARD and E. G. ROCHOW, *Angew. Chem. Internat. Edn.*, 1963, **2,** 325; *Z. anorg. Chem.*, 1964, **331,** 316.
61 W. FINK, *Helv. Chim. Acta*, 1964, **47,** 498.
62 *Idem, ibid.*, 1963, **46,** 720.
63 R. P. BUSH, C. A. PEARCE and N. C. LLOYD, *Chem. Comm.*, 1967, 1269.
64 L. W. BREED and R. L. ELLIOTT, *J. Organometallic Chem.*, 1968, **11,** 447.
65 U. WANNAGAT, *Angew. Chem. Internat. Edn.*, 1965, **4,** 605.
66 P. GEYMEYER and E. G. ROCHOW, *ibid.*, 592.
67 L. W. BREED, W. L. BUDDE and R. L. ELLIOT, *J. Organometallic Chem.*, 1966, **6,** 1676.
68 J. SILBIGER, J. FUCHS and N. GESUNDERHEIT, *Inorg. Chem.*, 1967, **6,** 39.
69 W. FINK, *Helv. Chim. Acta*, 1968, **51,** 954.
70 W. T. REICHLE, *Inorg. Chem.*, 1964, **3,** 402.
71 E. ELTENHUBER and K. RÜHLMANN, *Chem. Ber.*, 1968, **101,** 743.
72 W. FINK, *Helv. Chim. Acta*, 1966, **49,** 1408.
73 P. J. WHEATLEY, *J. Chem. Soc.*, 1962, 1721.
74 H. SCHMIDBAUR, *J. Amer. Chem. Soc.*, 1963, **85,** 2336.
75 R. P. BUSH and C. A. PEARCE, *J. Chem. Soc. (A)*, 1969, 808.
76 I. HAIDUC and H. GILMAN, *J. Organometallic Chem.*, 1969, **18,** P5.
77 O. J. SCHERER and D. BILLER, *Z. Naturforsch.*, 1967, **22b,** 1079.
78 A. STOCK and K. SOMIESKY, *Ber.*, 1921, **54,** 740.
79 S. D. BREWER and C. P. HABER, *J. Amer. Chem. Soc.*, 1948, **70,** 3888.
80 L. W. BREED and R. L. ELLIOT, *Inorg. Chem.*, 1964, **3,** 1622.
81 E. W. ABEL and R. P. BUSH, *J. Inorg. Nuclear Chem.*, 1964, **26,** 1685.
82 K. A. ANDRIANOV and G. YA. RUMBA, *Zhur. obshchei Khim.*, 1962, **32,** 1993; *Chem. Abs.*, 1963, **58,** 4592.
83 B. J. AYLETT, G. M. BURNETT, L. K. PETERSON and N. ROSS, *Soc. Chem. Ind. (London)*, Monograph 1961, **13,** 5.
84 R. SCHAEFFER, L. ROSS, M. THOMPSON and R. WELLS, *Chem. Abs.*, 1962, **57,** 16120.
85 U. WANNAGAT, E. BOGUSCH and R. BRAUN, *J. Organometallic Chem.*, 1969, **19,** 367; K. LIENHARD and E. G. ROCHOW, *Z. anorg. Chem.*, 1964, **331,** 307.
86 W. FINK, *Angew. Chem.*, 1961, **73,** 467; *Chem. Abs.*, 1962, **56,** 14726.
87 L. W. BREED, *Inorg. Chem.*, 1968, **7,** 1940.
88 L. W. BREED and R. L. ELLIOT, *ibid.*, 1963, **2,** 1069.
89 W. FINK, *Angew. Chem. Internat. Edn.*, 1969, **8,** 521.
90 *Idem, Helv. Chim. Acta*, 1962, **45,** 1081; *Chem. Abs.*, 1962, **57,** 9869i.
91 G. CHIOCOLLA and J. J. DALY, *J. Chem. Soc. (A)*, 1968, 1658.
92 U. WANNAGAT, E. BOGISCH and F. HÖFLER, *J. Organometallic Chem.*, 1967, **7,** 203.
93 M. YOKOI and K. YAMASAKI, *J. Amer. Chem. Soc.*, 1953, **75,** 4139.
94 R. PRINZ and H. WERNER, *Angew. Chem. Internat. Edn.*, 1967, **6,** 91.
95 G. R. WILLEY, *J. Amer. Chem. Soc.*, 1968, **90,** 3362.
96 R. N. LEWIS, *ibid.*, 1948, **70,** 1115.
97 G. S. SMITH and L. E. ALEXANDER, *Acta Cryst.*, 1963, **16,** 1015.
98 K. A. ANDRIANOV, B. A. ISMAILOV, A. M. KONONOV and G. V. KOTRELEV, *J. Organometallic Chem.*, 1965, **3,** 129.
99 C. R. KRUGER and E. G. ROCHOW, *Angew. Chem. Internat. Edn.*, 1962, **1,** 458; *Idem, J. Polymer Sci.*, 1964, **A2,** 3179.
100 E. LARSSON and L. BJELLERUP, *J. Amer. Chem. Soc.*, 1953, **75,** 995.
101 K. A. ANDRIANOV, G. A. KURAKOV, L. M. KHANANISVILI and T. A. LOMONOSOVA, *Zhur. obshchei Khim.*, 1963, **33,** 1294; *Chem. Abs.*, 1963, **59,** 10105.
102 M. M. MORGUNOVA, D. YA. ZHINKIN and M. V. SOBOLEVSKII, *Plasticheskie Massy*, 1963, **6,** 24; *Chem. Abs.*, 1963, **59,** 11674; 1964, **60,** 5664.
103 G. REDL and E. G. ROCHOW, *Angew. Chem. Internat. Edn.*, 1964, **3,** 516.

104 U. WANNAGAT and H. BURGER, *ibid.*, 1964, **3,** 446
105 J. DAVY, *Phil. Trans.*, 1812, **5,** 352.
106 M. F. LAPPERT and G. SRIVASTAVA, *Inorg. Nuclear Chem. Letters*, 1965, **1,** 53.
107 H. NÖTH, *Z. Naturforsch.*, 1961, **16b,** 618.
108 U. WANNAGAT, E. BOGUSCH and P. GEYMAYER, *Monatsh.*, 1964, **95,** 801; *Chem. Abs.*, 1964, **61,** 14701.
109 K. A. ANDRIANOV, V. V. ASTAKHIN and V. B. LOSEV, *Izvest. Akad. Nauk S.S.S.R., Otd. Khim. Nauk*, 1963, 950; *Chem. Abs.*, 1963, **59,** 7552; *Idem, ibid.*, 1964, **60,** 9305.
110 K. A. ANDRIANOV, G. V. KOTRELEV, B. A. KAMARITSKI, I. H. UNITSKI and N. I. SIDOROVA, *J. Organometallic Chem.*, 1969, **16,** 51.
111 K. A. ANDRIANOV and G. K. KOTRELEV, *ibid.*, 1967, **7,** 217.
112 M. M. SPRING and F. O. GUENTHER, *J. Amer. Chem. Soc.*, 1955, **77,** 3990, 3996.
113 W. FINK, *Chem. Ber.*, 1964, **97,** 1424.
114 J. J. DALY and W. FINK, *J. Chem. Soc.*, 1964, 4959.
115 A. E. FLOOD, *J. Amer. Chem. Soc.*, 1932, **54,** 1663.
116 I. RUIDISCH and M. SCHMIDT, *Angew. Chem. Internat. Edn.*, 1964, **3,** 231.
117 *Idem, ibid.*, 1964, **3,** 637.
118 I. SCHUMANN-RUIDISCH and B. JUTZI-MEBERT, *J. Organometallic Chem.*, 1968, **11,** 77.
119 W. EISENHUTH and J. R. VAN WAZER, *Inorg. Nuclear Chem. Letters*, 1967, **3,** 359.
120 M. V. GEORGE, P. B. TALUKDAR and H. GILMAN, *J. Organometallic Chem.*, 1966, **5,** 397.
121 K. JONES and M. F. LAPPERT, *J. Chem. Soc.*, 1965, 1944.
122 H. BREEDERVELD, *Rec. Trav. chim.*, 1962, **81,** 276.
123 J. SATGÉ, M. LESBRE and M. BAUDET, *Compt. rend.*, 1964, **259,** 4733; *Chem. Abs.*, 1965, **62,** 11842e.
124 T. A. GEORGE, K. JONES and M. F. LAPPERT, *J. Chem. Soc.*, 1965, 2157.
125 U. WANNAGAT and co-workers, *Monatsh.*, 1967, **97,** 1352; *Inorg. Nuclear Chem. Letters*, 1965, **1,** 13.
126 E. HENGGE, *Z. Naturforsch.*, 1965, **20b,** 397.
127 U. WANNAGAT and H. NIEDERPRÜM, *Angew. Chem.*, 1958, **70,** 745; *Z. anorg. Chem.*, 1961, **311,** 270.
128 U. WANNAGAT and O. BRANDSTÄTTER, *Angew. Chem. Internat. Edn.*, 1963, **2,** 263.
129 (a) N. N. SOKOLOV, *Zhur. obshchei Khim.*, 1959, **29,** 258; *Chem. Abs.*, 1959, **53,** 21622; 1960, **54,** 8603; (b) C. R. KRUGER and E. G. ROCHOW, *Inorg. Chem.*, 1963, **2,** 1295; *Angew. Chem. Internat. Edn.*, 1962, **1,** 455; (c) J. G. MURRAY and R. K. GRIFFITH, *J. Org. Chem.*, 1964, **29,** 1215; (d) L. W. BREED, M. E. WHITEHEAD and R. L. ELLIOT, *Inorg. Chem.*, 1967, **6,** 1254; (e) K.A. ANDRIANOV, and co-workers, *Izvest. Akad. Nauk S.S.S.R., Otd. Khim. Nauk*, 1963, 1847, 2045; *Chem. Abs.* 1964, **60,** 2999, 5534.
130 I. HAIDUC and H. GILMAN, *J. Organometallic Chem.*, 1969, **18,** P5.
131 K. A. ANDRIANOV, I. HAIDUC and L. M. KHANANASHVILI, *Izvest. Akad. Nauk S.S.S.R., Otd. Khim. Nauk*, 1963, 1701; *Chem. Abs.*, 1963, **59,** 15299.
132 J. SILBIGER and J. FUCHS, *Inorg. Chem.*, 1965, **4,** 1371.
133 H. GILMAN, H. N. BENEDICT and H. HARTZFELD, *J. Org. Chem.*, 1954, **19,** 419.
134 R. P. BUSH, N. C. LLOYD and C. A. PEARCE, *Chem. Comm.*, 1968, 1191.
135 I. IDRESTEDT and C. BROSSETT, *Acta Chem., Scand.*, 1964, **18,** 1879.
136 S. SUJISHI and S. WITZ, *J. Amer. Chem. Soc.*, 1954, **76,** 4631.
137 R. RUDMAN, W. C. HAMILTON, S. NOVICK and T. D. GOLDFARB, *ibid.*, 1967, **89,** 5157.
138 N. WIBERG and K. H. SCHMID, *Chem. Ber.*, 1967, **100,** 748.
139 E. W. ABEL and J. P. CROW, *J. Organometallic Chem.*, 1969, **17,** 337.
140 G. W. PARSHALL and R. V. LINDSAY, *J. Amer. Chem. Soc.*, 1959, **81,** 6273.
141 H. SCHUMANN and H. BENDA, *Angew. Chem. Internat. Edn.*, 1969, **8,** 989.
142 *Idem, ibid.*, 1968, **7,** 812.

143 *Idem*, *ibid.*, 813.
144 *Idem*, *J. Organometallic Chem.*, 1970, **21,** P12.
145 H. SCHUMANN, H. KOPF and M. SCHMIDT, *Angew. Chem. Internat. Edn.*, 1963, **2,** 546; *Chem. Ber.*, 1964, **97,** 2395.
146 *Idem*, *Chem. Ber.*, 1964, **97,** 1458.
147 E. W. ABEL and I. H. SABHERWAL, *J. Chem. Soc.* (*A*), 1968, 1105.
148 H. SCHUMANN and H. BENDA, *Angew. Chem. Internat. Edn.*, 1970, **9,** 76.
149 T. HARADA, *Sci. Papers Inst. Phys. Chem. Res., Tokyo*, 1939, **36,** 497.
150 D. L. ALLESTON, A. G. DAVIES and M. HANCOCK, *J. Chem. Soc.*, 1964, 5744.
151 E. W. ABEL, D. A. ARMITAGE and D. B. BRADY, *Trans. Faraday Soc.*, 1966, **62,** 3459.
152 YU. I. SMOLIN, *Chem. Comm.*, 1969, 395.
153 F. A. COTTON and G. WILKINSON, 'Advanced Inorganic Chemistry', John Wiley and Sons, Inc., New York, 2nd Edn., 1966, p. 469.
154 P. RAMDOHR and A. E. C. GORESEY, *Science*, 1970, **167,** 615.
155 A. WEISS and A. WEISS, *Naturwiss.*, 1954, **41,** 12; *Z. anorg. Chem.*, 1954, **276,** 95.
156 W. S. CHAMBERS and C. J. WILKINS, *J. Chem. Soc.*, 1960, 5088.
157 E. G. ROCHOW, *J. Amer. Chem. Soc.*, 1945, **67,** 963; U.S.P. 2,380,995/1945.
158 K. A. ANDRIANOV and V. V. SEVERNYI, *J. Organometallic Chem.*, 1964, **1,** 268, 340.
159 F. R. MAYO, *J. Polymer Sci.*, 1961, **55,** 65; T. C. WU and C. A. HIRT, *J. Organometallic Chem.*, 1968, **11,** 17.
160 L. H. VOGT, JR., J. F. BROWN, JR. and P. I. PRESCOTT, *Inorg. Chem.*, 1963, **2,** 189; *J. Amer. Chem. Soc.*, 1964, **86,** 1120.
161 J. F. BROWN, JR., L. H. VOGT, JR., A. KATCHMAN, J. W. EUSTANCE, K. M. KISER and K. W. KRANTZ, *J. Amer. Chem. Soc.*, 1960, **82,** 6194.
162 H. STEINFINK, B. POST and I. FANKUCHEN, *Acta Cryst.*, 1955, **8,** 420.
163 H. J. HICKTON, A. HOLT, J. HOMER and A. W. JARVIE, *J. Chem. Soc.* (*C*), 1966, 149.
164 A. G. DAVIES, P. G. HARRISON and T. A. G. SILK, *Chem. and Ind.*, 1968, 949.
165 M. KUMADA, M. ISHIKAWA and B. MURAI, *Kogyo Kagatu Zasshi*, 1963, **66,** 637.
166 H. GILMAN, W. H. ATWELL and F. K. CARTLEDGE, *Adv. Organometallic Chem.*, 1966, **4,** 31.
167 E. HENGGE, in 'Halogen Chemistry', ed. V. GUTMANN, 1967, Academic Press, London and New York, Vol. 2, p. 193.
168 M. SCHMIDT and H. SCHMIDBAUR, *Chem. Ber.*, 1960, **93,** 878; 1961, **94,** 2446.
169 H. H. ANDERSON, *J. Amer. Chem. Soc.*, 1950, **72,** 194; 1952, **74,** 237.
170 M. KUMADA and S. MAEDA, *Inorg. Chim. Acta*, 1967, **1,** 105.
171 R. S. TOBIAS and S. HUTCHESON, *J. Organometallic Chem.*, 1966, **6,** 535.
172 M. P. BROWN and E. G. ROCHOW, *J. Amer. Chem. Soc.*, 1960, **82,** 4166.
173 K. MOEDRITZER, *J. Organometallic Chem.*, 1966, **5,** 254.
174 I. RUIDISCH and M. SCHMIDT, *Chem. Ber.*, 1963, **96,** 821.
175 W. METLESICS and H. ZEISS, *J. Amer. Chem. Soc.*, 1960, **82,** 3321, 3324.
176 E. J. BULTEN and J. G. NOLTES, *Tetrahedron Letters*, 1966, **29,** 3471; K. KUHLEIN and W. P. NEUMANN, *Annalen*, 1967, **702,** 17; F. RIJKENS, E. J. BULTEN, W. DRENTH and G. J. M. VAN DER KERK, *Rec. Trav. chim.*, 1966, **85,** 1223.
177 P. G. HARRISON and J. J. ZUCKERMANN, *Inorg. Chem.*, 1970, **9,** 175.
178 G. STERR and R. MATTES, *Z. anorg. Chem.*, 1963, **322,** 319.
179 Y. KAWASAKI, T. TANAKA and R. OKAWARA, *J. Organometallic Chem.*, 1966, **6,** 95; Y. KAWASAKI and R. OKAWARA, *J. Inorg. Nuclear Chem.*, 1965, **27,** 1168.
180 J. C. POMMIER and J. VALADE, *Compt. rend.*, 1969, **268,** 633; *J. Organometallic Chem.*, 1968, **12,** 433.
181 A. G. DAVIES and P. G. HARRISON, *J. Chem. Soc.* (*C*), 1967, 298.
182 A. C. CHAPMAN, A. G. DAVIES, P. G. HARRISON and W. MCFARLANE, *J. Chem. Soc.* (*C*), 1970, 821.
183 H. KRIEGSMANN, H. HOFFMANN and S. PISCHTCHAU, *Z. anorg. Chem.*, 1962, **315,** 283.
184 R. OKAWARA and K. YASUDA, *J. Organometallic Chem.*, 1964, **1,** 356.

185 R. S. TOBIAS, *Organometallic Chem. Rev.*, 1966, **1,** 93.
186 C. A. KRAUS and R. H. BULLARD, *J. Amer. Chem. Soc.*, 1929, **51,** 3605.
187 E. AMBERGER and M. R. KULA, *Chem. Ber.*, 1963, **96,** 2562.
188 C. A. KRAUS and co-workers, *J. Amer. Chem. Soc.*, 1925, **47,** 2416; 1930, **52,** 4056; T. HARADA, *Bull. Chem. Soc. Japan*, 1927, **2,** 105; *Sci. Papers Inst. Phys. Chem. Res., Tokyo*, 1939, **36,** 504.
189 H. SCHMIDBAUR, *Chem. Ber.*, 1964, **97,** 830.
190 D. L. ALLESTON and A. G. DAVIES, *J. Chem. Soc.*, 1962, 2465; A. REICHLE and I. DAHLMANN, *Annalen*, 1964, **675,** 19; O. L. MAGILI and J. B. HARRISON, U.S.P. 3,152,156; *Chem. Abs.*, 1964, **61,** 16903.
191 A. J. BLOODWORTH, A. G. DAVIES and I. F. GRAHAM, *J. Organometallic Chem.*, 1968, **13,** 351.
192 C. WALLING and L. HEATON, *J. Amer. Chem. Soc.*, 1965, **87,** 48.
193 A. G. DAVIES and I. F. GRAHAM, *Chem. and Ind.*, 1963, 1622.
194 R. L. DANNLEY and W. A. AUE, *J. Org. Chem.*, 1965, **30,** 3845.
195 A. G. DAVIES and P. G. HARRISON, *J. Organometallic Chem.*, 1967, **7,** P13; 1967, **10,** P31.
196 A. STRECKER, *Annalen*, 1862, **123,** 365.
197 Y. TAKEDA, Y. HAJAKAWA, T. FUENO and J. FURKAWA, *Makromol. Chem.*, 1965, **83,** 234.
198 D. L. ALLESTON, A. G. DAVIES, M. HANCOCK and R. F. M. WHITE, *J. Chem. Soc.*, 1963, 5469.
199 T. HARADA, *Report Sci. Res. Inst. (Japan)*, 1948, **24,** 177.
200 *Idem, Sci. Papers Inst. Phys. Chem. Res. Tokyo*, 1947, **42,** 64.
201 D. L. ALLESTON, A. G. DAVIES and M. HANCOCK, *J. Chem. Soc.*, 1964, 5744.
202 R. OKAWARA and M. WADA, *J. Organometallic Chem.*, 1963, **1,** 81.
203 R. OKAWARA and co-workers, *ibid.*, 1965, **3,** 70; 1967, **8,** 261.
204 V. I. GOLDANSKII, E. F. MAKAROV, R. A. STUKAN, V. A. TRUKHTANOV and E. V. KHRAPOV, *Proc. Akad. Sci. (U.S.S.R.)*, 1963, **151,** 598.
205 A. G. DAVIES, P. G. HARRISON and P. R. PALAN, *J. Organometallic Chem.*, 1967, **10,** P33; *J. Chem. Soc. (C)*, 1970, 2030.
206 B. A. RADKIL, V. N. GLASHAKOVA and YU. A. ALEKSANDROV, *J. Gen. Chem. (U.S.S.R.)*, 1967, **37,** 195.
207 G. A. RAZUVAEV, O. A. SCHCHAPETKOVA and N. S. VYAZANKIN, *J. Gen. Chem. (U.S.S.R.)*, 1949, **19,** 2121.
208 D. SUKHANI, V. N. GUPTA and R. C. MEHROTRA, *J. Organometallic Chem.*, 1967, **7,** 85.
209 K. YAMADU, H. MATSUMOTO and R. OKAWARA, *ibid.*, 1966, **6,** 528.
210 K. DEHNICKE, *Z. anorg. Chem.*, 1961, **308,** 72.
211 D. MORAS, A. MITSCHLER and R. WEISS, *Chem. Comm.*, 1968, 26.
212 R. S. TOBIAS, J. OGRINS and B. A. NEVETT, *Inorg. Chem.*, 1962, **1,** 638.
213 V. A. MARONI and T. G. SPIRO, *J. Amer. Chem. Soc.*, 1967, **89,** 45; *Inorg. Chem.*, 1968, **7,** 183.
214 T. G. SPIRO, D. H. TEMPLETON and A. ZALKIN, *ibid.*, 1969, **8,** 857; T. G. SPIRO, V. A. MARONI and C. O. QUICKSALL, *ibid.*, 2524.
215 A. SCHMIDPETER and K. STOLL, *Angew. Chem. Internat. Edn.*, 1967, **6,** 252; 1968, **7,** 549.
216 W. KUCHEN and H. HERTEL, *ibid.*, 1969, **8,** 89.
217 H. TERCHMANN, *ibid.*, 1965, **4,** 785.
218 F. HUBER and F.-J. PADBERG, *Z. anorg. Chem.*, 1967, **351,** 1.
219 P. A. YEATS, B. F. E. FORD, J. R. SAMS and F. AUBKE, *Chem. Comm.*, 1969, 791.
220 M. SCHMEISSER and H. MÜLLER, *Angew. Chem.*, 1957, **69,** 781.
221 W. J. LILE and R. C. MENZIES, *J. Chem. Soc.*, 1950, 617.
222 R. C. EVANS, 'Crystal Chemistry', Cambridge University Press, 1952, p. 193.
223 J. GOUBEAU and H. GROSSE-RUYKEN, *Z. anorg. Chem.*, 1951, **264,** 230.
224 V. GUTMANN, P. HULMAYER and K. UTVARY, *Monatsh.*, 1961, **92,** 942.

225 Y. ETIENNE and co-workers, *Bull. Soc. chim. France*, 1953, **20,** 791; *Compt. rend.*, 1952, **234,** 1985; 1952, **235,** 966.
226 M. BLIX and W. WIRBLAUER, *Ber.*, 1903, **36,** 4218, 4220.
227 D. J. PANKHURST, C. J. WILKINS and P. W. CRAIGHEAD, *J. Chem. Soc.*, 1955, 3395.
228 M. MILLARD, K. S. STEELE and L. J. PAZDERNIK, *J. Organometallic Chem.*, 1968, **13,** P7.
229 W. T. REICHLE, *J. Org. Chem.*, 1961, **26,** 4634.
230 A. POLIS, *Ber.*, 1887, **20,** 3331.
231 W. T. REICHLE, *Inorg. Chem.*, 1962, **1,** 650.
232 M. SCHMIDT, H. J. DERSIN and H. SCHUMANN, *Chem. Ber.*, 1962, **95,** 1428.
233 R. K. INGHAM, S. D. ROSENBERG and H. GILMAN, *Chem. Rev.*, 1960, **60,** 459.
234 J. A. FORSTNER and E. L. MUETTERTIES, *Inorg. Chem.*, 1966, **5,** 552.
235 K. MOEDRITZER, *ibid.*, 1967, **6,** 1248.
236 M. KOMURA and R. OKAWARA, *J. Inorg. Nuclear Chem.*, 1966, **2,** 93.
237 J. C. J. BART and J. J. DALY, *Chem. Comm.*, 1968, 1207; C. DÖRFELT, E. F. PAULUS and H. SCHERER, *Angew. Chem. Internat. Edn.*, 1969, **8,** 288.
238 U. WANNAGAT and O. BRANDSTÄTTER, *Monatsh.*, 1963, **94,** 1090.
239 K. A. ANDRIANOV, I. HAIDUC, L. M. KHANANASHVILI and N. NEKHEEVA, *Zhur. obshchei Khim.*, 1962, **32,** 3447.
240 H. F. REIFF, B. R. LA LIBERTE, W. E. DAVIDSOHN and M. C. HENRY, *J. Organometallic Chem.*, 1968, **15,** 247.
241 E. W. ABEL, D. J. WALKER and J. N. WINGFIELD, *J. Chem. Soc.* (*A*), 1968, 2642.
242 E. W. ABEL and D. J. WALKER, *J. Chem. Soc.* (*A*), 1968, 2338.
243 A. G. DAVIES and P. G. HARRISON, *J. Organometallic Chem.*, 1967, **8,** P19.
244 S. MIDGAL, D. GERTNER and A. ZILKHA, *Canad. J. Chem.*, 1967, **45,** 2987.
245 R. C. POLLER, *Proc. Chem. Soc.*, 1963, 312.
246 R. C. POLLER and J. A. SPILLMAN, *J. Chem. Soc.* (*A*), 1966, 959.
247 M. SHINDO and R. OKAWARA, *Inorg. Nuclear Chem. Letters*, 1967, **3,** 75; Idem and Y. MATSUMURA, *J. Organometallic Chem.*, 1968, **11,** 299.
248 H. SCHUMANN, *Z. anorg. Chem.*, 1967, **354,** 192.
249 K. A. ANDRIANOV and V. V. SEVERNY, *J. Organometallic Chem.*, 1964, **1,** 268.
250 D. L. ALLESTON, A. G. DAVIES, M. HANCOCK and R. F. M. WHITE, *J. Chem. Soc.*, 1963, 5469.
251 A. G. DAVIES and P. G. HARRISON, *J. Organometallic Chem.*, 1967, **8,** P19; S. MIDZAL, D. GERTNER and A. ZILKHA, *Canad. J. Chem.*, 1967, **45,** 2987.
252 K. MOEDRITZER and J. R. VAN WAZER, *Inorg. Nuclear Chem. Letters*, 1966, **2,** 45; *Inorg. Chim. Acta*, 1967, **1,** 152; *J. Amer. Chem. Soc.*, 1968, **90,** 1708.
253 K. MOEDRITZER, *J. Organometallic Chem.*, 1966, **5,** 254.
254 E. O. SCHLEMPER and W. C. HAMILTON, *Inorg. Chem.*, 1966, **5,** 995.
255 R. HOPPE and W. DÄHNE, *Naturwiss.*, 1962, **49,** 254.
256 G. E. COATES, M. L. H. GREEN and K. WADE, 'Organometallic Compounds', Methuen and Co. Ltd., London, 3rd Edn., Vol. I, 1967, 494.
257 A. G. DAVIES, personal communication.
258 I. P. GOL'DSHTEIN, N. N. ZEINLYANSKII, O. P. SHAMAGINA, E. N. GUI'YANOVA, E. M. PANOV, N. A. SLOVOKHOTOVA and K. A. KOCHESCHKOV, *Doklady Chem.*, 1965, **163,** 715.

APPENDIX

Structure determinations on the alkaline earth disilicides MSi_2 show that the Si—Si bond lengths are similar, but the anions have fundamentally different structures for the 3 metals considered. $CaSi_2$ contains a 2-dimensional sheet, strontium a 3D array and barium, Si_4^{4-} anions.[1]

$IrSi_3$ contains planar sheets of silicon atoms, each with 4 nearest neighbours at 4 of the 6 apices of a hexagon (1 *trans* pair missing). The Si—Si bond distance is significantly shorter than that of a single bond.[2]

The compound once formulated as $LiSn_2$ has now been shown to be Li_2Sn_5. The anions containing pentagonal prisms involving ten tin atoms with Sn—Sn bonds of about 0.3 nm.[3]

Cyclopolysilanes and germanes

Bi-iso-butyl dichlorosilane and the anion of biphenyl yield deca-iso-butylcyclopentasilane.[4] Unlike $(Me_2Si)_5$, it is air stable and can be distilled at 250°C/4 mm. The *p*-tolylcyclo-polysilanes and germanes (poly = 4, 5 or 6) can be prepared from the dichloride and sodium in naphthalene.[5] Mixed rings containing silicon and germanium can be prepared from lithium and a Me_2SiCl_2/Me_2GeCl_2 mixture.[6] The $GeSi_5$ and Ge_2Si_4 rings have been separated and characterized, while other 6 membered, 5 and 7 membered have been characterized by v.p.c.

Permethylcyclopolysilanes produce good yields of dimethylchlorosilanes with HCl under ultraviolet light. Dimethylsilylene could well be present as an intermediate— indeed in the presence of ultraviolet light and diethylmethylsilane, dodecamethylcyclopentasilane yields the smaller cyclic homologues along with $Et_2MeSi(SiMe_2)_nH$ (n = 1, 2 and 4).[7] One phenyl group per silicon is cleaved from decaphenylcyclopentasilane with HI. The product $(PhISi)_5$ is used as a synthetic intermediate.[8]

The 'barralene' compound $MeSi(Si_2Me_4)_3SiMe$ results from Na–K alloy with Me_2SiCl_2 and $MeSiCl_3$.[9]

Cyclodisilazanes

Perchlorodisilyl cyclodisilazane has been prepared from $(Cl_3Si)_2NLi$, while dimethyldichlorosilane and hexamethylcyclotrisilazane gave mainly *N,N'*-bis(dimethylchlorosilyl)tetramethylcyclodisilazane $(ClMe_2SiNSiMe_2)_2$.[10] The homologue from phenylmethyldichlorosilane can be prepared similarly. The spiro compound I results from $(Me_3Si)_2NNa$ and $SiCl_4$.[11]

$SiMe_3$ $SiMe_3$ | N N | Me_2Si Si $SiMe_2$ | C N | Me_3Si H $SiMe_3$

I

F_3 Si | $Me_3P{=}N$ $N{=}PMe_3$ | Si F_3

m.p. 73–4°C.

II

While $Me_3P{=}N{-}SiMe_{3-n}F_n$ (n = 1, 2) are monomeric, $Me_3P{=}N{-}SiF_3$, is dimeric II, with 5-co-ordinate silicon.[12] Bis(trimethylphosphinimino) dimethylsilane and Me_2SiX_2 give compounds containing a similar cationic ring $(R_3PNSiMe_2)_2^{2+}$. With excess R_3M(M=Al, In), ionic adducts result incorporating the R_2M group into a 4-membered ring.

Other Si—N rings

A boat configuration is adopted for *N*,*N*′,*N*″-tris(trimethylsilyl) hexamethyl-cyclotrisilazane to reduce steric hindrance, induced by the valency sites at nitrogen occupying a planar configuration.[13] Dodecamethylcyclotetrasilazane has now been prepared by various stepwise routes involving substituted silazanes,[14] and also directly from dimethyldichlorosilane and methylamine. Linear chlorosilazanes can be readily synthesized from *N*,*N*′,*N*″-trimethylhexamethylcyclotrisilazane and dimethyldichlorosilane.

Unsymmetrical cyclotetrasilazanes and cycloheteropolysilazanes result from $Me_2Si(NMeSiMe_2NRLi)_2$ and silicon, germanium, boron and phosphorus dichlorides.[15] Chlorosilazanes and amides have also been used.[16] The mass spectra of cyclic silylhydrazines are also reported.[17]

The wide angle of 126° is recorded for the GeNGe linkage in $(Cl_2GeNMe)_3$, supporting multiple bonding,[18] though the Ge—N bond is not appreciably shorter than expected for a single bond.

Linear and cyclic tin–nitrogen compounds react with tin halides to give mixed amino–tin halides.[19] These are exceedingly moisture sensitive, and are difficult to characterize directly.

Further work has been carried out on the base-induced ring contractions of cyclosiloxazanes[20] with 2 oxygen and 2 nitrogen atoms as ring members.

Diphenyldichlorosilane and germane react with potassium derivatives of phenylphosphine to yield the 4 and 6 membered cyclosila- and germanaphosphanes.[21] Organotindichlorides and phenylphosphine give cyclic trimers.

Group IV—Oxygen compounds

The important features of silicon-oxygen chemistry have come to light in the past year. Heating octamethylcyclotetrasiloxane gives the cyclic trimer and pentamer.[22] Reaction rate determinations show that monomeric dimethyl silicone is formed as intermediate. Silylation of mineral silicates using hexamethyldisiloxane, HCl and water (Lent technique) gives volatile products readily separated and distinguished by g.l.c.[23]

Difluorosilylene readily adds to thionyl fluoride. Subsequent decomposition gives the fluorinated cyclodi- and trisiloxanes.[24] Cyclotetrasiloxane $(H_2SiO)_4$ can be prepared from copper oxide and di-iodosilane.[25] A significant feature of the structure is the wide Si—O—Si angle of 149°. Recent studies with $(PhSiO_{3/2})_n$ show it to be a random array of polycyclic cages[26] which give mainly low oligomers T_{8-14} at high dilution and not a long ladder as was first proposed (see p. 218).

The Mössbauer spectra of $SnOX_2$ supports a chain structure[27] rather than a cyclic one. The infrared and mass spectral data for $(Bu^n_3Sn)_2SO_4$ add weight to the proposals put forward based on the Mössbauer spectrum, that the compound is monomeric,[28] with a quadridentate sulphate group and 5-co-ordinate tin.

The crystal structure of dimethyltin bis(fluorosulphate) $Me_2Sn(SO_3F)_2$ shows it to be a 2D sheet polymer with tin 6-co-ordinate and a component (with 3 other tin atoms) of a 16 membered ring.[29] Dimeric $(Me_2Sn(NCS))_2O$ has the ladder structure already described.[30] The bridging Sn—O bonds appear stronger than those in the monomer chain. Similar products result when isomeric dibutyltin dichlorides are hydrolyzed. Most derivatives formulated as $XBu^n_2SnOSnBu^n_2X$ are dimeric and thought to possess the ladder structure, even when X = halogen. (See p. 233).

The products formed from ethylene carbonate and organoditin oxides have been used to prepare organic orthocarbonates and thiocarbonates.[31] The vibrational spectra of $Pb_4(OH)_4^{4+}$ is interpreted.[32]

The adducts formed from organotin halides and tin–sulphur rings are oils or low melting waxes.[33] Di-aryl trisulphides

$$R_{4-n}SnX_n + (R'_2SnS)_3 \longrightarrow X_{n-1}R_{4-n}Sn(SSnR'_2)X \qquad (n = 1 \text{ or } 2).$$

can also be conveniently prepared from cyclic tin sulphides and sulphenyl chlorides.[34]

Germanium disulphide and S^{2-} yield the ion $Ge_2S_6^{4-}$. This possesses a 4-membered ring structure with a small angle (86.1°) for the GeSGe linkage.[35] The metathiostannate ion is thought to have the same structure and not to be SnS_3^{2-}.

The crystal structure of dimethyltin dichloride shows no evidence of strong chlorine bridging[36] (see p. 255).

References

1 K. H. JANZON, H. SCHÄFER and A. WEISS, *Z. anorg. Chem.*, 1970, **372,** 87.
2 J. G. WHITE and E. F. HOCKINGS, *Inorg. Chem.*, 1971, **10,** 1934.
3 D. A. HANSEN and L. J. CHANG, *Acta Cryst.*, 1969, **25B,** 2392.
4 G. R. HUSK, R. WEXLER and B. M. KILCULLEN, *J. Organometallic Chem.*, 1971, **29,** C49.
5 M. RICHTER and W. P. NEUMANN, *ibid.*, 1969, **20,** 81.
6 E. CARBERRY and B. D. BOMBEK, *ibid.*, 1970, **22,** C43.
7 M. ISHIKAWA and M. KUMADER, *Chem. Comm.*, 1971, 507; 1970, 612.
8 E. HENGGE and H. MARKETZ, *Monatsch*, 1970, **101,** 528.
9 R. WEST and A. INDRIKSONS, *J. Amer. Chem. Soc.*, 1970, **92,** 6704.
10 U. WANNAGAT and co-workers, *Z. anorg. Chem.*, 1970, **375,** 157 and 1971, **381,** 288. L. W. BREED, R. L. ELLIOTT and J. C. WILEY JR., *J. Organometallic Chem.*, 1970, **24,** 315 and 1971, **31,** 179.
11 U. WANNAGAT, J. HERZIG and H. BÜRGER, *ibid.*, 1970, **23,** 373.
12 W. WOLFSBERGER, H. H. PICKEL and H. SCHMIDBAUR, *Chem. Ber.*, 1971, **104,** 1830; *ibid.*, *J. Organometallic Chem.*, 1971, **27,** 181 and **28,** 307; W. WOLFSBERGER and H. SCHMIDBAUR, *ibid.*, 301.
13 G. W. ADAMSON and J. J. DALY, *J. Chem. Soc.* (A), 1970, 2724.

14 U. WANNAGAT, R. BRAUN, L. GERSCHLER and H.-J. WISWAR, *J. Organometallic Chem.*, 1971, **26,** 321.
15 U. WANNAGAT and L. GERSCHLER, *Z. anorg. Chem.*, 1971, **383,** 249, *Annalen*, 1971, **744,** 111 and *Inorg. and Nuclear Chem. Letters*, 1971, **7,** 285.
16 I. GEISLER and H. NÖTH, *Chem. Ber.*, 1970, **103,** 2234.
17 K. G. DAS, P. S. KULKARNI, V. KALYAMARAMAN and M. V. GEORGE, *J. Organometallic Chem.*, 1970, **35.** 2140.
18 M. ZIEGLER and J. WEISS, *Z. Nat.*, 1971, **26b,** 735.
19 A. G. DAVIES and J. D. KENNEDY, *J. Chem. Soc.* (C), 1970, 759.
20 R. P. BUSH, N. C. LLOYD and C. A. PEARCE, *J. Chem. Soc.* (A), 1970, 1587.
21 H. SCHUMANN and H. BENDA, *Chem. Ber.*, 1971, **104,** 333.
22 I. M. T. DAVIDSON and J. F. THOMPSON, *Chem. Comm.*, 1971, 251.
23 J. GÖTZ and C. R. MASSON, *J. Chem. Soc.*, (A), 1970, 2683.
24 K. G. SHARP and J. L. MARGRAVE, *J. Inorg. Nuclear Chem.*, 1971, **33,** 2813.
25 C. GLIDEWELL, A. G. ROBRIETTE and G. M. SHELDRICK, *Chem. Comm.*, 1970, 931.
26 C. L. FRYE and J. M. KLOSOWSKI, *J. Amer. Chem. Soc.*, 1971, **93,** 4599.
27 H.-S. CHENG and R. H. HERBER, *Inorg. Chem.*, 1971, **10,** 1315.
28 R. H. HERBER and C. H. STAPFER, *Inorg. Nuclear Chem. Letters*, 1971, **7,** 617.
29 F. A. ALLAN, J. A. LERBSCHER and J. TROTTER, *J. Chem. Soc.*, (A) 1971, 2507.
30 Y. M. CHOW, *Inorg. Chem.*, 1971, **10,** 673, C. K. CHU and J. D. MURRAY, *J. Chem. Soc.* (A), 1971, 360. A. G. DAVIES, L. SMITH, P. J. SMITH and W. MCFARLANE, *J. Organometallic Chem.*, 1971, **29,** 245.
31 S. SAKAI, Y. FUJIMURA and Y. ISHII, *J. Organic Chem.*, 1970, **35,** 2344.
32 P. A. BULLMER and T. G. SPIRO, *Spec. Acta*, 1970, **26A,** 1641.
33 A. G. DAVIES and P. G. HARRISON, *J. Chem. Soc.* (C), 1970, 2035.
34 J. L. WARDELL and P. L. CLARK, *J. Organometallic Chem.*, 1971, **26,** 345.
35 B. KREBS, S. POHL and W. SCHIWY, *Angew. Chem. Internat. Edn.*, 1970, **9,** 897.
36 A. G. DAVIES, H. J. MILLEDGE, D. C. PUXLEY and P. J. SMITH, *J. Chem. Soc.* (A), 1970, 2862.

5

The Group V Elements

While there are many superficial similarities between the Group IV elements and those considered in this Chapter, the ability to form compounds in two oxidation states increases the range of cyclic compounds considered here. This point is readily illustrated with P_4O_6 and P_4O_{10}, and while the subject of cyclic mineral silicates is most extensive, that of cyclic phosphates is sparse. Indeed, research effort in Group V chemistry concerned with the synthesis of inorganic polymers has not been directed towards organo-oxides, as with silicon, but to P^V—N compounds, notably organo- and chloro-phosphazenes $[R_2P{=}N]_x$.

THE STRUCTURE OF THE ELEMENTS

As with the elements of Group IV, the allotropy of the post-nitrogen elements illustrate well the tendency to form cyclic structures. The polymorphic nature of phosphorus provides the more interesting structural contrasts, and comparison will be made with the heavier elements of the Group.

Both phosphorus and arsenic[1] occur as discrete tetra-atomic molecules in the vapour phase, and while the arsenic condensate is unstable at room temperature and readily forms a metallic polymorph, the phosphorus condensate, white phosphorus, is metastable. This contains discrete P_4 tetrahedra with a P—P bond distance of 0.221 nm, which corresponds to a single bond, as does the As—As bond in vaporized arsenic (0.244 nm).

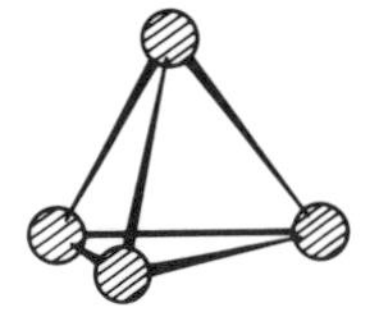

Fig. 5.1. The P_4 tetrahedron

The highly reactive nature of this tetra-atomic allotrope may result from the considerable strain imposed by bond angles of 60 degrees.

White phosphorus slowly changes into the red allotrope at room temperature, but this change can be accelerated by heat, or an iodine catalyst. Red phosphorus is polymeric, and amorphous to X-rays, and so is thought to contain randomly cross-linked P_4 tetrahedra.[2]

```
                  P
                  |
        P         P
       /|\       / \
    —P  |  P—P     P—P
       \|/       \ /
        P         P
                  |
                  P
```

Fig. 5.2. The P_4 units of amorphous red phosphorus

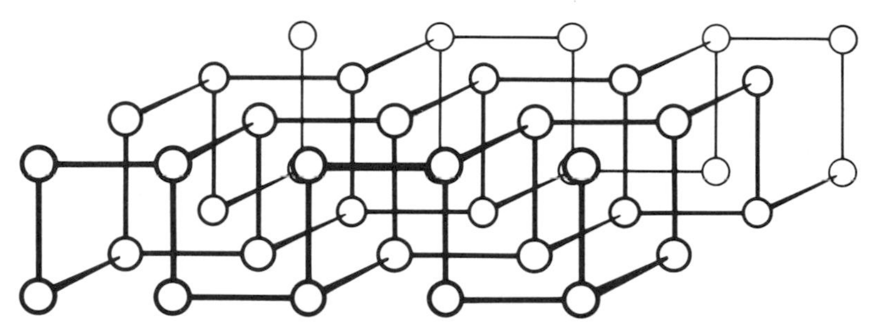

Fig. 5.3. The crystal structure of black phosphorus—portion of one layer[h]

Heating white phosphorus to 350°C for eight days with a Hg/Cu catalyst gives the black allotrope when seeded.[3] This is the most stable form of the element and can only be oxidized with difficulty. Indeed, it is best recrystallized from bismuth and nitric acid! Black phosphorus has a flaky character resembling graphite and a structure consistent with this. The individual sheets of this laminated structure comprise planar double layers of phosphorus atoms, in which three atoms of the upper layer and three of the lower one comprise the chair configuration of a 6-membered P_6 ring. It can also be considered as zig-zag chains of phosphorus atoms cross-linked through every alternate one. The P—P bonds are rather larger than in white phosphorus, and the weak forces between pairs of these double layers give the allotrope its flaky properties.[4] This structure closely resembles those of the sulphides and selenides of bivalent germanium and tin.[5]

Heating amorphous arsenic to 150°C in the presence of mercury yields a polymorph isostructural with black phosphorus, while antimony and bismuth occur as metallic, laminated forms of similar structure. With these two elements, however, the number of nearest neighbours is three and not seven, as with black phosphorus. These laminated forms of antimony and bismuth derive from the phosphorus structure by expansion along an axis perpendicular to the zig-zag and in the plane of the layer.[6]

A further structural variation occurs with Hittorf's violet phosphorus, which results when phosphorus glasses are annealed. This comprises tubes with a pentagonal cross-section built up of P_8 and P_9 units. The former is isostructural with realgar (As_4S_4), and these tubes are arranged in perpendicular sheets cross-linked through the P_9 residue.[7]

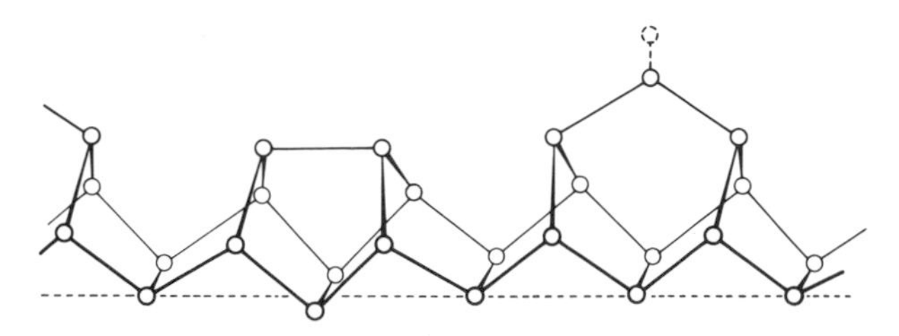

Fig. 5.4. Section of a "tube" of violet phosphorus[7]

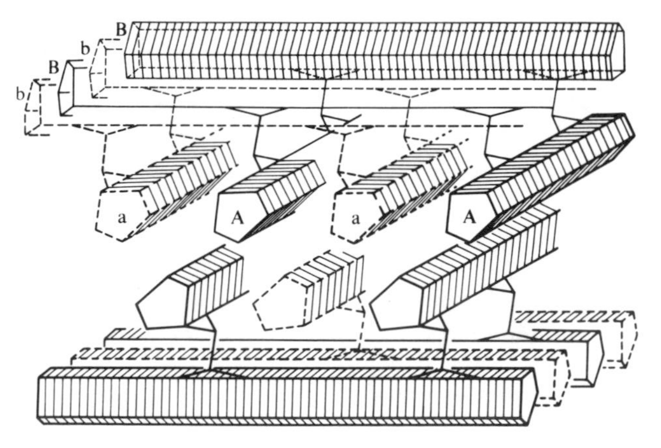

Fig. 5.5. Tubes of violet phosphorus, showing cross-linking[7]

This tubular array of phosphorus atoms is also encountered in potassium polyphosphide KP_{15}.[8] This forms as ruby needles when the elements are cooled from 650°C to 300°C, and are stable to air and moisture. This structure also occurs in $HgPbP_{14}$,[9] with the stippled P atom replaced by Pb and Hg bridging phosphorus atoms as shown in Figure 5.6.

There is little tendency among the elements of this Group to form cations comprising just the element, but it does occur. Bismuth monochloride was thought to result from the reduction of the trichloride with the metal.

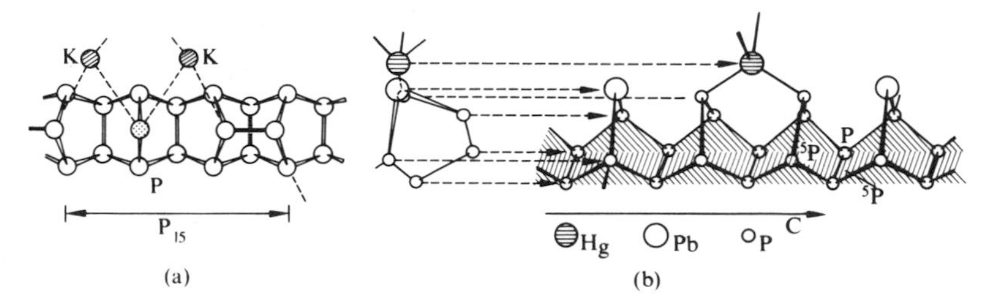

Fig. 5.6. The structure of (a) KP_{15} and (b) $MPbP_{14}$ (M = Zn, Cd, Hg)[8,9]

This "monochloride" is now formulated as $Bi_{24}Cl_{28}$ and comprises a mixture of ions,[10]

$$Bi_{24}Cl_{28} \equiv 2BiCl_5^{2-} + 2Cl^- + 2Bi_2Cl_8^{2-} + 2Bi_9^{5+}$$

of which Bi_9^{5+} is a face-bonded triangular prism cage of bismuth atoms resembling $B_9H_9^{2-}$ and isoelectronic with Pb_9^{4-} mentioned in the last Chapter (p. 155). In addition, the ions Bi_5^{3+} and Bi_8^{2+} have been isolated and possess the D_{3h} trigonal-bipyramidal structure of $B_3C_2H_5$ and the D_{2d} structure of $B_8H_8^{2-}$.[11,12]

CYCLOPENTAZOLES

The stability of the triple bond in N_2 is one of the factors rendering highly-catenated nitrogen compounds thermally unstable. Thus, tetramethyltetrazine explodes at its boiling point (130°C), releasing N_2 and dimethylamino radicals.[13]

$$Me_2N{-}N{=}N{-}NMe_2 \xrightarrow{130°C} 2Me_2N\cdot + N_2$$

Likewise, arylcyclopentazoles, with nitrogen as the sole ring member, are thermally unstable. However, they can be synthesized at low temperatures from aryldiazonium salts and the azide ion, and thermal stability increases with the electron donating power of the *para* substituent.

$$ArN_2^+ + N_3^- \longrightarrow ArN_5 \text{ (cyclic: ArN}\langle N{=}N{-}N{=}N\rangle)$$

Thus, the phenyl compound decomposes below 0° while *p*-dimethylaminophenylcyclopentazole is stable to 50°C.[14]

Evidence for the symmetrical nature of the ring stems from ^{15}N tracer experiments. Using $Na^{15}NO_2$ to prepare the aryldiazonium salt gives a pentazole which decomposes to nitrogen and the arylazide, both of which contain ^{15}N in half the molecules.

$$2p\text{-EtOC}_6H_4\overset{+}{N}\equiv{}^{15}N \xrightarrow{2N_3^-}$$

$$2p\text{-EtOC}_6H_4N\text{(cyclo-}{}^{15}N{=}N{-}N{=}N\text{)} \longrightarrow ArN^{15}NN + Ar^{14}N_3 + {}^{14}N_2 + {}^{15}NN$$

CYCLOPOLY-PHOSPHINES AND -ARSINES

Synthesis

These compounds have aroused considerable interest of late through their structures and their peculiar reactions. They are conveniently formulated as $[RM]_n$, where M is a phosphorus or arsenic atom and $n = 3, 4, 5$ or 6, and as well as being called cyclopolyphosphines degenerate names such as phosphoromethane $[MeP]_n$ and arsenobenzene $[PhAs]_n$ are often used.[a]

A convenient synthetic route involves the reduction of an organo-phosphorus or -arsenic dihalide with a metal. Bromides and sodium provide the most reactive combinations,[15] with lithium or magnesium giving a more controlled reduction,[16] *e.g.*,

$$4Bu^tPCl_2 + 8Na \longrightarrow [Bu^tP]_4 + 8NaCl$$

$$6p\text{-tolylAsCl}_2 + 12Na \longrightarrow [p\text{-tolylAs}]_6 + 12NaCl$$

$$RPCl_2 \xrightarrow{\text{Li or Mg}} [RP]_n \text{ (R = Et, Pr, Bu, Ph)}$$

Mercury appears to be the most convenient metal used to reduce perfluoro-organophosphorus and arsenic dihalides. Trifluoromethyl-phosphorus[17] and -arsenic di-iodide[18] give cyclic tetramers and pentamers on standing for a few days with mercury.

$$CF_3MI_2 + Hg \longrightarrow \frac{1}{n}[CF_3M]_n + HgI_2$$

Pentafluorophenylarsenic dichloride[19] gives the cyclic tetramer, while the cyclopentaphosphine[20] results from $C_6F_5PX_2$ (X = Br or I). Pentafluoroethylphosphorus di-iodide gives the cyclotetraphosphine $[C_2F_5P]_4$ along with the first reported neutral cyclotriphosphine $[C_2F_5P]_3$, (m.p. 23.5°C),[21] though this is disputed.[21a]

$$C_6F_5MX_2 + 2Hg \longrightarrow \frac{1}{n}[C_6F_5M]_n + Hg_2X_2$$

(M = P, $n = 5$, m.p. 156–161°C;
M = As, $n = 4$, m.p. 141°C)

As a point of historical interest, the first cyclopolyphosphine was prepared in 1877 from phenylphosphine and phenylphosphorus dichloride.[22] The product was thought to be the phosphorus analogue of azobenzene but has now been shown to be a cyclic oligomer. The formation of rings of various sizes has already been illustrated with the perfluoro compounds, and applies equally to phosphorobenzene.

This material has been isolated as the four compounds A, m.p. 150°C; B, m.p. 190°C; C, m.p. 252–6°C; and D, m.p. 262–85°C. A is a pentamer in the solid state and B is a hexamer occurring as four crystalline modifications. Both are tetrameric in solution, however, illustrating both the difficulties encountered in characterization and the lability of the P—P bond.[23]

Similar complications arise with the analogous reaction with *p*-chlorophenylphosphine and the dichloride.

$$p\text{-ClC}_6\text{H}_4\text{PH}_2 + p\text{-ClC}_6\text{H}_4\text{PCl}_2 \longrightarrow \frac{2}{x}[p\text{-ClC}_6\text{H}_4\text{P}]_x + 2\text{HCl}$$

A pentamer results from ether, a hexamer from benzene and a tetramer when no solvent is used. With $C_6F_5PCl_2$ and $C_6F_5PH_2$, only the tetramer has been reported.[24]

Lithium hydride effectively reduces methyl- and ethyl-phosphorus dichlorides to the cyclopolyphosphines,[25] while coupling diphenylarsine with phenylarsenic dichloride gives hexaphenylcyclohexa-arsine. This may well result through hydride–halogen exchange.[b]

$$\text{PhAsCl}_2 + \text{Ph}_2\text{AsH} \longrightarrow \text{Ph}_2\text{AsCl} + \tfrac{1}{6}[\text{PhAs}]_6 + \text{HCl}$$

However, arsenobenzene and arsenomethane are usually prepared by reducing the corresponding arsenic acid with hypophosphorous acid.[26]

$$\text{RAsO(OH)}_2 \longrightarrow [\text{RAs}]_n \ (\text{R} = \text{Ph}, n = 6; \text{R} = \text{Me}, n = 5)$$

The P^V acids are difficult to reduce, so P^{III} compounds are the normal precursors to cyclopolyphosphines. Tri-n-butylphosphine is a convenient reducing agent.[27]

$$\text{RPCl}_2 + \text{Bu}^n_3\text{P} \longrightarrow \frac{1}{n}[\text{RP}]_n + \text{Bu}^n_3\text{PCl}_2$$

Trialkylstibines will reduce both chlorophosphines and arsines.

$$\text{MePCl}_2 + \text{Et}_3\text{Sb} \longrightarrow \tfrac{1}{5}[\text{MeP}]_5 + \text{Et}_3\text{SbCl}_2$$

$$\text{PhAsCl}_2 + \text{Bu}^n_3\text{Sb} \longrightarrow \tfrac{1}{6}[\text{PhAs}]_6 + \text{Bu}^n_3\text{SbCl}_2$$

Organo-substituted difluorophosphines readily disproportionate on heating to RPF_4 and the cyclopolyphosphine.

$$2n\text{RPF}_2 \longrightarrow [\text{RP}]_n + n\text{RPF}_4$$
$$(\text{R} = \text{Me or Ph}, n = 5; \text{R} = \text{CF}_3, n = 4 \text{ and } 5)$$

The rate seems to decrease with the electronegativity of the substituent, CF_3PF_2 being stable for at least six months at room temperature while $MePF_2$ disproportionates steadily.[28] This again illustrates the instability of P^{III} relative to P^{I} and P^{V}.

Ring contraction of pentakis-(trifluoromethyl)cyclopentaphosphine readily occurs on heating to give the cyclic tetramer,[17] and indicates the possibility of forming the electron-deficient phosphinidene intermediate CF_3P. There is much chemical evidence to support the presence of these compounds, and this will be considered among the reactions of these rings.

Reactions

Generation of Phosphinidenes

The transient existence of these, the phosphorus analogues of carbenes, is now receiving widespread support both through physical data and the isolation of phosphinidene adducts.

Heating an equimolar mixture of methyl- and ethyl-substituted cyclopentaphosphines gives the mixed compounds, all of which have been detected by mass spectrometry.[29]

$$[MeP]_5 + [EtP]_5 \longrightarrow Me_nEt_{5-n}P_5 \ (n = 0\text{–}5)$$

Unsaturated hydrocarbons have been shown to react with carbenes, through direct addition to the multiple bond, to give bicyclic products. Phosphinidenes can be trapped in a similar manner. Thus, methyl- and phenyl-cyclopolyphosphines readily give phospholenes and diphosphorines when heated with substituted butadienes.

$$[R'P]_5 + \text{R-substituted butadiene} \xrightarrow{150\text{–}180°C} \text{phospholene} + \text{diphosphorine}$$

(R′ = Me, Ph)

phospholene diphosphorine

Only the diphosphorines results with $[EtP]_5$, and when all three cyclopolyphosphines are in turn irradiated with the diene. A similar insertion of two MeAs residues occurs with 2,3-dimethyl-buta-1,3-diene.[30]

Again, the isolation of di-*S*-ethylphenylphosphonodithioite, $(EtS)_2PPh$, when phenylcyclopolyphosphines are heated with diethyldisulphide supports a phosphinidene insertion into the S—S bond.[31]

$$\frac{1}{n}[PhP]_n + Et_2S_2 \xrightarrow{160°C} (EtS)_2PPh$$

Trifluoromethylphosphinidene, CF_3P, is believed to be formed under much milder conditions, possibly due to the strongly electron-withdrawing nature of the trifluoromethyl group.[32] Thus, tetrakis-(trifluoromethyl)-cyclotetraphosphine, $[CF_3P]_4$, reacts with trimethylphosphine at low temperatures.

$$[CF_3P]_4 + 4Me_3P \longrightarrow \underset{\textbf{(5-I)}}{4Me_3P{\rightarrow}PCF_3}$$

The reaction is reversed at room temperature, but both the cyclotetra- and penta-phosphines result in the ratio 6:1. The instability of this adduct contrasts with the thermal stability of the oxide, sulphide, carbene and nitrene adducts of trimethylphosphine.

This has led to the low-temperature oligomerization of these cyclopolyphosphines using Me_3P as a catalyst, both in solution and without a solvent. The pentamer $[CF_3P]_5$ is 50% converted to $[CF_3P]_4$ in two days at room temperature without a solvent and with 1 mol% Me_3P, in contrast to the usual method of heating at 260°C. The lattice energy of the tetramer (m.p. 66°C) probably renders the conversion thermodynamically feasible since the pentamer is a liquid (m.p. −33°C). In solution, however, such factors do not apply, and, both in ether and hexane, the tetramer tends to give mainly $[CF_3P]_5$ on standing with Me_3P.

Analogues of **(5-I)** probably form as intermediates in the reactions of primary and secondary phosphines with $[CF_3P]_4$. Thus, methylphosphine gives $[CH_3P]_5$ quantitatively after three days at room temperature.

$$MePH_2 + \tfrac{1}{4}[CF_3P]_4 \longrightarrow [MePH_2{\rightarrow}PCF_3] \longrightarrow CF_3PH_2 + \tfrac{1}{5}[MeP]_5$$

The basicity of the phosphine appears to play an important part in determining the extent to which this reaction proceeds, phenylphosphine giving 82% CF_3PH_2 while phosphine, even in excess, fails to react.[33]

Dimethylphosphine gives the expected products along with the mixed triphosphine.

$$Me_2PH + [CF_3P]_4 \longrightarrow CF_3PH_2 + Me_4P_2 + CF_3P(PMe_2)_2$$

This probably results from phosphinidene insertion into the diphosphine,[32] as the independent reaction with $[CF_3P]_4$ indicates.

$$Me_4P_2 + \tfrac{1}{4}[CF_3P]_4 \longrightarrow CF_3P(PMe_2)_2$$

It also readily inserts into cacodyl and dimethyldisulphide.[34]

$$(MeS)_2PCF_3 \xleftarrow[135°C]{Me_2S_2} [CF_3P]_4 \xrightarrow{Me_4As_2} CF_3P(AsMe_2)_2$$

Cleavage by Alkali Metals

The P—P bonds of cyclopolyphosphines are readily cleaved, but the products depend on the metal, its proportion and the solvent employed.

In dioxan, the alkali metals readily break one P—P bond of tetraphenylcyclotetraphosphine.

$$[PhP]_4 + 2M \xrightarrow[heat]{dioxan} M[PPh]_4M \ (M = Li, Na \text{ or } K).$$

Further breakdown occurs with more alkali metal, giving $M[PPh]_2M$ and M_2PPh. These phosphides readily react with halogenoalkanes yielding cyclopolyphosphines and phosphinocycloalkanes. K/Br exchange occurs between $K[RP]_nK$ and ethylene dibromide giving the cyclotetra- and penta-phosphine.

$$K[EtP]_nK \xrightarrow{(BrCH_2)_2} [EtP]_n \ (n = 4 \text{ or } 5).$$

With 1,4-dichlorobutane, $K[EtP]_2K$ coupling occurs to give the 1,2-diphosphinocyclohexane.[35]

$$K[EtP]_2K + Cl(CH_2)_4Cl \longrightarrow$$ (six-membered ring with P—P, each P bearing Et)

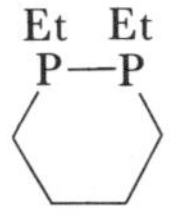

A peculiar product $K_2[PPh]_3{\cdot}2THF$ results from $[PPh]_4$ and potassium in tetrahydrofuran. The ^{31}P n.m.r. spectrum of these red crystals show equivalent phosphorus atoms, ruling out a linear anion and supporting a cyclic structure resembling the triphenylcyclopropenylium cation.

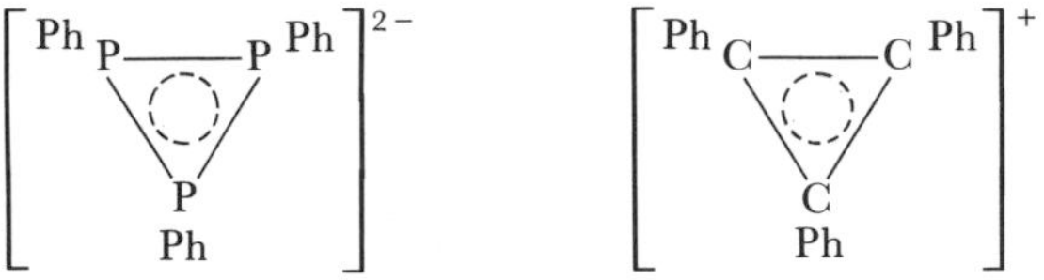

$[(PhP)_3]^{2-}$ and $[(PhC)_3]^{+}$ (three-membered rings with delocalized π-orbital)

Both ions are believed to possess a ring delocalized π-orbital which can function as an electron donor to transition metals.[36]

$K_2[EtP]_3$ results from ethylcyclopolyphosphines, but the AB_2 pattern of the ^{31}P n.m.r. spectrum supports a linear structure.

The alkali metal polyphosphides react readily with $MgBr_2$ to give cyclic magnesium derivatives.

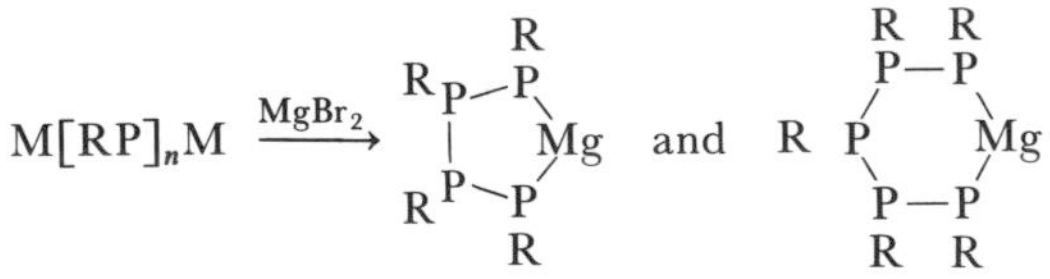

$$M[RP]_nM \xrightarrow{MgBr_2} [RP]_4Mg \text{ (five-membered ring) and } [RP]_5Mg \text{ (six-membered ring)}$$

(R = Me, Et; M = Li, Na, K; n = 4 or 5)

With dipyridyl, the cyclopolyphosphine is again formed along with a low valency magnesium–dipyridyl complex.[37]

$$[RP]_5Mg + \text{dipyridyl} \xrightarrow{THF} [RP]_5 + Mg(\text{dipy})_2 3THF$$

Halogenation

Controlled bromination of pentaphenylcyclopentaphosphine produces sym-diphenyldibromodiphosphine. This also results using phenylphosphorus dibromide as the brominating agent, and the low melting point range (118–120°C) supports the presence of only one conformer.[38]

$$2[PhP]_5 + 5Br_2 \longrightarrow 5Ph_2P_2Br_2 \longleftarrow 5PhPBr_2 + [PhP]_5$$

Iodine readily oxidizes the trifluoromethylcyclopolyphosphines $[CF_3P]_{4\text{ and }5}$ to CF_3PI_2,[17] while chlorine gives the pentavalent phosphorus halide CF_3PCl_4. This can be readily reduced by mercury to CF_3PCl_2, but no further.

Reactions with Sulphur and Selenium

All the P—P bonds of $[PhP]_5$ are readily broken by sulphur. The product $[PhPS]_3$ probably contains a 6-membered ring of alternate P and S atoms.[39] Selenium gives a cyclic tetramer $[PhPSe]_4$ probably existing as an 8-membered ring.[40]

While the P—P bonds in cyclopolyphosphines are thought to be single bonds, there is some support for weak delocalized π-bonding around the ring incorporating d orbitals. Neither oxygen nor an imino group can be inserted into tetrakis-(trifluoromethyl)cyclotetraphosphine but sulphur can. Thus, the monosulphide forms on heating and is itself quantitatively oxidized with excess sulphur at 200°C to $(CF_3PS)_3S_2$. The polysulphide can be reduced with mercury to the monosulphide and this, with trimethylphosphine, to the cyclotetraphosphine.[41]

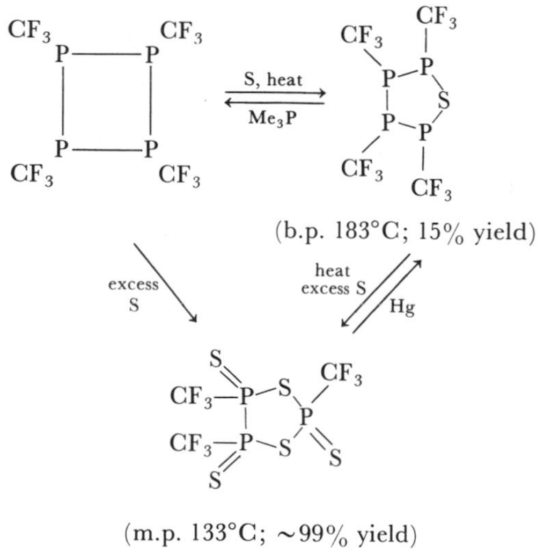

(b.p. 183°C; 15% yield)

(m.p. 133°C; ~99% yield)

The more extensive oxidation with trifluoromethyl cyclopolyphosphines compared with the aryl ones may be due to the strong electron-withdrawing nature of perfluoroalkyl substituents.

Addition to Alkynes

The addition of one or two phosphinidene residues to conjugated dienes has already been noted (p. 272). Two or three trifluoromethylphosphinidene groups readily add to hexafluorobut-2-yne, using $[CF_3P]_4$ and $[CF_3P]_5$.

CF_3—C=C—CF_3 ring with P(CF_3)—P(CF_3) and CF_3C=CCF_3 ring with P(CF_3)—P(CF_3)—P(CF_3)

(5-II) and **(5-III)**

At 170°C, iodine readily oxidizes pentakis-(trifluoromethyl)-1,2,3-triphosphocyclopent-4-ene **(5-III)** to CF_3PI_2 and tetrakis-(trifluoromethyl)-1,2-diphosphocyclobut-3-ene **(5-II)**.[42]

$$\textbf{(5-III)} + I_2 \longrightarrow \textbf{(5-II)} + CF_3PI_2$$

Trifluoromethylcyclopolyphosphines and arsines are readily reduced by mercury in the presence of hydriodic acid.[43]

$$\frac{1}{n}[CF_3M]_n \xrightarrow{Hg/HI} CF_3MH_2 \ (M = P \text{ or } As)$$

Alcoholysis

The products vary with the ring size of the cyclopolyphosphine. The trifluoromethyl tetramer gives esters of trifluoromethylphosphonous acid and $[CF_3PH]_2$, while the pentamer gives a complex mixture of trifluoromethyl-substituted mono-, di- and tri-phosphines.[44] The mechanism probably involves P—P cleavage, followed by nucleophilic attack at the P—OH group so formed. Such a mechanism is also proposed for alcoholysis.

$$[CF_3P]_n + ROH \longrightarrow H[PCF_3]_{n-1}P(CF_3)OR \ (n = 4 \text{ or } 5)$$

$$H[PCF_3]_{n-1}P(CF_3)OR \longrightarrow H[PCF_3]_{n-1}H + CF_3P(OR)_2$$

$$[C_6F_5P]_4 \longrightarrow C_6F_5P(OR)_2$$

Quaternization of Phosphorus in Cyclopolyphosphines

While monophosphines readily form phosphonium salts with alkyl halides, cyclopolyphosphines readily undergo ring cleavage. Thus, while $[Bu^tP]_4$ readily forms a 1:1 phosphonium salt with methyl iodide, phenylcyclopolyphosphines are rapidly broken.

$$[Bu^tP]_4 + MeI \longrightarrow [(Bu^tP)_3P(Bu^t)Me]^+ I^-$$

Benzyl chloride and iodide break all P—P bonds in tetraphenylcyclotetraphosphine, while methyl and ethyl iodide give only diphosphines.[45]

$$[PhP]_4 + 7PhCH_2X \longrightarrow Ph(PhCH_2)_3P^+X^- + 2Ph(PhCH_2)_2PX_2 + PhPX_2 \quad (X = Cl \text{ or } I)$$

$$[PhP]_4 + 5RI \longrightarrow \underset{I\quad\quad I}{\overset{Ph\quad\quad Ph}{P{-}P}} + PhR_3P^+I^- + PhR_2PI_2$$

(R = Me or Et)

$$\xrightarrow{2PhCH_2I} PhPI_2 + Ph(PhCH_2)_2PI_2$$

Sym-diphenyldi-iododiphosphine also results from tetraphenylcyclotetraphosphine and phenylphosphorus di-iodide.

$$[PhP]_4 + 4PhPI_2 \longrightarrow 4[PhPI]_2$$

No quaternization has yet been reported for perfluoro organo-substituted rings, as would be expected for such a strongly electron-withdrawing group as CF_3.

Transition Metal Complexes

Though tris-(trifluoromethyl)phosphine complexes of nickel carbonyl are well-characterized, those of $[CF_3P]_4$ are of uncertain composition. The alkyl- and aryl-cyclopolyphosphine complexes are better characterized, though some structures still remain unresolved.[46]

These complexes have been mainly limited to cyclopolyphosphines, but hexaphenylcyclohexa-arsine does give three complexes with molybdenum carbonyl, namely $[PhAs]_4Mo(CO)_4$, $[PhAs]_4[Mo(CO)_4]_2$ and $[PhAs]_6Mo(CO)_4$.[46] The puckered structure of $[CF_3P]_4$ tends to support the idea that opposite atoms of the $[PhAs]_4$ ring chelate to the metal, while with the chair conformer of $[PhAs]_6$, $As_{(1)}$ and $As_{(3)}$ co-ordinate.

A variety of products also result from the reaction of cyclopolyphosphines with transition metals, the products depending on the conditions used. Thus, phosphorobenzene $[PhP]_5$ gives a monodentate phosphine complex with nickel carbonyl in the absence of solvent, but in ether ring contraction occurs, and the cyclotetraphosphine complex is isolated.

$$[PhP]_4Ni(CO)_3 \xleftarrow[Et_2O]{Ni(CO)_4} [PhP]_5 \xrightarrow{Ni(CO)_4} [PhP]_5Ni(CO)_3$$

This ring contraction in solution has already been mentioned (p. 273).

Phosphorobenzene reacts with iron pentacarbonyl at 170°C. The complex $[PhP]_4[Fe(CO)_4]_2$ is thought to involve alternate phosphorus atoms bonding to the $Fe(CO)_4$ residue. With molybdenum and tungsten carbonyls, monodentate cyclopentaphosphine complexes result $[PhP]_5M(CO)_5$ (M = Mo or W).[47]

Alkyl-substituted cyclopolyphosphines also undergo ring size variation during the formation of transition metal complexes. Thus, while tetraethylcyclotetraphosphine maintains ring size in its complexes with tungsten, palladium and copper, expansion occurs with molybdenum.[48]

$$(MeCN)_3W(CO)_3 \xrightarrow{[EtP]_4} [EtP]_4W(CO)_4$$

$$(PhCN)_2PdCl_2 \xrightarrow{[EtP]_4} [EtP]_4PdCl_2$$

$$CuCl \xrightarrow{[EtP]_4} [EtP]_4CuCl$$

$$Mo(CO)_6 + [EtP]_4 \longrightarrow [EtP]_5Mo(CO)_4$$

A structure determination shows that phosphorus atoms in 1- and 3-positions of the cyclopentaphosphine chelate in *cis* positions on the metal carbonyl residue.

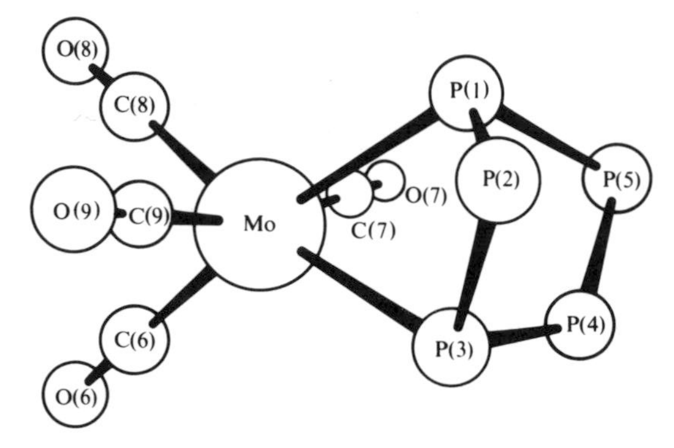

Fig. 5.7. The configuration of $[EtP]_5Mo(CO)_4$; the ethyl groups are not shown[49]

The P—P distances (0.221 nm) are all close to that expected for a single bond, and the infrared spectrum[49] shows four peaks in the metal–carbonyl stretching region, at 2014 (s), 1929 (s), 1917 (vs) and 1905 cm^{-1} (vs), typical of a *cis*-substituted metal hexacarbonyl.

C_{2v}	E	C_2	σ_v	σ'_v
Γ_{CO}	4	0	2	2

$$\Gamma_{CO} = A'_1 + A''_1 + B_1 + B_2 \text{ (all infrared active).}$$

With cupric bromide, pentamethylcyclopentaphosphine yields the cuprous complex $[MeP]_5CuBr$, while with the chloride both reduction and ring contraction occurs.

$$[MeP]_4CuCl \xleftarrow{CuCl_2} [MeP]_5 \xrightarrow{CuBr_2} [MeP]_5CuBr$$

Again, this reaction is thought to proceed via a phosphinidene intermediate MePCuCl.[48]

In the Section on anionic polyphosphides (p. 274), $[PhP]_3^{2-}$ was described. This is a two-electron π-donor, and complexes with metal carbonyls as a monodentate ligand under the influence of ultraviolet light,[36] *e.g.*,

$$Ni(CO)_4 + [PhP]_3^{2-} \xrightarrow{u.v.} [(PhP)_3Ni(CO)_3]^{2-}$$

As would be expected, the C—O stretching frequencies are all low (1865, 1925, 1950 cm^{-1}) due to strong back-bonding from the metal to the carbonyl groups.

Among the products isolated from the reaction of cobalt octacarbonyl with pentamethylcyclopenta-arsine (at 200°C under a pressure of 100 atm of carbon monoxide) is the peculiar air-stable compound $As_3Co(CO)_3$.[50] The infrared spectrum gives a non-degenerate A_1 vibration at 2083 cm^{-1} and the E one at 2039 cm^{-1}, as expected for a molecule of C_{3v} symmetry. This molecule can be conveniently considered as As_4 with one arsenic atom replaced by the trivalent $Co(CO)_3$ residue. Indeed it represents the last member of the series $M_4(CO)_{12}$ (M = Ir), $X[Co(CO)_3]_3$ (X = RC, S, Se), $As_2Co_2(CO)_6$, $As_3Co(CO)_3$ and $As_4(P_4)$.

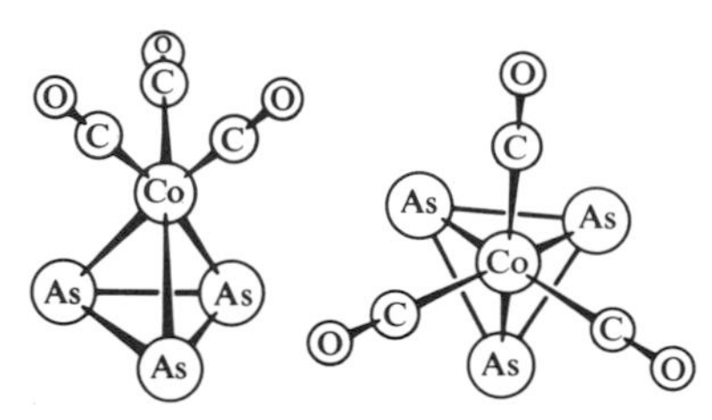

Fig. 5.8. The idealized C_{3v} molecular configuration of $As_3Co(CO)_3$[50]

The proposal that As_4 is unstable because of inter-orbital coulombic repulsions receives support from the structural and stability data of both $As_3Co(CO)_3$ and $As_2Co_2(CO)_6$. The $Co(CO)_3$ residue is an effective electron sink, removing charge from the As_3 and As_2 residues. Consequently, both compounds are more stable than As_4 and have shorter bond lengths. In $As_3Co(CO)_3$, the As—As distance of 0.2372 nm is shorter than a single bond value (0.242 nm), while that in $As_2Co_2(CO)_6$ is 0.2273 nm, the shortest reported to date.

Structural determinations

The disconcerting way in which the ring size of these cyclopolyphosphines and arsines vary with both synthetic conditions and dissolution has prompted much structural work, especially through crystallography and n.m.r. spectroscopy.

Tetrakis-(trifluoromethyl)cyclotetraphosphine possesses a folded 4-membered ring with the average P—P distance of 0.2213 nm.

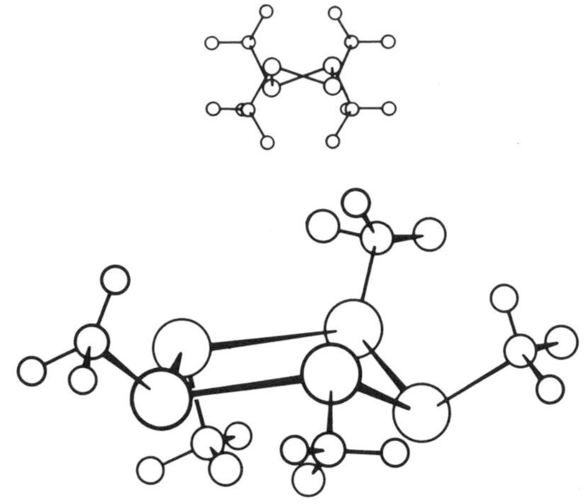

Fig. 5.9. The molecular conformations of $[CF_3P]_{4\ \mathrm{and}\ 5}$[51]

This agrees closely with that of the pentamer (0.2223 nm) where there is more ring strain through the two eclipsed CF_3 groups.[51] Again, the ring is non-planar, which leads to peculiar ^{31}P and ^{19}F n.m.r. spectra. The ^{31}P n.m.r. spectrum gives a doublet for the pentamer (signal ratio 2:3) indicating non-equivalence of the phosphorus atoms. Similarly, two ^{19}F bands are observed at 42°C (ratio 2:3), which exhibit a complex structure, due probably to a combination of ^{31}P—^{19}F coupling and restricted rotation of the P—C bonds.[52] This complex structure vanishes on heating, but the two separate peaks are maintained, and supports a mechanism of ring structure change which will give an apparent rotation of the ring-puckering. Such a mechanism probably involves restricted rotation about both P—P and P—C bonds.

The P_4 ring in tetracyclohexylcyclotetraphosphine is again non-planar, and each C_6H_{11} ring is in the chair conformer with the P atom occupying an equatorial position.[53]

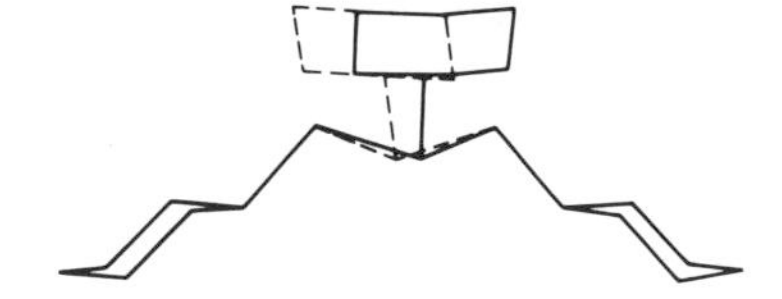

Fig. 5.10. The structure of $(C_6H_{11}P)_4$.[53] (From J. W. BART, *Acta. Cryst.*, 1969, **B25**, 762.)

The crystal structures of the phosphorobenzene pentamer and hexamer have been studied. The pentamer (compound A, m.p. 150°C) possesses an almost planar polyphosphine ring,[54] as distinct from that of $[CF_3P]_5$.

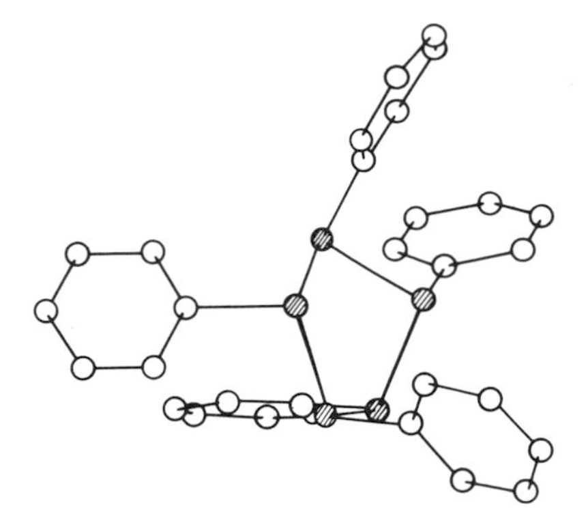

Fig. 5.11. The structure of pentaphenyl-cyclopentaphosphine[54]

The P—P distances vary little (0.2207–0.2223 nm), and agree closely with those of the trifluoromethylcyclopolyphosphines.

The trigonal form of phosphorobenzene B, $[PhP]_6$, (m.p. 190°C) possesses the chair conformer of the P_6 ring with all phenyl groups equatorial and arranged like a paddle-wheel. The P—P distance (0.2237 nm) is not significantly different from the other cyclopolyphosphines.[55]

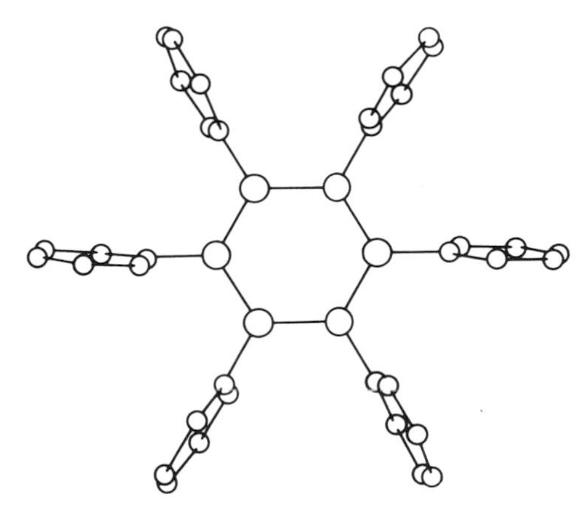

Fig. 5.12. The structure of hexaphenyl-cyclohexaphosphine[55]

Though the molecular structure is the same in the triclinic polymorph, two molecules are unrelated by symmetry in the unit cell, whereas in the trigonal form they are related by a two-fold axis.

These provide the third example of phosphorus catenated in 6-membered rings. The P—P distances are comparatively long in black phosphorus (0.228 nm), while in $Cs_6[(PO_2)_6]$ they are 0.222 nm.[56]

The structure of the yellow, low-melting form of pentameric arsenomethane (m.p. 12°C) comprises a puckered 5-membered ring with a mean As—As bond length of 0.2428 nm.[57]

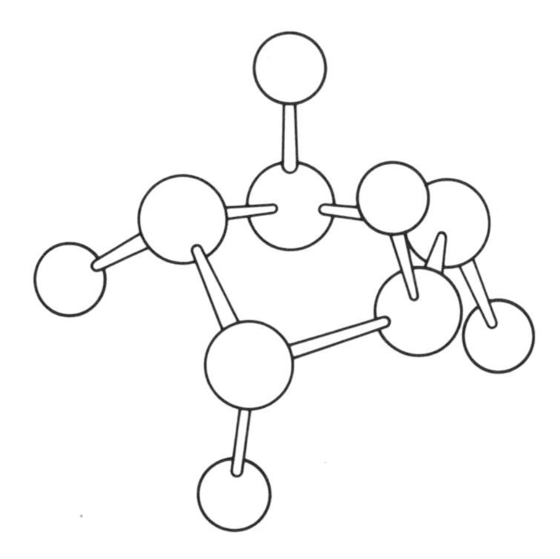

Fig. 5.13. The structure of arsenomethane[57]

This is close to that expected for a single bond, and the ring structure closely resembles that of $[CF_3P]_5$. The As—As distance in arsenobenzene $[PhAs]_6$ is significantly larger (0.2456 nm) though shorter than in arsenic metal.[58] As with hexameric phosphorobenzene, the phenyl groups are equatorial on the chair conformer of the As_6 ring.[b]

The n.m.r. spectra of $[MeAs]_5$ and $[CF_3As]_5$ are both temperature-dependent, and the inequivalence of the substituents can be resolved, as with the cyclopentaphosphine, by assuming the progressive rotation of a time-averaged perpendicular symmetry plane yielding a time-averaged 5-fold rotation axis. This proceeds one atom at a time, thereby changing the position of ring puckering in a progressive manner and is known as a "click-stop" process.[59]

Phosphorus cage compounds

Both phosphorus trichloride and phosphoryl chloride react with dilithium phenylphosphide in the ratio 2:3 to give compounds assumed to have a cage structure, and formulated as $P_4[PPh]_6$ and $(PO)_4[PPh]_6$.

$$4PCl_3 + 6Li_2PPh \longrightarrow 12LiCl + P_4[PPh]_6$$

Molecular weight measurements support this, so the structure probably resembles the adamantane cage **(5-IV)**.

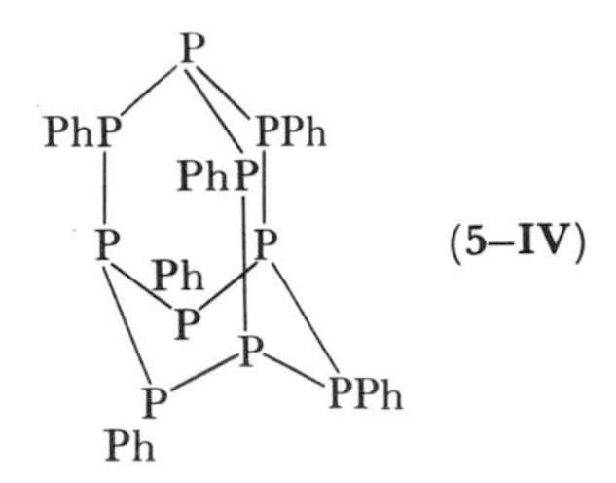

(5–IV)

Fig. 5.14. The proposed structure of $P_4[PPh]_6$

(5-IV) oxidizes iodomethane to iodine presumably in a manner resembling that with tetraphenyldiphosphine.[60]

$$Ph_4P_2 + 4MeI \longrightarrow 2Ph_2PMe_2^+I^- + I_2$$

$$Ph_6P_{10} + 34MeI \longrightarrow 6PhPMe_3^+I^- + 4Me_4P^+I^- + 12I_2$$

NITROGEN COMPOUNDS

There is a wide variety of cyclic Group V–nitrogen compounds, but the majority can be conveniently classified under three general formulae:

cyclopolyphospha-III-azanes $[XP—NY]_n$

cyclopolyphospha-V-azanes $[X_3P—NY]_n$

cyclopolyphospha-V-azenes $[X_2P{=}N]_n$

Indeed, most of the cyclic Group V–nitrogen compounds are of phosphorus; arsenic and antimony compounds will be considered under the appropriate class of phosphorus compounds.

The cyclopolyphospha-III-azanes and their arsenic analogues

This ring system was first reported in 1894 with the isolation of *N,N'*-diphenyl-*P,P'*-dichlorocyclodiphospha-III-diazane from aniline and phosphorus trichloride.[61]

$$2PhNH_2 + 2PCl_3 \longrightarrow PhN{<}(PCl)_2{>}NPh + 4HCl$$

Cl
P
PhN NPh
P
Cl (5-V)

This is surprisingly stable in cold water.

Aminated derivatives of **(5-V)** result when an excess of a primary amine is used[62] and also when bis-(benzylamino)pentafluorophenylphosphine is heated.[63] Pentafluorobenzene is lost.

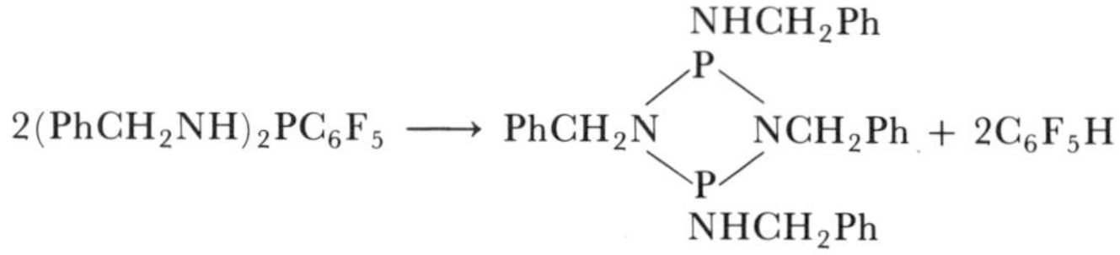

$$10RNH_2 + 2PCl_3 \longrightarrow RN\langle P(NHR) \rangle_2 NR + 6RNH_3Cl$$

(4-membered ring: RN and NR bridged by two P atoms, each P bearing NHR)

$$2(PhCH_2NH)_2PC_6F_5 \longrightarrow PhCH_2N\langle P(NHCH_2Ph) \rangle_2 NCH_2Ph + 2C_6F_5H$$

(4-membered ring: $PhCH_2N$ and NCH_2Ph bridged by two P atoms, each P bearing $NHCH_2Ph$)

This 4-membered ring also results in t-butylaminophosphorus dichloride and base, or lithium t-butyltrimethylsilylamide.[64]

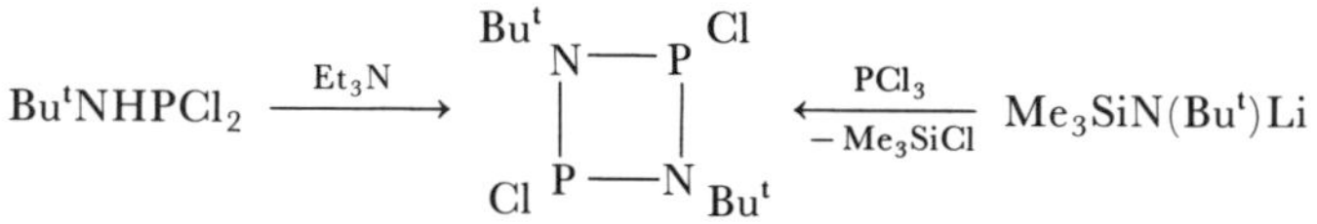

$$Bu^tNHPCl_2 \xrightarrow{Et_3N} [Bu^tN-PCl]_2 \xleftarrow[-Me_3SiCl]{PCl_3} Me_3SiN(Bu^t)Li$$

(4-membered ring: Bu^tN—P(Cl), P(Cl)—NBu^t; m.p. 42–4°C)

The analogous As^{III}—N rings are conveniently prepared by transamination and subsequent deamination, though they can be obtained directly from an arsenic halide and the amine.

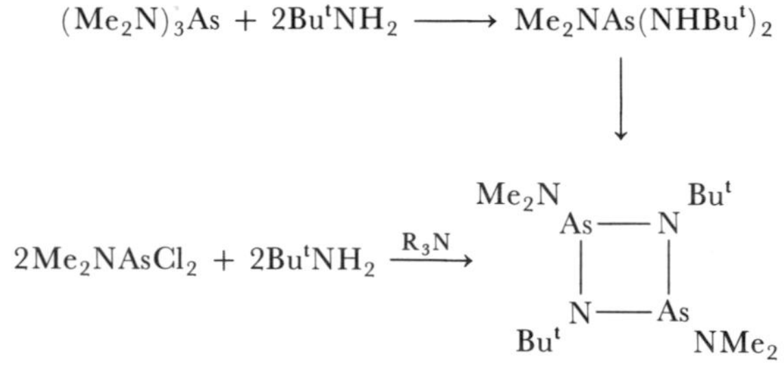

$$(Me_2N)_3As + 2Bu^tNH_2 \longrightarrow Me_2NAs(NHBu^t)_2$$

$$\downarrow$$

$$2Me_2NAsCl_2 + 2Bu^tNH_2 \xrightarrow{R_3N} [Me_2NAs-NBu^t]_2$$

(4-membered ring: Me_2NAs—NBu^t, Bu^tN—AsNMe_2)

This ring system also results using aniline, benzamide and *N,N*-dimethylhydrazine, and all compounds can be readily distilled or sublimed.[65] Attempts to isolate the *N,N'*-di-n-butyl ring resulted in disproportionation.

$$3[Me_2NAsNBu^n]_2 \xrightarrow{\Delta} 2(Me_2N)_3As + As_4(NBu^n)_6$$

$As_4(NBu^n)_6$ is believed to have the cage structure characteristic of urotropine, $(CH_2)_6N_4$, and will be discussed later.

Since small rings (*e.g.*, 4-membered) reduce steric hindrance between substituents, *N,N'*-di-t-butyl-*As,As'*-di-t-butylamino-cyclodiarsa-III-azane is formed from arsenic trichloride and excess t-butylamine.

$$2AsCl_3 + 10Bu^tNH_2 \longrightarrow [Bu^tHN\text{-}As(\mu\text{-}NBu^t)_2As\text{-}NHBu^t] \xrightarrow{2HCl} [ClAs(\mu\text{-}NBu^t)_2AsCl]$$

(m.p. 61–4°C)

With smaller substituents, however, larger rings result.

Ethylaminotrimethylsilane and arsenic trichloride readily yield the cyclotriarsa-III-azane,[66]

$$Me_3SiNHEt + AsCl_3 \longrightarrow Me_3SiCl + [Cl_2AsNHEt] \xrightarrow{-HCl} \tfrac{1}{3}[EtNAsCl]_3$$

while the *N*-methyl homologue is conveniently synthesized from HCl and the As—N cage $As_4(NMe)_6$.[67]

$$3As_4(NMe)_6 + 12HCl \longrightarrow 4[ClAsNMe]_3 + 6MeNH_2$$

A structure determination shows the 6-membered As—N ring to be slightly puckered with the As—N bonds varying in length from 0.176 to 0.186 nm, and with the As—Cl bonds about 0.226 nm.[68]

Eight-membered As—N and Sb—N rings are well-characterized, with the syntheses summarized below.[69]

$$PhNH_2 + AsI_3 \longrightarrow [PhNAsI]_4$$

$$PhAsCl_2 + NH_3 \xrightarrow{C_6H_6} [PhAsNH]_4$$

$$RNH_2 + SbX_3 \longrightarrow [RNSbX]_4 \ (R = \text{alkyl}, X = \text{I or OEt})$$

The phospha-V-azanes

Much attention has been paid to cyclic P^V—N compounds. These can be conveniently divided into two classes, those formulated with a P—N single bond and those with a P—N double bond. The former, the phospha-V-azanes $[X_3P\text{—}NY]_n$, have no arsenic or antimony analogues unlike the phospha-V-azenes $[X_2P\text{=}N]_n$, where a limited amount of work has been conducted on $[Ph_2As\text{=}N]_4$.

The phospha-V-azanes $[X_3P\text{—}NY]_n$ normally occur with $n = 2$, and are conveniently synthesized directly from PCl_5 and a primary amine.

$$2PCl_5 + 2MeNH_2 \longrightarrow [Cl_3PNMe]_2 + 4HCl$$

The simplest, *P,P'*-hexachloro-*N,N'*-dimethylcyclodiphospha-V-diazane, has been widely used as a synthetic intermediate.[70]

An interesting development using sulphonyl amines leads to a wider variety of products through secondary reactions. With aminomethanesulphonic acid, PCl_5 chlorinates the acid as well as forming the phosphinimine. This readily loses SO_2 yielding the cyclodiphospha-V-azane.

$$H_2NCH_2SO_3H \xrightarrow{PCl_5} H_2NCH_2SO_2Cl \xrightarrow{PCl_5} Cl_3P{=}NCH_2SO_2Cl \xrightarrow{-SO_2} [Cl_3P{-}NCH_2Cl]_2$$

A little tris-chloromethylamine $(ClCH_2)_3N$ is also formed, an amine best synthesized from PCl_5 and $N(CH_2SO_3Na)_3$.

$$N(CH_2SO_3Na)_3 + 2PCl_5 \longrightarrow N(CH_2Cl)_3 + 3SO_2 + 3POCl_3 + 3NaCl$$

Chlorosulphonamide, $ClSO_2NH_2$, forms the thermally unstable $Cl_3P{=}NSO_2Cl$ which decomposes to phosphoryl chloride and sulphanuric chloride.[71]

$$3Cl_3P{=}NSO_2Cl \xrightarrow{\Delta} 3Cl_3PO + [NSOCl]_3$$

Phosphoryl and thiophosphoryltriamides readily deaminate on heating to give dimeric products possessing the 4-membered ring.[62, 72]

$$2XP(NHR)_3 \longrightarrow [(RNH)(X{=})P{-}NR]_2 \text{ (4-membered ring)} + 2RNH_2 \quad (X = O \text{ or } S).$$

This readily occurs with both aliphatic and aromatic amines, and there is little evidence for side reactions, except with the aniline derivative $(PhNH)_3PO$ which forms a cross-linked polymer $[C_6H_4PON]_x$.

The deamination of $PhP(S)(NHR)_2$ (R = H, alkyl, phenyl) is considerably more complicated, the products depending on R and the deamination temperature. Up to 160°C, deamination of the alkyl- and aryl-amino compounds produces the 4-membered ring as expected, while $PhP(S)(NH_2)_2$ yields the less strained 6-membered ring as steric hindrance is less.

$$3PhP(S)(NH_2)_2 \xrightarrow{160^\circ C} [Ph(S{=})P{-}NH]_3 \text{ (6-membered ring)}$$

$$2PhP(S)(NHR)_2 \xrightarrow{160^\circ C} [Ph(S{=})P{-}NR]_2 \text{ (4-membered ring)} \quad (R = Me, Et, Pr^n, Bu^n, Ph)$$

Above 180°C, the deamination of bis-(*N*-ethylamino)phenylphosphine sulphide is more complicated.

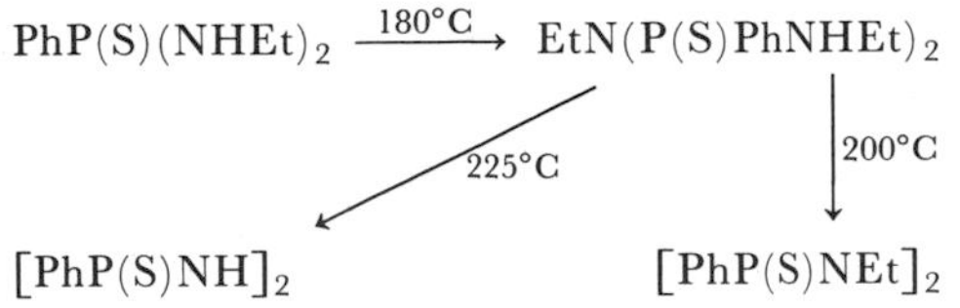

$$PhP(S)(NHEt)_2 \xrightarrow{180^\circ C} EtN(P(S)PhNHEt)_2$$

$$EtN(P(S)PhNHEt)_2 \xrightarrow{225^\circ C} [PhP(S)NH]_2$$

$$EtN(P(S)PhNHEt)_2 \xrightarrow{200^\circ C} [PhP(S)NEt]_2$$

At 265°C, a compound analyzing for $[PhP(S)NPh]_2$ is formed from bis-(*N*-phenylamino)phenylphosphine sulphide.

Ph_2 P, S, N, N, S, P Ph_2 (six-membered ring)

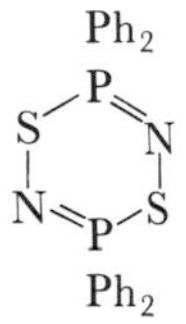

Fig. 5.15.

The 6-membered ring structure is supported by hydrolysis, which gives diphenylphosphinic acid. It also forms when *N*-phenylaminodiphenylphosphine sulphide is heated.

Bis-(imidazol-1-yl)phenylphosphine oxide readily undergoes transamination with aromatic diamines to give linear polymers involving the P_2N_2 4-membered ring.[73] The properties of these polymers can be varied by using mixtures of the diamines.

(R = *p*-phenylene, *p*-diphenyl,2,6-pyridyl)

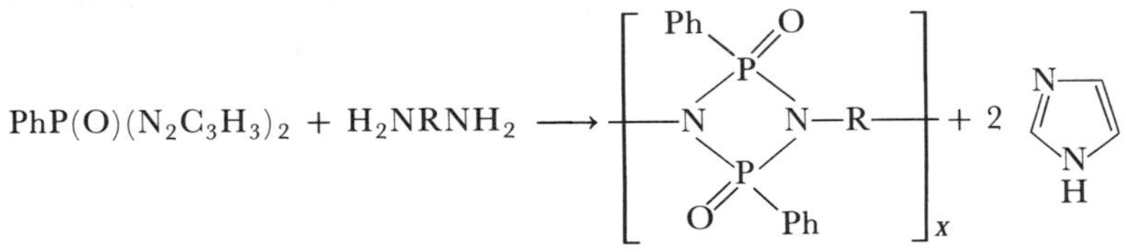

$$PhP(O)(N_2C_3H_3)_2 + H_2NRNH_2 \longrightarrow [-N(P(O)Ph)_2N-R-]_x + 2\ C_3H_4N_2$$

Trimethylsilyl compounds have been widely used as intermediates in the synthesis of phospha-V-azanes, in particular the fluoro ones. *N*-methylhexamethyldisilazane reacts vigorously with phosphorus pentafluoride and phenylphosphorus tetrafluoride to give the *P*-fluorophospha-V-azanes.

$$RPF_4 + (Me_3Si)_2NMe \longrightarrow [RPF_2NMe]_2 + 2Me_3SiF \quad (R = Ph\ or\ F)$$

The contrast in the physical properties of the products is quite remarkable, the hexafluoro compound boiling at 92°C, while the diphenyltetrafluorocyclodiphospha-V-azane melts at 162°C. Diphenylphosphorus trifluoride gives a monomeric product, $Ph_2P(F){=}NMe$.[74]

Reactions of Cyclodiphospha-V-diazanes

Ring substitution. The chlorine atoms of *P*-hexachlorocyclodiphospha-V-diazanes can be readily substituted using H_2S or liquid SO_2, while amino- and thio-silanes conveniently displace the last chlorine atoms of (**5-VI**).[75]

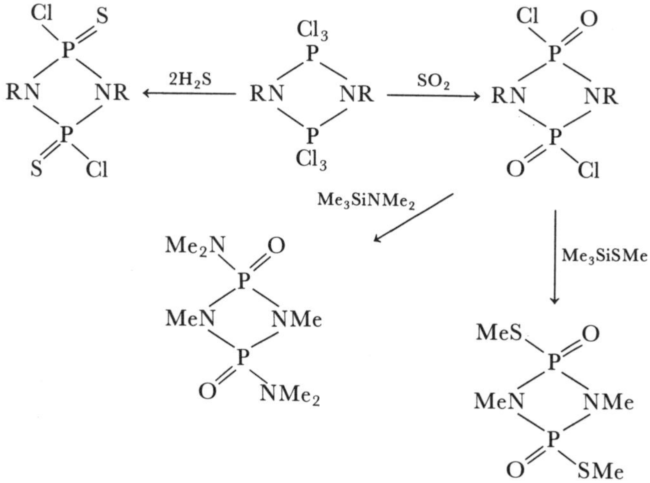

$[Cl(S)PNR]_2 \xleftarrow{2H_2S} [Cl_3PNR]_2 \xrightarrow{SO_2} [Cl(O)PNR]_2$ (**5-VI**)

$[Cl(O)PNR]_2 \xrightarrow{Me_3SiNMe_2} [Me_2N(O)PNMe]_2$

$[Cl(O)PNR]_2 \xrightarrow{Me_3SiSMe} [MeS(O)PNMe]_2$

Ring opening. Ammonia will readily substitute at chlorine in $[Cl_3PNMe]_2$, but in addition, ring opening occurs to give the salt $[P_2N_7H_{10}Me_2]^+Cl^-$.

$$[Cl_3PNMe]_2 + 10NH_3 \longrightarrow [P_2N_7H_{10}Me_2]^+Cl^- + 5NH_4Cl$$

A structure determination of the iodide shows the central PNP skeleton with $P\hat{N}P$ 130 degrees and localized multiple bonding in these P—N bonds, both being shorter than the terminal P—N bonds.

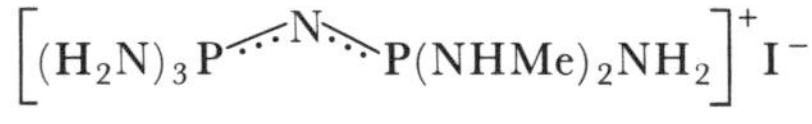

$$\left[(H_2N)_3P\cdots N\cdots P(NHMe)_2NH_2\right]^+ I^-$$

(**5-VII**)

Fig. 5.16.

The most peculiar feature is the presence of both methylamino groups on the one phosphorus atom.[76]

The reaction between $[ClP(S)NMe]_2$ and ammonia might be expected to give a similar product but several distinct differences arise.

$$[ClP(S)NMe]_2 + NH_3 \longrightarrow P_2N_5H_7Me_2S_2 + 2NH_4Cl$$

It is a non-electrolyte, and the X-ray structure shows N—Me bridging, in contrast to (**5-VII**). In addition, the bridging P—N bonds are longer

(0.171 nm) than the terminal P—N bonds (0.159–0.169 nm), the former being close to that expected for a single bond. The P—S bond distance (0.195 nm) also indicates multiple-bond character.

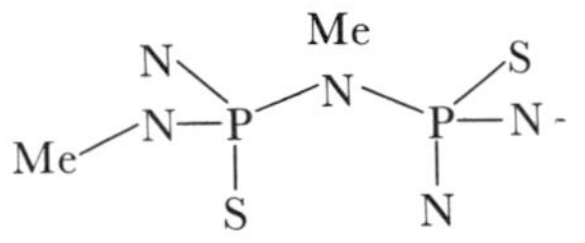

Fig. 5.17. The skeleton of $P_2N_5H_7Me_2S_2$

Heating dimeric phospha-V-azanes with phenylisocyanate provides a useful route to mixed carbodi-imides and probably proceeds via the monomeric intermediate.[77]

$$[Cl_3PNR]_2 \rightleftharpoons [Cl_3PNR] \xrightarrow{R'NCO} R'NCNR + POCl_3$$
$$(R = Me, Ph;\ R' = Ph)$$

$[PhCl_2PNMe]_2$ and $[Ph_2ClPNMe]_2$ react similarly, while carbon dioxide and disulphide give isocyanates and isothiocyanates.

$$[Cl_3PNR]_2 + 2CX_2 \longrightarrow 2RNCX + 2PXCl_3\ (X = O \text{ or } S)$$

With methylisocyanate, however, a different reaction course is observed.

$$[ClP(X)NMe]_2 + MeNCO \longrightarrow \quad (5\text{-VIII})$$
$$(X = Cl_2, O, S)$$

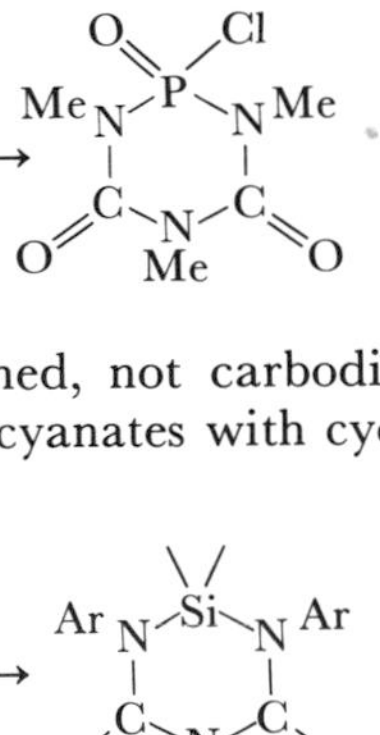

The heterocycle (**5-VIII**) is formed, not carbodi-imides.[75] This closely resembles the reactions of arylisocyanates with cyclosilazanes, mentioned in the last Chapter (p. 185), *i.e.*,

$$\frac{1}{n}[Me_2SiNH]_n + 2ArNCO \longrightarrow$$
$$(n = 3, 4)$$

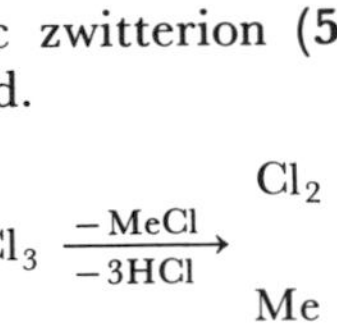

$[Cl_3PNMe]_2$ reacts with methylammonium chloride in the presence of boron trichloride.[78] The cyclic zwitterion (**5-IX**) results through a peculiar cleavage of the C—N bond.

$$[Cl_3PNMe]_2 + MeNH_3Cl + BCl_3 \xrightarrow[-3HCl]{-MeCl} \quad (5\text{-IX})$$

Phosphorus pentachloride will also cleave the C—N bond, giving the perchlorodiphospha-V-azene cation.

$$2PCl_5 + MeNH_3Cl \xrightarrow[-3HCl]{-MeCl} [Cl_3P{-}N{=}PCl_3]^+Cl^-$$

This has also been isolated as an intermediate in the synthesis of chlorocyclophospha-V-azenes. With an excess of $MeNH_3Cl$ and BCl_3, the zwitterion (**5-IX**) results.

$$[Cl_6P_2N]^+Cl^- + BCl_3 + 2MeNH_3Cl \longrightarrow (\textbf{5-IX}) + 6HCl$$

Chlorination of $[Cl_3PNMe]_2$ under the influence of u.v. light breaks the ring and substitutes at the protons of the methyl group, giving the monomeric perchloromethylphosphinimine.

$$[Cl_3PNMe]_2 + 6Cl_2 \longrightarrow 2Cl_3P{=}NCCl_3 + 6HCl$$

Structure and Bonding

Several peculiar structural features have already emerged from the four structures determined to date. $Ph_2FP{=}NMe$ is a monomer, but $[PhF_2PNMe]_2$ is dimeric. This is thought to be due to the strong electron-withdrawing nature of the F atom rendering phosphorus in PhF_2PNMe a better acceptor than Ph_2FPNMe.[c] However, the dimeric nature of $Ph_2ClPNMe$ tends to refute this. The P—F bond length in $Ph_2FP{=}NMe$ is very short (0.1488 nm), while the P—N bond length (0.1641 nm) corresponds to a double bond,[79] which supports the presence of $p_\pi{-}d_\pi$ bonding in the P—N and P—F bonds.[80]

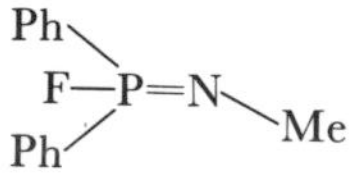

Fig. 5.18.

The dimeric phospha-V-azanes possess 5-co-ordinate phosphorus atoms with trigonal-bipyramidal stereochemistry and, as with PCl_5, the axial bonds are longer than the equatorial ones. $[PhF_2PNMe]_2$ possesses a planar P_2N_2 ring. The substituent orientation is trigonal-bipyramidal around each phosphorus, with one P—N and one P—F bond in the axial positions.

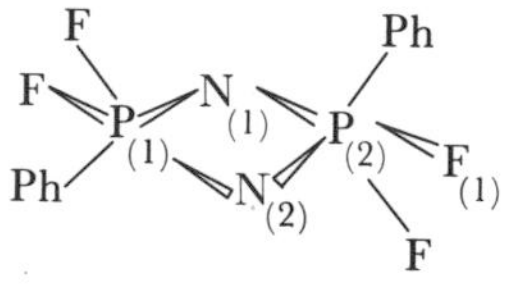

Fig. 5.19. Dimeric $[PhF_2PNMe]_2$ represented as two bridging trigonal-bipyramids

The less-hindered geometrical isomer with *trans* phenyl groups is the one obtained. $P_{(2)}$—$N_{(1)}$ and $P_{(2)}$—$F_{(1)}$ are axial bonds whose lengths (0.178 and 0.162 nm) are close to single bond lengths. The $P_{(2)}$—$N_{(2)}$ bond (0.164 nm) appears to be a double bond, and $P_{(2)}$—$F_{(2)}$ is shorter (0.157 nm) than $P_{(2)}$—$F_{(1)}$.[81]

Similar features arise in the structure of the *N*-methyltrichlorophosphinimine dimer. Again, both axial and equatorial P—N bonds occur in the P_2N_2 ring (0.1769 and 0.1635 nm), while the two P—Cl bond distances (0.2133 and 0.2026 nm) are close to those obtained for PCl_5.[82]

Replacing two Cl atoms by an S atom gives a pseudo-tetrahedral stereochemistry around the phosphorus atom of 2,4-dithio-2,4-dichloro-1,3-dimethylcyclodiphospha-V-azane, $P_2Cl_2S_2N_2Me_2$.[83] So both P—N bonds of the P_2N_2 ring are equal in length (0.167 nm) and close to that observed for a double bond.

Azides

Recent interest in metalloid azides has led to the isolation of dimeric azidotetrachlorostibine. This can be readily synthesized from triorganosilyl-, chloro-, or hydrogen azides and is a yellow solid, bridged through the nitrogen atom alpha to the antimony atom.[84]

$$2R_3SiN_3 + 2SbCl_5 \longrightarrow \underset{\text{(m.p. 126–7°C)}}{[Cl_4SbN_3]_2} + 2R_3SiCl$$

Excessive heating results in the loss of chlorine and nitrogen, while hydrolysis is slow. While pyridine readily breaks the azide bridge to give the 6-co-ordinate complex $C_5H_5NSbCl_4N_3$ with the two nitrogen ligands *trans* to each other, triphenylphosphine reduces Sb^V.

$$[Cl_4SbN_3]_2 + 4Ph_3P \longrightarrow [Ph_3P{=}N{-}N{=}N{-}PPh_3]^+SbCl_4^-$$

The product, bis-(triphenylphosphine)triazenium tetrachloroantimonite possesses equivalent P atoms, and loses N_2 at 100°C.[85]

$$[Ph_3PN_3PPh_3]^+SbCl_4^- \xrightarrow{100°C} [Ph_3P{=}N{-}PPh_3]^+SbCl_4^- + N_2$$

A most unexpected azidation of carbon tetrachloride also occurs with azidotetrachlorostibine.

$$3[Cl_4SbN_3]_2 + 2CCl_4 \longrightarrow 2[C(N_3)_3]^+SbCl_6^- + 4SbCl_5$$

The pale yellow product is sensitive to shock, and has a vibrational spectrum characteristic of the triazidomethyl cation, which, possesses C_{3v} point group symmetry,[86] with carbon and the 3 α-N atoms planar.

Cage compounds

Compounds with an atomic skeleton resembling that of adamantane, $(CH)_4(CH_2)_6$, or urotropine, $(CH_2)_6N_4$, are becoming increasingly common in Inorganic Chemistry, and the elements of Group V provide varied examples. Oxides and sulphides will be discussed later in the Chapter. Imides are known with trivalent phosphorus, arsenic and antimony, $M_4(NR)_6$ (M = P, As, Sb).

The phosphorus cage readily forms from PCl_3 and methylamine in the absence of solvent, and is a crystalline solid melting at 225°C.[87]

$$4PCl_3 + 18MeNH_2 \longrightarrow P_4(NMe)_6 + 12MeNH_3Cl$$

It is also formed from *N*-methyl-bis-(dichlorophosphino)amine and *N*-methylhexamethyldisilazane.[88]

$$2MeN(PCl_2)_2 + 4(Me_3Si)_2NMe \longrightarrow P_4(NMe)_6 + 8Me_3SiCl$$

$$\underset{\text{(excess)}}{MeNH_3Cl + PCl_3} \xrightarrow{Cl_2C{=}CCl_2} MeN(PCl_2)_2$$

The arsenic analogues can also be synthesized directly,[69, 89] but in addition they have been formed by processes involving trans-amination[67] and As—N skeletal rearrangements.[65]

$$4AsCl_3 + 18RNH_2 \longrightarrow As_4(NR)_6 + 12RNH_3Cl$$
$$(R = Me, Pr^i, Bu^n)$$

$$4(Me_2N)_3As \xrightarrow{Bu^nNH_2} As_4(NBu^n)_6 \xleftarrow{heat} \begin{array}{c} Me_2NAs—NBu^n \\ | \quad\quad | \\ Bu^nN—AsNMe_2 \end{array}$$

$$As_4(NPr^n)_6 \xrightarrow{ArNH_2} As_4(NAr)_6 \text{ (Ar = Ph, p-tolyl)}$$

Aromatic amines react with either antimony tri-iodide or triethoxide, again yielding what is probably the Sb—N adamantane cage.[69]

$$4SbX_3 + 6ArNH_2 \longrightarrow Sb_4(NAr)_6 + 12HX$$
$$(X = I \text{ or } OEt)$$

Reactions

While oxidation of $P_4(NMe)_6$ at 170°C gives a polymeric residue $[P_2O_2(NMe)_3]_n$,[90] the cage structure is maintained on heating with sulphur.[89]

$$P_4(NMe)_6 \xrightarrow[\text{sulphur}]{185°C} P_4S_4(NMe)_6$$

The phosphorus and arsenic cages react with methyl iodide in significantly different ways. A 1:1 phosphonium adduct forms with $P_4(NMe)_6$, while with excess iodomethane aminophosphonium salts[89] are obtained.

$$P_4(NMe)_6 \xrightarrow{1:1} [MeP_4(NMe)_6]^+I^-$$

$$P_4(NMe)_6 \xrightarrow{\text{excess MeI}} [MeP(NMe_2)_3]^+I^-$$

With the arsenic cage, however, two adjacent As—N bonds are cleaved to form trimethylamine and the di-iodide $As_4I_2(NMe)_5$, (m.p. 162°C).

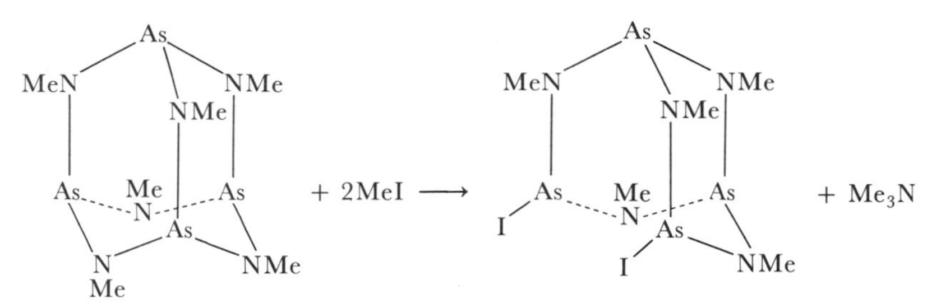

This opened cage can be formed directly, or by mixing $As_4(NR)_6$ with AsI_3 in suitable proportions. It equilibrates with the 8-membered As—N ring in the presence of AsI_3.[91]

$$4AsI_3 + 15RNH_2 \longrightarrow As_4I_2(NR)_5 + 10RNH_3I$$
(R = alkyl, $PhCH_2$, Ph)

$$4AsI_3 + 5As_4(NR)_6 \longrightarrow 6As_4I_2(NR)_5$$

$$4As_4I_2(NR)_5 + 4AsI_3 \rightleftharpoons 5[IAsNR]_4$$

Hydrogen chloride will also cleave in a step-wise manner, giving first the cage dichloride and subsequently the 6-membered ring, supporting the adamantane cage structure.[67]

$$As_4(NMe)_6 + 2HCl \longrightarrow As_4(NMe)_5Cl_2 \longrightarrow [MeNAsCl]_3$$

Carbon disulphide will also readily attack the imino bridge of $As_4(NMe)_6$ in a manner rather similar to its reaction with dimeric phospha-V-azanes.

$$As_4(NMe)_6 + 4CS_2 \longrightarrow As_4(NMe)_2S_4 + 4MeNCS$$

Four of the six imino groups can be replaced in this way.

The lone pairs of electrons on both phosphorus and arsenic will readily co-ordinate to borane and metal carbonyls, as would be expected from the interactions of these cages with iodomethane. Borane successively adds to all phosphorus atoms, but only the fully-complexed cage can be conveniently isolated.[92]

$$P_4(NMe)_6 + \frac{n}{2}B_2H_6 \longrightarrow [H_3B]_nP_4(NMe)_6 \ (n = 1\text{–}4)$$

Both the phosphorus and arsenic cages complex with nickel carbonyl but again, although all species are indicated in solution, only the tetrakis-(tricarbonylnickel) adduct can be fully characterized.[93]

$$M_4(NMe)_6 + nNi(CO)_4 \longrightarrow [(CO)_3Ni]_nM_4(NMe)_6$$
(n = 1–4, M = P or As)

While it is reasonable to suppose that the adamantane cage will predominate among compounds with this formulation, it is not inevitable, as was thought with $(MeSi)_4S_6$ mentioned in the last Chapter (p. 245). This was subsequently shown to have the adamantane structure, and although the 4-membered As—N ring is well characterized, $As_4(NMe)_6$ has structure (**5-X**) and not (**5-XI**).

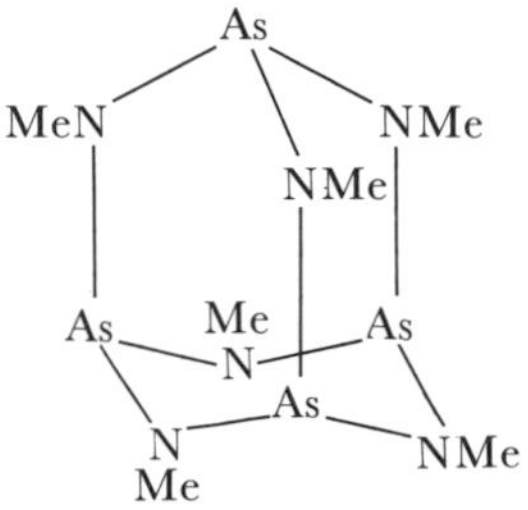

(5-**X**)

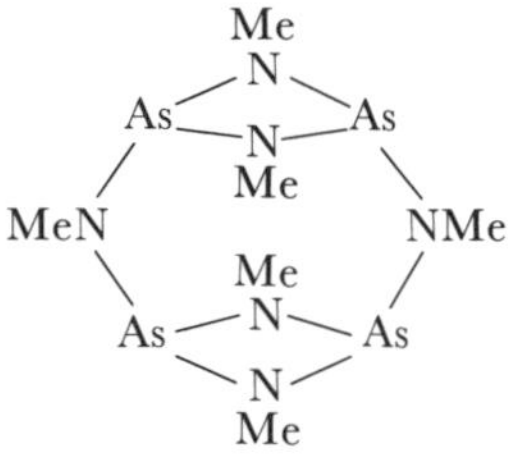

(**5-XI**)

The As—N bond length (0.187 nm) corresponds with that expected for a single bond.[94]

Hydrazines will readily trans-aminate aminophosphines and arsines. 1,2-dimethylhydrazine hydrochloride and tris-(dimethylamino)phosphine give the "barralene" compound (**5-XII**), (m.p. 116–7°C).

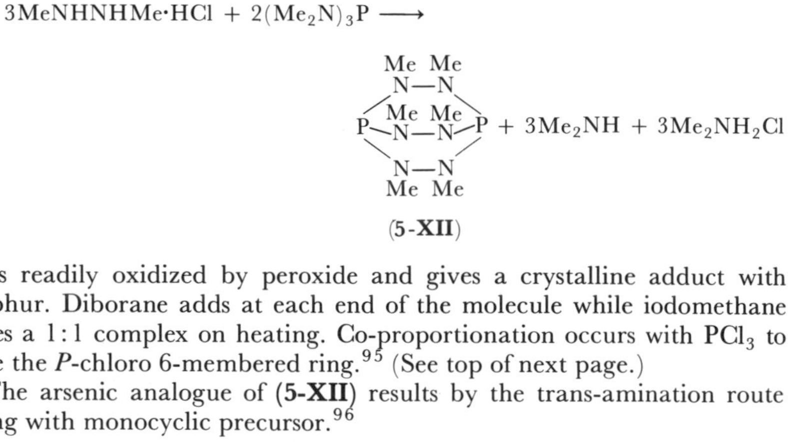

$$3MeNHNHMe\cdot HCl + 2(Me_2N)_3P \longrightarrow (\textbf{5-XII}) + 3Me_2NH + 3Me_2NH_2Cl$$

(5-**XII**)

It is readily oxidized by peroxide and gives a crystalline adduct with sulphur. Diborane adds at each end of the molecule while iodomethane gives a 1:1 complex on heating. Co-proportionation occurs with PCl_3 to give the *P*-chloro 6-membered ring.[95] (See top of next page.)

The arsenic analogue of (**5-XII**) results by the trans-amination route along with monocyclic precursor.[96]

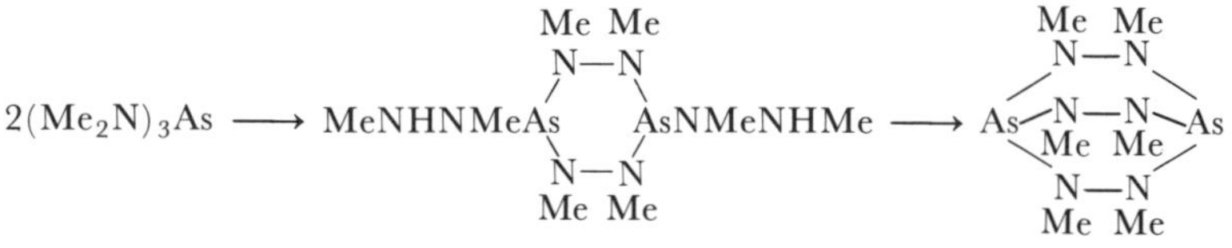

$$2(Me_2N)_3As \longrightarrow MeNHNMeAs(NMeNMe)_2AsNMeNHMe \longrightarrow As(NMeNMe)_3As$$

(m.p. 97–9°C)

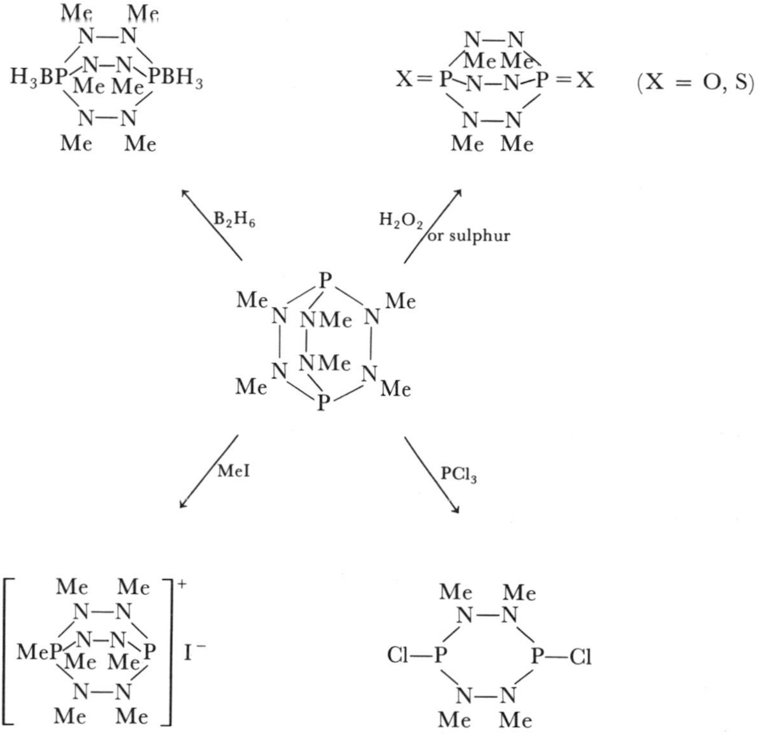

Me Me
N—N
H_3BP N—N PBH_3
Me Me
N—N
Me Me

N—N
Me Me
X = P—N—N—P = X (X = O, S)
N—N
Me Me

B_2H_6

H_2O_2 or sulphur

P
Me N NMe N Me
Me N NMe N Me
P

MeI

PCl_3

[Me Me
N—N
MeP N—N P
Me Me
N—N
Me Me]$^+$ I^-

Me Me
N—N
Cl—P P—Cl
N—N
Me Me

The synthesis of dimeric *P*-chlorinated phospha-V-azanes from PCl_5 and methylammonium chloride has already been outlined. The presence of a trace of water in the hydrochloride gives a second product analyzing for $(MeN)_6(P_4Cl_8)$, which is reported to show but a single ^{31}P n.m.r. signal.[70] This naturally led to structural proposals based on the adamantane cage which proved misleading as 5-co-ordinate phosphorus in such a cage is more hindered. The X-ray analysis showed a third structural isomer for the formula X_4Y_6 (where X is trivalent, Y divalent) based on linked 4-membered P_2N_2 rings (**5-XIII**).

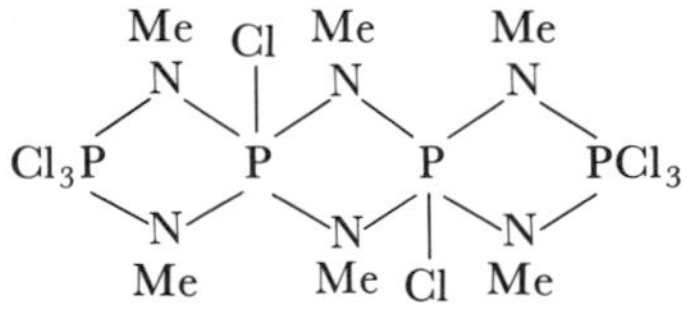

Me Cl Me Me
N N N
Cl_3P P P PCl_3
N N N
Me Me Cl Me

(**5-XIII**)

This is, in effect, four interconnected trigonal-bipyramidal centres of two types, *viz.* Cl_3PN_2 (terminal) and $ClPN_4$ (central). The bond lengths of these groupings are illustrated, showing the long axial bonds, which compare quite closely with those in $[Cl_3PNMe]_2$. The narrow N—P—N angle ($\sim$80 degrees) is also common to both compounds.[97]

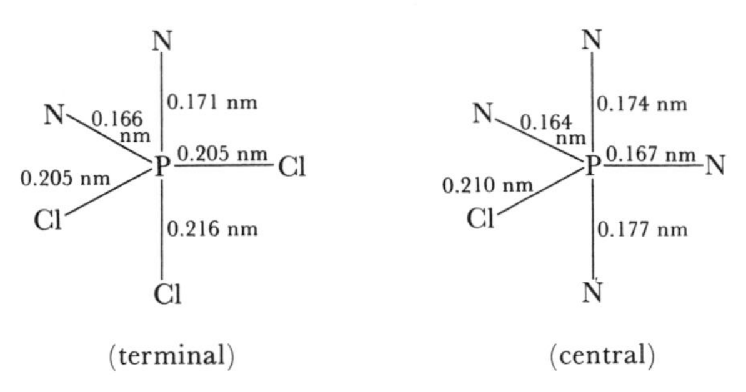

Fig. 5.20. The trigonal-bipyramidal environments of phosphorus in $(MeN)_6P_4Cl_8$

Bi- and tri-cyclic systems also result from *N*,*N*′-dimethylsulphamide and PCl_5 according to the sequence shown.[98]

$$(MeNH)_2SO_2 + PCl_5 \longrightarrow [Cl_3PNMe]_2$$

C_5H_5N ↓ : $O_2S(NMe)_2PCl(NMe)_2SO_2$ (m.p. 170–171.5°C)

$(MeNH)_2SO_2$ ↓ : $O_2S(NMe)_2PCl(NMe)_2PCl(NMe)_2SO_2$ (m.p. 221°C)

Cl_2 ⇄ $(MeNH)_2CO$: $OC(NMe)_2PCl(NMe)_2PCl(NMe)_2CO$

AgF ↓ : $[O_2S(NMe)_2]_2PF$ (m.p. 184–8°C)

The phospha-V-azenes

In Inorganic Chemistry, this is the most widely studied class of phosphorus compound.[c] They are conveniently formulated as $[R_2P{=}N]_n$ where n can vary from three to several thousand. In general, the small rings can be readily polymerized, and this has led to much industrial interest in the formation of inorganic, heat-resistant polymers.[d]

Synthesis

Phosphazenes were first prepared in 1895 by H. N. Stokes. The synthesis involved heating ammonium chloride with phosphorus pentachloride in sealed tubes between 150°C and 200°C.[99] The high pressure of HCl raised an explosion hazard, and subsequent modifications using a bubbler led to 90% conversion of PCl_5 to chlorophosphazenes, about a half being trimer and tetramer.

$$n\mathrm{NH_4Cl} + n\mathrm{PCl_5} \longrightarrow [\mathrm{Cl_2PN}]_n + 4n\mathrm{HCl}$$

The reaction is now usually conducted in a solvent such as sym-tetrachloroethane, and heating to 135°C for 20 h gives the expected mixture $[Cl_2PN]_n$. About 40% of the final product may be extracted with petrol and consists of a mixture of trimer and tetramer (ratio 3:1). Modifying the solvent, reaction temperature and the order of reactant addition has a most pronounced effect on the products.[100]

Slow addition of PCl_5 to a suspension of NH_4Cl in refluxing $(Cl_2CH)_2$ produces mainly cyclophosphazenes, while excess PCl_5 present all the time gives "end-capped" polymers $Cl_4P—[NPCl_2]_n—NPCl_3$ (n = 10–20). These products are oils which, like the larger rings, are relatively soluble in benzene.[101]

Bromophosphazenes can be prepared by a similar method, yields again varying with the conditions employed. Cyclic trimer and tetramer result at 150°C in sym-tetrachloroethane, while in $(Br_2CH)_2$ at 170°C the tetramer and higher homologues predominate. The yields of cyclic bromophosphazenes decrease appreciably above 160°C, while below 130°C large quantities of the adducts $NPBr_2PBr_5$ and $NPBr_2PBr_7$ form.[102]

Mixed bromochlorophosphazenes result from mixed starting materials and only the trimers have been isolated.[103] Fluorinated phosphazenes cannot be prepared directly but result from the fluorination of polymeric phosphorus nitride with NF_3 or CF_3SF_5. Alternatively, they can be obtained by the fluorination of chlorophosphazenes with potassium fluorosulphite.[104]

$$[\mathrm{Cl_2PN}]_4 + 8\mathrm{KSO_2F} \longrightarrow [\mathrm{F_2PN}]_4 + 8\mathrm{KCl} + 8\mathrm{SO_2}$$

While chlorophosphazenes are useful precursors to the substituted rings, organophosphazenes can often be conveniently prepared directly from an organophosphorus(V) halide. Dimethylphosphazenes can be readily prepared this way, the principal products being linear polymers which can be cyclized by heating with triethylamine and ammonium chloride in chloroform. The trimer and tetramer are both water-soluble, while the analogous ethyl compounds $[Et_2PN]_{3\text{ and }4}$ result when diethylphosphorus trichloride reacts with excess ammonia at $-78°C$, followed by heating at $250°C$.[105]

Fully-phenylated phosphazenes are formed from Ph_2PCl_3 and either ammonium chloride or liquid ammonia.

$$4nNH_3 + nPh_2PCl_3 \longrightarrow [Ph_2PN]_n + 3nNH_4Cl$$

The cyclic trimer and tetramer result in high yield when the intermediates formed in liquid NH_3 are heated to $275°C$. One such intermediate, $Ph_4P_2N_3H_4Cl$, has been isolated, and is readily converted at high temperature to the cyclic trimer and tetramer in good yield.[106]

Mixed phenylchloro- and phenylbromo-phosphazenes can be prepared similarly from the phenylphosphorus tetrahalide. Both *cis* and *trans* isomers have been isolated for the planar trimer, while three isomers of $[PhClPN]_4$ are known.[107] The isolation of two geometrical isomers of $Ph_3Br_2ClP_3N_3$, when the reaction below was performed in $(Cl_2CH)_2$, indicates solvent participation in the reaction.[103]

$$nNH_4Br + nPhPBr_2 + nBr_2 \longrightarrow [PhBrPN]_n + 4nHBr$$

$$\downarrow (Cl_2CH)_2$$

$$Ph_3Br_2ClP_3N_3$$

Aryl-substituted phosphazenes can also be prepared from halogenophosphazenes by a Friedel–Crafts reaction with benzene and aluminium chloride.[108] The reaction is slow, and substitution normally occurs geminally, with the final product being $[Ph_2PN]_3$.

This substitution is believed to involve the complex $[Cl_5P_3N_3]^+AlCl_4^-$ which attacks the benzene ring. Secondary arylation occurs on the substituted phosphorus atom.

$$[Cl_5P_3N_3]^+AlCl_4^- + C_6H_6 \longrightarrow PhCl_5P_3N_3 + AlCl_3 + HCl$$

Non-geminally substituted phosphazenes result if non-geminal bis-(dimethylamino)tetrachlorotriphosphatriazene is arylated and then deaminated.

```
      Cl Cl                       Cl Cl                      Cl Cl
       \/                          \/                         \/
     N=P-N        C6H6           N=P-N        2HCl          N=P-N
     |    ||     ------>         |    ||     ------>        |    ||
Cl-P=N-P-Cl      AlCl3      Ph-P=N-P-Ph                Ph-P=N-P-Ph
   |     |                     |     |                    |     |
   N     N                     N     N                    Cl    Cl
  Me2   Me2                   Me2   Me2
```

The thermal decomposition of phosphorus(III) azides gives phospha-V-azenes. Lithium and sodium azides undergo a metathetical reaction with PBr_3, $PhPCl_2$ or Ph_2PCl to yield thermally-unstable monoazides, *e.g.*,

$$PBr_3 + NaN_3 \xrightarrow{165°C} \frac{1}{n}[Br_2PN]_n + N_2 + NaBr$$

The phosphazenes formed are normally polymeric, and only with Ph_2PCl is any cyclic oligomer formed (33% tetramer).[109]

This method has also been used to prepare phenylated arsa-V-azenes,[110] which are difficult to synthesize, because of the instability of arsenic(V) tetra- and penta-chlorides and bromides.

$$Ph_2AsCl + LiN_3 \longrightarrow Ph_2AsN_3 \longrightarrow \tfrac{1}{4}[Ph_2AsN]_4 + N_2$$

The oxidation of phosphorus(III) compounds using chloramine readily yields phosphorus(V) amines and phosphazenes. With ammonia in $(Cl_2CH)_2$, diphenylchlorophosphine gives the solvated phosphazene trimer.[106] This reaction is thought to proceed through the isolable intermediate $[Ph_2P(NH_2)NPPh_2(NH_2)]^+Cl^-$, which is readily pyrolyzed to the trimer and tetramer.[111]

$$[Ph_2P(NH_2)NPPh_2(NH_2)]^+Cl^- \longrightarrow \frac{2}{n}[Ph_2PN]_n + NH_4Cl$$

With pure chloramine $Ph_2P(NH_2)ClNHPPh_2Cl_2$ results,

$$Ph_2PCl + NH_2Cl \longrightarrow Ph_2P(NH_2)Cl_2 \longrightarrow Ph_2P(NH_2)ClNHPPh_2Cl_2$$

which either loses HCl directly to give cyclic phosphazenes, or with ammonia giving the salt **(5-XIV)**

$$Ph_2P(NH_2)ClNHPPh_2Cl_2 \longrightarrow \underset{\textbf{(5-XIV)}}{[Ph_2P(NH_2)NPPh_2(NH)]^+Cl^-}$$

This intermediate is also formed in the hydrazine oxidation and ammonation of Ph_2PCl.

$$Ph_2PCl \xrightarrow[-HCl-NH_4Cl]{H_3NNH_2^+Cl^-} Ph_2P(Cl)NP(NH)Ph_2 \xrightarrow{NH_3} [Ph_2P(NH_2)NPPh_2(NH_2)]^+Cl^-$$

$$Ph_2P(Cl)NP(NH)Ph_2 \xrightarrow{-HCl} [Ph_2PN]_{3\text{ and }4} \xleftarrow{-NH_4Cl} [Ph_2P(NH_2)NPPh_2(NH_2)]^+Cl^-$$

In a similar manner, oxidizing diphenylchloroarsine with a chloramine–ammonia mixture in chloroform gives the unstable intermediate $[Ph_2As(NH_2)NAsPh_2(NH_2)]^+Cl^-\cdot CHCl_3$, which decomposes to $[Ph_2AsN]_3$ on standing. This also results from the direct oxidation of Ph_2AsCl with chlorine, followed by ammonolysis.[112]

Chlorination of $(CF_3)_2PNH_2$, followed by dehydrochlorination, gives trifluoromethylphosphazenes, as the trimer, tetramer and as polymers.[113]

$$(CF_3)_2PNH_2 + Cl_2 \longrightarrow (CF_3)_2P(NH_2)Cl_2 \xrightarrow{R_3N} [(CF_3)_2PN]_n$$

Unsymmetrical phosphazenes are generally synthesized in a step-wise manner. Thus, the 1,3-triamide salt $[Ph_2P(NH_2)NPPh_2(NH_2)]^+Cl$ can be readily cyclized with PCl_5 or $PhPCl_4$, though the product from the latter is only obtained in low yield.

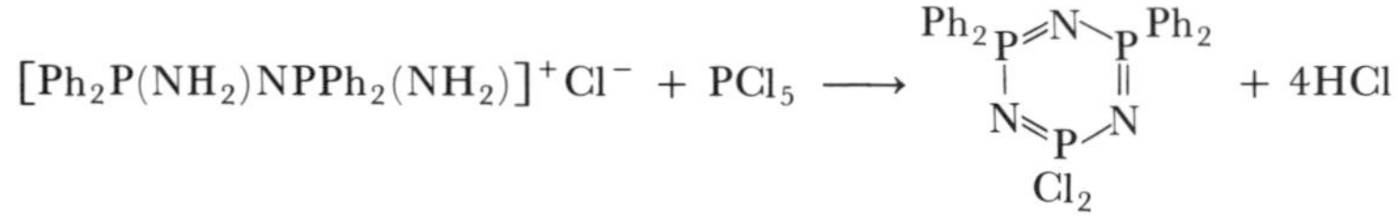

$$[Ph_2P(NH_2)NPPh_2(NH_2)]^+Cl^- + PCl_5 \longrightarrow \text{cyclo-}(Ph_2P{=}N{-}PPh_2{=}N{-}PCl_2{=}N) + 4HCl$$

(88% yield)

Rather unexpectedly, quantities of cyclotetraphosphazenes were also obtained.[114]

1,1-Bis-amino-3,3,5,5-tetrachlorotriphosphatriazene will successively condense with two moles of PCl_5 and one of $(Me_3Si)_2NMe$ to give a spiro compound.[115]

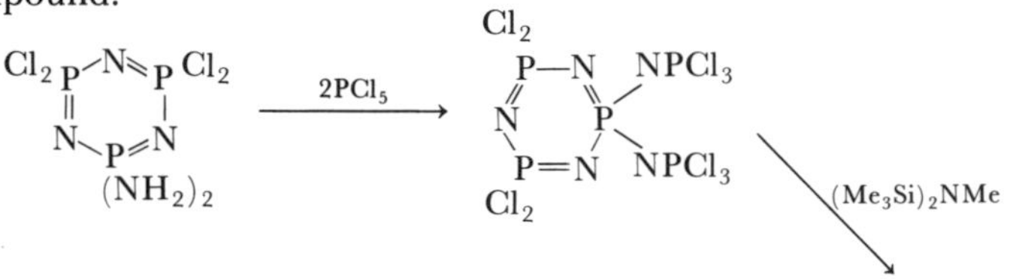

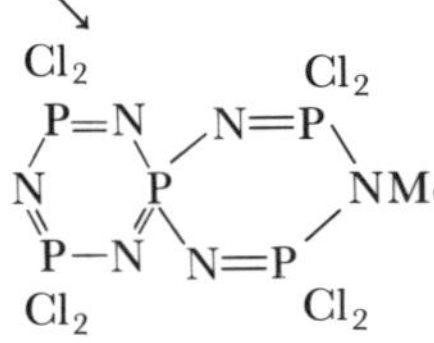

$$\text{cyclo-}(Cl_2P{-}N{=}PCl_2{-}N{=}P(NH_2)_2{-}N) \xrightarrow{2PCl_5} \text{cyclo-}(Cl_2P{-}N{=}PCl_2{-}N{=}P(NPCl_3)_2{-}N) \xrightarrow{(Me_3Si)_2NMe} \text{spiro compound}$$

(spiro product: $Cl_2P{=}N$, N, P, $P{-}N$, Cl_2 ring joined at P to $N{=}PCl_2{-}NMe{-}PCl_2{=}N$ ring)

The equilibrium

$$>P{-}X{-}H \rightleftharpoons {-}\overset{H}{\underset{|}{P}}{=}X$$

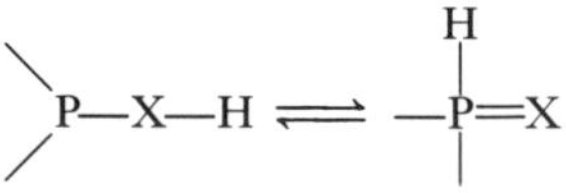

normally exists to the right when X = O and to the left when X = NR. In triphosphazenes, however, the right-hand form can be stabilized. Condensing 1-amino-3-iminodiphosphazene with triphenylphosphite or phenylphosphonites should yield cyclotriphosphazadienes (**5-XV**), but the properties are indicative of a phosphazatriene (**5-XVI**).

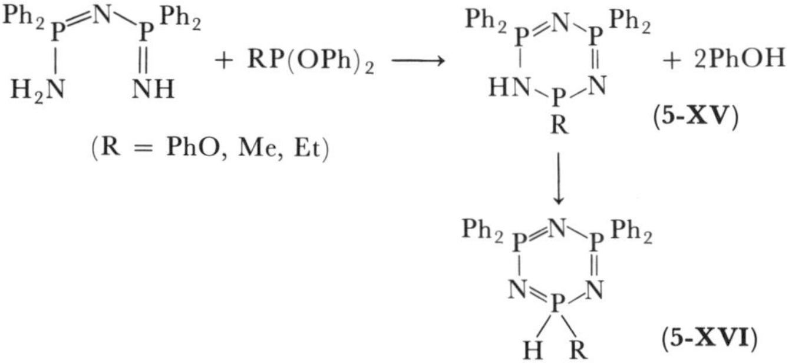

$$Ph_2P(NH_2){=}N{-}PPh_2{=}NH + RP(OPh)_2 \longrightarrow \text{cyclo-}(Ph_2P{=}N{-}PPh_2{=}N{-}PR{-}NH) + 2PhOH$$

(R = PhO, Me, Et) (**5-XV**)

↓

$$\text{cyclo-}(Ph_2P{=}N{-}PPh_2{=}N{-}PHR{=}N)$$

(**5-XVI**)

They are not oxygen-sensitive, have infrared absorptions typical of the P—H group and an n.m.r. coupling constant typical of a $N{=}P^{V}$—H group.[116]

Boron halides give cyclic compounds with $[Ph_2P(NH_2)NPPh_2(NH_2)]^+Cl^-$ but the single valency of three for boron precludes the tautomerism just discussed with phosphorus.

$$[Ph_2P(NH_2){=}N{-}PPh_2(NH)]^+Cl^- \xrightarrow{RBX_2} [\text{cyclo-}Ph_2P{=}N{-}PPh_2{-}NH{-}B(R){-}NH]^+X^- \xrightarrow{-HX} \text{cyclo-}Ph_2P{=}N{-}PPh_2{=}N{-}B(R){-}NH$$

(R = Cl, Br, Ph; X = Cl, Br)

These boraphosphanitrile rings are all crystalline solids and can be considered as mixed borazine/phosphazene compounds.[117]

The use of the nitrogen–halogen bond in oxidizing phosphines to phosphazenes is apparent in the reaction of the *N*-bromo diaza analogue of dimethylsulphone and bis-(diphenylphosphino)amine.[118]

$$Me_2S(NBr)_2 + (Ph_2P)_2NH \xrightarrow{-HBr} [\text{cyclo-}Me_2S{=}N{-}PPh_2{=}N{-}PPh_2{=}N]^+ Br^-$$

The wide academic and industrial interest in the synthesis of cyclic and linear polyphosphazenes, and the variation of the products with reaction conditions, has naturally led to a detailed investigation of the mechanism of phosphazene formation from ammonium chloride and phosphorus pentachloride.

There are two schools of thought regarding the mechanism and both successfully account for the listed experimental facts:[101]

(a) In boiling s-tetrachloroethane, the cyclic trimer, tetramer and high polymer phosphazenes result along with "end-capped" phosphazenes $Cl_3PN[Cl_2PN]_nPCl_4$ ($n \sim 10$).

(b) Maintaining an excess of PCl_5 gives a preponderance of linear compounds, while with ammonium chloride cyclic products predominate.

(c) Linear and cyclic phosphazenes are readily interconverted.

$$[Cl_2PN]_n \underset{(Cl_2CH)_2/\text{heat}/NH_4Cl}{\overset{PCl_5/\text{heat}}{\rightleftarrows}} Cl_3PN[Cl_2PN]_{n-1}PCl_4$$

The important step in the formation of these compounds appears to be the formation of $Cl_3P{=}NH$ as an intermediate.[e] Though this has never been isolated, both $Cl_3P{=}NPh$[119] and $Me_3P{=}NH$[120] are known. Both routes to this are quite plausible, the one involving the decomposition of $NH_4^+PCl_6^-$ and the other, ammonolysis of PCl_4^+.

$$NH_4Cl + PCl_5 \longrightarrow NH_4^+PCl_6^- \longrightarrow NH_2PCl_4 \longrightarrow NHPCl_3$$

$$2PCl_5 \rightleftharpoons PCl_4^+ + PCl_6^-$$

$$NH_4Cl \rightleftharpoons NH_3 + HCl$$

$$NH_3 + PCl_4^+ \longrightarrow [H_3N—PCl_4]^+ \xrightarrow{-HCl} [Cl_3PNH_2]^+ \longrightarrow Cl_3P{=}NH$$

The intermediate can then form the products by the two reaction routes proposed.

The first involves reaction with itself, or with more PCl_5, to give $H[NPCl_2]_nCl$ or $Cl_4P[NPCl_2]_nCl$. Loss of HCl or PCl_5 leads to cyclophosphazenes.

$$HNPCl_3 \xrightarrow[-HCl]{HNPCl_3} H[NPCl_2]_nCl \xrightarrow{-HCl} [NPCl_2]_n$$

$$HNPCl_3 \xrightarrow[-HCl]{PCl_5} Cl_4P[NPCl_2]_2Cl \xrightarrow[NH_4Cl]{PCl_5} Cl_4P[NPCl_2]_nCl$$

$$[NPCl_2]_n \underset{PCl_5}{\overset{-PCl_5}{\rightleftharpoons}} Cl_4P[NPCl_2]_nCl$$

Alternatively, Cl_3PNH may react with PCl_4^+, forming $[Cl_3P{=}NPCl_3]^+$, which is readily attacked by ammonia.[121]

$$[Cl_3P{=}N—PCl_3]^+ + NH_3 \longrightarrow Cl_3P{=}N—PCl_2{=}NH$$

Chain-lengthening occurs with PCl_4^+ or $[Cl_3PNPCl_3]^+$.

$$Cl_3PNPCl_2NH \xrightarrow{PCl_4^+} [Cl_3PNPCl_2NPCl_3]^+ \xleftarrow{Cl_3PNH} [Cl_3PNPCl_3]^+$$

$$Cl_3PNPCl_2NH \xrightarrow{[Cl_3PNPCl_3]^+} [Cl_3P(NPCl_2)_2NPCl_3]^+$$

While chain-lengthening and subsequent polymerization readily occurs, cyclization requires a terminal NH group. This forms with NH_4Cl.

$$[Cl_3P(NPCl_2)_2NPCl_3]^+ + Cl_3PNH \longrightarrow [Cl_3P(NPCl_2)_3NPCl_3]^+ + HCl$$

$$[Cl_3PNPCl_2NPCl_3]^+ + NH_4Cl \longrightarrow Cl_3PNPCl_2NPCl_2NH \xrightarrow{-HCl} [NPCl_2]_3$$

Yields of the tetramer are higher than expected, and this may be due to the dimerization of the unstable dimer $[NPCl_2]_2$, formed from Cl_3PNPCl_2NH.

The isolation of $[Cl_3PNPCl_3]^+[PCl_6]^-$, $[Cl_3PNPCl_2NPCl_3]^+[PCl_6]^-$, $Cl_3PNPCl_2NPCl_4$ and its PCl_5 adduct adds credence to the more complex second mechanism.[119, 122]

Properties

Extensive interest in the phosphazenes, particularly the chloro ones, stems from their potential as inorganic polymers. Thus, above 250°C, a cyclic chlorophosphazene can be converted to transparent rubbery compounds within hours. These comprise linear and branched polymers soluble in benzene along with insoluble cross-linked ones, with molecular weights up to 2×10^6. Hydrolysis in moist air is slow.

Polymerization occurs best between 270°C and 300°C; above 350°C, depolymerization and decomposition commences. Increasing the pressure favours the formation of high polymers but at a slower rate. Organo-oxygen compounds and metals will catalyze the polymerization, and free-radical initiators appear to have little effect. Indeed, oxygen is reported to inhibit the polymerization. This information, along with the pronounced increase in conductivity, is consistent with a mechanism involving the removal of a chloride ion.[123]

$$[Cl_2PN]_3 \xrightarrow[\text{catalyst}]{\text{heat or}} [Cl_5P_3N_3]^+Cl^-$$

$$[Cl_5P_3N_3]^+ + [Cl_2PN]_3 \longrightarrow$$

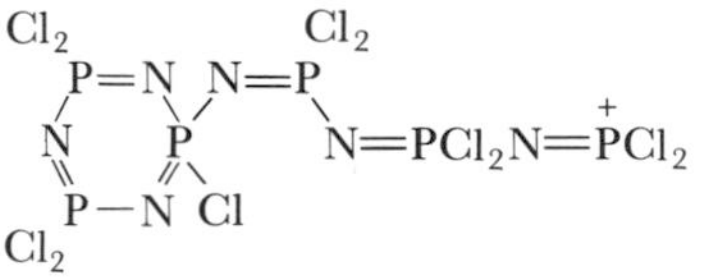

Cl₂ Cl₂
P=N N=P
N P N=PCl₂N=P⁺Cl₂
P—N Cl
Cl₂

The $=\overset{+}{P}Cl_2$ group can then induce electrophilic cross-linking and branching to give the wide variety of polymers encountered. Termination results on cooling, probably through *P*-chlorination.

$$—N{=}\overset{+}{P}Cl_2Cl^- \xrightleftharpoons{25°C} —N{=}PCl_3$$

Polymerization of the tetramer $[Cl_2PN]_4$ is slower than the trimer, possibly reflecting a greater shielding of the nitrogen atoms to electrophilic attack in this larger, puckered ring.

While the Cl_2PN unit remains intact on polymerization in the absence of moisture, an isomeric rearrangement of the $(RO)_2PN$ group occurs when alkoxyphosphazenes are heated to 200°C. The *N*-alkylphosphazanes are readily formed with simple alkoxy groups, but not with phenoxy or fluoroalkoxy ones.[124]

$$[(RO)_2PN]_3 \xrightarrow{200°C} [ROP(O)NR]_3 \ (R = Me, Et, Pr^n, Pr^i, PhCH_2)$$

$$[(RO)_2PN]_4 \xrightarrow{200°C} [ROP(O)NR]_4 \ (R = Me, Et)$$

The thermal stability of the fluoroalkoxyphosphazenes may be due to the strongly electron-withdrawing nature of the group. This will increase the strength of the P—N bond and lower the electron density at nitrogen, thereby reducing nucleophilic attack at the α-carbon atom of the alkyl group. The stability of phenoxy compounds may be similarly explained, and increasing the electron-withdrawing power of the *para*-substituent also increases thermal stability ($NO_2 \sim Cl > H > Me \sim OMe$).

The majority of the reactions of phosphazenes studied involve the interaction with some protic compound, whether this be active, as with water, or comparatively inert, as with benzene.

The halogenophosphazenes are hydrolyzed much more slowly than the phosphorus(V) halides. This is thought to occur in stages, involving hydroxyphosphazene and its tautomeric phosphazane.[99, 125]

$$[X_2PN]_3 \longrightarrow [(HO)_2PN]_3 \rightleftharpoons [HOP(O)NH]_3$$

The rate varies with pH, and under alkaline conditions, the trimetaphosphinate ion $[O_2PNH]_3^{3-}$ can be isolated. The P—N bond distance of 0.168 nm is about 0.01 nm larger than in phosphazenes, supporting the proposed tautomerism.[126]

This ion hydrolyzes further in acid to cyclic trimetaphosphate, and subsequently chain polyphosphates, pyro- and mono-phosphate.

$$[O_2PNH]_3^{3-} \xrightarrow{3H_3O^+} [PO_3]_3^{3-}$$

In the light of this, the slow hydrolysis of polymeric $[Cl_2PN]_n$ by atmospheric moisture probably involves initial cleavage of the P—Cl bond followed by cross-linking. This is marked by loss of elasticity in the polymer.

$$2\,Cl{-}\underset{|}{\overset{\overset{|}{N}}{\overset{\|}{P}}}{-}Cl + H_2O \longrightarrow Cl{-}\underset{|}{\overset{\overset{|}{N}}{\overset{\|}{P}}}{-}O{-}\underset{|}{\overset{\overset{|}{N}}{\overset{\|}{P}}}{-}Cl + 2HCl$$

Further hydrolysis results in chain cleavage, and the formation of ammonia and phosphate.

Fluorophosphazenes hydrolyze more readily, while organo-substituted phosphazenes are more stable to water than halogeno-substituted ones.[127] They are degraded by acid and base, however. Thus, while the alkyl esters are fairly stable to water, they can be decomposed by hot hydrochloric acid. It is therefore rather surprising that fluoroalkoxy-fluorophosphazenes $[CF_3(CF_2)_nCH_2OP(F)N]_n$ are resistant both to strong aqueous acid and base.[128]

Alkoxy- and aryloxy-cyclophosphazenes are normally prepared by the alcoholysis of halogenophosphazenes in the presence of base. Fully-substituted derivatives usually result, but partial substitution can be arranged. The non-geminal isomer of diphenoxytetrachlorocyclotriphosphatriazene results from phenol and triethylamine in non-polar solvents, while in acetone, tri-substitution occurs. With the phenoxide ion, substitution is partial in benzene and ether but complete in THF or monoglyme.[129]

While this substitution is non-geminal, probably for steric reasons, simple alkoxide substitution is often geminal, while with catechols it is exclusively so. With alkanediols, however, dehydration to the cyclic ether occurs.[130]

$$N{=}PCl_2 + HO(CH_2)_nOH \longrightarrow N{=}PO + 2HCl + \left[(CH_2)_nO\right] \quad (n = 3, 4, 5)$$

Non-geminal substitution with phenol can also be explained on electronic grounds, oxygen–phosphorus π-bonding decreasing its electrophilic character relative to the other phosphorus atoms. Full substitution occurs in polar solvents, probably through solvation weakening the P—Cl bonds still present.

The mercaptide anion substitutes geminally, and the extent depends on the solvent and size of the nucleophile. This is thought to be due to the initial thiolysis "softening" the substituted phosphorus atom and thereby encouraging further attack. Unlike the alkoxyphosphazenes, these thio derivatives do not isomerize on heating.[131]

These ideas are still speculative, however, as a comprehensive study of substitution has only been done with primary and secondary amines, though again much work still remains to be done.

Only the relatively unhindered primary amines will completely substitute hexachlorophosphazene at room temperature. Others only completely substitute on heating. With hindered secondary amines, complete substitution is impossible, even under the severest conditions. The tetrameric chlorophosphazene can be more readily substituted, and partial substitution can be effected with a limited amount of amine. Aziridinylphosphazenes, formed from ethylene imine, act as effective chemosterilants for houseflies.

The pattern of substitution has been closely studied for amines and occurs both geminally and non-geminally. The pattern appears to depend upon the basicity of the amine, the strongly nucleophilic ones substituting non-geminally.[132]

$$N_3P_3Cl_5(NR_2) \longrightarrow N_3P_3Cl_4(NR_2)_2 \text{ (geminal: } Cl_2P{=}N{-}P(NR_2)_2\text{)} \quad \text{or} \quad N_3P_3Cl_4(NR_2)_2 \text{ (non-geminal: } PCl(NR_2) \text{ at two P atoms)}$$

Partial replacement not only gives rise to positional, but also to *cis–trans* isomerism. Thus, di-substitution of $[Cl_2PN]_3$ with dimethylamine gives the non-geminal *cis*-bis-(dimethylamino) isomer.

$$[Cl_2PN]_3 + 4Me_2NH \longrightarrow (Me_2N)ClP(NPCl_2N)PCl(NMe_2)\ (\text{ring}) + 2Me_2NH_2Cl$$

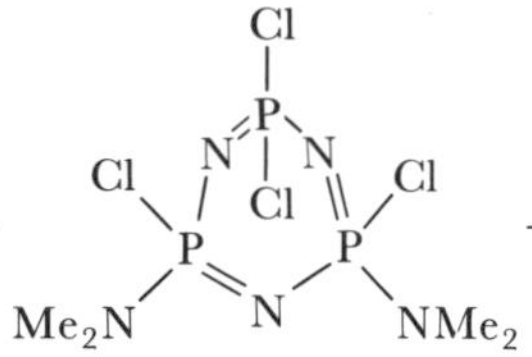

Cl
P
N N
Cl Cl Cl
P P
Me_2N N NMe_2

With tri-substitution, both *cis* and *trans* isomers result. This non-geminal substitution can be conveniently explained through N—P π-bonding decreasing the electrophilic character of the aminated phosphorus atom relative to the unsubstituted ones. Indeed, in $[(Me_2N)_2PN]_4$, the P—NMe_2 residues are planar, with the P—N bond about 0.007 nm shorter than expected for a single bond, showing that basic amino groups will reduce the electrophilic character of the phosphorus atom to which they are bonded. Arylamines, which are weaker bases, tend to substitute geminally, as does t-butylamine.

However, the group bonded to the ring does not always determine the position of substitution of a second amine molecule. Thus, while the t-butylamino group directs Bu^tNH_2 geminal, $EtNH_2$ goes non-geminal. Similarly, the ethylamino group directs $EtNH_2$ non-geminal and Bu^tNH_2 geminal.

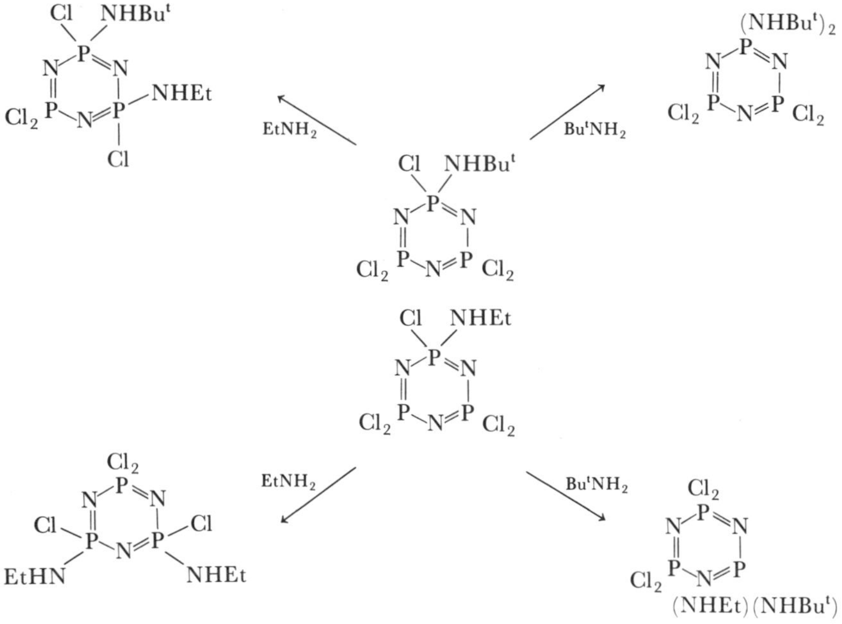

Cl $NHBu^t$
P
N N
Cl_2 P N P NHEt
Cl

$EtNH_2$

Bu^tNH_2

$(NHBu^t)_2$
P
N N
Cl_2 P N P Cl_2

Cl $NHBu^t$
P
N N
Cl_2 P N P Cl_2

Cl NHEt
P
N N
Cl_2 P N P Cl_2

$EtNH_2$

Bu^tNH_2

Cl_2
P
N N
Cl P N P Cl
EtHN NHEt

Cl_2
P
N N
Cl_2 P N P
$(NHEt)(NHBu^t)$

Consequently a mechanistic analysis of the substitution patterns of the phosphazene rings is far from complete and must consider more deeply the role of the incoming nucleophile as well as the groups already bonded to the ring.[133]

Reactions with Grignard reagents. The synthesis of arylphosphazenes directly from phenylphosphorus halides, or by a Friedel–Crafts reaction has already been mentioned (p. 298). Arylation with a Grignard reagent is more effective, however, especially with the triphosphazene.

$$[Cl_2PN]_3 + 6PhMgBr \longrightarrow [Ph_2PN]_3$$

With the tetramer $[Cl_2PN]_4$, two tetraphenyl and two octaphenyl products result, depending upon the conditions employed.

Above 100°C, 1,1,5,5-tetraphenyl-3,3,7,7-tetrachlorocyclotetraphosphatetrazene forms, while at room temperature the predominant product involves ring contraction. This is rather similar to the base-catalyzed contractions of cyclosiloxazanes. Hydrolysis of the two phosphazenes gives products expected of the two structures.[134]

$[Cl_2PN]_4$

100°C (PhMgBr) → cyclic $Ph_2P{-}N{=}PCl_2$ / N / N / $Cl_2P{=}N{-}PPh_2$ (m.p. 212.5°C)

room temp. (PhMgBr) → cyclic $Cl_2P{=}N{-}P(Ph)(NPPh_3){=}N{-}PCl_2{=}N$ (m.p. 181°C)

Hydrolysis

$$2Ph_2P(O)OH + 2H_3PO_4 \qquad Ph_3PO + PhP(O)(OH)_2 + 2H_3PO_4$$

Phenylation readily occurs with phenyl-lithium, while the alkyl Grignard reagent successfully yields organophosphazenes. Simple alkylation with lithium alkyls is impossible, and only limited amounts of simple phenylphosphazenes result from diphenylmagnesium and $[Cl_2PN]_3$. The products do include 7% of a compound analyzing for $Ph_2Cl_8P_6N_6$. The structure, with two rings coupled by a P—P bond, is supported by a strong infrared absorption at 1215 cm^{-1} typical of the cyclotriphosphazene ring and hydrolysis products including phosphoric and phenylphosphoric acids as the only phosphorus residues.[135]

$$[Cl_2PN]_3 \xrightarrow[\text{dioxan}]{Ph_2Mg} \begin{array}{c} Cl_2 \quad\quad Ph \quad\quad Cl_2 \\ P{=}N / \quad N{-}P \\ N \quad\; P{-}P \quad\; N \\ P{-}N \quad / N{=}P \\ Cl_2 \quad\quad Ph \quad\quad Cl_2 \end{array} \xrightarrow{\text{hydrolysis}}$$

$$4H_3PO_4 + 2PhP(O)(OH)_2$$

Such coupling also occurs with thiophosphoryl chloride and alkyl Grignard reagents, to give tetra-alkyldiphosphine disulphides.

$$2PSCl_3 \xrightarrow{RMgX} R_2P(S)P(S)R_2$$

Adducts and complexes. The ring nitrogen atoms are only weakly basic, and protonation only occurs if the side groups supply electrons. Under these circumstances, salt formation occurs with hydrogen halides, and normally the nitrogen of the ring is protonated and not the electron-rich exocyclic group. With amino groups, the $[PN]_3$ ring is a better base than $[PN]_4$, but with OR substituents, this order is reversed.

Basicity increases with the inductive effect of R, with CF_3CH_2O < $PhCH_2O$ < MeO < EtO ~ Bu^nO < Pr^iO, and this readily explains the stability of $[(CF_3CH_2O)_2PN]_n$ to heat and subsequent isomerization. Alkylthiophosphazenes show a similar trend.

Perchloric acid also forms adducts with the chlorinated tri- and tetra-phosphazenes,[136] and a structure determination of the hydrochloride of $(Pr^iNH)_4Cl_2P_3N_3$ has now been determined and exhibits several interesting features.

$$\begin{array}{c} Cl \\ \vdots \\ H \\ | \\ (Pr^iNH)_2P \;\;\; N \;\;\; P(NHPr^i)_2 \\ N \quad P \quad N \\ Cl \quad\; Cl \end{array}$$

Fig. 5.21. The hydrochloride of $(Pr^iNH)_4Cl_2P_3N_3$

Protonation occurs on the endocyclic nitrogen as predicted, while the N—PCl_2 distances of 0.156 nm are in accordance with reported values. The exocyclic P—N bonds are about 0.161 nm, shorter than in $[(Me_2N)_2PN]_4$, while the ring bonds involving the protonated nitrogen are 0.167 nm in length. This appears to indicate a reversal in the π-bonding roles of the two kinds of nitrogen atom, the exocyclic ones fulfilling the stronger π-bonding role to phosphorus when the ring nitrogen is protonated.[137]

Various other Lewis acids form adducts with phosphazenes, notably aluminium halides,[138] sulphur trioxide,[139] iodine monochloride,[140] tin and titanium tetrachlorides.[141] The bonding in these complexes varies quite considerably. Chlorophosphazenes give ionic products with aluminium chloride, as does hexaphenylcyclotriphosphatriazene with ICl, both in the pure state and in concentrated HCl.

$$[Cl_2PN]_3 + AlCl_3 \longrightarrow [Cl_5P_3N_3]^+[AlCl_4]^-$$

$$[Ph_2PN]_3 + ICl \xrightarrow{HCl} [Ph_6P_3N_3H]^+[ICl_2]^-$$

The adduct $Cl_6P_3N_3{\cdot}3SO_3$ is thought to involve *N*-co-ordination, as with $[Me_2PN]_3$ and the tetrachlorides of tin and titanium.

Methylphosphazenes are readily alkylated with methyl or ethyl iodide at the endocyclic nitrogen atom.[142]

$$[Me_2PN]_4 + EtI \longrightarrow [Me_8EtP_4N_4]^+I^-$$

With hexakis-(dimethylamino)cyclotriphosphazene, alkylation with trimethyloxonium tetrafluoroborate occurs, rather surprisingly, on the amino group, since the nitrogen of the ring is more basic and less sterically hindered.[143]

$$[(Me_2N)_2PN]_3 + 2Me_3O^+BF_4^- \longrightarrow$$

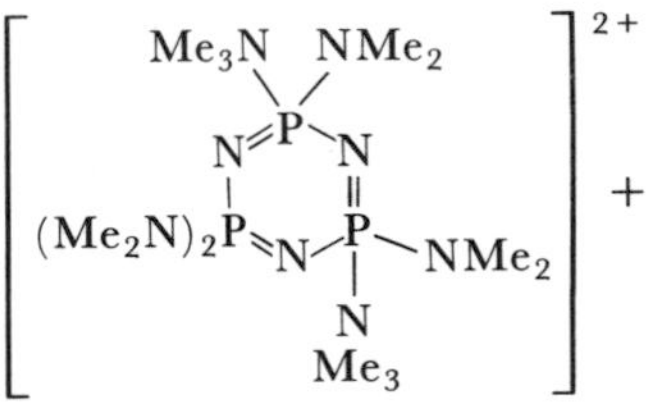

```
 ┌                                     ┐ 2+
 │          Me3N  NMe2                 │
 │             \  /                    │
 │            N=P\N                    │
 │            |    ||                  │     + 2BF4⁻ + 2Me2O
 │   (Me2N)2P=N/P\NMe2                 │
 │                |                    │
 │                N                    │
 │               Me3                   │
 └                                     ┘
```

This tends to indicate that exocyclic charge transfer from nitrogen to phosphorus is not as great as anticipated from the P—N bond lengths of $[(Me_2N)_2PN]_{4\ \text{and}\ 6}$, which are about 0.01 nm shorter than expected for a single bond (0.178 nm).

There has been much speculation and experimentation since the isolation and characterization of π-bonded aromatic transition metal compounds, *e.g.*, $C_6H_6Cr(CO)_3$, as to whether purely inorganic "aromatic" systems would bond in a similar way. The isolation of boron–nitrogen analogues of benzene and the cyclopentadienyl anion π-bonded to metal–carbonyl residues shows this to be possible with rings involving atoms of similar size, but, to date, only σ-complexes of the phosphazene ring are known.

Two *cis* carbonyl groups in $Mo(CO)_6$ can be displaced by $[Me_2PN]_4$. Nitrogen atoms in positions 1 and 5 are thought to co-ordinate.

$$Mo(CO)_6 + [Me_2PN]_4 \longrightarrow [Me_2PN]_4Mo(CO)_4 + 2CO$$

The analogous reaction with $[Me_9P_4N_4]^+I^-$ gives an ionic product resulting from iodide attack on the carbonyl.[144]

$$[Me_9P_4N_4]^+I^- + Mo(CO)_6 \longrightarrow [Me_9P_4N_4]^+Mo(CO)_5I^-$$

Theoretical and Structural Aspects of Phosphazene Chemistry

The simple formulation for the structure of a phosphonitrilic halide involves alternate single and double bonds around the ring. While this is obviously in accordance with the valencies of phosphorus and nitrogen, it does nothing to explain the approximate equivalence of all the P—N bond lengths, irrespective of whether the phosphazene ring is planar. The situation with the C—C bonds in benzene is explained by resonance or, through a molecular orbital treatment, by delocalization of electron pairs over the whole ring in *A*- and *E*-type orbitals.

The non-planar nature of some of the P—N rings renders this kind of description only partly valid, while the presence of low-lying d orbitals, and their ability to participate in π-bonding, obviously introduces complications.

So the bonding of the single —N=PCl_2— group must be first considered. The four σ-bonds at phosphorus can be formed through sp^3-hybridized orbitals on phosphorus and sp^2 on nitrogen. The shortening of the P—N bond in these rings from the value of 0.178 nm expected for a single bond indicates considerable multiple bonding which must involve the d orbitals of phosphorus. The short P—N bonds in $[Cl_2PN]_3$ and $[F_2PN]_3$ can be conveniently accounted for in this way.[f]

With nitrogen sp^2-hybridized and $P\hat{N}P$ 120 degrees, π-bonding between nitrogen and phosphorus involves the p orbital of nitrogen and d_{xz} of phosphorus.

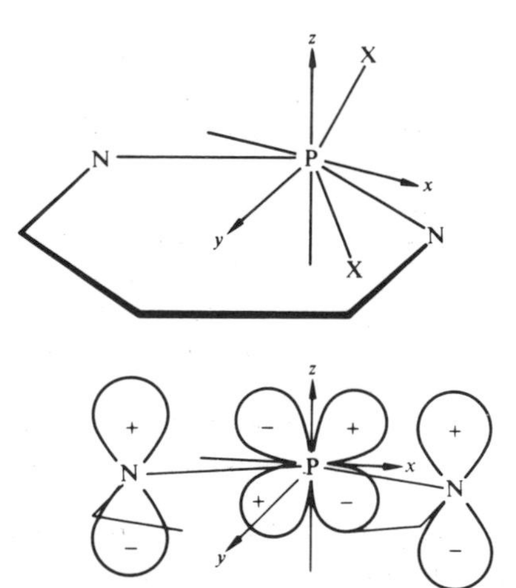

Fig. 5.22. π_a-Bonding in the phosphazene ring[c,f]

Such bonding is conveniently represented as π_a, since it is antisymmetric to reflection in the molecular plane.

π-Bonding in the plane of the ring is also possible, however, and in the case of the planar trimers just mentioned, would involve the d_{xy} and $d_{x^2-y^2}$ orbitals on phosphorus with the sp^2-hybridized lone-pair of nitrogen. This π-bonding is symbolically represented as π_s as it is symmetric to reflection in the molecular plane.

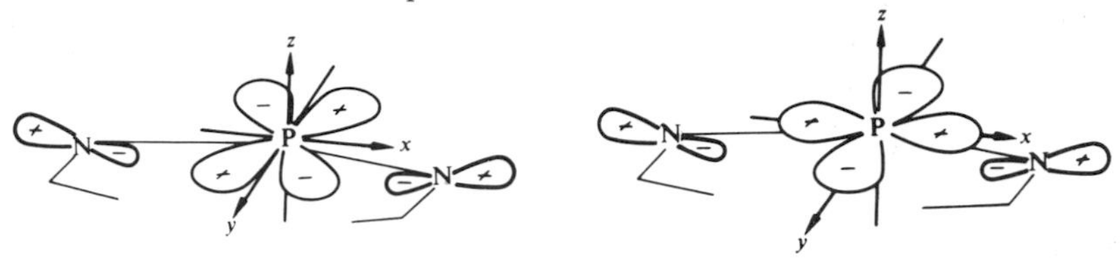

Fig. 5.23. π_s-Bonding in the phosphazene ring[c,f]

Overlap integrals readily show that π_s-bonding using $d_{x^2-y^2}$ is the stronger. Widening the P—N—P angle decreases π-bonding since the large positive lobe of the nitrogen lone-pair overlaps better, while the decrease in s character increases the size and overlap of the negative lobe of the lone pair with the $d_{x^2-y^2}$ orbital of phosphorus.

So both π_a- and π_s-bonding will shorten the P—N bond, and overlap integrals show that both types of bonding are effective even when the ring is non-planar. This means that an NPN residue can be considered in isolation with three atom centres, or "islands", of π_a- and π_s-delocalization.[145] This is similar to the π-allyl group encountered in metal complexes, *e.g.*, bis-(π-allyl)nickel $(\pi\text{-}C_3H_5)_2Ni$, and will lead to equivalent P—N bonds.

A careful comparison of the structures of the compounds $[F_2PN]_{3 \text{ and } 4}$, $[Cl_2PN]_{3,\ 4 \text{ and } 5}$ readily illustrate these points,[146] and the variation of the P—N bond strength shows up through infrared and n.m.r. spectral data. The important ring parameters are shown in Table 5.1.

Table 5.1. Structural Data for Cyclic Phospha-V-azenes

Compound	P—N (nm)	N$\hat{P}$N (degrees)	P$\hat{N}$P (degrees)	Structure
$[F_2PN]_3$	0.1560	119.4	120.3	planar
$[F_2PN]_4$	0.151	122.7	147.0	planar
$[Cl_2PN]_3$	0.1595	120.0	119.0	planar
$[Cl_2PN]_4$	0.1570	121.2	131.3	puckered
$[Cl_2PN]_5$	0.1521	118.4	148.6	nearly planar
$[Me_2PN]_4$	0.1596	119.8	131.9	puckered

The shortening of the P—N bonds from the length of 0.178 nm encountered for a P—N single bond[147] supports the presence of π-bonding, and their equivalence is reminiscent of the borazoles. With the larger rings, however, a variation in the P—N—P angles around the ring is reflected in the P—N bond lengths. Increasing the angle produces a decrease in the P—N distance, supporting the theory of increased π_s-bonding. The other ring parameters, notably $N\hat{P}N$, $X\hat{P}X$ and the P—X bond length, remain reasonably constant.

Considering the two fluorides first, both are planar, but the tetramer, with its wide P—N—P angle and more effective π_s-bonding, has the shorter P—N bond length. This is also reflected in its poor solubility in sulphuric acid, due to poor basicity. These fluorinated phosphazenes complex with good acceptors such as SbF_5.

$$[F_2PN]_n + 2SbF_5 \longrightarrow [F_2PN]_n{\cdot}2SbF_5 \ (n = 3\text{–}6)$$

As hexamethylcyclotriphosphazene will only form a 1:1 complex with SbF_5, through nitrogen co-ordination, it seems likely that the fluorophosphazene complexes involve fluoride co-ordination.[142, 148] Ionization is unlikely as the complexes are volatile and the fluorophosphazene can be recovered after hydrolysis of SbF_5. The infrared spectra of $F_6P_3N_3$ and $F_6P_3N_3{\cdot}2SbF_5$ are quite similar, the major differences arising through a lowering of the symmetry of the molecule. The degenerate $\nu_{(PF_2)_s}$ is split, while the strong degenerate P—N stretching band at 1300 cm^{-1} is increased by 20 cm^{-1}. This normally disappears completely on nitrogen co-ordination.

With the three cyclic chlorides, again bond shortening occurs with an increase in the P—N—P angle. The chemical shift increases from $[Cl_2PN]_3$ to $[Cl_2PN]_4$ and steadies out at the pentamer and above. This reflects an increase in the electron density around the phosphorus as the ring size increases. In a complementary way, the change in the OH stretching frequency when the chlorophosphazene hydrogen bonds through the nitrogen lone-pair shows that the trimer is the strongest base, hydrogen-bonding the most strongly and hence producing the biggest shift.

Table 5.2. N.m.r. and Infrared Spectral Shifts Relating to Ring Size in the Cyclic Chlorophospha-V-azenes

$[Cl_2PN]_n$	$n =$	3	4	5	6	7
Shift (ppm) relative to 85% H_3PO_4		−20	7	17	16	18
$\Delta\nu_{OH}$		135	125	115	105	95

Protonation of chlorophosphazenes occurs readily in sulphuric acid, and the ^{31}P n.m.r. absorptions are all reduced by this *N*-protonation to the value of the neutral trimer. Again, these facts fit the theory of π_s-bonding which is weak in the trimeric chloride, as would be expected for a P—N—P angle of 120 degrees.

The differences in the heats of formation of the $NPCl_2$ unit from a trimer and higher polymers show that in the case of the trimer, this is less, as expected for a weaker P—N bond. A steady increase occurs above the tetramer.[149]

Table 5.3. Variation in Heat of Formation of P—N *Linkage in Cyclic Chlorophospha-V-azenes*

n in $[Cl_2PN]_n$	3	4	5	6	7
$\bar{E}(P—N)_n - \bar{E}(P—N)_3$ (kJ mol^{-1})	0	1.63	2.26	2.51	2.59

The significant difference in the structure of the trimeric fluoride and chloride is the shorter P—N bond of the former. This difference is probably best understood by comparing it with the difference in axial P—Cl bond lengths in PCl_5 and PCl_2F_3. The 0.014 nm shortening on substitution of fluorine is probably due to the stronger electron-withdrawing nature of fluorine contracting the phosphorus orbitals. This would shorten the P—Cl bond. This conveniently explains the difference with the trimeric phosphazenes, though the effect (only 0.0025 nm) is probably masked by π-bonding.

However, extending to other phosphazenes shows that the idea is probably a realistic one. Thus, $[Me_2PN]_4$ has a P—N bond length 0.026 nm longer than in $[Cl_2PN]_4$, though the P—N—P bond angles are similar. As π_s-bonding is similar with each compound, the increase may well be due to a swelling of the phosphorus orbitals by the less electronegative methyl group.[146]

In *gem*-diphenyltetrafluorocyclotriphosphazene,[150] the P—N bonds associated with the Ph_2P group are 0.162 nm, while the others are much shorter, at 0.154–0.156 nm, again supporting the idea that electronegative groups lead to shorter P—N bonds, all other things being equal.

The tetrameric phosphazene $[Me_2PN]_4$ can be readily protonated. The ion can be stabilized as a copper or cobalt salt, both compounds forming from the anhydrous metal chloride in methyl ethyl ketone.

$$\underset{\text{(blue)}}{[(Me_2PN)_4H^+]_2CoCl_4^{2-}} \xleftarrow{CoCl_2} [Me_2PN]_4 \xrightarrow{CuCl_2} \underset{\text{(yellow)}}{[Me_2PN]_4HCuCl_3}$$

The copper complex is not ionic, the copper being bonded with a ring nitrogen atom. The cobalt complex is a conductor, involving the phosphazenium ion and $CoCl_4^{2-}$.[151]

In $[Me_2PN]_4HCuCl_3$, the proton and $CuCl_3$ residue are bonded to nitrogen atoms on opposite sides of the tub-shaped 8-membered ring. See Figure 5.24.

Copper is surrounded in a square-planar manner by one nitrogen, and three chlorine atoms. The P—N bond lengths vary in pairs as listed 0.163,

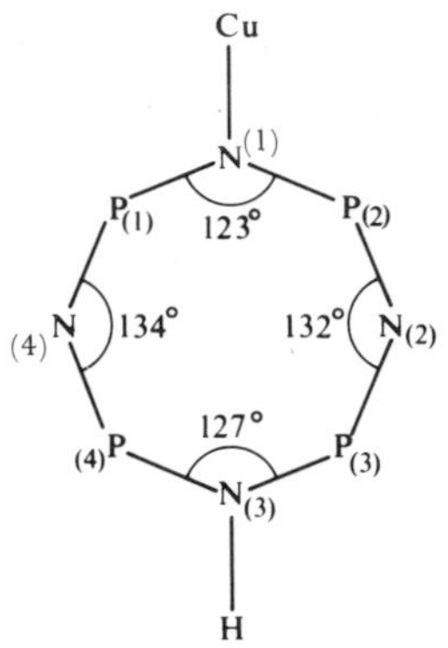

Fig. 5.24. Bond angles of the $[PN]_4$ ring in $[Me_2PN]_4HCuCl_3$

0.160, 0.156 and 0.167 nm from $N_{(1)}$—$P_{(1)}$, $N_{(1)}$—$P_{(2)}$ pair to the $N_{(3)}$—$P_{(3)}$, $N_{(3)}$—$P_{(4)}$ pair (0.167 nm). Complexing the nitrogen atoms with Cu or H will decrease π_s-bonding but not π_a-bonding, since the p_z orbital on nitrogen is not involved in complexing. While π_s-bonding may well still occur in the $CuNP_2$ skeleton (P—N 0.163 nm) it is less likely in the HNP_2 skeleton. The proton is a better acceptor, and the P—N length (0.167 nm) is close to that observed in $[NHPO_2]_3^{3-}$ and $(NH)_4P_4O_8H_4$, where only π_a-bonding can occur. The bond angles at nitrogen in these phosphinates are also close to those observed at $N_{(1)}$ and $N_{(3)}$.

With the $[(Me_2PN)_4H]^+$ cation, a similar variation in the P—N bond length occurs to that observed in the copper complex. The mean value of 0.1695 nm for the HNP_2 residue supports the presence of only weak π-bonding (π_a only).

OXYGEN DERIVATIVES

Anionic compounds

Although boron, silicon and phosphorus all form strong bonds with oxygen, only with the former two elements is the existence of minerals with oxyanions of cyclic structure apparent. With phosphorus, orthophosphate minerals predominate, though cyclic polyphosphates can be readily synthesized.[g]

The structural chemistry of these metaphosphates closely parallels that of the metasilicates, with structural determinations on the cyclic anions $P_3O_9^{3-}$, $P_4O_{12}^{4-}$ and $P_6O_{18}^{6-}$. In addition, chromatographic work supports the existence of the series $[PO_3]_n^{n-}$ (n = 3 to 8).

The progenitor of all trimetaphosphates appears to be the sodium salt.[152] This results from heating NaH_2PO_4 to between 500°C and 600°C and the well-formed crystals are now manufactured on a large scale.

$$3NaH_2PO_4 \longrightarrow Na_3P_3O_9 + 3H_2O$$

The same phosphate also results on treating sodium pyrophosphate with ammonium chloride.

$$3Na_4P_2O_7 + 6NH_4Cl \longrightarrow 2Na_3P_3O_9 + 6NaCl + 6NH_3 + 6H_2O$$

The majority of metaphosphates are soluble in water, particularly those of the alkali metals, of silver, copper and calcium. They are, however, less strong complexing agents than the chain phosphates, and in line with this, form strong acids, unlike the chain phosphoric acids, which dissociate less readily.

These ring phosphates are stable in neutral solution, but low or high pH solutions degrade them to simpler phosphates.

The most convenient synthetic route to tetrametaphosphates involves either treating the soluble form of P_2O_5 with sodium bicarbonate, or by heating equimolar proportions of NaH_2PO_4 and H_3PO_4 at 400°C.[153]

$$2P_2O_5 + 4NaHCO_3 \xrightarrow{0°C} Na_4P_4O_{12} + 2H_2O + 4CO_2$$

$$2NaH_2PO_4 + 2H_3PO_4 \xrightarrow{400°C} Na_2H_2P_4O_{12} + 4H_2O$$

Tetrametaphosphate salts can readily be precipitated using heavy metal ions, and hydrolyze under basic conditions more slowly than trimetaphosphate.[154] Above 400°C, $Na_4P_4O_{12}$ is irreversibly transformed into the trimetaphosphate.

Two-dimensional paper chromatography employing both acid and basic media shows that vitreous sodium metaphosphate (Graham's salt) breaks down into metaphosphates with up to eight phosphorus atoms in the ring. The hexamer has also been independently synthesized by pyrolyzing orthophosphoric acid and lithium carbonate.

$$6H_3PO_4 + 3Li_2CO_3 \longrightarrow Li_6P_6O_{18} + 3CO_2 + 9H_2O$$

The molecular structure of the trimetaphosphate ion in the solid state confirms the symmetry group proposed from vibrational spectra of the solid as C_{3v}.

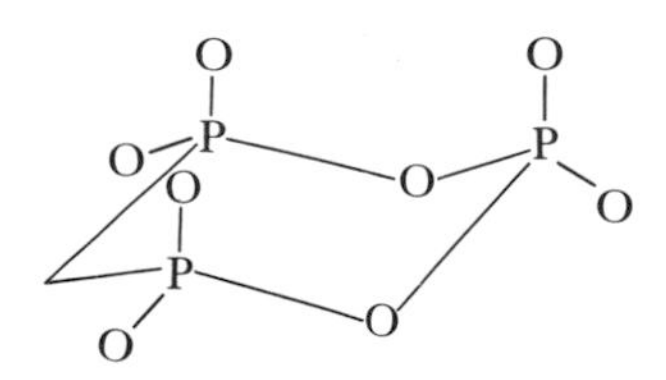

Fig. 5.25. The chair structure of trimetaphosphate

It crystallizes as a chair structure with P—O bond lengths of 0.1615 nm (endocyclic) and 0.1484 nm (exocyclic). Even the larger one is about 0.01 nm shorter than would be anticipated for a single bond, supporting the presence of p_π–d_π bonding in all P—O bonds.[155] In solution however, the vibrational spectra are consistent with a D_{3h} structure.[156]

With the tetrametaphosphate ion a more complicated picture arises. There appear to be two different configurations for this anion, the "boat" and "chair" forms, both of which are sufficiently stable in solution to be isolated as different salts, though the "boat" form slowly converts into the "chair" form.

The crystal structure of the "chair" form of $(NH_4)_4P_4O_{12}$ gives equal ring (0.1607 nm) and terminal (0.1479 nm) P—O bonds. The vibrational spectroscopic data also supports a C_{2h} structure.[156,157]

The structure of the hexametaphosphate ion also shows equivalence in the two kinds of P—O bonds, both in close agreement with those observed for the tri- and tetra-homologues.

As with the trimetaphosphate, a "chair" configuration is indicated for trimeta-arsenate in the solid state.[158] This is best synthesized from sodium dihydrogen arsenate.[159]

$$3NaH_2AsO_4 \longrightarrow Na_3As_3O_9 + 3H_2O$$

There appears to be no conclusive evidence yet for the existence of higher meta-arsenates and this tendency for the larger atoms of this group to form the smaller rings is reflected in the structures of the complex fluoro-oxo anions of arsenic and antimony. Complex hydroxyfluorides readily lose HF to give oxo-bridged dimers.[160]

$$2K^+MF_5OH^- \longrightarrow 2K^+\left[F_4M\begin{matrix}O\\ \\ O\end{matrix}MF_4\right]^{2-} \quad (M = As, Sb)$$

The mixed oxychloride $[SbCl_4OPOCl_2]_2$ is a dimeric solid formed in two ways from antimony pentachloride, using either a $Cl_2O/POCl_3$ mixture or pyrophosphoryl chloride $Cl_2P(O)\cdot OPOCl_2$.[161] The structure is thought to involve an 8-membered ring, and with pyridine this is

$$POCl_3 + Cl_2O + 2SbCl_5 \longrightarrow$$
$$P_2O_3Cl_4 + 2SbCl_5 \longrightarrow$$
$$Cl_4Sb(-O-P(=O)Cl_2-O-)_2SbCl_4 \quad \text{(m.p. 147°C)}$$

progressively cleaved to give initially the 6-co-ordinate complex (**5-XVII**) and then the complex salt.

$$[SbCl_4OPOCl_2]_2 + 2py \longrightarrow 2pySbCl_4OPOCl_2 \quad (\textbf{5-XVII})$$

$$pySbCl_4OPOCl_2 + py \longrightarrow [py_2SbCl_4]^+PO_2Cl_2^-$$

Oxides of group V

Trivalent State

Though the oxides of nitrogen are non-cyclic and often bonded through $p_\pi-p_\pi$ bonds, those of the heavier elements involve cyclic and cage structures involving single bonds supplemented by $p_\pi-d_\pi$ bonding, as invoked to explain the short bonds in the oxo-anions.

Phosphorus trioxide results when white phosphorus is burned at 50°C in enriched air (75% O_2) at 90 mm Hg pressure.

$$P_4 + 3O_2 \longrightarrow P_4O_6 \text{ (m.p. 24°C, b.p. 175°C)}$$

Excess phosphorus can readily be converted to the involatile red form by irradiation.[162] The structure is based on the adamantane cage and electron diffraction studies show the P—O bonds to be 0.165 nm long. This is shorter than a single bond, while the wide angle at oxygen (127 degrees) supports $p_\pi-d_\pi$ bonding in this bond.[163]

In the vapour phase, both As_4O_6 and Sb_4O_6 possess cage structures, while two polymorphic forms exist in the solid state. The high-temperature forms, arsenolite (As_4O_6) and senarmontite (Sb_4O_6), possess M_4O_6 molecules in a cubic array. With the low-temperature forms of these two oxides, however, polymeric arrays of atoms are observed. In monoclinic claudetite, a laminated structure comprising puckered As_6O_6 rings (*cf.* orpiment As_2S_3) is found, while orthorhombic valentinite possesses Sb_3O_3 rings.

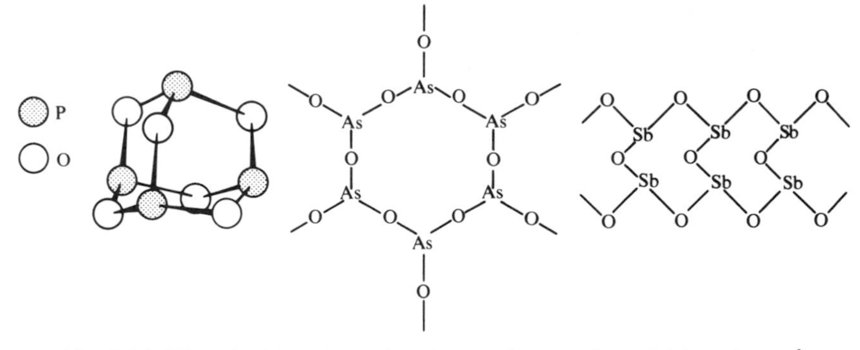

Fig. 5.26. The trioxides of (a) phosphorus, (b) arsenic and (c) antimony[h]

The bond lengths of 0.180 nm (As—O) and 0.20 nm (Sb—O) have not been sufficiently refined to allow any conclusions to be drawn regarding the presence of multiple bonding.[h]

With the heavier elements, there is a strong tendency to form hydroxide cage compounds, as was seen with tin and lead. With bismuth, solutions of composition $[Bi(OH)_2]^+$ have been shown to contain the ion $Bi_6(OH)_{12}^{6+}$. This is believed to comprise an octahedron of bismuth atoms edge-bridged by hydroxide groups.[164]

Reactions of Phosphorus Trioxide

This readily acts as a quadridentate ligand. It displaces CO from nickel carbonyl as do the cage amides,[165]

$$4Ni(CO)_4 + P_4O_6 \longrightarrow [Ni(CO)_3]_4P_4O_6 + 4CO$$

and structural data shows each phosphorus to co-ordinate to a carbonyl residue.[166]

Controlled oxidation of P_4O_6 with sulphur gives $P_4O_6S_4$ with terminal P—S groups and a structure resembling P_4O_{10}.[167] The P—O bonds are similar in length to those of the metaphosphate rings and of P_4O_{10}, but a little shorter than those of P_4O_6. The P—S bond distance (0.186 nm) is shorter than the terminal P—S bonds in phosphorus sulphides by about 0.01 nm.

Oxidation with chlorine or bromine is violent and gives the phosphoryl halides. Iodine slowly gives as yet unidentified products, but heating under pressure with carbon disulphide gives orange-red prisms of diphosphorus tetraiodide through disproportionation.

$$5P_4O_6 + 8I_2 \longrightarrow 4P_2I_4 + 3P_4O_{10}$$

Shaking phosphorus trioxide with an excess of cold water gives phosphorus acid exclusively, but, unless mixing occurs, disproportionation results with the formation of phosphoric acid and polymeric phosphines. Hot water gives phosphine and what is probably red phosphorus.

Phosphorus Pentoxide

This oxide, the anhydride of phosphoric acid, can be obtained in a very pure state by the carefully controlled burning of phosphorus. It occurs in various structural forms which appear to be related to its reactivity, as will be seen from the reactions with water.

In the vapour state, electron diffraction studies show a structure analogous to P_4O_6, but with four additional terminal P—O bonds. These are much shorter (0.139 nm) than the bridge-bonds of the P_4O_{10} cage, which are in close agreement with P—O bond lengths of P_4O_6.

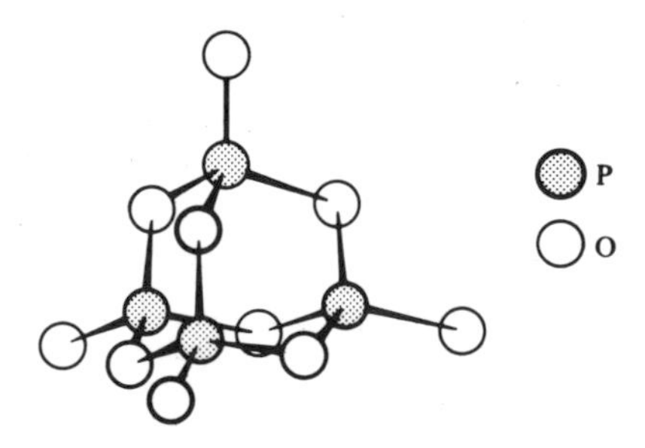

Fig. 5.27. The molecular structure of P_4O_{10}[h]

Condensing this vapour gives crystals of a metastable rhombohedral polymorph which has a high vapour pressure and is readily attacked by water. It comprises distinct P_4O_{10} molecules, but if melted and maintained 500°C for some days, transforms to a less volatile orthorhombic form (O) which is only slowly attacked by water. X-ray diffraction studies indicate an infinite array of PO_4 tetrahedra sharing three of their oxygen atoms. These are built into rings involving ten PO_4 units, and the phosphorus atoms lie in one of the simplest of three-dimensional tri-connected networks.

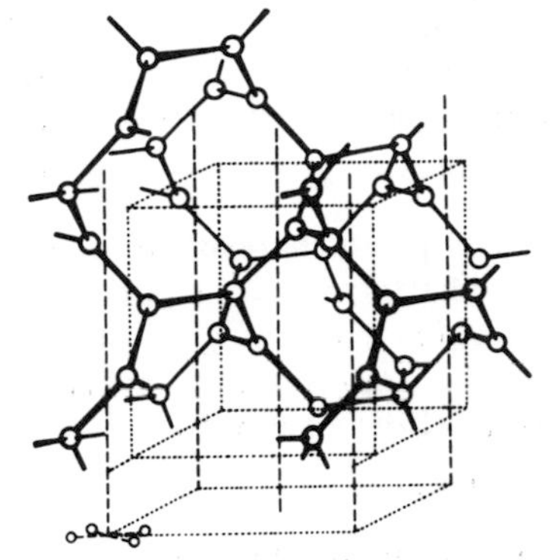

Fig. 5.28. Orthorhombic $P_4O_{10}(O)^h$

A second orthorhombic form (O′) comprises a two-dimensional array of PO_4 groups, involving a 12-membered ring lattice.

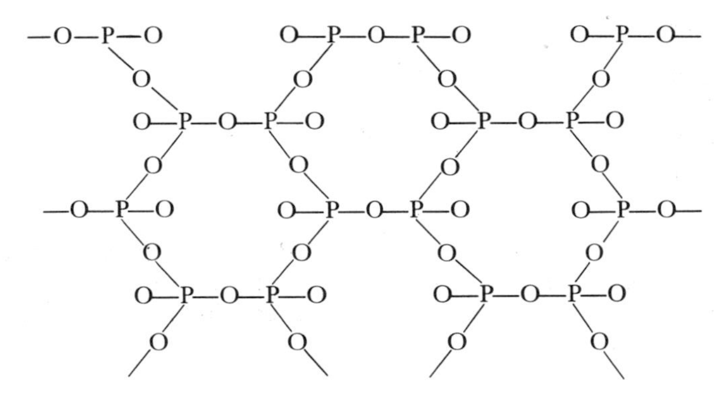

Fig. 5.29. Orthorhombic $P_4O_{10}(O')^h$

This form dissolves more rapidly in water than the O-form, as would be expected for the more open of these two polymorphs.

Indeed, the rate at which water is absorbed is the same for all three polymorphs initially, but decreases rapidly for the two orthorhombic forms.

The avidity of P_4O_{10} for water renders it an excellent drying and dehydrating agent. Both nitric and sulphuric acids will give anhydrides, while organic amides yield nitriles and methylaryl carbinols their respective styrenes. Ammonia reacts readily to give ammonium salts of condensed phosphates which often contain the P—NH—P linkage. Bases react in a similar manner to give phosphates. Volatile halides result when P_4O_{10} is heated to 350°C with calcium fluoride or sodium chloride. These include PF_3 as well as various oxyhalides.

Esters of phosphoric acid result with alcohols and, unlike the products of alcoholysis of P_4S_{10}, these do not decompose further.

$$P_4O_{10} + 6ROH \longrightarrow 2ROPO(OH)_2 + 2(RO)_2PO{\cdot}OH$$

With diethyl ether, various products result from P_4O_{10}, including tetrameric $[EtO{\cdot}P(O)O]_4$, indicating a change of ring size on reaction.[168]

Oxides analyzing for P_4O_7 and P_4O_8 are formed when the tri- and pentoxide are mixed in the appropriate proportions. Again, these have cage structures with two or three terminal P—O groups missing.[169]

$$3P_4O_6 + P_4O_{10} \longrightarrow 4P_4O_7$$

Phosphorus peroxide is formed when the pentoxide and oxygen are subjected to a glow discharge. The product analyzes for P_4O_{11}, and hydrolysis is found to give no H_2O_2 but peroxydiphosphoric acid and pyrophosphate. The structure proposed therefore involves a $P—O_2—P$ bridge in place of one P—O—P bridge.[170]

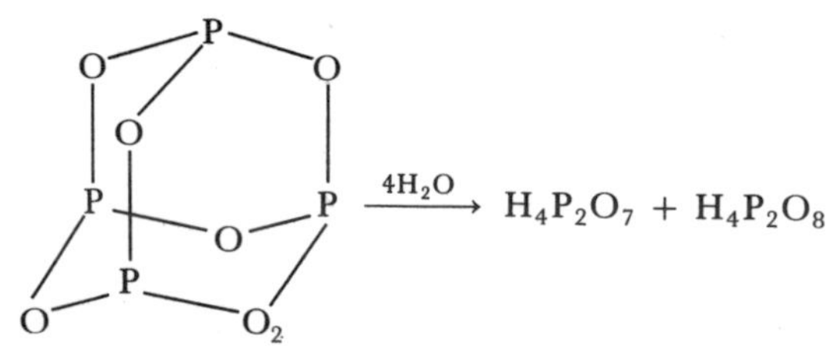

So despite these varied P:O ratios, the P_4O_6 cage holds intact except on peroxide formation, and no P—P products have been isolated. With P—S compounds, the isolation of compounds with P—P bonds appears to be the rule rather than the exception.

THE BINARY PHOSPHORUS SULPHIDES

Below about 100°C, no reaction takes place between white phosphorus and sulphur, but the phase diagram shows that solid solutions are formed. Above 100°C, however, a vigorous reaction occurs between these elements and five compounds are thought to be formed, four of which have subsequently been characterized, namely P_4S_3, P_4S_5, P_4S_7 and P_4S_{10}.[g]

Only P_4S_{10} has a structural analogue among the oxides, and this is prepared by heating the elements and slowly cooling the melt. Further crystallization from hot carbon disulphide gives pale yellow crystals, an X-ray diffraction study showing the expected adamantane cage with the two P—S distances 0.209 nm (bridging) and 0.196 nm (terminal). This bridging distance is close to the value expected for a phosphorus–sulphur single bond.

Alcoholysis gives two thiophosphate esters, the mono one reacting further with the alcohol, unlike the situation with P_4O_{10}.[171]

$$P_4S_{10} + 6ROH \longrightarrow 2(RO)_2P(S)SH + 2ROP(S)(SH)_2 \xrightarrow{2ROH} 2(RO)_2P(S)SH + 2H_2S$$

Anions such as fluoride and azide break the P_4S_{10} cage giving thio-anions including $F_2PS_2^-$, $(FPS_2)_2S^{2-}$, $(N_3)_2PS_2^-$ and $(N_3PS_2)_2S^{2-}$. Diazidodithiophosphate will also substitute the P_4S_{10} cage giving the anion $P_4S_9N^-$. Phosphorus pentasulphide is used as a sulphurating agent for organic carbonyl compounds, notably amides,[172]

$$10RCONH_2 + P_4S_{10} \longrightarrow 10RCSNH_2 + P_4O_{10}$$

while the thio-esters are used as high-pressure lubricants, oil additives and flotation agents.

Tetraphosphorus trisulphide, or phosphorus sesquisulphide, results when the elements are heated till a reaction just starts. This is exothermic and maintains the reaction temperature. P_4S_3 can be recrystallized from CS_2 or distilled in an atmosphere of CO_2. The structure involves sulphur insertion into three P—P bonds of the P_4 tetrahedron, as shown, giving it C_{3v} point group symmetry.

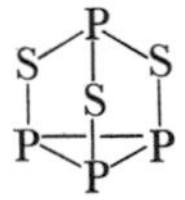

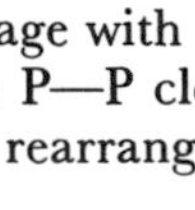

Fig. 5.30. The structure of P_4S_3

The P—P (0.2235 nm) and P—S bonds (0.2090 nm) are close to single bonds, and the reaction of this cage with iodine shows the lability of these bonds. A 1 : 1 addition in which P—P cleavage has occurred is probably the first step, followed by a cage rearrangement.

$$P_4S_3 + I_2 \longrightarrow \quad \longrightarrow$$ (m.p. 118–120°C)

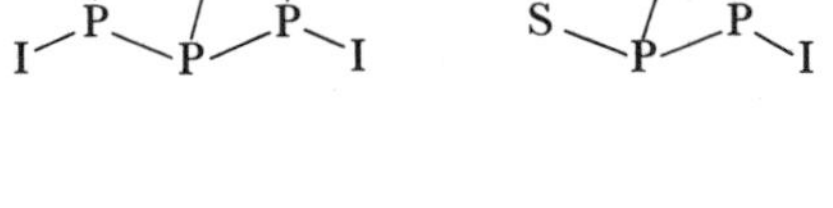

It is interesting to compare this reaction with iodine oxidation of P_4O_6, where P—P bonds were encountered in the major product, P_2I_4.[173]

The ^{31}P n.m.r. of P_4S_3 shows the presence of the two types of phosphorus, the unique phosphorus atom co-ordinating readily with zero-valent transition metals, as did P_4O_6. Thus, heating the norbornadiene complexes of the Group VI metal carbonyls with P_4S_3 in CS_2 gives the di- and tri-substituted *cis* complexes.

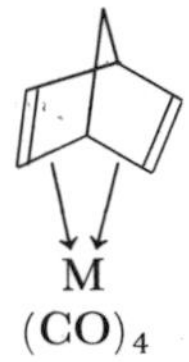

$$\text{(norbornadiene)M(CO)}_4 + P_4S_3 \xrightarrow{CS_2} cis\text{-}[P_4S_3]_2M(CO)_4 + cis\text{-}[P_4S_3]_3M(CO)_3$$

A large downfield shift occurred in the ^{31}P n.m.r. spectrum of the unique phosphorus, supporting trithiophosphite co-ordination. A large excess of sulphur will oxidize the di-substituted complex to P_4S_7.[174]

Bis-(π-allyl)nickel and P_4S_3 give the expected tetraphosphite nickel complex $[P_4S_3]_4Ni$ and, while iodine oxidizes this to $P_4S_3I_2$ and nickel di-iodide, PF_3 will not replace the thiophosphite.

$$(\pi\text{-allyl})_2Ni + 4P_4S_3 \longrightarrow [P_4S_3]_4Ni \xrightarrow[PF_3]{60°C} \text{no reaction observed}$$

$$[P_4S_3]_4Ni \xrightarrow{I_2} 4P_4S_3I_2 + NiI_2$$

The sesquiselenide is conveniently prepared from the elements and charcoal in heptane heated under reflux. Heating to 207°C after adding tetralin, followed by filtration, gives the selenide as orange crystals.

$$P_4 + 3Se \longrightarrow P_4Se_3 \text{ (m.p. 245°C)}$$

Exposing a 1:2 mixture of P_4S_3 and sulphur in carbon disulphide to diffuse sunlight is sufficient to give P_4S_5, whose structure can be conveniently constructed from that of P_4S_3 by cleaving and oxidizing the P_3 triangle.[175]

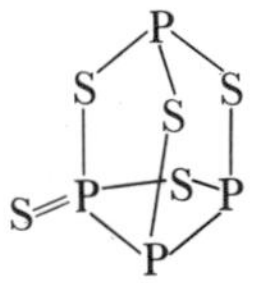

Fig. 5.31. The structure of P_4S_5

While P—P and the terminal P—S bond lengths agree closely with those reported, the bridging P—S bond distances vary from 0.208–0.219 nm. Some are therefore very labile, and this is reflected in the disproportionation at its melting point into the more thermally stable and more symmetrical P_4S_3 and P_4S_7. Indeed, both of these can be distilled at atmospheric pressure in an inert atmosphere.

Tetraphosphorus heptasulphide is the least soluble of the compounds, and is prepared by heating stoichiometric quantities of the elements. The pale yellow prisms produced possess the C_{2v} point group symmetry.[176]

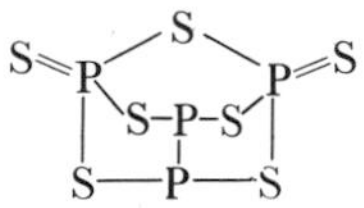

```
        S
S=P/      \P=S
  | \S-P-S/ |
  |    |    |
  S----P----S
```

Fig. 5.32. The structure of P_4S_7

The terminal and bridging P—S bond lengths are all very similar (0.195 nm terminal; 0.208 nm bridging), and closely agree with values in other phosphorus–sulphur compounds. The P—P bond is remarkably long (0.235 nm).

Alcoholysis in CS_2 at 20°C gives the esters $(RO)_2P(S)SH$, $(RO)_2P(S)H$, $(RO)_2P(S)SR$, hydrogen sulphide and phosphine. The initial reaction is thought to involve the formation of $P_{(2)}$ and $P_{(1)}$ esters by nucleophilic attack at the P—S bonds.

$$P_4S_7 + 7ROH \longrightarrow 2(RO)_2P(S)SH + 2H_2S + (RO)_2PP(OR)SH$$

Further alcoholysis occurs on standing, with P—P cleavage. All products were characterized by thin layer chromatography.[177]

Bromination in CS_2 gives the thiobromides $P_2S_6Br_2$ (m.p. 118°C) and $P_2S_5Br_4$ (m.p. 90°C).

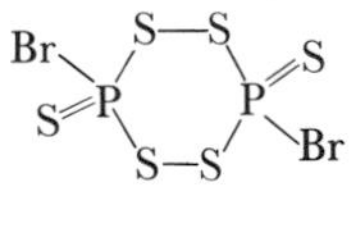

```
Br     S—S
  \   /   \  =S
 S=P       P
    \     / \
     S—S     Br
```

Fig. 5.33. The structure of $P_2S_6Br_2$

The former possesses a 6-membered puckered P_2S_4 symmetrical ring, with terminal Br and S atoms on each phosphorus.[178] The bond lengths are close to the values expected for this structure.

Though $P_4S_4O_6$ has been known for some time, $P_4O_4S_6$ has eluded synthesis till recently. This is probably because the P_4S_6 cage, analogous to P_4O_6, is unknown and so cannot be used for oxidation, as P_4O_6 can. However, the P_4S_6 cage can be stabilized with pentavalent phosphorus. $P_4O_4S_6$ can be prepared in good yield from phosphoryl chloride and hexamethyldisilthiane. The pale yellow crystals melted at 290–295°C.[179]

$$4POCl_3 + 6(Me_3Si)_2S \longrightarrow P_4O_4S_6 + 12Me_3SiCl$$

Attempts to prepare P_4S_6 the same way, using PCl_3 gave a powder, but an X-ray powder photograph supported the presence of P_4S_7 and possibly lower sulphides.

The complex structures of the phosphorus sulphides does not extend to those of the heavier elements. Most possess polymeric lattices and only As_4S_4 forms a molecular crystal. This occurs naturally as the mineral realgar and has a structure comprising a square of sulphur atoms and a pseudo-tetrahedral array of arsenic atoms.[180,h]

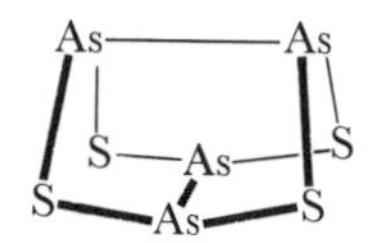

Fig. 5.34. The structure of realgar As_4S_4

It can be readily sublimed or distilled, and dissolves in alkali with the formation of thioarsenites and the precipitation of arsenic. Exposure to sunlight readily changes it into orpiment, As_2S_3, which possesses the layer structure already discussed for claudetite (As_2O_3) (p. 317). A molecular structure exists in the vapour phase (As_4S_6).

The structures of antimony and bismuth trisulphides are more complicated than that of orpiment. They possess two weakly connected double chains, each chain comprising M_2S_3 units as shown.

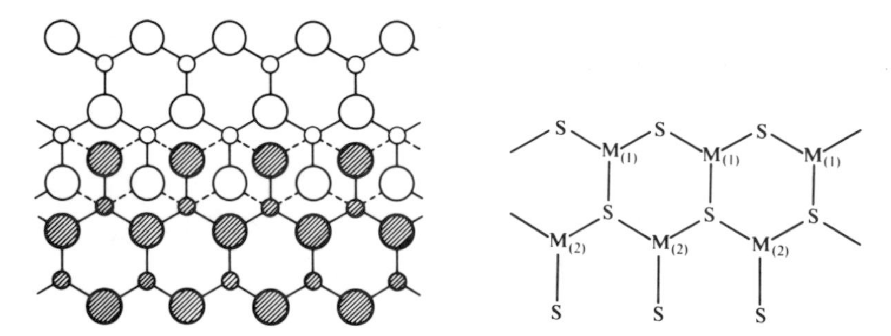

Fig. 5.35. The double chain structure of Sb_2S_3. Large circles represent sulphur and all atoms of one chain are shaded. Dotted lines represent weak Sb—S bonds between double chains[h]

With $M_{(1)}$ and $M_{(2)}$ antimony, the $M_{(1)}$—S distances are all about 0.250 nm, whereas the shortest $M_{(2)}$—S distance is 0.238 nm, with two at 0.267 nm and two at 0.283 nm (dotted lines).[181]

The majority of the mixed thio- and seleno-halides of antimony and bismuth are known, and all but BiSeCl are isostructural. The structure comprises an infinite puckered ladder with Sb—S bonds, each "rung"

0.249 nm, and side bonds, 0.267 nm long, giving the cation $[SbS]_n^{n+}$. The halide ions are held between the chains, with two Sb and one S atom as nearest neighbours.

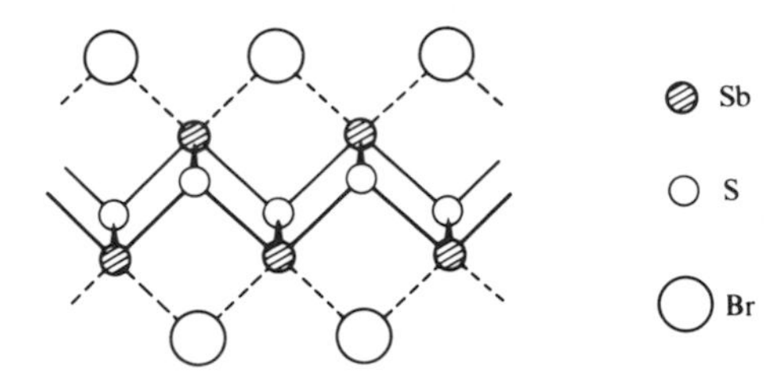

Fig. 5.36. The structure of SbSBr[h]

ORGANO-SUBSTITUTED CYCLIC PHOSPHORUS–SULPHUR COMPOUNDS

Trimeric phenylphosphorus sulphide is conveniently prepared from phenylphosphorus dichloride and methyl silthianes, whether they be linear or cyclic.

$$\underset{(n = 2 \text{ or } 3)}{(Me_3Si)_2S/[Me_2SiS]_n} \xrightarrow{PhPCl_2} Me_3SiCl/Me_2SiCl_2 + \underset{(\text{m.p. } 148^\circ\text{C})}{[PhPS]_3}$$

The same ring size is maintained in the product irrespective of starting material.[179] It is also formed from pentaphenylcyclopentaphosphine and sulphur in CS_2.[182]

Hexamethylcyclotrisilthiane gives tetrameric phenylarsenic sulphide with $PhAsCl_2$,[183] while hydrogen sulphide[184] gives the trimer.

$$\underset{(\text{m.p. } 152^\circ\text{C})}{[PhAsS]_3} \xleftarrow{H_2S} PhAsCl_2 \xrightarrow{[Me_2SiS]_3} \underset{(\text{m.p. } 175\text{–}6^\circ\text{C})}{[PhAsS]_4}$$

The tetramer is also formed from phenylarsine and thionyl compounds.[185]

$$3PhAsH_2 \xrightarrow{2SOCl_2} PhAsO + \tfrac{1}{2}[PhAsS]_4 + H_2O + 4HCl$$

$$3PhAsH_2 \xrightarrow{2PhNSO} 2PhNH_2 + \tfrac{1}{2}[PhAsS]_4 + PhAsO + H_2O$$

Cyclic phosphorus(V)–sulphur compounds are generally prepared from organophosphorus dichlorides and disulphides.[186]

$$2RPCl_2 + 2R_2S_2 \longrightarrow [RP(S)(\mu\text{-}S)]_2 + 4RCl \quad (R = Me, Et, Ph)$$

The structure of these compounds involves tetrahedral orientation of the groups about phosphorus, and a planar P_2S_2 ring with a small angle of 85 degrees at sulphur.[187]

Tri-n-butylphosphine will reduce these perthiophosphonic anhydrides to cyclopolyphosphines. The reaction proceeds via a ring-opening process involving a stabilized phosphonium salt.

$$[ArPS_2]_2 + 2Bu^n_3P \xrightarrow{C_6H_6} 2Bu^n_3P^+\text{—}P(\bar{S})(=S)\text{—}Ar \xrightarrow{Bu^n_3P} [ArP]_n + Bu^n_3PS$$

Tertiary amines will form analogous inner salts which undergo displacement reactions with Bu^n_3P. This, in turn, can be displaced by primary and secondary amines, giving outer ammonium salts.[188]

$$[ArPS_2]_2 \xrightarrow{Et_2NH} Et_2NH_2^+\,ArP(=S)(S^-)\text{—}NEt_2$$

$$[ArPS_2]_2 \xrightarrow{C_5H_5N} ArP(=S)(S^-)\text{—}\overset{+}{N}C_5H_5$$

$$[ArPS_2]_2 \xrightarrow{Bu^n_3P} ArP(=S)(S^-)\text{—}\overset{+}{P}Bu^n_3$$

$$ArP(=S)(S^-)\text{—}\overset{+}{N}C_5H_5 \xrightarrow{Bu^n_3P} ArP(=S)(S^-)\text{—}\overset{+}{P}Bu^n_3 \xrightarrow{Et_2NH} Et_2NH_2^+\,ArP(=S)(S^-)\text{—}NEt_2$$

Alcohols and thiols also readily cleave the anhydride ring forming esters of arylphosphonic acid.

$$[ArPS_2]_2 + R'XH \longrightarrow ArP(S)(SH)XR' \quad (X = O \text{ or } S)$$

Methyl bromide and the azide ion open the ring giving thiophosphonyl compounds.[189]

$$[RPS_2]_2 \xrightarrow{2MeBr} RP(S)(SMe)Br$$

$$[RPS_2]_2 \xrightarrow{N_3^-} RPS_2N_3^- \xrightarrow{Ph_3P} [Ph_3P{=}NPRS_2]^-$$

Ring-substitution occurs between phenylisocyanate and phenyldithiophosphonic anhydride. Carbon oxysulphide is eliminated and the mixed ring (m.p. 183°C) formed. Phenylisothiocyanate reacts similarly.

$$[PhPS_2]_2 + PhNCO \longrightarrow \text{cyclo-}[Ph(S{=})P\text{—}N(Ph)\text{—}P(=S)Ph\text{—}S\text{—}] + COS$$

Nitriles will also form heterocyclic compounds with thiophosphonic anhydride rather similar to those formed with sulphur trioxide.[190]

$$\tfrac{1}{2}[RPS_2]_2 + 2R'_2NCN \longrightarrow$$

Bibliography

General

a S. MAIER, *Prog. Inorg. Chem.*, 1963, **5,** 27.

b W. R. CULLEN, *Adv. Organometallic Chem.*, 1966, **4,** 145.

c N. L. PADDOCK, *Quart. Rev.*, 1964, **18,** 168.

d H. R. ALLCOCK, 'Heteroatom Ring systems and polymers', Academic Press, New York, 1967.

e N. L. PADDOCK and H. T. SEARLE, *Adv. Inorg. Chem. Radiochem.*, 1959, **1,** 347.

f D. P. CRAIG and N. L. PADDOCK, *J. Chem. Soc.*, 1962, 4118.

g J. R. VAN WAZER, 'Phosphorus and its Compounds', Interscience, New York, 1958, Vol. 1.

h A. F. WELLS, 'Structural Inorganic Chemistry', Oxford University Press, 3rd Edn., 1962.

References

1 L. R. MAXWELL, S. B. HENDRICKS and V. M. MOSLEY, *J. Chem. Phys.*, 1935, **3,** 699.

2 L. PAULING and M. SIMONETTA, *J. Chem. Phys.*, 1952, **20,** 29.

3 H. KREBS, H. WEITZ and K. H. WORMS, *Z. anorg. Chem.*, 1955, **280,** 119.

4 A. BROWN and S. RUNDQUIST, *Acta. Cryst.*, 1965, **19,** 684.

5 S. N. DUTTA and G. A. JEFFREY, *Inorg. Chem.*, 1965, **4,** 1363.

6 A. F. WELLS, 'Structural Inorganic Chemistry', Oxford University Press, 3rd Edn., 1962, p. 662.

7 H. THURN and H. KREBS, *Angew. Chem. Internat. Edn.*, 1966, **5,** 1047.

8 H. G. VON SCHNERING and H. SCHMIDT, *ibid.*, 1967, **6,** 356.

9 H. KREBS and TH. LUDWIG, *Z. anorg. Chem.*, 1958, **294,** 257.

10 A. HERSHAFT and J. D. CORBETT, *Inorg. Chem.*, 1963, **2,** 979; J. D. CORBETT and R. E. RUNDLE, *ibid.*, 1964, **3,** 1408.

11 N. J. BJERRUM and G. PEDRO SMITH, *Inorg. Nuclear Chem. Letters*, 1967, **3,** 165; *Inorg. Chem.*, 1967, **6,** 1968.

12 J. D. CORBETT, *Inorg. Nuclear Chem. Letters*, 1967, **3,** 173; *Inorg. Chem.*, 1968, **7,** 198.

13 J. S. WATSON, *J. Chem. Soc.*, 1956, 3677.

14 I. UGI, H. PESHINGER and L. BEHRINGER, *Chem. Ber.*, 1958, **91,** 2324; 1959, **92,** 1864.

15 K. ISSLEIB and M. HOFFMANN, *Chem. Ber.*, 1966, **99,** 1320.

16 W. KUCHEN and W. GRÜNEWALD, *Chem. Ber.*, 1965, **98,** 480; *Angew. Chem. Internat. Edn.*, 1963, **2,** 399; W. A. HENDERSON JR., M. EPSTEIN and F. S. SEICHTER, *J. Amer. Chem. Soc.*, 1963, **85,** 2462.

17 W. MAHLER and A. B. BURG, *J. Amer. Chem. Soc.*, 1957, **79,** 251; 1958, **80,** 6161.

18 A. H. COWLEY, A. B. BURG and W. R. CULLEN, *ibid.*, 1966, **88,** 3178.
19 M. GREEN and D. KIRKPATRICK, *Chem. Comm.*, 1967, 58; *J. Chem. Soc.* (*A*), 1968, 483.
20 A. H. COWLEY and R. P. PINNELL, *J. Amer. Chem. Soc.*, 1966, **88,** 4533.
21 A. H. COWLEY, T. A. FURTSCH and D. S. DIERDORF, *Chem. Comm.*, 1970, 523.
21a P. S. ELMES, M. E. REDWOOD and B. O. WEST, *ibid.*, 1970, 1120.
22 H. KÖHLER and A. MICHAELIS, *Ber.*, 1877, **10,** 807.
23 L. MAIER, *Angew. Chem. Internat. Edn.*, 1967, **6,** 887; E. FLUCK and K. ISSLEIB, *Z. Naturforsch.*, 1966, **21b,** 736.
24 M. FILD, I. HOLLENBERG and O. GLEMSER, *Naturwiss.*, 1967, **54,** 89.
25 M. BANDLER and K. HAMMERSTRÖM, *Z. Naturforsch.*, 1965, **20b,** 810.
26 A. E. GODDARD, in 'Textbook of Inorganic Chemistry', ed. J. N. FRIEND, C. GRIFFIN, London, 1930, Vol. XI, Part II.
27 J. C. SUMMERS and H. H. SISLER, *Inorg. Chem.*, 1970, **9,** 862; J. F. NIXON, *J. Inorg. Nuclear Chem.*, 1965, **27,** 1281.
28 H. G. ANG and R. SCHMUTZLER, *J. Chem. Soc.* (*A*), 1969, 702.
29 V. SCHMIDT, R. SCHRÖER and H. ACHENBACH, *Angew. Chem. Internat. Edn.*, 1966, **5,** 316.
30 V. SCHMIDT and co-workers, *ibid.*, 1966, **5,** 1038; *Chem. Ber.*, 1968, **101,** 1381.
31 V. SCHMIDT and CH. OSTERROHT, *Angew. Chem. Internat. Edn.*, 1965, **4,** 457.
32 A. B. BURG and W. MAHLER, *J. Amer. Chem. Soc.*, 1961, **83,** 2388.
33 A. H. COWLEY, *ibid.*, 1967, **89,** 5990.
34 A. H. COWLEY and D. S. DIERDORF, *ibid.*, 1969, **61,** 6609.
35 K. ISSLEIB and K. KRECH, *Chem. Ber.*, 1965, **98,** 2545.
36 K. ISSLEIB and E. FLUCK, *Angew. Chem. Internat. Edn.*, 1966, **5,** 587.
37 K. ISSLEIB, CH. ROCKSTROH, I. DUCHEK and E. FLUCK, *Z. anorg. Chem.*, 1968, **360,** 77.
38 M. BANDLER, O. GEHLEN, K. KIPKER and P. BACKES, *Z. Naturforsch.*, 1967, **22b,** 1354.
39 M. BANDLER, K. KIPKER and W. W. VALPERTZ, *Naturwiss.*, 1967, **54,** 43.
40 H. L. KRAUSS and H. JUNG, *Z. Naturforsch.*, 1960, **15b,** 545.
41 A. B. BURG, *J. Amer. Chem. Soc.*, 1966, **88,** 4299; A. B. BURG and D. M. PARKER, *ibid.*, 1970, **92,** 1898.
42 W. MAHLER, *ibid.*, 1964, **86,** 2306.
43 R. G. CAVELL and R. C. DOBBIE, *J. Chem. Soc.* (*A*), 1967, 1308.
44 A. B. BURG and L. K. PETERSON, *Inorg. Chem.*, 1966, **5,** 943.
45 H. HOFFMANN and R. GRÜNEWALD, *Chem. Ber.*, 1961, **94,** 186.
46 H. G. ANG, J. S. SHANNON and B. O. WEST, *Chem. Comm.*, 1965, 11; H. G. ANG and B. O. WEST, *Austral. J. Chem.*, 1967, **20,** 1133.
47 G. W. A. FOWLES and D. K. JENKINS, *Chem. Comm.*, 1965, 61; A. H. COWLEY and R. P. PINNELL, *Inorg. Chem.*, 1966, **5,** 1459.
48 C. S. CUNDY, M. GREEN, F. G. A. STONE and D. E. A. TAUNTON-RIGBY, *Inorg. Nuclear Chem. Letters*, 1966, **2,** 233; *J. Chem. Soc.* (*A*), 1968, 1776.
49 M. A. BUSH and P. WOODWARD, *J. Chem. Soc.* (*A*), 1968, 1221.
50 A. S. FOUST, M. S. FOSTER and L. F. DAHL, *J. Amer. Chem. Soc.*, 1969, **91,** 5631, 5633.
51 C. J. SPENCER and W. N. LIPSCOMB, *Acta Cryst.*, 1961, **14,** 250; *Idem* and P. G. SIMPSON, *ibid.*, 1962, **15,** 509; G. J. PALENIK and J. DONOHUE, *Acta Cryst.*, 1962, **15,** 564; J. DONOHUE, *ibid.*, 1962, **15,** 708.
52 E. J. WELLS, H. P. K. LEE and L. K. PETERSON, *Chem. Comm.*, 1967, 894.
53 J. W. BART, *Acta Cryst.*, 1969, **B25,** 762.
54 J. J. DALY, *J. Chem. Soc.*, 1964, 6147.
55 *Idem*, *ibid.*, 1965, 4789; 1966, 428.
56 J. WEISS, *Z. anorg. Chem.*, 1960, **306,** 30.
57 J. H. BURNS and J. WASER, *J. Amer. Chem. Soc.*, 1957, **79,** 859.

58 K. HEDBERG, E. W. HUGHES and J. WASER, *Acta Cryst.*, 1961, **14,** 369.
59 E. J. WELLS, R. C. FERGUSON, J. G. HALLETT and L. K. PETERSON, *Canad. J. Chem.*, 1968, **46,** 2733.
60 E. WIBERG, M. VAN GHEMEN and G. MÜLLER-SCHIEDMAYER, *Angew. Chem. Internat. Edn.*, 1962, **2,** 646.
61 A. MICHAELIS and G. SCHROETER, *Ber.*, 1894, **27,** 490.
62 R. R. HOLMES and J. A. FORSTNER, *Inorg. Chem.*, 1963, **2,** 380.
63 M. G. BARLOW, M. GREEN, R. N. HASZELDINE and H. G. HIGSON, *J. Chem. Soc.* (*C*), 1966, 1592.
64 O. J. SCHERER and P. KLUSMANN, *Angew. Chem. Internat. Edn.*, 1969, **8,** 752.
65 H. J. VETTER, H. STRAMETZ and H. NÖTH, *ibid.*, 1963, **2,** 218; H. J. VETTER and H. NÖTH, *ibid.*, 1963, **2,** 663.
66 E. W. ABEL, D. A. ARMITAGE and G. R. WILLEY, *J. Chem. Soc.*, 1965, 57.
67 H. J. VETTER, H. NÖTH and W. JAHN, *Z. anorg. Chem.*, 1964, **328,** 144.
68 J. WEISS and W. EISENHUTH, *Z. Naturforsch.*, 1967, **20b,** 455.
69 D. HASS, *Z. anorg. Chem.*, 1963, **325,** 139; W. IPATIEW, G. RAZNWAJEW and W. STROMSKI, *Chem. Ber.*, 1929, **62,** 598; D. HASS, *Z. anorg. Chem.*, 1964, **332,** 287.
70 M. BECKE-GOEHRING and L. LEICHNER, *Angew. Chem. Internat. Edn.*, 1964, **3,** 590.
71 T. MOELLER and A. H. WESTLAKE, *J. Inorg. Nuclear Chem.*, 1967, **29,** 957.
72 E. H. M. IBRAHIM and R. A. SHAW, *Angew. Chem. Internat. Edn.*, 1967, **6,** 556; Idem, *Chem. Comm.*, 1967, 244.
73 L. PARTS, M. L. NIELSEN and J. T. MILLER JR., *Inorg. Chem.*, 1964, **3,** 1261.
74 R. SCHMUTZLER, *Angew. Chem. Internat. Edn.*, 1965, **4,** 500; *Z. Naturforsch.*, 1964, **19b,** 1101; *Chem. Comm.*, 1965, 19.
75 H. P. LATSCHA and P. B. HORMUTH, *Z. anorg. Chem.*, 1968, **359,** 81; M. GREEN, R. N. HASZELDINE and G. S. A. HOPKINS, *J. Chem. Soc.* (*A*), 1966, 1766.
76 V. GUTMANN, K. UTVARY and M. BERGMANN, *Monatsh.*, 1966, **97,** 1745; M. L. ZIEGLER and J. WEISS, *Z. anorg. Chem.*, 1968, **361,** 136; M. ZIEGLER, *Angew. Chem. Internat. Edn.*, 1967, **6,** 369.
77 H. ULRICH and A. A. SAYIGH, *ibid.*, 1962, **1,** 595; P. B. HORMUTH and H. P. LATSCHA, *Z. anorg. Chem.*, 1969, **365,** 26.
78 M. BECKE-GOEHRING and H. J. MÜLLER, *Z. anorg. Chem.*, 1968, **362,** 51.
79 G. W. ADAMSON and J. C. J. BART, *J. Chem. Soc.* (*A*), 1970, 1452.
80 A. C. CHAPMAN, W. S. HOLMES, N. L. PADDOCK and H. T. SEARLE, *J. Chem. Soc.*, 1961, 1825.
81 J. WESLEY COX and E. R. COREY, *Chem. Comm.*, 1967, 123.
82 L. G. HOARD and R. A. JACOBSON, *J. Chem. Soc.* (*A*), 1966, 1203; H. HESS and D. FORST, *Z. anorg. Chem.*, 1966, **342,** 240.
83 J. WEISS and G. HARTMANN, *Z. Naturforsch.*, 1966, **21b,** 891.
84 N. WIBERG and K. H. SCHMID, *Chem. Ber.*, 1967, **100,** 741; V. MÜLLER and K. DEHNICKE, *Z. anorg. Chem.*, 1967, **350,** 113.
85 N. WIBERG and K. H. SCHMID, *Angew. Chem., Internat. Edn.*, 1967, **6,** 953.
86 V. MÜLLER and K. DEHNICKE, *ibid.*, 1966, **5,** 841.
87 R. R. HOLMES and J. A. FORSTNER, *J. Amer. Chem. Soc.*, 1960, **82,** 5509.
88 J. F. NIXON, *Chem. Comm.*, 1967, 669; R. JEFFERSON, J. F. NIXON and T. M. PAINTER, *ibid.*, 1969, 622.
89 H. NÖTH and H. J. VETTER, *Naturwiss.*, 1961, **48,** 553; R. R. HOLMES and J. A. FORSTNER, *Inorg. Chem.*, 1963, **2,** 377.
90 R. R. HOLMES and J. A. FORSTNER, *Inorg. Chem.*, 1962, **1,** 89.
91 D. HASS, *Z. anorg. Chem.*, 1963, **326,** 192.
92 J. G. RIESS and J. R. VAN WAZER, *Bull. Soc. chim. France*, 1966, 1846.
93 *Idem*, *J. Organometallic Chem.*, 1967, **8,** 347.
94 J. WEISS and W. EISENHUTH, *Z. anorg. Chem.*, 1967, **350,** 9.
95 D. S. PAYNE, H. NÖTH and G. HENNIGER, *Chem. Comm.*, 1965, 327.

96 H. NÖTH and N. REGNET, *Z. Naturforsch.*, 1965, **20b,** 604.
97 J. WEISS and G. HARTMANN, *Z. anorg. Chem.*, 1967, **351,** 152.
98 M. BECKE-GOEHRING and H. J. WALD, *ibid.*, 1969, **371,** 88; M. BECKE-GOEHRING and H. SCHWIND, *ibid.*, 1970, **372,** 285.
99 H. N. STOKES, *J. Amer. Chem. Soc.*, 1895, **17,** 257; 1897, **19,** 78; *Chem. Ber.*, 1895, **28,** 437.
100 R. SCHENCK and G. RÖMER, *Ber.*, 1924, **57,** 1343.
101 L. G. LUND, N. L. PADDOCK, J. E. PROCTER and H. T. SEARLE, *J. Chem. Soc.*, 1960, 2542.
102 K. JOHN and T. MOELLER, *J. Inorg. Nuclear Chem.*, 1961, **22,** 199; *J. Amer. Chem. Soc.*, 1960, **82,** 2647.
103 R. G. RICE, L. N. DAASCH, J. R. HOLDEN and E. J. KOHN, *J. Inorg. Nuclear Chem.*, 1958, **5,** 190; G. E. COXON and D. B. SOWERBY, *J. Chem. Soc.* (*A*), 1967, 1567.
104 F. SEEL and J. LANGER, *Angew. Chem.*, 1956, **68,** 461.
105 H. T. SEARLE, *Proc. Chem. Soc.*, 1959, 7; A. J. BILBO, *Z. Naturforsch.*, 1960, **15b,** 330.
106 C. P. HABER, D. L. HERRING and E. A. LAWTON, *J. Amer. Chem. Soc.*, 1958, **80,** 2116; H. H. SISLER, H. S. AHUJA and N. L. SMITH, *Inorg. Chem.*, 1962, **1,** 84.
107 R. A. SHAW and C. STRATTON, *Chem. and Ind.*, 1960, 839; *J. Chem. Soc.*, 1962, 5004.
108 D. G. ACOCK, R. A. SHAW and F. B. G. WELLS, *J. Chem. Soc.*, 1964, 121.
109 D. L. HERRING, *Chem. and Ind.*, 1960, 717; D. L. HERRING and C. M. DOUGLAS, *Inorg. Chem.*, 1965, **4,** 1012; K. L. PACIOREK and R. KRATZER, *ibid.*, 1964, **3,** 594.
110 W. T. REICHLE, *Tetrahedron Letters*, 1962, 51; *J. Organometallic Chem.*, 1968, **13,** 529.
111 I. T. GIBSON and H. H. SISLER, *Inorg. Chem.*, 1965, **4,** 273.
112 H. H. SISLER and C. STRATTON, *ibid.*, 1966, **5,** 2003.
113 G. TESI, C. P. HABER and C. M. DOUGLAS, *Proc. Chem. Soc.*, 1960, 219.
114 C. D. SCHMULBACH and C. DERDERIAN, *J. Inorg. Nuclear Chem.*, 1963, **25,** 1395; D. L. HERRING and C. M. DOUGLAS, *Inorg. Chem.*, 1964, **3,** 428.
115 W. LEHR, *Angew. Chem. Internat. Edn.*, 1967, **6,** 982.
116 A. SCHMIDPETER and J. EBELING, *ibid.*, 1968, **7,** 209.
117 F. G. SHERIF and C. D. SCHMULBACH, *Inorg. Chem.*, 1966, **5,** 322; M. BERMANN and K. UTVARY, *J. Inorg. Nuclear Chem.*, 1969, **31,** 271.
118 R. APPEL, D. HANSSGEN and B. ROSS, *Z. Naturforsch.*, 1967, **22b,** 1354.
119 I. N. ZHMUROVA and A. V. KIRSANOV, *J. Gen. Chem. U.S.S.R.*, (English Transl.), 1960, **30,** 3044.
120 H. SCHMIDBAUR and G. JONAS, *Chem. Ber.*, 1967, **100,** 1120.
121 M. BECKE-GOEHRING and E. FLUCK, *Angew. Chem. Internat. Edn.*, 1962, **1,** 281; M. BECKE-GOEHRING and W. LEHR, *Z. anorg. Chem.*, 1964, **327,** 128.
122 M. BECKE-GOEHRING and W. LEHR, *Chem. Ber.*, 1961, **94,** 159; *Z. anorg. Chem.*, 1963, **325,** 287; W. L. GROENEVELD, J. H. VISSER and A. M. J. H. SENTER, *J. Inorg. Nuclear Chem.*, 1958, **8,** 245; O. GLEMSER and E. WYSZOMIRSKII, *Naturwiss.*, 1961, **48,** 25; K. NIEDENZU and G. MAGIN, *Z. Naturforsch.*, 1965, **20b,** 604.
123 H. R. ALLCOCK and R. J. BEST, *Canad. J. Chem.*, 1964, **42,** 447.
124 B. W. FITZSIMMONS and R. A. SHAW, *Proc. Chem. Soc.*, 1961, 258; B. W. FITZSIMMONS, C. HEWLETT and R. A. SHAW, *J. Chem. Soc.*, 1964, 4459.
125 H. N. STOKES, *J. Amer. Chem. Soc.*, 1895, **17,** 275; 1896, **18,** 629, 780; A. BESSON and G. ROSSET, *Compt. rend.*, 1906, **143,** 37.
126 R. OLTHOF, T. MIGCHELSEN and A. VOS, *Acta Cryst.*, 1965, **19,** 596.
127 F. SEEL and J. LANGER, *Z. anorg. Chem.*, 1958, **295,** 316.
128 B. W. FITZSIMMONS and R. A. SHAW, *Chem. and Ind.*, 1961, 109; *J. Chem. Soc.*, 1964, 1735; R. RÄTZ, H. SCHROEDER, H. ULRICH, E. KOBER and C. GRUNDMANN, *J. Amer. Chem. Soc.*, 1962, **84,** 551.

129 G. ALLEN, D. J. OLDFIELD, N. L. PADDOCK, F. PALLO, J. SERREGI and S. M. TODD, *Chem. and Ind.*, 1965, 1032.
130 A. J. MATUSZKO and M. S. CHANG, *J. Org. Chem.*, 1966, **31,** 2004.
131 A. P. CARROLL and R. A. SHAW, *Chem. and Ind.*, 1962, 1908; *J. Chem. Soc.*, 1966, 914.
132 R. KEAT and R. A. SHAW, *J. Chem. Soc.*, 1965, 2215; *Idem* and B. C. SMITH, *J. Chem. Soc.*, 1965, 5032; G. J. BULLEN, *ibid.*, 1962, 3193; M. BECKE-GOEHRING, K. JOHN and E. FLUCK, *Z. anorg. Chem.*, 1959, **302,** 103.
133 R. KEAT and R. A. SHAW, *Angew. Chem. Internat. Edn.*, 1968, **7,** 212.
134 M. BIDDLESTONE and R. A. SHAW, *Chem. Comm.*, 1965, 205.
135 *Idem*, *ibid.*, 1968, 407.
136 H. BODE, K. BÜTOW and G. LIENAU, *Chem. Ber.*, 1948, **81,** 547.
137 N. V. MAIN and A. J. WAGNER, *Chem. Comm.*, 1968, 658.
138 I. I. BEZMAN and C. T. FORD, *Chem. and Ind.*, 1963, 163; Idem and F. E. DICKSON, *Inorg. Chem.*, 1964, **3,** 77; G. E. COXON and D. B. SOWERBY, *J. Chem. Soc.* (*A*), 1969, 3012.
139 M. BECKE-GOEHRING, H. HOHENSCHUTZ and R. APPEL, *Z. Naturforsch.*, 1954, **9b,** 678.
140 R. D. WHITAKER, J. C. CARLETON and H. H. SISLER, *Inorg. Chem.*, 1963, **2,** 420.
141 M. F. LAPPERT and G. SRIVASTAVA, *J. Chem. Soc.* (*A*), 1966, 210.
142 G. ALLEN, J. DYSON and N. L. PADDOCK, *Chem. and Ind.*, 1964, 1832; J. N. RAPKO and G. R. FEISTEL, *Chem. Comm.*, 1968, 474.
143 D. FEAKINS, W. A. LAST and R. A. SHAW, *J. Chem. Soc.*, 1964, 4464; Idem and S. N, NABI, *J. Chem. Soc.* (*A*), 1966, 1831.
144 J. DYSON and N. L. PADDOCK, *Chem. Comm.*, 1966, 191.
145 M. J. S. DEWAR, E. A. C. LUCKEN and M. A. WHITEHEAD, *J. Chem. Soc.*, 1960, 2423.
146 A. W. SCHLUETER and R. A. JACOBSEN, *J. Amer. Chem. Soc.*, 1966, **88,** 2051.
147 E. HOBBS, D. E. C. CORBRIDGE and B. RAISTRICK, *Acta. Cryst.*, 1953, **6,** 621.
148 T. CHIVERS and N. L. PADDOCK, *J. Chem. Soc.* (*A*), 1969, 1687.
149 J. K. JAQUES, M. F. MOLE and N. L. PADDOCK, *J. Chem. Soc.*, 1965, 2112.
150 C. W. ALLEN, J. B. FAUGHT, T. MOELLER and I. C. PAUL, *Inorg. Chem.*, 1969, **8,** 1719.
151 J. TROTTER and S. H. WHITLOW, *J. Chem. Soc.* (*A*), 1970, 455, 460.
152 R. N. BELL, *Inorg. Synth.*, 1950, **III,** 103.
153 E. J. GRIFFITH, *J. Amer. Chem. Soc.*, 1954, **76,** 5892.
154 I. A. BROVKINA, *Zhur. obshchei Khim.*, 1952, **22,** 1917.
155 H. M. ONDIK, *Acta Cryst.*, 1965, **18,** 226.
156 W. P. GRIFFITH and K. J. RUTT, *J. Chem. Soc.* (*A*), 1968, 2331.
157 D. A. KOSTER and A. J. WAGNER, *J. Chem. Soc.* (*A*), 1970, 435.
158 W. P. GRIFFITH, *J. Chem. Soc.* (*A*), 1967, 905.
159 L. KOLDITZ, B. NUSSBÜCKER and M. SCHÖNHERR, *Z. anorg. Chem.*, 1965, **335,** 189.
160 L. KOLDITZ, *Angew. Chem. Interact. Edn.*, 1965, **4,** 361; *Z. Chem.*, 1967, **7,** 240.
161 K. DEHNICKE and R. SCHMITT, *Z. anorg. Chem.*, 1968, **358,** 1.
162 L. WOLF and H. SCHMAGER, *Chem. Ber.*, 1929, **62,** 771; J. J. MANLEY, *J. Chem. Soc.*, 1922, **121,** 331.
163 G. C. HAMPTON and A. J. STOSICK, *J. Amer. Chem. Soc.*, 1938, **60,** 1814; L. R. MAXWELL, S. B. HENDRICKS and L. S. DEMING, *J. Chem. Phys.*, 1937, **5,** 626.
164 W. H. NELSON and R. S. TOBIAS, *Inorg. Chem.*, 1963, **2,** 985.
165 J. G. REISS and J. R. VAN WAZER, *J. Amer. Chem. Soc.*, 1965, **87,** 5506; *Idem*, *ibid*, 1966, **88,** 2166.
166 E. D. PERRON, P. J. WHEATLEY and J. G. RIESS, *Acta Cryst.*, 1966, **21,** 288.
167 F. C. MIJLHOFF, J. POSTHUME and C. ROMERS, *Rec. Trav. chim.*, 1967, **86,** 257.
168 G. BURKHARDT, M. P. KLEIN and M. CALVIN, *J. Amer. Chem. Soc.*, 1965, **87,** 591.
169 K. H. JOST, *Acta Cryst.*, 1966, **21,** 34; B. BEAGLEY, D. W. J. CRUICKSHANK, T. G. HEWITT and A. HAALAND, *Trans. Faraday Soc.*, 1967, **63,** 836.

170 P. W. SCHENK and H. VIETZKE, *Angew Chem. Internat. Edn.*, 1962, **1,** 48.
171 F. A. COTTON and G. WILKINSON, 'Advanced Inorganic Chemistry', Interscience, New York, 2nd Edn., p. 501.
172 W. WALTER and K. D. BODE, *Angew. Chem. Internat. Edn.*, 1966, **5,** 451.
173 M. BAUDLER, H. W. VALPERTZ and K. KEPLER, *Chem. Ber.*, 1967, **100,** 1766.
174 R. JEFFERSON, K. F. KLEIN and J. F. NIXON, *Chem. Comm.*, 1969, 536.
175 S. VAN HOUTEN and E. H. WIEBENGA, *Acta Cryst.*, 1957, **10,** 156.
176 D. T. DIXON, F. W. B. EINSTEIN and B. R. PENFOLD, *ibid.*, 1965, **18,** 221.
177 H. PETSCHUK and E. STEGER, *Angew. Chem. Internat. Edn.*, 1964, **3,** 314.
178 J. M. ANDREWS, J. E. FERGUSSON and C. J. WILKINS, *J. Inorg. Nuclear Chem.*, 1963, **25,** 829; F. W. B. EINSTEIN, B. R. PENFOLD and Q. T. TAPSELL, *Inorg. Chem.*, 1965, **4,** 186.
179 E. W. ABEL, D. A. ARMITAGE and R. P. BUSH, *J. Chem. Soc.*, 1964, 5584.
180 CHIA-SI LU and J. DONOHUE, *J. Amer. Chem. Soc.*, 1944, **66,** 818.
181 G. D. CHRISTOFFERSON and J. D. MCCULLOUGH, *Acta Cryst.*, 1959, **12**, 14.
182 M. BANDLER, K. KIPKER and W. W. VALPERTZ, *Naturwiss.*, 1967, **54,** 43.
183 E. W. ABEL and D. A. ARMITAGE, *J. Organometallic Chem.*, 1966, **5,** 326.
184 A. E. KRETOV and A. YA. BERLIN, *J. General Chem.* (U.S.S.R.), 1931, **1,** 411; *Chem. Abs.*, 1932, **26,** 2415.
185 L. AUCHUTZ and H. WIRTH, *Chem. Ber.*, 1956, **89,** 411.
186 M. BAUDLER and H. W. VALPERTZ, *Z. Naturforsch.*, 1967, **22b,** 222.
187 J. J. DALY, *J. Chem. Soc.*, 1964, 4065.
188 E. FLUCK and H. BINDER, *Z. anorg. Chem.*, 1967, **354,** 113; *Angew. Chem. Internat. Edn.*, 1966, **5,** 666.
189 *Idem, ibid.*, 1967, **6,** 260; H. W. ROESKY and D. BORMANN, *Chem. Ber.*, 1968, **101,** 630.
190 A. SCHMIDPETER and N. SCHINDLER, *Angew. Chem. Internat. Edn.*, 1968, **7,** 943.

APPENDIX

White phosphorus (P_4) has been co-ordinated to rhodium, using a tris(phosphine)rhodium(I) chloride.[1] It bonds as a monodentate ligand, the P_4 cage vibrations showing a reduction in symmetry on bonding. It is readily displaced by aliphatic phosphines and CO.

Melts of $BiCl_3$, $HfCl_4$ and Bi give black crystals of $Bi^+(Bi_9^{5+})(HfCl_6^{2-})_3$. The polyatomic cation has C_{3h} symmetry in the form of a face-bonded triangular prism and the lattice stabilizes Bi^+ cations.[2]

Antimony can be stabilized as Sb_n^{n+} in fluorosulphuric acid. The product $SbSO_3F$ is slowly oxidized to Sb(III) by the solvent, which is reduced to polysulphur cations and sulphur.[3] With arsenic[4] and antimony pentafluorides, the metal yields $SbMF_6$ and MF_3. This illustrates the caution that must be exercised in using weakly oxidizing sulphur solvents.

$$2Sb + 3AsF_5 \longrightarrow 2SbAsF_6 + AsF_3$$

Cyclopolyphosphines and arsines

Tetrakis(pentafluorophenyl)cyclotetraphosphine possesses a nonplanar P_4 ring with the substituents equatorial.[5] $(CF_3As)_4$ has a similar structure.[6]

The synthesis and properties of cyclopenta- and cyclohexa-arsines are summarized,[7] the alkyl derivatives reacting with the group VI metal carbonyls to produce various mono- and bimetallic cyclo-arsine complexes.[8] The complex $[Fe(CO_3)]_2(AsMe)_4$ has been shown to possess a linear tetra-arsine as the ligand,[9] showing that ring opening has occurred. Likewise the pentafluorophenyl-cyclotetraphosphine and arsines[10] are opened by $Fe(CO)_5$ yielding $(C_6F_5As)_2Fe(CO)_4$, $Fe_3(CO)_9(PC_6F_5)_2$ and $[Fe(CO)_3]_2(PC_6F_5)_4$.

Trigermanylarsine reacts with dimethylchloro-arsine to give the diarsine $Me_2AsAs(GeH_3)_2$. This dissociates on warming into dimethylgermanylarsine and pentagermanylcyclopenta-arsine.[11] The silyl analogue can be prepared the same way.

Phosphorus–nitrogen rings

Tri(chloromethyl)amine has already been reported in connection with the reactions of PCl_5 with aminomethanesulphonic acid to yield a cyclodiphospha-V-azane. It results in quantitative yields as a crystalline solid, m.pt. 93°C from hexamethylenetetramine and PCl_5.[12]

$$N_4(CH_2)_6 + 3PCl_5 \longrightarrow N(CH_2Cl)_3 + (NCH_2PCl_4)_3.$$

The tri-cyclic compound below results from *N*,*N'*-dimethylurea and *P*,*P'*-hexachloro-*N*,*N'*-dimethylcyclodiphospha-V-azane,[13]

$$(Cl_3PNMe)_2 + 2(MeNH)_2CO \longrightarrow$$

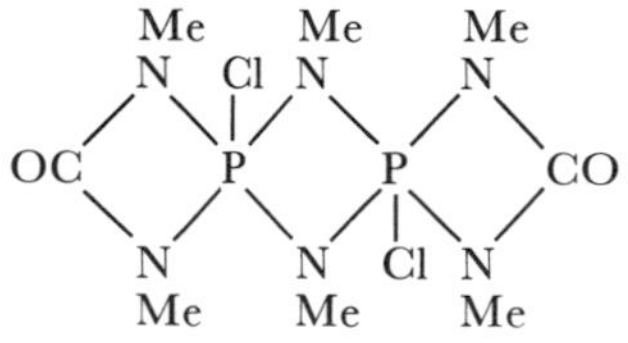

```
      Me     Me     Me
      N  Cl  N      N
     / \ | /   \  /   \
   OC    P       P      CO --Cl2--> (Cl3PNMe)2
     \ /   \   / | \  /
      N      N  Cl  N
      Me     Me     Me
```

to which it reverts on chlorine oxidation.

The cyclopolyphospha-V-azenes

The cyclodiphosphazene $(Bu^tNPCl)_2$ possesses a slightly puckered ring with P—N bond lengths indicating some multiple bonding.[14]

Much attention has been focussed on the synthesis, reactions and structure of these compounds. A critical survey of the conditions used for the synthesis of chlorophosphazenes is made.[15] This reports a convenient synthesis of $N_3P_3Cl_6$.

Extensive work on ring substitution concentrates on organic nucleophiles. With lithium alkyls, the fluoropha-V-azenes can be monosubstituted and polysubstituted either geminally or non-geminally. The mode of substitution is explained and the product from monosubstitution of $N_3P_3F_6$ by phenylacetylene gives a cobalt carbonyl complex.

With phenylmagnesium bromide, $N_4P_4Cl_8$ undergoes substitution and ring contraction. Diphenylmagnesium gives similar products while with $N_3P_3Cl_6$, ring substitution, substituent lengthening, ring coupling and polymerization occurs.[17]

Products involving substituent lengthening through amination and phosphorylation can be readily prepared using amines or disilazanes and PCl_5.[18] Substitution to yield spirophosphazenes occurs readily with diols[19] and chelating amines.[20] Optical isomers of 1,1-dichloro-trans-3,5-bis(*p*-tolyl)-3,5-diphenylcyclo-triphosphazene have been prepared.[21]

The kinetics of the reaction of $N_3P_3Cl_6$ with Me_2NH in tetrahydrofuran are measured. The mechanism proposed involves co-ordination as the first step, followed by solvent encouraged substitutions.

The π-bonding theory proposed for these rings has led to many structures being determined to ascertain the influence of substitution and co-ordination. The rings of $(NPX_2)_3$ (X = Cl, Br) are almost planar with similar P—N bond lengths of about 0.158 nm.[23] The fused tricyclic compound $P_6N_7Cl_9$ was first isolated by Stokes in 1897 and subsequently shown to

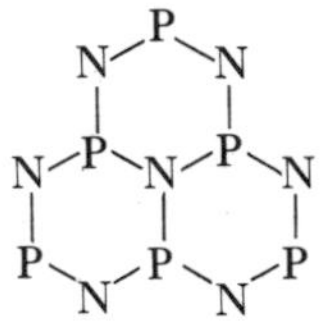

```
        P
      N   N
      |   |
  N - P - N - P - N
  |       |       |
  P       P       P
    N   N   N
```

be present as the decomposition product of the higher cyclophosphazenes. It has the structure shown in which the central nitrogen atom is bonded to the 3 phosphorus atoms in a planar arrangement.[24] These P—N bonds are long (0.1723 nm) and the proposal of weak σ- and strong π-bonds is supported by the lack of basicity. The planes of the rest of the ring do not contain the N of this NP_3 residue, so the molecule resembles a flattened triangular prism that has been truncated.

A detailed structure determination of $N_3P_3Cl_2(NHPr^i)_4 \cdot HCl$ is presented,[25] while the flattened saddle of $F_6Me_2P_4N_4$ and saddle conformer (D_{2d}) of $(F_2PNMe_2PN)_2$ support the structural trend[26] from a planar $(F_8P_4N_4)$ to a tub conformation $(Me_8P_4N_4)$. $(PhClPN)_4$ adopts a flattened crown structure[27] while $(MeO)_8P_4N_4$ resembles a distorted saddle.[28] The conformation adopted probably depends on the extent to which MeO π-bonds to phosphorus, and the orientations adopted to reduce steric clash. $(PhO)_6P_3N_3$ is nearly planar.[29]

$(Me_2N)_{12}P_6N_6$ forms a complex $(Me_2N)_{12}P_6N_6CuCl^+CuCl_2^-$ with $CuCl_2$ in MeCOEt. Copper is stabilized in 2 oxidation states and in the cation, Cu^{II} bonds in a square pyramid to 4 of the 6 nitrogen atoms of the hexaphosphazene ring, with chlorine in the apical position.[30] P—N bonds involving these nitrogen atoms are longer (0.162 nm) than the others (0.155 nm), since π_s-bonding is reduced.

The X-ray structure determination of orientated fibres of polymeric difluorophophazene shows 2 conformers present.[31] It is elastomeric at lower temperatures (−90°C) than the chloride (−63°C) or bromide (−15°C). This implies backbone torsional mobility which is consistent with low intramolecular interactions.

Miscellaneous nitrogen rings

Boron and aluminium have both been incorporated into the 6-membered P—N ring, as has sulphur VI.[32] Diphenylphosphorus trichloride oxidizes hydrazine hydrochloride and is itself reduced to the P_2N_3 ring indicated.[33]

$$Ph_2PCl_3 + H_2NNH_3^+Cl^- \longrightarrow$$

$$N_2 + [Ph_2PCl{=}N{-}PClPh_2]^+Cl^- + \left[Ph_2P{=}\overset{+}{N}{=}PPh_2 \;\; (HN{-}NH \text{ bridging the two P}) \right] Cl^-$$

Phenylarsenic dichloride is oxidized by ClN_3 to $PhAsCl_3N_3$ which slowly loses Cl_2 and N_2 at room temperature, to give the cyclotriarsazene.[34]

$$3PhAsCl_3N_3 \longrightarrow (PhClAsN)_3 + 3Cl_2 + 3N_2$$

The phosphazene $OPR_2(NPR_2)_3Cl$ readily cyclizes to the oxa-cyclotetraphosphazene which shows 2 kinds of phosphorus atom in its n.m.r. spectrum.[35]

$$OPR_2(NPR_2)_3X \longrightarrow \text{ring: } N{=}PR_2{-}N{=}PR_2{-}\overset{+}{O}{=}PR_2{-}N{=}PR_2{-}N \text{ (cyclic)}$$

Oxygen rings

The molecular structure of P_4O_6 and P_4O_8 has been elucidated by electron diffraction.[36] With P_4O_6, the P—O bond length is 0.1638 nm. The P^{III}—O—P^{III} has the same bond lengths but with P^{V}—O—P^{III}, P^{III}—O is 0.1666 nm and P^{V}—O (bridge), 0.1595 nm, supporting stronger π-bonding. Thallium cyclotetrametaphosphate is isomorphous with $Tl_4(AsO_3)_4$ and possesses the expected eight-membered ring.[37] The Tl—O distances indicate weak interaction.

Methoxytetrachlorostibine is dimeric and possesses methoxide bridges[38] while tetrameric fluorotetrachlorostibine is bridged by fluorine.[39] Antimony pentafluoride is also tetrameric and possesses an 8-membered Sb_4F_4 ring. The SbFSb angles are very wide (141–170°).

Sulphur and selenium derivatives

The reaction of iodine with P_4S_3 has been further investigated.[40] β-$P_4S_3I_2$ is isolated first, m.p. 107°C, and has a structure involving the cleavage of one P—P bond of the P_4S_3 cage and is isostructural with $P_4Se_3I_2$. It isomerises at 125°C into α-$P_4S_3I_2$.

$$P_4S_3 + I_2 \longrightarrow \beta\text{-isomer} \xrightarrow{125^\circ} \alpha\text{-isomer}$$

β-isomer (C_s) m.p. 107° α-isomer (C_2) m.p. 120°

The mass spectra of phosphorus sulphides[41] and selenides give breakdown patterns dominated by the ions P_3S^+ (isoelectronic with P_4), $P_3S_4^+$ (isoelectronic with P_4S_3) and PS^+.

P_4Se_3, the Raman spectrum of which is reported,[42] can be oxidized by bromine to P_4Se_5.[43] This is isostructural with P_4S_5. As_4S_3 has the same molecular structure[44] as P_4S_3 and P_4Se_3.

The structure of the $As_4S_6^{2-}$ anion in $(C_5H_{12}N)_2^+As_4S_6^{2-}$ closely resembles that of As_4S_4, with one As—As bond cleaved and the extra sulphur atoms bonded to each of these arsenic atoms.[45]

References

1 A. P. GINSBERG and W. E. LINDSELL, *J. Amer. Chem. Soc.*, 1971, **93,** 2082.
2 R. M. FRIEDMAN and J. D. CORBETT, *Chem. Comm.*, 1971, 422.
3 R. J. GILLESPIE and O. C. VAIDYA, *ibid.* 1972, 40.
4 P. A. W. DEAN and R. J. GILLESPIE, *ibid.*, 1970, 853.
5 F. SANZ and J. J. DALY, *J. Chem. Soc.* (A), 1971, 1083.
6 N. MANDEL and J. DONOHUE, *Acta Cryst.*, 1971, **B27,** 476.
7 P. S. ELMES, S. MIDDLETON and B. O. WEST, *Aust. J. Chem.*, 1970, **23,** 1559.
8 P. S. ELMES and B. O. WEST, *ibid.*, 2247.
9 B. M. GATEHOUSE, *Chem. Comm.*, 1969, 948.
10 P. S. ELMES, P. LEVERETT and B. O. WEST, *ibid.*, 1971, 747.
11 J. N. ANDERSON and J. E. DRAKE, *Chem. Comm.*, 1971, 1372.
12 E. FLUCK and P. MEISER, *Angew. Chem., Int. Edn.*, 1971, **10,** 653.
13 M. BECKE-GOEHRING and H. SCHWIND, *Z. anorg. Chem.*, 1970, **372,** 285.
14 K. W. MUIR and J. F. NIXON, *Chem. Comm.*, 1971, 1405.
15 J. EMSLEY and P. B. UDY, *J. Chem. Soc.* (A), 1971, 768.
16 E. NIECKE, H. THAMM and O. GLEMSER, *Z. Nat.*, 1971, **26b,** 366; N. L. PADDOCK, T. N. RANGANATHAN and S. M. TODD, *Canad. J. Chem.*, 1971, **49,** 164, T. CHIVERS, *Inorg. Nuclear Chem. Letters*, 1971, **7,** 827.
17 M. BIDDLESTONE and R. A. SHAW, *J. Chem. Soc.* (A), 1970, 1750; *idem, ibid.*, 1971, 2715.
18 H. W. ROESKY, W. GROSSE-BOWING and E. NIECKE, *Chem. Ber.*, 1971, **104,** 653.
19 H. R. ALLCOCK and E. J. WALSH, *Inorg. Chem.*, 1971, **10,** 1643.

20 T. CHIVERS and R. HEDGELAND, *Inorg. Nuclear Chem. Letters*, 1971, **7,** 767.
21 C. D. SCHMULBACH, C. DERDERIAN, O. ZECK and S. SAHURI, *Inorg. Chem.*, 1971, **10,** 195.
22 J. M. E. GOLDSMIDT and E. LICHT, *J. Chem. Soc.* (A), 1971, 2429.
23 G. J. BULLEN, *J. Chem. Soc.* (A), 1971, 1450; H. ZOER and A. J. WAGNER, *Acta Cryst.*, 1970, **B26,** 1812.
24 W. HARRISON, R. T. OAKLEY, N. L. PADDOCK and J. TROTTER, *Chem. Comm.*, 1971, 357; C. E. BRION and N. L. PADDOCK, P. σHEm. GOC. (A), 1968, 388.
25 N. V. MANI and A. J. WAGNER, *Acta Cryst.*, 1971, **B27,** 51.
26 W. C. MARSH and J. TROTTER, *J. Chem. Soc.* (A), 1971, 569 and 573.
27 G. J. BULLEN and P. A. TUCKER, *Chem. Comm.*, 1970, 1185.
28 G. B. ANSELL and G. J. BULLEN, *J. Chem. Soc.* (A), 1971, 2498.
29 W. C. MARSH and J. TROTTER, *ibid.*, 169.
30 *Idem*, *ibid.*, 1482.
31 H. R. ALLCOCK, G. F. KERNOPSKI, R. L. KUGEL and E. G. STROH, *Chem. Comm.*, 1970, 985.
32 H. BINDER, *Z. Nat.*, 1971, **26b,** 616; H. VOLLMER and M. BECKE-GOEHRING, *ibid.*, 1971, **380,** 314.
33 W. HANBOLD, D. KAMMEL and M. BECKE-GOEHRING, *ibid.*, 1971, **380,** 23.
34 V. KREIG and J. WEIDLEM, *Angew. Chem. Internat. Edn.*, 1971, **10,** 516.
35 A. SCHMIDPETER and K. STOLL, *ibid.*, 131.
36 D. W. J. CRUICKSHANK, T. G. HEWITT and K. H. JOST, *Trans. Faraday Soc.*, 1969, **65,** 1219.
37 J. K. FAWCETT, V. KOEMAN, S. C. NYBURG and R. J. O'BRIEN, *Chem. Comm.*, 1970, 1213.
38 VON H. PREISS, *Z. anorg. Chem.*, 1971, **380,** 65.
39 I. R. BEATTIE, K. M. S. LIVINGSTONE, G. A. OZIN and D. J. REYNOLDS, *J. Chem. Soc.*, (A), 1969, 958; A. J. EDWARDS and P. TAYLOR, *Chem. Comm.*, 1971, 1376.
40 G. J. PENNEY and G. M. SHELDRICK, *J. Chem. Soc.* (A), 1971, 1100; G. W. HUNT and A. W. CORDES, *Inorg. Chem.*, 1971, **10,** 1935.
41 G. J. PENNEY and G. M. SHELDRICK, *J. Chem. Soc.* (A), 1971, 243.
42 V. A. MARCONI and R. V. SCHABLASKE, *J. Inorg. Nuclear Chem.*, 1971, **33,** 3182.
43 G. J. PENNEY and G. M. SHELDRICK, *J. Chem. Soc.* (A), 1971, 245.
44 H. J. WHITFIELD, *ibid.*, 1970, 1800.
45 MISS E. J. PORTER and G. M. SHELDRICK, *J. Chem. Soc.* (A), 1971, 3130.

6

Sulphur, Selenium and Tellurium

THE ELEMENTS

One of the striking features of the chemistry of sulphur,[a] introduced at school level, is that of allotropy. This can be conveniently demonstrated by determining the transition temperature between the rhombic and monoclinic allotropes.[1] Both possess different crystalline arrangements of the S_8 molecules, comprising a puckered, 8-membered ring. This change represents only one of the transition points observed in the phase diagram of sulphur.[2]

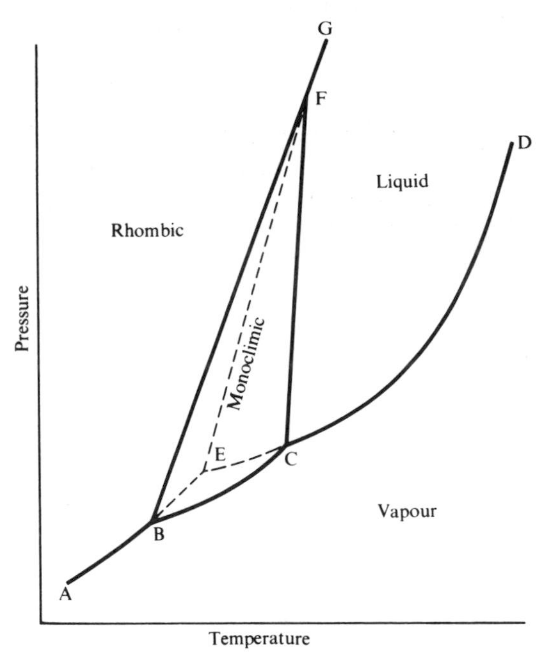

Fig. 6.1. Phase equilibrium diagram of sulphur (schematic)[2]

Rhombic sulphur is the stable allotrope at room temperature and atmospheric pressure. The area *BCF* gives the conditions for stable monoclinic sulphur relative to rhombic (*BF*), liquid (*CF*) or sulphur vapour (*BC*). If rhombic sulphur is slowly heated the transition into monoclinic is observed at 95.5°C (*B*). This melts at 119.25°C (*C*). Rapid heating causes the rhombic sulphur to melt at 114.5°C (*E*) before the transition into monoclinic sulphur, and the curve *EF* shows the effect of pressure on the melting point of rhombic sulphur.

Cooling sulphur slowly at atmospheric pressure gives large crystals of monoclinic sulphur which revert to small crystals of rhombic below 95.5°C. So the large crystals of rhombic sulphur found naturally are believed to have been formed directly from liquid sulphur by cooling at pressures above those at triple point *F*, *i.e.*, about 1300 atm.

$$S_{liq.} \longrightarrow S_{monoclinic} \longrightarrow S_{rhombic} \ (\Delta G = 0.402 \text{ kJ g atom}^{-1})$$

(large crystals) (small crystals)

The S_8 ring is puckered with S—S bond distance 0.204 nm. With the bond angles at sulphur 108 degrees, the ring possesses a crown structure, both in the rhombic and monoclinic polymorphs.

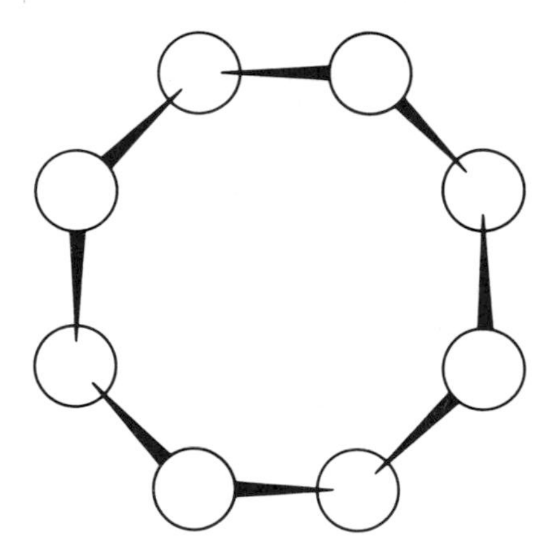

Fig. 6.2. The structure of the S_8 molecule in orthorhombic sulphur[1]

A third crystalline form of sulphur, Engel's sulphur S_ρ, is normally made by pouring a solution of sodium thiosulphate into conc. HCl and extracting with toluene. Molecular weight and structure determinations show this allotrope to be hexa-atomic sulphur S_6. It is less stable thermally than rhombic sulphur and melts irreversibly.

Both this, and various other polyatomic sulphur rings have now been synthesized in an integrated manner using various polythio residues. Thus, S_6 has been prepared by the elimination of hydrogen chloride from a hydrogen polysulphide and a dichlorosulphane. The crystals are a deeper yellow than either rhombic or monoclinic sulphur.[3]

$$2H_2S + 2S_2Cl_2 \xrightarrow[Et_2O]{dark} \underset{(8\%)}{S_6}$$

$$H_2S_4 + S_2Cl_2 \longrightarrow \underset{(30\%)}{S_6}$$

Titanocene dichloride reacts with ammonium pentasulphide or sodium polysulphides. The titanocene pentasulphide formed[4] can also be synthesized from titanocene di(hydrosulphide) and dichlorosulphanes.[5]

$$cp_2TiCl_2 + (NH_4)_2S_5 \longrightarrow \underset{(m.p.\ 201\text{–}2°C)}{cp_2TiS_5} + 2NH_4Cl$$

$$cp_2TiCl_2 + 2H_2S + 2Et_3N \longrightarrow cp_2Ti(SH)_2 + 2Et_3NHCl$$

$$cp_2Ti(SH)_2 \xrightarrow{S_nCl_2} cp_2TiS_5$$

With SCl_2, titanocene pentasulphide gives S_6 in high yield, along with S_{12} (11%).[6]

$$cp_2TiS_5 + SCl_2 \longrightarrow cp_2TiCl_2 + \frac{1}{n} S_{6n} \quad (n = 1 \text{ or } 2)$$

Unlike ρ-sulphur, cyclododecasulphur is a paler yellow and more thermally stable than most other forms of sulphur. It melts at 148°C and is irreversibly converted into monoclinic sulphur. The crystal structure shows the expected 12-membered ring with S—S bond distances of 0.2055 nm. It can also be made from hydrogen polysulphides and is encountered in 0.1% quantities in quenched liquid sulphur, independent of the temperature.[7]

$$H_2S_8 + S_4Cl_2 \longrightarrow S_{12} + 2HCl$$

$$2H_2S_4 + 2S_2Cl_2 \longrightarrow S_{12} + 4HCl$$

Cyclodecasulphur S_{10} has been prepared by a similar HCl condensation reaction between H_2S_6 and S_4Cl_2,[3] and also from titanocene pentasulphide and sulphuryl chloride.[6]

$$H_2S_6 + S_4Cl_2 \xrightarrow{dark} S_{10}$$

$$cp_2TiS_5 + SO_2Cl_2 \longrightarrow cp_2TiCl_2 + \tfrac{1}{2}S_{10} + SO_2$$

The expected product incorporating the SO_2 group into a five-atom sulphur ring is thermally unstable, like most compounds with the —S—SO_2—S— grouping, and readily loses SO_2. It forms intense yellow rhombic platelets which are decomposed by X-rays, and which readily polymerize above 60°C.

S_7 results in a 23% yield from cp_2TiS_5 and sulphur monochloride.[6]

$$cp_2TiS_5 + S_2Cl_2 \longrightarrow cp_2TiCl_2 + S_7$$

This crystallizes from toluene as deep yellow needles which melt reversibly at 39°C. X-ray diffraction shows the ring to possess a chair structure.[8] Rhombic sulphur can be made similarly using S_3Cl_2.

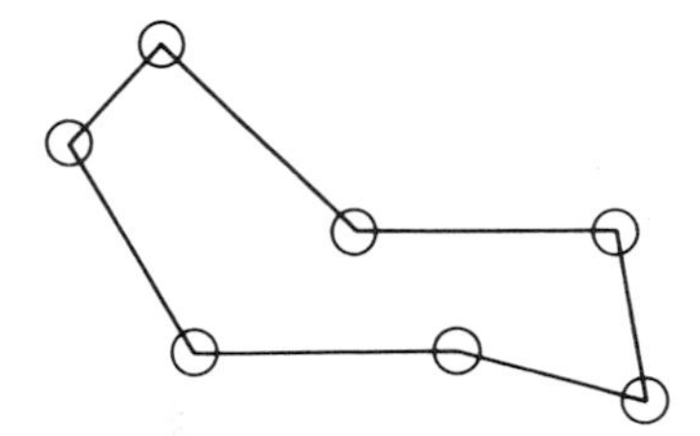

Fig. 6.3. Cycloheptasulphur showing the chair conformation[8]

Cyclononasulphur, S_9, cannot be made by this method directly as the S—Cl bond in S_4Cl_2 is not reactive enough to cleave the Ti—S bond.[8a] A catalytic amount of hydrogen chloride is required.

$$\mathrm{cp_2TiS_5 + HCl \longrightarrow cp_2Ti(Cl)S_5H \xrightarrow{S_4Cl_2} HCl + cp_2Ti(Cl)S_9Cl \longrightarrow cp_2TiCl_2 + S_9}$$

The product results in 30% yield, and is unstable towards heat and light, melting, like S_6 and S_{10}, irreversibly above 50°C into polymeric sulphur.

Heating mixtures of sulphur and selenium in various ratios, and examining the products by mass spectrometry, shows the existence of mixed sulphur–selenium rings containing eight atoms.[9] These include S_7Se, S_6Se_2, S_5Se_3 and S_4Se_4, while S_7Cl_2 and H_2Se give S_7Se in high yield. With tellurium, S_7Te has been shown to exist, while $TeCl_4$ and polysulphanes give Cl_2TeS_7 in high yield as orange crystals which melt at 110–112°C. The ring is the same shape as that of S_8.[10]

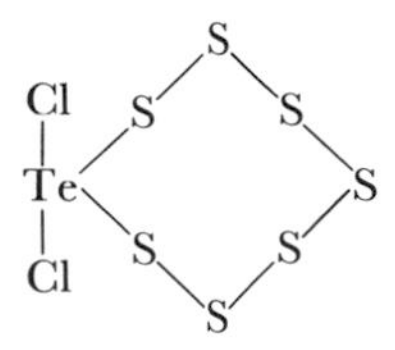

Fig. 6.4. The proposed structure of Cl_2TeS_7[10]

Aspects of the allotropy of sulphur are apparent in the chemistry of elemental selenium. However, the inevitable increase in metallic character as a Group is descended results in the non-metallic aspects of selenium chemistry occurring over a lower temperature range.

Both rhombic and monoclinic selenium result on evaporation of the dark red solution in CS_2 below 72°C. Both contain the Se_8 ring with Se—Se bonds 0.234 nm long.[11] Both forms revert to the grey form on heating, this allotrope containing an infinite chain of selenium atoms arranged in a spiral.[12] It has a metallic appearance and possesses conducting properties when light-activated. It is therefore a useful semi-conductor, like the only allotrope of tellurium.

CATIONIC DERIVATIVES OF SULPHUR, SELENIUM AND TELLURIUM

The three elements sulphur, selenium and tellurium can be oxidized to polyatomic cations using mild oxidizing agents, including derivatives of sulphuric acid and arsenic pentafluoride.

Selenium dissolves in sulphuric acid to give green solutions containing the cation Se_8^{2+}.

$$8Se + 5H_2SO_4 \longrightarrow Se_8^{2+} + 2H_3O^+ + 4HSO_4^- + SO_2$$

The acid is reduced to SO_2.[13] The structure of this cation, resembles the Se_8 ring, with the Se—Se bond 0.232 nm in length.

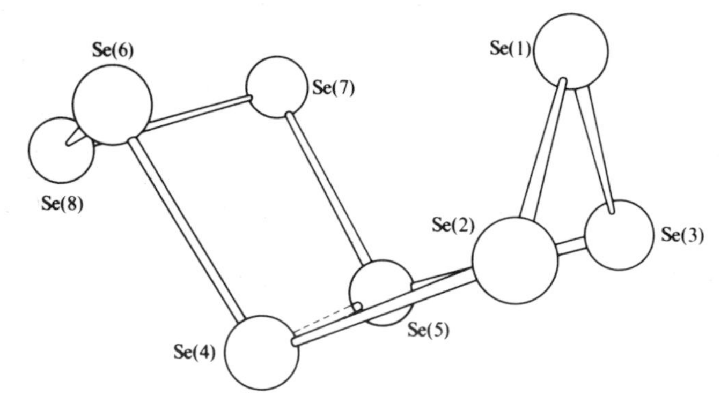

Fig. 6.5. The structure of Se_8^{2+} [14]

Nevertheless there are two outstanding differences. One is the way $Se_{(1)}$ has flipped across the $Se_{(2)}$—$Se_{(3)}$—$Se_{(4)}$—$Se_{(5)}$ quadrilateral, and the other, the strong interaction between $Se_{(4)}$ and $Se_{(5)}$ (0.284 nm) compared with $Se_{(2)}$—$Se_{(3)}$ and $Se_{(6)}$—$Se_{(7)}$ (0.33 nm).[14] Further oxidation might be expected to give Se_8^{4+}, with $Se_{(8)}$ flipped and a strong $Se_{(1)}$—$Se_{(8)}$ interaction, giving the ion a structure analogous to As_4S_4.

However, oxidation of Se_8^{2+} with selenium dioxides gives Se_4^{2+}.[15]

$$7Se_8^{2+} + 4SeO_2 + 24H_2SO_4 \longrightarrow 15Se_4^{2+} + 8H_3O^+ + 24HSO_4^-$$

An X-ray structure determination of this yellow cation shows it to have a planar structure with equal Se—Se bonds of length 0.228 nm.[16] This structure is also supported by the non-coincidental nature of the infrared and Raman peaks (i.r. 306 cm^{-1}; Raman 188, 326 cm^{-1}), which indicates a centre of symmetry to the ion.[17]

Tellurium forms a similar ion in fluorosulphuric acid. This is red, diamagnetic and has a Raman spectrum (219 cm^{-1}, 139 cm^{-1}) similar to that of Se_4^{2+}.

$$4Te + 4HSO_3F \longrightarrow Te_4^{2+} + SO_2 + HF + H_3O^+ + 3SO_3F^-$$

This ion also forms using antimony pentafluoride at $-23°C$ in liquid SO_2. This is soluble, but a yellow insoluble residue forms which, after heating, analyzes for $TeSbF_6$ and probably contains the cation Te_n^{n+}.[18]

Sulphur is readily oxidized by AsF_5[19] in HF at $-78°C$ giving the two cations S_8^{2+} and S_{16}^{2+}.

$$S_8 + 3AsF_5 \longrightarrow \underset{\text{(blue)}}{S_8^{2+}(AsF_6^-)_2} + AsF_3$$

$$2S_8 + 3AsF_5 \longrightarrow \underset{\text{(red)}}{S_{16}^{2+}(AsF_6^-)_2} + AsF_3$$

These ions also result with $S_2O_6F_2$ and SbF_5, but oxidation will go further to colourless S_4^{2+}.[20]

$$\tfrac{1}{2}S_8 + S_2O_6F_2 \longrightarrow S_4^{2+}(SO_3F^-)_2$$

Cationic derivatives of the Group VI elements are becoming increasingly common with all three tetra-atomic cations now known, and having similar absorption spectra.[20,21]

SULPHUR–NITROGEN COMPOUNDS

Nitrogen derivatives of sulphur are known for a wide range of oxidation states, from bivalent to hexavalent, and involve a variety of ring sizes. Those based on the S_8 ring and bivalent sulphur will be considered first.

Sulphur(II)–Nitrogen Rings[b,f]

Heptasulphur imide was first prepared in 1923 as a by-product of the reaction of sulphur monochloride and ammonia.[22] In CS_2 below $-10°C$, it is the major product.[23]

$$7S_2Cl_2 + 16NH_3 \longrightarrow \underset{\text{(m.p. 113.5°C)}}{2S_7NH} + 14NH_4Cl$$

It forms a variety of metallic derivatives, both of alkali metals (green sodium salt) and otherwise (mercury forms a yellow imide). Metallation also readily occurs with lithium alkyls, S_7N^-, readily attacking alkyl halides.[24] The N—H bond of the parent imide will react directly with acyl halides and formaldehyde, while with SCl_2 in the presence of pyridine, or sulphuric acid, ring-coupling occurs.[25]

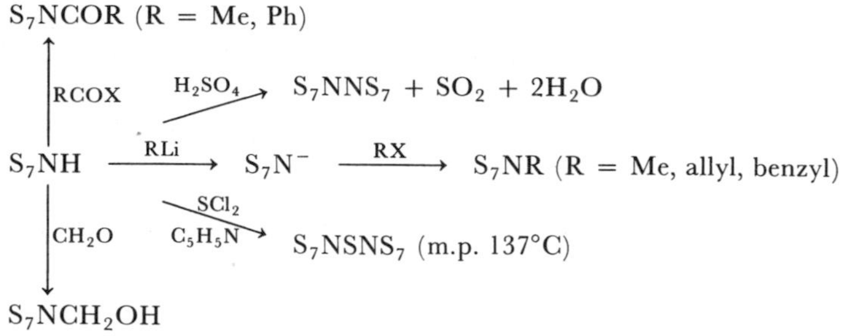

S_7NCOR (R = Me, Ph)

↑ RCOX

$S_7NH \xrightarrow{H_2SO_4} S_7NNS_7 + SO_2 + 2H_2O$

$S_7NH \xrightarrow{RLi} S_7N^- \xrightarrow{RX} S_7NR$ (R = Me, allyl, benzyl)

$S_7NH \xrightarrow[C_5H_5N]{SCl_2} S_7NSNS_7$ (m.p. 137°C)

↓ CH_2O

S_7NCH_2OH

The structure of heptasulphur imide, like that of sulphur, involves an 8-membered ring. The S—S bonds are the same length as in S_8, with the S—N bond 0.173 nm and close to a single bond.[26]

The substitution of two NH groups in the S_8 ring can give three structural isomers (excluding the case with the N—N bond). All three isomers result as side-products in the reaction of sulphur monochloride with ammonia.

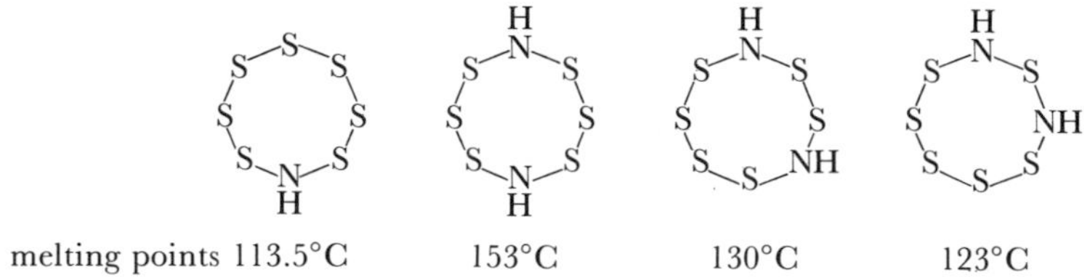

melting points 113.5°C   153°C   130°C   123°C

Fig. 6.6. The sulphur imides

The more symmetrical 1,5-diazo compound has the highest melting point, while in the 1,4 isomer bond lengths are close to those of S_7NH.[27] The 1,3 isomer reacts with S_5Cl_2 in the presence of base to give the bicyclic sulphur nitride $S_{11}N_2$. Cyclization is encouraged by using CS_2 as the solvent.[28]

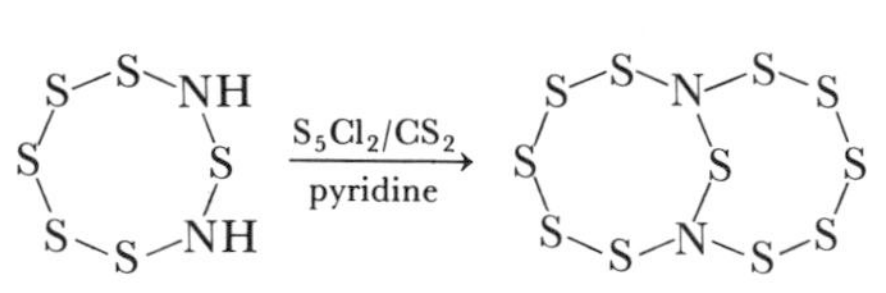

$\xrightarrow[\text{pyridine}]{S_5Cl_2/CS_2}$

The methyl analogue of the 1,5 isomer results from S_3Cl_2 and methylamine under high dilution in petroleum,[29] and the method has been extended and used to prepare the 6-membered rings from sulphur monochloride. Again, these compounds are low-melting crystalline solids.[30]

$$2S_2Cl_2 + 4RNH_2 \longrightarrow RN\langle S{-}S, S{-}S \rangle NR + 2RNH_3Cl$$

Reacting primary amines with sulphur dichloride affords the two tri-imide isomers along with the tetrasulphur 1,3,5,7-tetra-imide.[31] The parent tetra-imide $S_4N_4H_4$ is best prepared by reducing tetrasulphur tetranitride with stannous chloride.[32] It possesses an 8-membered ring structure involving both the four sulphur and four nitrogen atoms mutually orientated in a square plane, with the S—N bond distance 0.167 nm.[33]

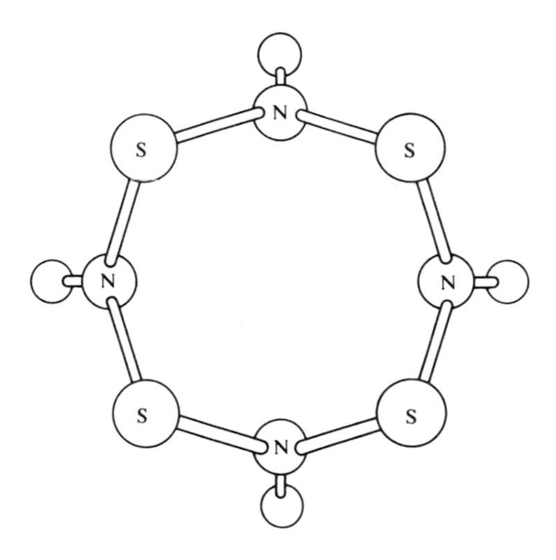

Fig. 6.7. The $S_4N_4H_4$ molecule[33]

The structure is very similar to that of rhombic sulphur, and the S—N bonds are shorter than expected for single bonds. It disproportionates on heating, and is oxidized to tetrameric thionylimide by air at 120°C[34] and by mercuric acetate, while with $LiAlH_4$ it forms what is probably $Li^+[S_4N_4Al]^-$.

$$2S_4N_4 + 4S + 4NH_3 \xleftarrow[N_2]{120°C} 3S_4N_4H_4 \xrightarrow[120°C]{O_2} 3[HNSO]_4$$

Tetrasulphur Tetranitride S_4N_4

This is by far the widest studied of cyclic sulphur–nitrogen compounds.[c,d] It is normally prepared by passing sulphur monochloride over hot ammonium chloride,

$$6S_2Cl_2 + 4NH_4Cl \longrightarrow S_4N_4 + 8S + 16HCl$$

but better yields result using ammonia and sulphur tetrafluoride or sulphur monochloride and the azide ion.[35]

$$12SF_4 + 64NH_3 \xrightarrow{-90°C} 3S_4N_4 + 48NH_4F$$

$$2N_3^- + S_2Cl_2 \longrightarrow S_2(N_3)_2 \longrightarrow \tfrac{1}{2}S_4N_4 + 2N_2$$

It is an orange crystalline compound that sometimes detonates at its melting point of 178°C.

The structure involves a puckered 8-membered ring of alternating sulphur and nitrogen atoms, with the nitrogen ones arranged in a plane, the sulphur ones in a distorted tetrahedron.[36]

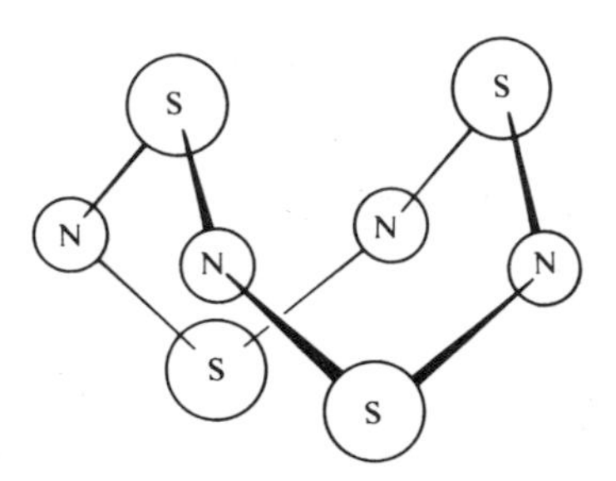

Fig. 6.8. The structure of tetrasulphur tetranitride[e]

The close proximity of the sulphur atoms (0.258 nm) indicates a weak bond which probably involves the overlap of sulphur p orbitals. On a molecular orbital scheme, this will be the highest occupied orbitals, with the lowest unoccupied ones a double degenerate *E* level. This readily allows for the formation of the anions $S_4N_4^-$ to $S_4N_4^{4-}$. This latter ion probably possesses a structure resembling $S_4N_4H_4$.[37] All the S—N bonds in S_4N_4 are equal in length (0.1616 nm) and shorter than a single bond, supporting the presence of delocalized p_π–d_π bonding around the ring in a manner similar to that proposed with the phosphazenes. Tetraselenium tetranitride, which results from various selenium(IV) compounds and ammonia, has a similar structure to S_4N_4 involving analogous Se—Se trans-annular interactions.[38]

Tetrasulphur tetranitride can be readily reduced with stannous chloride as already mentioned (p. 345), and the tetrasulphur tetra-imide so formed can be readily oxidized back to S_4N_4 using chlorine.[c]

$$S_4N_4H_4 + 2Cl_2 \longrightarrow S_4N_4 + 4HCl$$

Further chlorination oxidizes S_4N_4 to trithiazyl trichloride $[NSCl]_3$, while AgF_2 yields tetrathiazyl tetrafluoride $[NSF]_4$.[39]

Lewis acids readily form adducts with S_4N_4, through co-ordination of the nitrogen atoms. While the BF_3 adduct readily dissociates on warming, the 1:1 complex, $S_4N_4{\cdot}BCl_3$, can be sublimed. BCl_3 and $SbCl_5$ both displace BF_3 from $S_4N_4{\cdot}BF_3$, while antimony pentachloride forms a double adduct with $S_4N_4{\cdot}BCl_3$. This loses BCl_3 at 90°C under vacuum.[40]

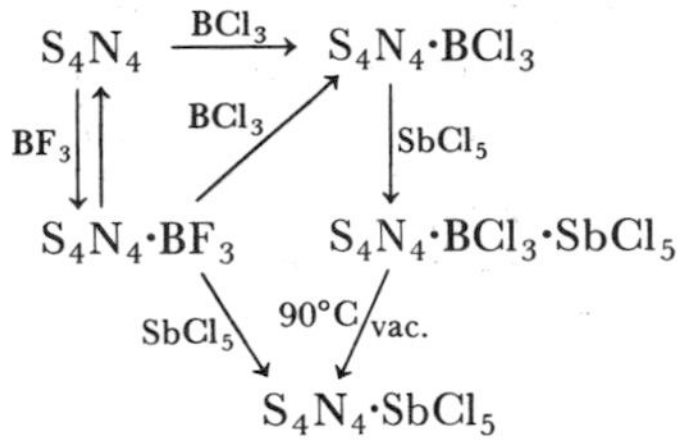
S4N4
BCl3
S4N4·BCl3
BF3
BCl3
SbCl5
S4N4·BF3
S4N4·BCl3·SbCl5
SbCl5
90°C/vac.
S4N4·SbCl5

While SbF_5 gives a 4 : 1 adduct at room temperature, this decomposes at 145°C into thiazyl fluoride. This oxidation occurs more readily with SF_4.

Cyclic olefines form 2 : 1 adducts with tetrasulphur tetranitride.[41] These are Diels–Alder-like complexes, involving the addition of norbornene or norbornadiene to a 1,3-delocalized part of the S_4N_4 ring.

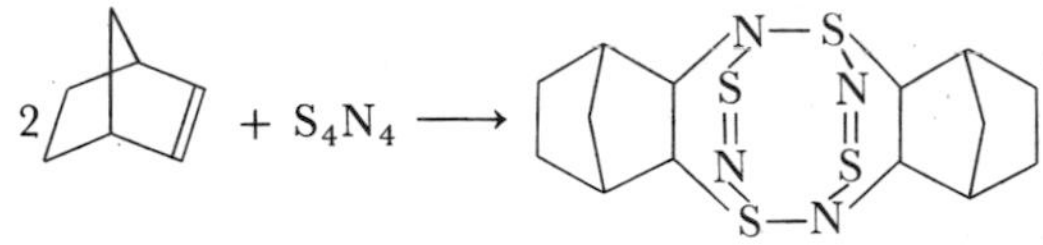
2
+ S4N4

With cyclopentadiene, a 4 : 1 complex results through the addition of C_5H_6 to the 2 : 1 complex.

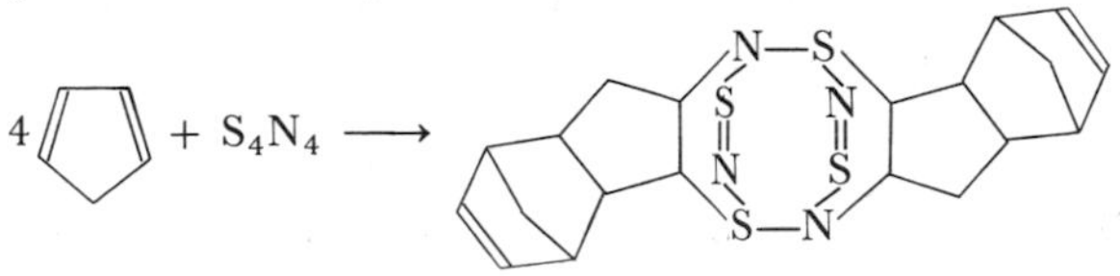
4
+ S4N4

Passing S_4N_4 vapour over silver wool produces S_2N_2, which can be conveniently stored in methylene chloride at 20°C, though out of solution it explodes at 30°C.[42] The molecule is almost square with S—N bonds of the length found in S_4N_4. Like S_4N_4, it complexes with excess $SbCl_5$ giving a 1 : 2 adduct with ring parameters similar to those observed in S_2N_2. This readily reacts with more S_2N_2, yielding the 1 : 1 complex $S_2N_2{\cdot}SbCl_5$ and finally $S_4N_4{\cdot}SbCl_5$. Thus, $SbCl_5$ catalyzes the dimerization of S_2N_2.[43]

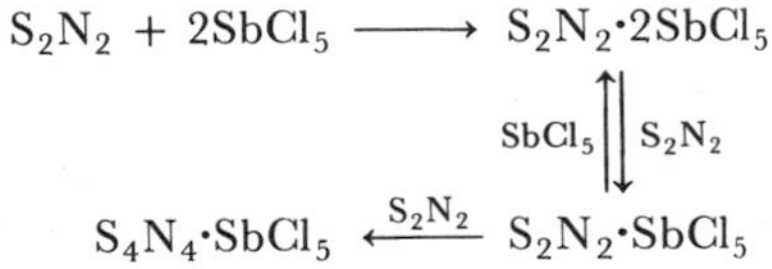
S2N2 + 2SbCl5
S2N2·2SbCl5
SbCl5
S2N2
S4N4·SbCl5
S2N2
S2N2·SbCl5

This dissociation of S_4N_4 also occurs in liquid ammonia, amines of S_2N_2 resulting.

$$S_4N_4 + \text{liq. } NH_3 \rightleftharpoons 2S_2N_2(NH_3)_n$$

Thionyl chloride will oxidize S_4N_4 into $S_5N_5^+$ in the presence of $AlCl_3$. This ion is a 10-membered heart-shaped ring, and resembles a butterfly with partly opened wings. The tetrachloraluminate melts at 181°C, and the S—N bond lengths of 0.155 nm indicate considerable multiple bonding, supported by the wide valence angles at nitrogen.[44]

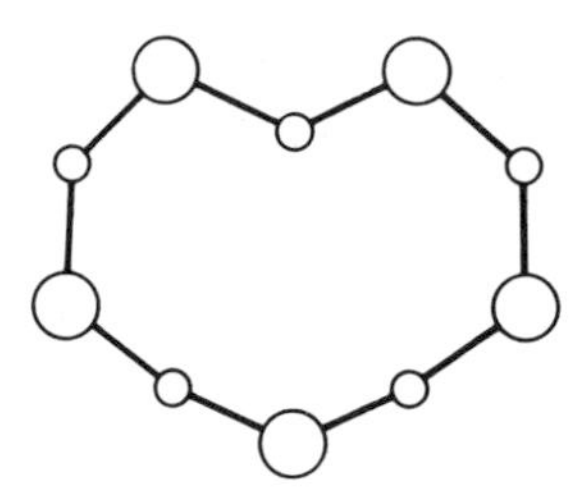

Fig. 6.9. The projection of the $S_5N_5^+$ ion (large circles represent sulphur atoms)[44]

Both phosphorus trichloride and phenylphosphorus dichloride disrupt the S_4N_4 ring, and give diphosphazenium chlorides along with polymeric phosphazenes and thiophosphorus chlorides.[45]

$$S_4N_4 + 10PCl_3 \longrightarrow 2[Cl_3PNPCl_3]^+Cl^- + \frac{2}{n}[NPCl_2]_n + 4SPCl_3$$

$$\begin{aligned} S_4N_4 + PhPCl_2 \longrightarrow & [PhCl_2PNPPhCl_2]^+Cl^- \\ & + [PhCl_2PN(PPhClN)_nPPhCl_2]^+Cl^- + PhPSCl_2 \\ & + PhCl_2PNP(S)ClPh \end{aligned}$$

HALOGEN-SUBSTITUTED SULPHUR–NITROGEN RINGS

While sulphur–nitrogen rings involving sulphur in a low oxidation state ((II) or (III)) are normally stabilized by non-oxidizing substituents (*e.g.*, alkyl, or hydrogen), those involving sulphur in higher oxidation states normally have electronegative oxidizing groups bonded to sulphur, *e.g.*, oxygen or a halogen.[e]

These compounds naturally fall into two groups, those involving sulphur(IV), stabilized by a halogen, and those rings containing sulphur(VI) stabilized by oxygen and a halogen. Tetrasulphur tetranitride plays a prominent part in the synthesis of these compounds, and while trithiazyl trichloride can be prepared from chlorine,[46] fluorine is too vigorous an oxidizing agent giving F_5SNSF_2 and not polythiazyl polyfluorides. The syntheses of these various sulphur(IV) rings are tabulated below. (See top of next page.)

Tetrathiazyl tetrafluoride can only be prepared by fluorinating S_4N_4 with silver difluoride.[39] This probably does not proceed through ring cleavage since thiazyl fluoride NSF only trimerizes.[47] The skeletal ring

$$N_3S_3Cl_3 \longleftarrow N_4S_4Cl_4 \qquad N_4S_4F_4 \text{ (m.p. 153°C, dec}^n\text{ above 128°C)}$$

(m.p. 168°C)

$$S_4N_4 \xrightarrow{Cl_2/CS_2} N_4S_4Cl_4 \qquad S_4N_4 \xrightarrow{AgF_2} N_4S_4F_4$$

$$S_4N_4 \xrightarrow{S_2Cl_2} S_4N_3^+Cl^-$$

$$N_3S_3Cl_3 \xrightarrow{AgF_2|CCl_4} N_3S_3F_3$$

$$S_4N_4 \xrightarrow{HgF_2} NSF \qquad S_4N_4 \xrightarrow{F_2} F_5SNSF_2$$

$$N_3S_3F_3 \text{ (m.p. 74°C)} \longleftarrow NSF \qquad F_5SNSF_2 \text{ (b.p. 43°C)}$$

of $N_4S_4F_4$ is flatter than that of tetrasulphur tetranitride and the S—N bonds alternate in length from 0.166 nm to 0.154 nm.[48] The shorter corresponds to a double-bond length, and hence little delocalization exists as is also observed with cyclo-octatetraene. This contrasts with the delocalized picture in S_4N_4, and is possibly due to the induced positive charge on sulphur causing orbital contraction, tighter bonding and hence a weaker tendency to delocalize.

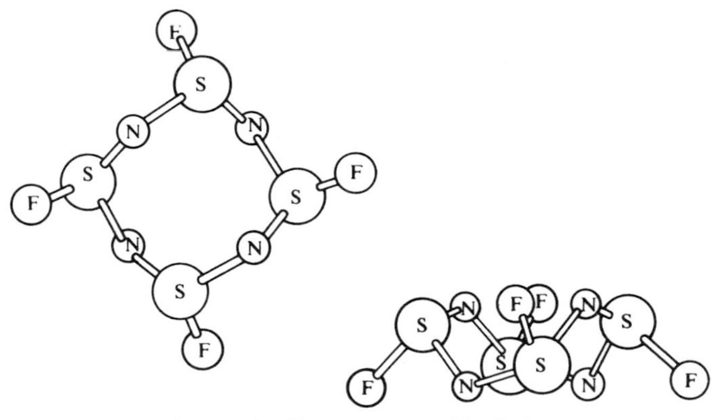

Fig. 6.10. The structure of $N_4S_4F_4$[e]

Like S_4N_4, however, it will co-ordinate with boron trifluoride.[49] The solitary ^{19}F n.m.r. signal is close in position to that of SF_6, a situation expected as the two molecules have S—F bonds of similar length (0.160 nm).

Trithiazyl trifluoride cannot be prepared from S_4N_4, but results when the trichloride is fluorinated[46] or by allowing thiazyl fluoride to trimerize at room temperature under pressure. It is readily hydrolyzed in basic solution to ammonium fluoride and sulphite,

$$N_3S_3F_3 + 9H_2O \longrightarrow 3NH_4F + 3H_2SO_3$$

while a trace of water gives the gaseous products NSF, HNSO, SO_2, SOF_2, SF_4 and NH_3.

The trichloride, which was first isolated in 1931 directly from tetrasulphur tetranitride, possesses an essentially planar 6-membered ring structure with all S—N bond lengths equal (0.1605 nm) and corresponding to a bond order of about 1.5.[50] This is in contrast with the fluorides, where the S—N bonds alternate in length.

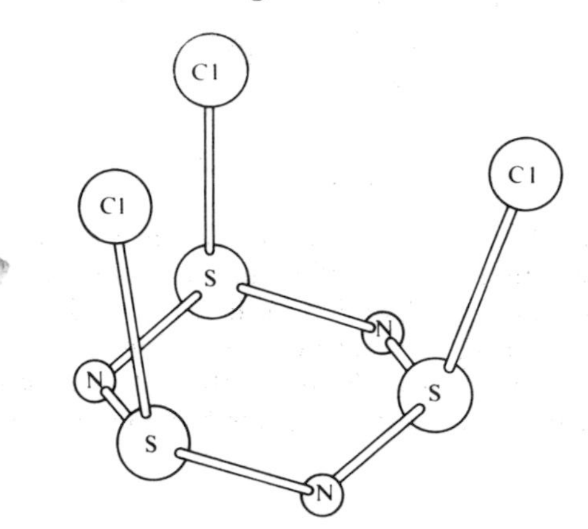

Fig. 6.11. The structure of trithiazyl trichloride[e]

Again, basic hydrolysis gives S^{IV} products, while with tetrasulphur tetra-imide co-proportionation of S^{II} and S^{IV} gives tetrasulphur tetra-nitride.[51] With dimethylsulphoxide, a peculiar reaction involving the formation of the dithiazonium ion occurs.[52]

$$[NSCl]_3 + 6Me_2SO \longrightarrow 3[Me_2S{=}N{=}SMe_2]^+Cl^- + 3SO_2$$

It appears that the sulphur of the ring is released as SO_2.

Thiotrithiazyl chloride, $S_4N_3^+Cl^-$, can be prepared from S_4N_4 and sulphur halides,[35,53] notably sulphur monochloride and thionyl chloride. However, it is better prepared directly from lithium azide and S_2Cl_2, which involves S_4N_4 as an intermediate.

$$12LiN_3 + 8S_2Cl_2 \longrightarrow 4S_4N_3^+Cl^- + 12LiCl + 12N_2$$

Molecular weight and X-ray crystal structure determinations show thio-trithiazyl chloride to be ionic, with the cation $S_4N_3^+$ planar with alternate S and N atoms and an S—S bond. This is probably a single bond (0.206 nm) unlike the S—N bonds, whose multiple bonding is inferred from the short bond distances (0.152 to 0.160 nm).[54]

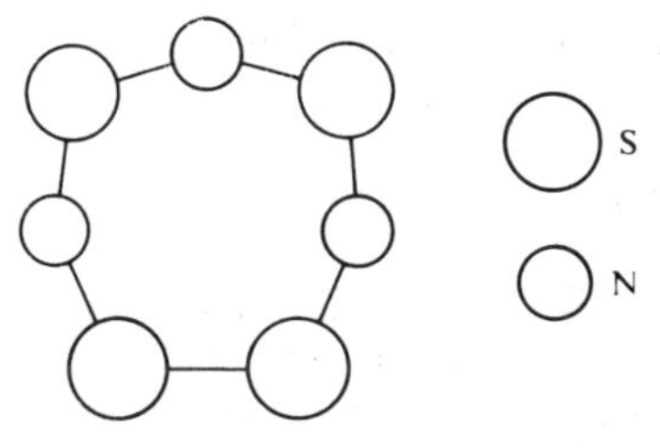

Fig. 6.12. The structure of the $S_4N_3^+$ cation[e]

The ionic nature of this compound is also supported by the ease with which the chloride can be converted into the nitrate, bromide or thiocyanate,[55] while with aluminium azide (explosive), ring expansion into S_4N_4 readily occurs.[56]

$$S_4N_3^+Cl^- + NH_4SCN \xrightarrow{HCO_2H} S_4N_3^+SCN^- + NH_4Cl$$

$$S_4N_3^+Cl^- + \tfrac{1}{3}Al(N_3)_3 \longrightarrow S_4N_4 + N_2 + \tfrac{1}{3}AlCl_3$$

Tetrasulphur tetranitride is also the major decomposition product resulting when $S_4N_3^+Cl^-$ is heated to 170°C, hydrolyzed or allowed to react with ammonia.[55,57]

If thiotrithiazyl chloride is prepared from thionyl chloride in the presence of selenium monochloride, a compound formulated as $SeS_2N_2Cl_2$ is also formed. The infrared spectrum shows peaks characteristic of $S_4N_3^+$ and $SeCl_6^{2-}$, so this material probably comprises the selenotrithiazyl cation $[(SeS_3N_3)^+]_2SeCl_6^{2-}$.[58]

Though the sulphanuryl chloride isomers can be isolated by oxidizing trithiazyl trichloride with SO_3,[51] they are more efficiently prepared by the thermal decomposition of trichlorophosphazosulphuryl chloride, $ClSO_2NPCl_3$. This is formed from imidosulphonic acid and PCl_5.[59]

$$NH_2SO_3H + 2PCl_5 \longrightarrow 3HCl + POCl_3 + \underset{(\text{m.p. } 33\text{–}5°C)}{ClSO_2NPCl_3} \xrightarrow{\text{heat}} POCl_3 + [NSOCl]_3$$

The two isomers of $[NSOCl]_3$ can be conveniently separated, since the more symmetrical α-form has a higher dipole moment and melting point (144–5°C) than the β-isomer (m.p. 46–7°C). Filtering the mixture at 50°C partially separates the two isomers. The filtrate contains mainly the β-isomer, which remains in solution on recrystallizing from cyclohexane. The α-isomer separates as rhombic prisms. Evaporating much of the solvent and cooling to 0°C produces long needles of the β-isomer, which can be readily converted to the α-form in polar solvents such as acetonitrile or ether. Both compounds can be further purified by sublimation, the less polar β-isomer subliming at room temperature (0.005 mm) while the α-isomer has to be heated to 80°C.[60] (See Figure 6.13.)

The S_3N_3 ring of the α-isomer has a chair conformation with the S_3 and N_3 planes 0.025 nm apart.[61] All S—N bonds are the same length (0.1571 nm) and significantly shorter than those of trithiazyl trichloride, due possibly to additional orbital contraction due to the extra electronegative group. This bond shortening due to π-bonding is also reflected in the S—O bond length which is close to that of sulphuryl chloride (~0.141 nm).

Halogen-exchange on α-sulphanuryl chloride with KF in CCl_4 produces two isomeric sulphanuryl fluorides which also result from ammonia and OSF_4.[62] These can be separated by gas chromatography and readily

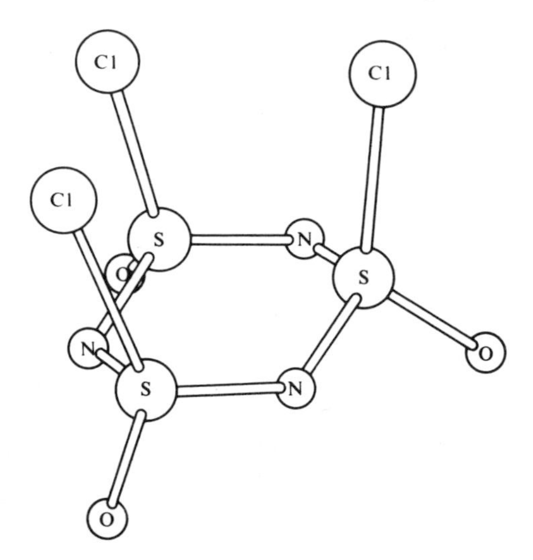

Fig. 6.13. The structure of α-sulphanuryl chloride[e]

distinguished, the *cis* isomer (m.p. 17.4°C), giving a single ^{19}F n.m.r. signal, while the *trans* isomer (m.p. −12.5°C) gives a multiplet.[63]

Attempts to substitute these rings normally produces cleavage. However, the azide ion will oxidize benzenesulphinyl chloride to 1,3,5-triphenyl-1,3,5,2,4,6-trithiatriazene-1,3,5-trioxide. This is believed to be the first reported non-halogeno sulphanuryl derivative, and is formed via benzenesulphinyl azide.[64]

$$PhSOCl + N_3^- \xrightarrow{-30°C} PhSON_3 \xrightarrow{-20°C} [PhS(O)N] \longrightarrow$$

Examples of S—N rings containing sulphur in different oxidation states include $S_3N_3Cl_3O$ and $[Me_2SN_2SR]_2^{2+}$. Chlorine oxidation of sulphur(II) bis-thionylamine $S(NSO)_2$ gives the former,[65] which undergoes halogen-exchange at the two S^{IV} atoms with silver difluoride.

$$3S(NSO)_2 + 6Cl_2 \longrightarrow 3SOCl_2 + 2 \qquad \xrightarrow{AgF_2}$$

The *N*-bromo derivative of dimethylsulphur-di-imide oxidizes not only dialkyl disulphides[66] to the disulphonium heterocyclic ion $[Me_2SN_2SR]_2^{2+}$,

$$Me_2S(NBr)_2 + R_2S_2 \longrightarrow [\text{cyclo-}(RS)_2(NSMe_2N)_2]^{2+}\ 2Br^-$$

but also $(Ph_2P)_2NH$ to the sulphur-substituted phosphazene.[67]

$$Me_2S(NBr)_2 + (Ph_2P)_2NH \longrightarrow [Me_2S(N{-}PPh_2)_2N]^+\ Br^- + HBr$$

Likewise, the parent sulphur-di-imide forms S^{VI}-substituted cyclosilazanes with sym-dichlorotetramethyldisilazane by HCl elimination and transamination.[68]

$$Me_2S(NH)_2 + (ClSiMe_2)_2NH \longrightarrow Me_2S(N{-}SiMe_2)_2NH \xrightarrow{Me_2S(NH)_2} \text{cyclo-}(Me_2Si)_2(NSMe_2N)_2$$

(m.p. 114°C) (m.p. 154°C)

OXYGEN DERIVATIVES OF SULPHUR AND SELENIUM

While the dioxides of these elements possess no cyclic characteristics, the more reactive forms of the trioxides do. Both trioxides have asbestos-like structures involving MO_4 chains (M = S or Se), along with a reactive crystalline oligomer. Sulphur trioxide is trimeric (m.p. 16.8°C, b.p. 44.8°C) while selenium trioxide is a tetramer.[69]

$$(SO_3)_3 \qquad (SeO_3)_4$$

The bond lengths within both rings are a little shorter than expected for single bonds, while the exocyclic bonds (S—O, 0.14 nm; Se—O, 0.155 nm) show considerable π-bonding.[70]

Both sulphur and selenium trioxides complex with organic donor molecules, those of selenium being the less stable because of the strong oxidizing nature of the oxide. Arsenic trifluoride will also complex with sulphur trioxide, producing various adducts whose composition depends upon the proportions of factors used.[71] The ^{19}F n.m.r. spectrum of $2AsF_3 \cdot 3SO_3$ shows three peaks typical of O_4SF, $OAsF_2$ and O_2AsF groups (ratio 3:2:1). This is in accordance with structure **(6-I)**, involving fused 6-membered rings.

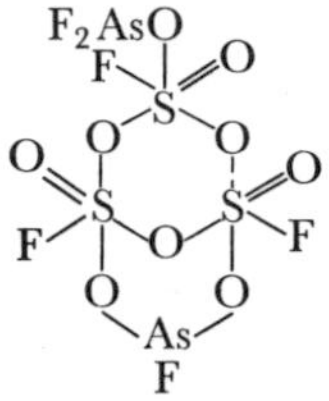

$2AsF_3 \cdot 3SO_3$
(6–I)

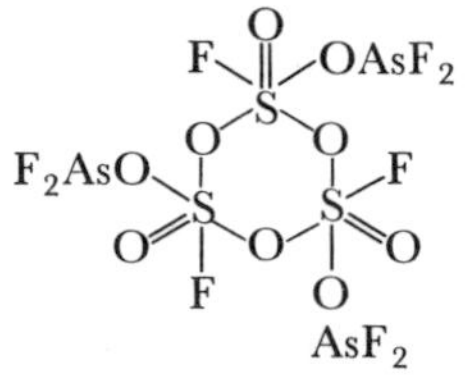

$3AsF_3 \cdot 3SO_3$
(6–II)

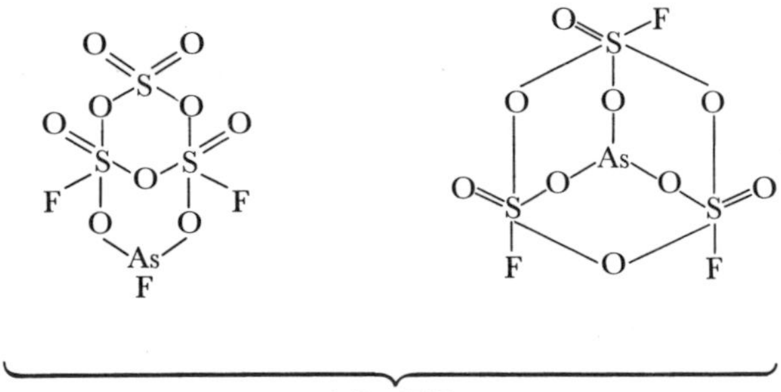

$AsF_3 \cdot 3SO_3$
(6–III)

Fig. 6.14. Structures proposed from the ^{19}F n.m.r. spectra for the $AsF_3 \cdot SO_3$ mixtures

The structure **(6-II)** of the equimolar adduct $3AsF_3 \cdot 3SO_3$ gives a spectrum involving two peaks (O_4SF : $OAsF_2$, 1:2), while $AsF_3 \cdot 3SO_3$ is thought to exist as a mixture of the two isomers of **(6-III)** in similar proportions since ^{19}F n.m.r. spectroscopy gives two peaks (O_4SF : O_2AsF about 6:1).

THE GROUP VI TETRAHALIDES

Halides, more than any other class of compound, show structural variations dependent upon the conditions under which these measurements are made. Thus, phosphorus pentachloride is monomeric, ionic ($PCl_4^+ PCl_6^-$) or dimeric in vapour, solid or solution.

The X-ray crystal structure determination of tellurium tetrachloride shows it to be tetrameric,[72] confirming the molecular weight determination in benzene. The Te_4Cl_{16} molecule comprises four $TeCl_3^+$ cations (C_{3v}) held in a cube by four Cl^- ions.

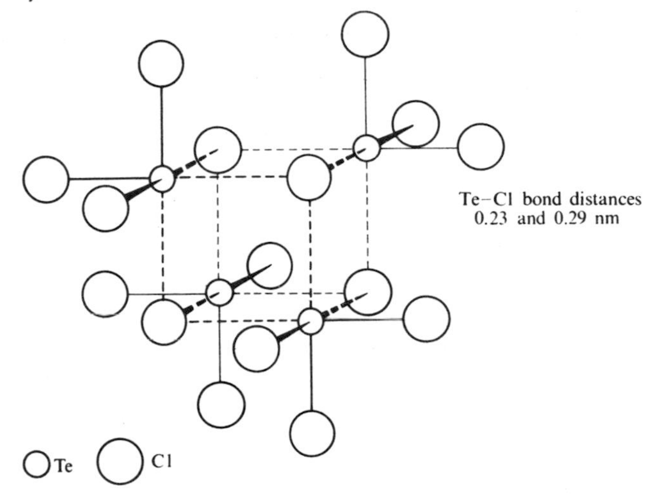

Fig. 6.15. Te_4Cl_{16} structural unit in solid $TeCl_4$[72]

The terminal Te—Cl bonds in the $TeCl_3^+$ residues are single, while the ionic Te—Cl bonds of the cube are 0.06 nm longer, supporting the partially-ionic structure of $TeCl_4$ as $TeCl_3^+Cl^-$ previously proposed. Selenium tetrachloride and tellurium tetrabromide probably have a similar crystal structure.

Bibliography

General

a B. MEYER, 'Elemental Sulphur', John Wiley and Sons, Inc., New York, 1965.
b M. BECKE-GOEHRING, *Inorg. Chem. Radiochem.*, 1960, **2,** 159.
c M. BECKE-GOEHRING, *Quart. Rev.*, 1956, **10,** 437.
d C. W. ALLEN, *J. Chem. Educ.*, 1967, **44,** 38.
e O. GLEMSER and M. FELD, in 'Halogen Chemistry', ed. V. GUTMANN, 1967, Academic Press, London and New York, Vol. 2, p. 1.
f H. G. HEAL, in 'Inorganic Sulphur Chemistry', ed. G. NICKLESS, Elsevier, London 1968, p. 459.

References

1 S. C. ABRAHAM, *Acta Cryst.*, 1955, **8,** 661; 1961, **14,** 311; A. CARON and J. DONOHUE, *ibid.*, 1965, **18,** 562; D. E. SANDS, *J. Amer. Chem. Soc.*, 1965, **87,** 1395.

2 S. GLASSTONE, 'Textbook of Physical Chemistry', Macmillan and Co. Ltd., London, 2nd Edn., 1960, p. 471.
3 M. SCHMIDT and E. WILHELM, *Inorg. Nuclear Chem. Letters*, 1965, **1,** 39; M. SCHMIDT, Elemental Sulphur, *Chem. Phys.*, 1965, 327; J. DONOHUE, A. CARON and E. GOLDISH, *J. Amer. Chem. Soc.*, 1961, **83,** 3748.
4 H. KÖPF, B. BLOCK and M. SCHMIDT, *Chem. Ber.*, 1968, **101,** 272; H. KÖPF and B. BLOCK, *ibid.*, 1969, **102,** 1504.
5 H. KÖPF, *Chem. Ber.*, 1969, **102,** 1509.
6 M. SCHMIDT, B. BLOCK, H. D. BLOCK, H. KÖPF and E. WILHELM, *Angew. Chem. Internat. Edn.*, 1968, **7,** 632.
7 M. SCHMIDT and E. WILHELM, *ibid.*, 1966, **5,** 964; A. KUTOGLU, E. HELLNER and J. BUCHLER, *ibid.*, 1966, **5,** 965; M. SCHMIDT and H. D. BLOCK, *ibid.*, 1967, **6,** 955; M. SCHMIDT, G. KNIPPSCHILD and E. WILHELM, *Chem. Ber.*, 1968, **101,** 381.
8 I. KAWADA and E. HELLNER, *Angew. Chem. Internat. Edn.*, 1970, **9,** 379.
8a M. SCHMIDT and E. WILHELM, *Chem. Comm.*, 1970, 1111.
9 R. COOPER and J. V. CULKA, *J. Inorg. Nuclear Chem.*, 1965, **27,** 755; 1970, **32,** 1857.
10 *Idem, ibid.*, 1967, **29,** 1877; J. WEISS and M. PUPP, *Angew. Chem. Internat. Edn.*, 1970, **9,** 463.
11 R. D. BURBANK, *Acta Cryst.*, 1951, **4,** 140.
12 M. SCHMIDT, in 'Inorganic Polymers', ed. F. G. A. STONE and W. A. G. GRAHAM, Academic Press, New York, 1962, p. 98
13 J. BARR, R. J. GILLESPIE, R. KAPOOR and K. C. MALHOTRA, *Canad. J. Chem.*, 1969, **47,** 149.
14 R. C. MCMILLAN, D. J. PRINCE and J. D. CORBETT, *Chem. Comm.*, 1969, 1438.
15 J. BARR, D. B. CRUMP, R. J. GILLESPIE, R. KAPOOR and R. K. UMMAT, *Canad. J. Chem.*, 1968, **46,** 3607.
16 I. D. BROWN, D. B. CRUMP, R. J. GILLESPIE and D. P. BANTRY, *Chem. Comm.*, 1968, 853.
17 R. J. GILLESPIE and G. P. PEZ, *Inorg. Chem.*, 1969, **8,** 1229.
18 J. BARR, R. J. GILLESPIE, R. KAPOOR and G. P. PEZ, *J. Amer. Chem. Soc.*, 1968, **90,** 6855; J. BARR, R. J. GILLESPIE, G. P. PEZ, P. K. UMMAT and O. C. VAIDYA, *ibid.*, 1970, **92,** 1081; R. C. PAUL, J. K. PURI and K. C. MALHOTRA, *Chem. Comm.*, 1970, 776.
19 R. J. GILLESPIE and J. PASSMORE, *Chem. Comm.*, 1969, 1333.
20 J. BARR, R. J. GILLESPIE and P. K. UMMAT, *ibid.*, 1970, 264.
21 P. J. STEPHENS, *ibid.*, 1969, 1496.
22 A. K. MACBETH and H. GRAHAM, *Proc. Roy. Irish Acad.*, 1923, **36,** 31.
23 J. R. ALFORD, D. C. H. BIGG and H. G. HEAL, *J. Inorg. Nuclear Chem.*, 1967, **29,** 1538.
24 B. A. OLSEN, F. P. OLSEN and E. M. TINGLE, *Chem. Comm.*, 1968, 554; B. A. OLSEN and F. P. OLSEN, *Inorg. Chem.*, 1969, **8,** 1736.
25 M. BECKE-GOEHRING and W. KOCH, *Z. Naturforsch.*, 1952, **7b,** 634; P. MACHMER, *ibid.*, 1969, **24b,** 1056.
26 J. WEISS and H. L. NEUBERT, *Acta Cryst.*, 1965, **18,** 815.
27 J. C. VAN DE GRAMPEL and A. VOS, *ibid.*, 1969, **25A,** 611.
28 H. G. HEAL, M. S. SHAHID and H. GARCIA-FERNANDEZ, *Chem. Comm.*, 1969, 1063.
29 R. C. BRASTED and J. S. POND, *Inorg. Chem.*, 1965, **4,** 1163; W. I. GORDON and H. G. HEAL, *J. Inorg. Nuclear Chem.*, 1970, **32,** 1863.
30 M. BECKE-GOEHRING and H. JENNE, *Chem. Ber.*, 1959, **92,** 1149.
31 H. GARCIA-FERNANDEZ, *Compt. rend.*, 1965, **260,** 6107; H. GARCIA-FERNANDEZ and C. RÉRAT, *ibid.*, 1966, **262C,** 1866; B. D. STONE and M. L. NIELSEN, *J. Amer. Chem. Soc.*, 1957, **79,** 1264; 1959, **81,** 3580.
32 H. WÖLBLING, *Z. anorg. Chem.*, 1908, **57,** 281.
33 E. W. LUND and S. R. SVENDSEN, *Acta Chem. Scand.*, 1957, **11,** 940; R. L. SASS and J. DONOHUE, *Acta Cryst.*, 1958, **11,** 497.

34 Y. SASAKI and F. P. OLSEN, *Inorg. Nuclear Chem. Letters*, 1967, **3,** 351; E. FLUCK and M. BECKE-GOEHRING, *Z. anorg. Chem.*, 1957, **292,** 229; A. MEUWSEN and M. LÖSEL, *Z. anorg. Chem.*, 1953, **271,** 217; M. BECKE-GOEHRING and G. ZIRKER, *Z. Naturforsch.*, 1955, **10b,** 58.
35 W. L. JOLLY and M. BECKE-GOEHRING, *Inorg. Chem.*, 1962, **1,** 76; B. COHEN, T. R. HOOPER and R. D. PEACOCK, *J. Inorg. Nuclear Chem.*, 1966, **28,** 920; O. GLEMSER, A. HAAS and H. REINKE, *Z. Naturforsch.*, 1965, **20b,** 809.
36 D. CLARK, *J. Chem. Soc.*, 1952, 1615; J. (BANUS) MASON, *J. Chem. Soc. (A)*, 1969, 1567.
37 A. G. TURNER and F. S. MORTIMER, *Inorg. Chem.*, 1966, **5,** 906; R. A. MEINZER, D. W. PRATT and R. J. MYERS, *J. Amer. Chem. Soc.*, 1969, **91,** 6623.
38 J. JANDER and V. DOETSCH, *Chem. Ber.*, 1960, **93,** 561; H. BÄRNIGHAUSEN, T. VON VOLKMANN and J. JANDER, *Angew. Chem. Internat. Edn.*, 1965, **4,** 72.
39 O. GLEMSER, *ibid.*, 1963, **2,** 530; O. GLEMSER, H. SCHROEDER and H. HAESELER, *Naturwiss.*, 1955, **42,** 44; *Z. anorg. Chem.*, 1955, **279,** 28.
40 K. J. WYNNE and W. L. JOLLY, *Inorg. Chem.*, 1967, **6,** 107; B. COHEN, T. R. HOOPER, D. HUGILL and R. D. PEACOCK, *Nature*, 1965, **207,** 748.
41 M. BECKE-GOEHRING and D. SCHLAFER, *Z. anorg. Chem.*, 1968, **356,** 234.
42 J. R. W. WARN and D. CHAPMAN, *Spectrochim. Acta*, 1966, **22,** 1379; R. L. PATTON and W. L. JOLLY, *Inorg. Chem.*, 1969, **8,** 1389.
43 R. L. PATTON and K. N. RAYMOND, *ibid.*, 1969, **8,** 2426; J. T. NELSON and J. J. LAGOWSKI, *ibid.*, 1967, **6,** 1292.
44 A. J. BANISTER, P. J. DAINTY, A. C. HAZELL, R. G. HAZELL and J. G. LOMBORG, *Chem. Comm.*, 1969, 1187.
45 O. GLEMSER and E. WYSZOMEIRSKI, *Naturwiss.*, 1961, **48,** 25; E. FLUCK and R. N. REIMISCH, *Z. anorg. Chem.*, 1964, **328,** 165.
46 A. MEUWSEN, *Chem. Ber.*, 1931, **64,** 2311; H. SCHROEDER and O. GLEMSER, *Z. anorg. Chem.*, 1959, **298,** 78; B. COHEN, T. R. HOOPER and R. D. PEACOCK, *Chem. Comm.*, 1966, 32; J. NELSON and H. G. HEAL, *Inorg. Nuclear Chem. Letters*, 1970, **6,** 429.
47 O. GLEMSER, H. MEYER and A. HAAS, *Chem. Ber.*, 1964, **97,** 1704.
48 G. A. WIEGERS and A. VOS, *Acta Cryst.*, 1963, **16,** 152; D. P. CRAIG and N. L. PADDOCK, *J. Chem. Soc.*, 1962, 4118.
49 O. GLEMSER and H. LÜDEMANN, *Angew. Chem.*, 1958, **70,** 190.
50 G. A. WIEGERS and A. VOS, *Acta Cryst.*, 1961, **14,** 462.
51 M. BECKE-GOEHRING and H. MATZ, *Z. Naturforsch.*, 1954, **9b,** 567.
52 M. BECKE-GOEHRING and H. P. LATSCHA, *Angew. Chem. Internat. Edn.*, 1962, **1,** 551.
53 E. DEMARCAY, *Compt. rend.*, 1880, **91,** 854, 1066; A. MEUWSEN, *Chem. Ber.*, 1932, **65,** 1724; O. GLEMSER, A. HAAS and H. REMKE, *Z. Naturforsch.*, 1965, **20b,** 809.
54 J. WEISS, *Z. anorg. Chem.*, 1964, **333,** 314.
55 M. BECKE-GOEHRING and H. P. LATSCHA, *Z. Naturforsch.*, 1962, **17b,** 125.
56 M. BECKE-GOEHRING and G. MAGIN, *Z. Naturforsch.*, 1965, **20b,** 493.
57 A. MEUWSEN and O. JACOB, *Z. anorg. Chem.*, 1950, **263,** 200.
58 A. J. BANISTER and J. S. PADLEY, *J. Chem. Soc.*, 1967, 1437.
59 A. W. KIRSANOV, *Chem. Abs.*, 1951, **45,** 1503a; 1952, **46,** 6984b.
60 A. VANDI, T. MOELLER and T. L. BROWN, *Inorg. Chem.*, 1963, **2,** 899.
61 A. C. HAZELL, G. A. WIEGERS and A. VOS, *Acta Cryst.*, 1966, **20,** 186.
62 T. MOELLER and A. OUCHI, *J. Inorg. Nuclear Chem.*, 1966, **28,** 2147; F. SEEL and G. SIMON, *Z. Naturforsch.*, 1964, **19b,** 354.
63 S. M. WILLIAMSON, *Progr. Inorg. Chem.*, 1966, **7,** 69.
64 T. J. MARICICH, *J. Amer. Chem. Soc.*, 1968, **90,** 7179.
65 D. SCHLÄFER and M. BECKE-GOEHRING, *Z. anorg. Chem.*, 1968, **362,** 1.
66 R. APPEL, D. HÄNSSGEN and W. MÜLLER, *Chem. Ber.*, 1968, **101,** 2855.
67 R. APPEL, D. HÄNSSGEN and B. ROSS, *Z. Naturforsch.*, 1967, **22b,** 1354.
68 R. APPEL, L. SIEKMANN and H. O. HOPPEN, *Chem. Ber.*, 1968, **101,** 2861.

69 F. C. MIJLHOFF, *Acta Cryst.*, 1965, **18,** 795.
70 F. A. COTTON and G. WILKINSON, 'Advanced Inorganic Chemistry', Interscience, London, 2nd Edn., 1966, p. 543.
71 R. J. GILLESPIE and J. V. OUBRIDGE, *Proc. Chem. Soc.*, 1960, 308.
72 B. BUSS and B. KREBS, *Angew. Chem. Internat. Edn.*, 1970, **9,** 463.

APPENDIX

The elements

Visible absorption spectra show that the red colour developed on heating liquid sulphur is probably due to S_3 (thiozone) and S_4^1.

In oleum, sulphur gives the cations S_8^+, S_{16}^{2+}, S_8^{2+}, S_4^{2+} depending upon the conditions employed.[2] Such ions also result in disulphuric acid, and both arsenic and antimony pentafluoride. Arsenic pentafluoride will also react with disulphur difluoride[3] to give $S_2F^+AsF_6^-$, which decomposes on heating to S_8^{2+} and S_{16}^{2+}.

Selenium[4] and AsF_5 give Se_8^{2+}, the crystal structure of which has been determined for the $AlCl_4^-$ salt, isolated from a melt of selenium, its tetrachloride and aluminium chloride.[5] It possesses a short *trans*-annular bond of 0.284 nm. The tellurium cation Te_4^{2+} can also be prepared this way,[6] or from the metalloid in oleum, $S_2O_6F_2$, A_sF_5 or SbF_5.

Sulphur (II)—nitrogen rings

Heptasulphur imide can be prepared from S_8 and triphenylarsine imide Ph_3AsNH.[7] S_7NH couples with polysulphur dichlorides to give $(S_7N)_2S_n$ (n = 3, 5)[8] while formaldehyde will add to the three NH bonds of 1,3,5,$(HN)_3S_5$. The e.s.r. spectra of S_7NH and $S_4N_4H_4$ have been measured after loss of hydrogen.[9]

S_4N_2 (m.p. 23.5°C), prepared either from $Hg_5(NS)_8$ and S_2Cl_2 or from S_4N_4 and sulphur, is not to the thio analogue of dinitrogen tetroxide,[10] but contains the sulphur di-imide residue N=S=N.

Tetrasulphur tetranitride

$(CF_3)_2NO$ will add to S_4N_4 to give $(NSON(CF_3)_2)_4$. With $Hg(ON(CF_3)_2)_2$ and NSF monomeric $NSON(CF_3)_2$ forms which spontaneously trimerises.[11] Hydrogen chloride yields $(NSCl)_3$. While benzylamine reduces S_4N_4 to sulphur and aminopolysulphides,[12] sulphuric acid gives a variety of products, including sulphur and SO_2. Triphenylphosphine causes ring contraction of S_4N_4 to $Ph_3PNS_3N_3$. The structure shows the S_2N_3 residue not containing the sulphur bonded to the Ph_3PN side-chain to be planar, with variable S—N bond lengths.[13] Molecular orbital calculations are consistant with a structure for S_4N_4 involving *trans*-annular S—S bonds.[14] This leads to predictions about the structure of $S_4N_4^{2-}$ and the mode of reaction with olefins.

S_4N_4 complexes with carbonyls of iron, cobalt and molybdenum. These readily decompose and this is substantiated by the high carbonyl stretching frequency.[15] $(Ph_3P)_2Ir(CO)Cl$ yields $(Ph_3P)Ir(CO)ClN_4S_4$. Ionic structures are proposed for the adducts with $SbCl_5$, $SeCl_4$, $TeCl_4$ and BCl_3.[16]

Halogenated sulphur–nitrogen rings

$S_3N_3Cl_3$ dissociates on heating into monomeric NSCl[17] and with $SCl_2/AlCl_3$ yields ionic $N(SCl)_2^+ AlCl_4^-$. The melting point diagram for *cis*- and *trans*-sulphanuric fluoride gives a eutectic point at −32.5°C.[18] With BCl_3, FSO_2NSO undergoes halogen exchange, loses SO_2 on warming, then reacts with $ClSO_2NSO$ to give the heterocyclic compound I.[19]

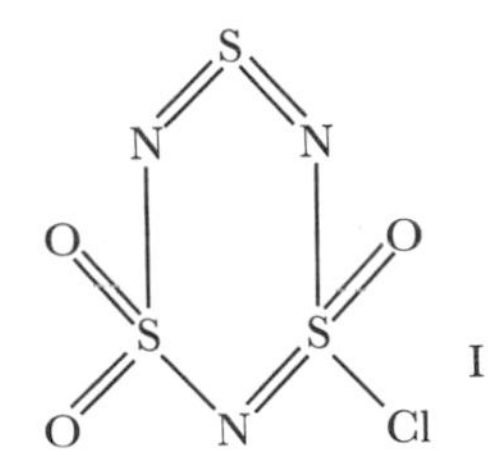

I

Sulphanuryl chlorides and fluorides[20] can be readily organo-substituted and aminated, while the structure of $PhF_2S_3N_3O_3$ shows the phenyl group *trans* to the 2 fluorine atoms.[21] Tellurium tetrachloride catalyses the dimerization of $Ar_2C{=}CH_2$ and forms adducts with olefins.[22]

References

1 B. MEYER, T. STROGER-HANSEN, D. JENSEN and T. V. OOMMEN, *J. Amer. Chem. Soc.*, 1971, **93**, 1034.

2 W. F. GIGGENBACH, *Chem. Comm.*, 1970, 852; R. A. BEAUDET and P. J. STEPHENS, *ibid.*, 1971, 1083; M. STILLINGS, M. C. R. SYMONS and J. G. WILKINSON, *ibid.*, 1971, 372; R. J. GILLESPIE, J. PASSMORE, P. K. UMMAT and O. C. VAIDYA, *Inorg. Chem.*, 1971, **10**, 1327; R. C. PAUL, J. K. PURI and K. C. MALHOTRA, *Inorg. Nuclear Chem. Letters*, 1971, **7**, 729.

3 F. SEEL, V. HARTMANN, I. MOLNAR, R. BUDENZ and W. GOMBLER, *Angew. Chem. Int. Edn.*, 1971, **10**, 186.

4 R. J. GILLESPIE and P. K. UMMAT, *Canad. J. Chem.*, 1970, **48**, 1239.

5 R. K. MCMILLAN, D. J. PRINCE and J. D. CORBETT, *Inorg. Chem.*, 1971, **10**, 1749.

6 D. J. PRINCE, J. D. CORBETT and B. GARBISCH, *ibid.*, 1970, **9**, 2731, N. J. BJERRUM, *ibid.*, 1965, J. BARR, R. J. GILLESPIE, G. P. PEZ, P. K. UMMAT and O. C. VAIDYA, *ibid.*, 1971, **10**, 362.

7 R. APPEL, *Angew. Chem.*, 1963, **73**, 220.

8 H. GARCIA-FERNANDEZ, H. G. HEAL and M. S. SHAHIB, *Compt. Rendu*, 1971, **272**, 60.

9 P. MACHMER, D. A. C. MCNEIL and M. C. R. SYMONS, *Trans. Faraday Soc.*, 1970, **66**, 1309.

10 J. NELSON and H. G. HEAL, *J. Chem. Soc.* (A), 1971, 136.

11 H. J. EMELÉUS, R. A. FORDER, R. J. PENLET and G. M. SHELDRICK, *Chem. Comm.*, 1970, 1483.

12 Y. SASAKI and F. P. OLSEN, *Canad. J. Chem.*, 1971, **49,** 271; S. A. LIPP and W. L. JOLLY, *Inorg. Chem.*, 1971, **10,** 33.
13 E. M. HOLT and S. L. HOLT, *Chem. Comm.*, 1970, 1704.
14 R. GLEITER, *J. Chem. Soc.* (A), 1970, 3174.
15 D. A. BROWN and F. FRIMMEL, *Chem. Comm.*, 1971, 579, B. J. MCCORMICK and B. M. ANDERSON, *J. Inorg. Nuclear Chem.*, 1970, **32,** 3414.
16 R. C. PAUL, C. L. ARORA, J. KISHORE and K. C. MALHOTRA, *Austr. J. Chem.*, 1971, **24,** 1637.
17 R. L. PATTON and W. L. JOLLY, *Inorg. Chem.*, 1970, **9,** 1079, O. GLEMSER and J. WEGENER, *Inorg. Nuclear Chem. Letters*, 1971, **7,** 623.
18 F. SEEL, K. VELLEMAN and E. HEINRICH, *Z. anorg. Chem.*, 1971, **382,** 61.
19 H. W. ROESKY, *Angew. Chem. Internat. Edn.*, 1971, **10,** 266.
20 A. J. BANISTER and B. BELL, *J. Chem. Soc.* (A), 1970, 1659; Preparative Inorganic Chemistry, 1971, **6,** 63.
21 D. E. ARRINGTON, T. MOELLER and I. C. PAUL, *ibid.*, 2627.
22 D. ELMALEH, S. PATAI and Z. PAPPOPORT, *J. Chem. Soc.* (C), 1971, 3100; D. KOBELT and E. F. PAULUS, *J. Organometallic Chem.*, 1971, **27,** C63.

7

The Halogens

The fact that the halogen elements are normally monovalent considerably restricts the number of cyclic interhalogen compounds. However, a few such compounds do exist.

While the high boiling point and Trouton's constant for bromine trifluoride and iodine pentafluoride indicate association, no dimeric form has ever been characterized. Indeed, the conductivity of these compounds supports dissociation.[1]

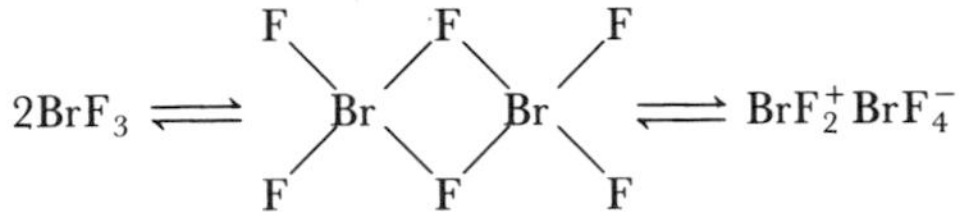

$$2BrF_3 \rightleftharpoons F_2Br(\mu\text{-}F)_2BrF_2 \rightleftharpoons BrF_2^+ BrF_4^-$$

Iodine trichloride was first prepared by Gay-Lussac and by Davy in 1814,[2] and is now conveniently synthesized by adding finely-powdered iodine to an excess of liquid chlorine.[3] The yellow crystals melt in a sealed tube at 101°C but will readily sublime at −12°C. The crystal structure shows the compound to be dimeric and planar with a 4-membered I_2Cl_2 ring.[4] The bridging I—Cl bonds (0.270 nm) are longer than the terminal ones (0.238 nm) and this difference is probably best explained through multi-centre bonding.

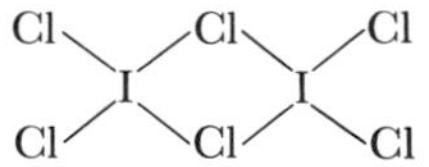

$$Cl_2I(\mu\text{-}Cl)_2ICl_2$$

Fig. 7.1. Dimeric iodine trichloride

Iodine trichloride forms adducts with aluminium trichloride, antimony pentachloride and sulphur tetrachloride. While those of aluminium and sulphur are ionic and conductors, the antimony one is a non-conductor and may have a cyclic structure.[5] (See top of next page.)

Tetraphenyl-tin and -lead are readily chlorinated by ICl_3, with cleavage of the metal–carbon bond.

$$ICl_3 + Ph_4M \xrightarrow{CHCl_3} Ph_2MCl_2$$

$$ICl_2^+AlCl_4^- \xleftarrow{AlCl_3} ICl_3 \xrightarrow{SCl_4} SCl_3^+ICl_4^-$$

(m.p. 105°C)

$$ICl_3 \xrightarrow{SbCl_5} ISbCl_8$$

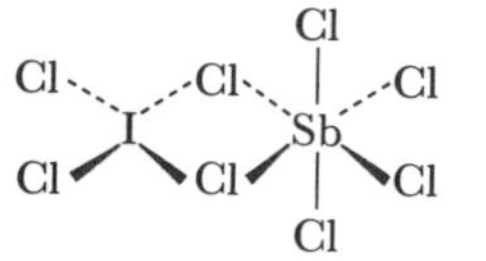

$ISbCl_8$ (m.p. 83.5°C)

With phenylsilicon trichloride only ring-substitution occurs, with essentially mono-substitution at −40°C.[6]

Hydrocarbons are readily chlorinated with both iodine trichloride and phenyliodine dichloride in the presence of ultraviolet light. The radicals $Cl_2I\cdot$ and $PhClI\cdot$ are thought to be involved.

$$Me_2CHCHMe_2 \longrightarrow ClCH_2CHMeCHMe_2 \text{ and } Me_2CClCHMe_2$$

Phenyliodine dichloride can be prepared directly from iodobenzene and chlorine.[7] It is a lemon-yellow crystalline solid which readily photolyzes to *p*-chloro-iodobenzene, and functions in reactions as a chlorine carrier. Thus, sulphides can be readily oxidized to sulphoxides in high yield in aqueous pyridine with no subsequent oxidation to the sulphone.[8] The substituent can be alkyl, aryl or heterocyclic.

$$PhICl_2 + R_2S \xrightarrow[-40°\text{to}-20°C]{C_2H_5N/H_2O} R_2SO$$

The structure of phenyliodine dichloride indicates a T-shaped monomer with the phenyl ring perpendicular to the linear Cl—I—Cl system.[9] The I—Cl bond distances (0.245 nm) are close to those found in ICl_3, while the intermolecular I—Cl distance (0.340 nm) supports weak association.

Diphenyliodine chloride is dimeric according to the crystal structure. The 4-membered I_2Cl_2 ring lies perpendicular to all the phenyl rings and has I—Cl bond lengths of between 0.31 and 0.32 nm.[7] While these are longer than normal covalent bonds, they are shorter than the sum of the van der Waals radii. Indeed the I—I distance of 0.334 nm indicates an interaction between these atoms.

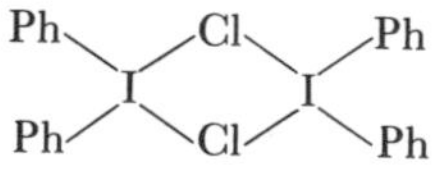

Fig. 7.2. Dimeric diphenyliodine monochloride

Bibliography

References

1 L. STEIN, in 'Halogen Chemistry', ed. V. GUTMANN, 1967, Academic Press, London and New York, Vol. 1, p. 133.
2 J. L. GAY-LUSSAC, *Ann. chim. et phys.*, 1814, **91,** 5; H. DAVY, *Trans. Roy. Soc.*, 1814, **104,** 487.
3 *Inorg. Synth.*, 1939, **I,** 167.
4 K. H. BOSWIJK and E. H. WIEBENGA, *Acta Cryst.*, 1954, **7,** 514.
5 G. VONK and E. H. WIEBENGA, *Rec. Trav. chim.*, 1959, **78,** 913; S. N. NABI, *Pakistan J. Sci. Res.*, 1964, **16,** 138.
6 Z. M. MANULKIN, *Uzbek. khim. Zhur.*, 1960, **2,** 66; *Chem. Abs.*, 1961, **55,** 12330b; G. V. MOTSAREV and A. YA-YAKUBOVICH, *Zhur. obshchei Khim.*, 1965, **35,** 1056; *Chem. Abs.*, 1965, **63,** 11604c.
7 D. F. BANKS, *Chem. Rev.*, 1966, **66,** 243.
8 G. BARBIERI, M. CINQUINI, S. COLONNA and F. MONTANARI, *J. Chem. Soc.* (*C*), 1968, 659.
9 E. M. ARCHER and T. G. D. VAN SCHALKWYK, *Acta Cryst.*, 1953, **6,** 88.

Formulae Index

Specific Index of Compounds

GROUP VI

General Index